ASTRONOMY AND ASTROPHYSICS LIBRARY

Series Editors: I. Appenzeller, Heidelberg, Germany
G. Börner, Garching, Germany
M. A. Dopita, Canberra, Australia
M. Harwit, Washington, DC, USA
R. Kippenhahn, Göttingen, Germany
J. Lequeux, Paris, France
A. Maeder, Sauverny, Switzerland
V. Trimble, College Park, MD, and Irvine, CA, USA

Springer
Berlin
Heidelberg
New York
Hong Kong
London
Milan
Paris
Tokyo

Physics and Astronomy ONLINE LIBRARY

http://www.springer.de/phys/

Reinhard Schlickeiser

Cosmic Ray Astrophysics

With 164 Figures

 Springer

Professor Dr. Reinhard Schlickeiser
Ruhr-Universität Bochum
Theoretische Physik IV
44780 Bochum, Germany

Cover picture: The magnetic field structure in the external edge-on galaxy NGC 4631 derived from radio polarization measurements at $\lambda = 20$ cm wavelength by Hummel et al. (1988 [232])

Library of Congress Cataloging-in-Publication Data. Schlickeiser, R. (Reinhard), 1952– . Cosmic ray astrophysics/ R. Schlickeiser. p. cm. – (Astronomy and astrophysics library, ISSN 0941-7834). Includes bibliographical references. ISBN 3540664653 (hardcover: alk. paper) 1. Cosmic rays. I. Title. II. Series. QC485.S334 2000 523.01'97223–dc21 99-089515

First Edition 2002
Corrected Second Printing 2003
ISSN 0941-7834
ISBN 3-540-66465-3 Springer-Verlag Berlin Heidelberg New York

This work is subject to copyright. All rights are reserved, whether the whole or part of the material is concerned, specifically the rights of translation, reprinting, reuse of illustrations, recitation, broadcasting, reproduction on microfilm or in any other way, and storage in data banks. Duplication of this publication or parts thereof is permitted only under the provisions of the German Copyright Law of September 9, 1965, in its current version, and permission for use must always be obtained from Springer-Verlag. Violations are liable for prosecution under the German Copyright Law.

Springer-Verlag Berlin Heidelberg New York
a member of BertelsmannSpringer Science+Business Media GmbH

http://www.springer.de

© Springer-Verlag Berlin Heidelberg 2002
Printed in Germany

The use of general descriptive names, registered names, trademarks, etc. in this publication does not imply, even in the absence of a specific statement, that such names are exempt from the relevant protective laws and regulations and therefore free for general use.

Data conversion by LeTeX, Leipzig
Cover design: *design & production* GmbH, Heidelberg

Printed on acid-free paper   SPIN: 10893926   55/3141/ba - 5 4 3 2 1 0

To Wernhild, Christina, Frank, Anna and Heinz

Preface

This book is based on a series of lectures that were given at the universities of Bonn and Bochum to undergraduate and graduate students from the departments of physics and astronomy. The book is intended as an introduction to the physics and astrophysics of cosmic rays.

I am grateful to many colleagues, friends and students for their support and encouragement. I am especially grateful to the undergraduate and graduate students for whom I have been thesis advisor: their work appears throughout these pages. The list includes Ulrich Achatz, Markus Böttcher, Andre Crusius-Wätzel, Wolfgang Dröge, Rüdiger Dung, Uwe Jaekel, Meinhard Kneller, Harald Lesch, Hartmut Mause, Martin Pohl, Olaf Stawicki, Adriane Steinacker, Jürgen Steinacker and Ralf Weyer.

Moreover, I want to thank Rainer Beck, Elly Berkhujsen, Ed Churchwell, Karl Mannheim, Wolfgang Reich, Manfred Simon and Richard Wielebinski for carefully reading parts of the manuscript and for their valuable suggestions. I especially wish to thank Ian Lerche for his meticulous reading of the whole book and valuable comments, corrections and suggestions to improve the presentation. Needless to say, all remaining errors are my own.

Special thanks go to my frequent collaborators Charles Dermer, Horst Fichtner, James A. Miller, Michal Ostrowski and Rami Vainio for sharing their insights and special knowledge with me. The final editing of the manuscript was ably performed by Angelika Schmitz who maintained her very high standards of technical precision. Bernd Neubacher, G. Hutschenreiter and W. Fußhöller provided invaluable help with the illustrations. Last but not least I want to thank Springer-Verlag, in particular Wolf Beiglböck, for the years-long patience and the repeated encouragements to finish a difficult manuscript. This book would not have been written without the support of my wife Wernhild and my children Christina and Frank.

Bochum, May 2001 *Reinhard Schlickeiser*

Contents

1. **Introduction** 1
   1.1 The Start of Cosmic Ray Research 1
   1.2 The Scope of Cosmic Ray Research 1
   1.3 Organization of the Book 3

2. **Cosmic Rays as Part of the Universe** 7
   2.1 Our Present View of the Structure of the Universe 7
   2.2 Interstellar Medium 10
   2.3 Cosmic Photon Fields 17
   2.4 Cosmic Magnetic Fields 19
   2.5 Intergalactic Medium 23
   2.6 Interlude 24

3. **Direct Observations of Cosmic Rays** 25
   3.1 The Cosmic Ray Landscape 26
   3.2 Solar Modulation 29
   3.3 Solar and Heliospheric Cosmic Rays 36
       3.3.1 Solar Flare Cosmic Ray Particles 38
       3.3.2 The Anomalous Cosmic Ray Component 46
   3.4 Extrasolar Cosmic Rays 46
       3.4.1 Elemental Composition 46
       3.4.2 Isotopic Composition 57
       3.4.3 Anisotropy 64
       3.4.4 Time History from Cosmogenic Nuclei 69
       3.4.5 Ultrahigh Energy Cosmic Rays 70

4. **Interactions of Cosmic Ray Electrons** 73
   4.1 Synchrotron Radiation 74
       4.1.1 Synchrotron Power 74
       4.1.2 Emission and Transfer of Synchrotron Radiation ... 76
       4.1.3 Synchrotron Radiation from Cosmic Ray Electrons
             in the Interstellar Medium 78
       4.1.4 Synchrotron Energy Loss of Relativistic Electrons. 80
   4.2 Inverse Compton Scattering 80

		4.2.1	Inverse Compton Power	81
		4.2.2	Inverse Compton Scattering of Target Photons in the Interstellar Medium by Cosmic Ray Electrons	84
		4.2.3	Inverse Compton Energy Loss of Relativistic Electrons	88
	4.3	Triplet Pair Production		89
	4.4	Nonthermal Relativistic Electron Bremsstrahlung		93
		4.4.1	Nonthermal Bremsstrahlung Power	93
		4.4.2	Nonthermal Bremsstrahlung of Cosmic Ray Electrons in the Interstellar Medium	95
		4.4.3	Bremsstrahlung Energy Loss of Relativistic Electrons	97
	4.5	Ionization and Coulomb Interactions		98
	4.6	Total Energy Loss Rate		99
	4.7	Continuum Radiation Processes of Relativistic Electrons		102
5.	**Interactions of Cosmic Ray Nuclei**			**105**
	5.1	Relativistic Kinematics of Inelastic Collisions		105
		5.1.1	Threshold Energy	106
		5.1.2	The Energy of One Particle Seen from the Rest System of Another	107
	5.2	Interactions Between Cosmic Ray Nuclei and Cosmic Photons		107
		5.2.1	Photo-Pair Production	109
		5.2.2	Photo-Hadron Production	109
		5.2.3	Photo-Disintegration	112
	5.3	Interactions Between Cosmic Ray Nuclei and Matter		114
		5.3.1	Gamma-Ray, Electron, Positron and Neutrino Source Functions	115
		5.3.2	Pion Production Spectra	116
		5.3.3	π^0 Decay γ-Rays	120
		5.3.4	Secondary Electrons and Positrons	123
		5.3.5	Secondary Neutrinos	125
		5.3.6	Energy Loss by Pion Production	128
		5.3.7	Excitation of Nuclei	128
		5.3.8	Coulomb and Ionization Interactions	130
		5.3.9	Catastrophic Losses from Fragmentation and Radioactive Decay	135
		5.3.10	Total Energy Loss Rate from Interactions with Matter	138
		5.3.11	Ionization and Heating Rate of Interstellar Matter by Cosmic Rays	139
		5.3.12	Continuum Radiation Processes of Relativistic Nuclei	140

6. Indirect Observations of Cosmic Rays ... 143
- 6.1 Clues from Radio Astronomy ... 143
  - 6.1.1 Free–Free Emission ... 143
  - 6.1.2 Radio Continuum Surveys ... 145
- 6.2 Clues from γ-Ray Astronomy ... 154
  - 6.2.1 γ-Ray Emission Above 10 MeV ... 155
  - 6.2.2 Continuum γ-Ray Emission Below 10 MeV ... 162
  - 6.2.3 γ-Ray Line Emission Below 10 MeV from the Orion Region ... 163
  - 6.2.4 Interlude ... 164
- 6.3 γ-Ray Point Sources ... 165
  - 6.3.1 γ-Ray Observations of Active Galactic Nuclei ... 165
  - 6.3.2 Nonthermal γ-Ray Emission Processes in the Jets of Active Galactic Nuclei ... 172
  - 6.3.3 Interlude ... 176
  - 6.3.4 Galactic γ-Ray Sources ... 177

7. Immediate Consequences of Galactic Cosmic Ray Observations ... 179
- 7.1 Cosmic Ray Energetics ... 179
- 7.2 Global Cosmic Ray Source Energetics ... 180
- 7.3 Cosmic Ray Scattering, Confinement and Isotropy ... 182

8. Statistical Mechanics of Charged Particles ... 183
- 8.1 Basic Equations ... 183
  - 8.1.1 Plasma Parameter ... 183
  - 8.1.2 Kinetic Description of Plasmas ... 187
  - 8.1.3 Magnetohydrodynamics of Plasmas ... 190
- 8.2 The Relativistic Vlasov Equation and Maxwell Equations ... 194
- 8.3 Kinetic Theory of Plasma Waves ... 195
  - 8.3.1 Linearization of Kinetic Plasma Equations ... 195
  - 8.3.2 Solution of the Linearized Vlasov Equation ... 196
  - 8.3.3 Dispersion Relation ... 197
  - 8.3.4 The Landau Contour ... 200
  - 8.3.5 Polarization Vector ... 203
  - 8.3.6 Conductivity Tensor in a Homogeneous Hot Magnetized Plasma ... 204
  - 8.3.7 Energy Transfer in Dispersive Media ... 208

9. Test Wave Approach 1. Waves in Cold Magnetized Plasmas ... 211
- 9.1 Plasma Waves in a Cold Moving Magnetized Plasma ... 211
- 9.2 Plasma Waves in a Cold Magnetized Plasma at Rest ... 213
  - 9.2.1 Cold Unmagnetized Plasma at Rest ... 215
  - 9.2.2 Cold Electron-Proton Plasma ... 217

	9.2.3	Propagation Along the Static Magnetic Field 219
	9.2.4	Propagation Across the Static Magnetic Field 224
	9.2.5	Oblique Propagation of Low-Frequency Waves 228
	9.2.6	Oblique Propagation of High-Frequency Waves 231
	9.2.7	More Rigorous Treatment of Dispersion Relation and Polarization of Low-Frequency Waves 234
	9.2.8	Faraday Rotation................................. 238

10. Test Wave Approach 2.
Waves in Hot Magnetized Isotropic Plasmas 241
10.1 Conductivity Tensor for Isotropic Distribution Functions 241
10.2 The Longitudinal Mode 244
    10.2.1 Superluminal Waves 247
    10.2.2 Subluminal Waves 249
10.3 Illustrative Example: Longitudinal Waves
    in an Equilibrium Electron Plasma 250
    10.3.1 Superluminal Waves 252
    10.3.2 Subluminal Waves 253
    10.3.3 Non-relativistic Plasma Temperatures 254
    10.3.4 Infinitely Large Speed of Light, $c \to \infty$ 257
    10.3.5 Solution of the Non-relativistic Dispersion Relation .. 257
    10.3.6 Landau Damping of Subluminal Solutions 262
    10.3.7 Weak Damping Approximation.................... 262
10.4 The Transverse Modes 263
    10.4.1 Collisionless Damping of Transverse Oscillations..... 265
    10.4.2 An Illustrative Example 266

11. Test Wave Approach 3.
Generation of Plasma Waves.............................. 269
11.1 Two-Stream Instability 269
11.2 Cosmic-Ray-Induced Instabilities 273
    11.2.1 Parallel Propagating Waves...................... 274
    11.2.2 Growth Rates at Arbitrary Angles of Propagation ... 283
    11.2.3 Cosmic Ray Self-Confinement in Galaxies 286
11.3 Competition of Wave Growth and Damping................ 287
    11.3.1 Damping by Ion-Neutral Collisions 287
    11.3.2 Nonlinear Landau Damping...................... 287
    11.3.3 Wave Cascading................................. 289
    11.3.4 Super-Alfvénic Cosmic Ray Propagation?........... 290

12. Test Particle Approach 1.
Hierarchy of Transport Equations 293
12.1 Quasilinear Theory 293
12.2 Quasilinear Fokker-Planck Coefficients 298
    12.2.1 Unperturbed Particle Orbits 298

   12.2.2 Fokker-Planck Coefficients
      for Plasma Wave Turbulence 302
  12.3 The Diffusion Approximation 306
   12.3.1 Cosmic Ray Anisotropy 309
   12.3.2 The Diffusion-Convection Transport Equation 310

13. Test Particle Approach 2. Calculation of Transport Parameters 313
  13.1 Linearly Polarized Alfvén Waves.......................... 313
   13.1.1 Alfvén Wave Fokker-Planck Coefficients 315
   13.1.2 Magnetic Turbulence Tensors 315
   13.1.3 Slab Linearly Polarized Alfvén Turbulence 318
   13.1.4 Isotropic Linearly Polarized Alfvén Waves 319
  13.2 Parallel Propagating Magnetohydrodynamic Waves 320
   13.2.1 Undamped Waves 322
   13.2.2 Undamped Non-dispersive Alfvén Waves 322
   13.2.3 Influence of Damping and Dissipation 337
   13.2.4 Numerical Test of Quasilinear Theory 340
  13.3 Magnetosonic Waves 342
   13.3.1 Fast Mode Wave Fokker-Planck Coefficients 343
   13.3.2 Isotropic Fast Mode Waves 345
   13.3.3 Cosmic Ray Mean Free Path
      from Fast Mode Waves........................... 357
   13.3.4 Cosmic Ray Momentum Diffusion Coefficient
      from Fast Mode Waves........................... 359
   13.3.5 Fast Mode Time Scale Relation 361
  13.4 Cosmic Transport Parameters
   from Fast Mode Waves and Slab Alfvén Waves 361
   13.4.1 Modifications Due to Slab Alfvén Waves............ 361
   13.4.2 Transport Parameters in the Case of Admixture
      of Slab Alfvén Waves to Fast Mode Waves 362

14. Acceleration and Transport Processes of Cosmic Rays 365
  14.1 Structure of the Cosmic Ray Transport Equation 365
   14.1.1 Inclusion of Momentum Losses 367
   14.1.2 Continuous and Catastrophic Momentum Losses 368
  14.2 Steady-State and Time-Dependent Transport Equations 370
  14.3 Scattering Time Method:
   Separation of Spatial and Momentum Problem 371
   14.3.1 Formal Mathematical Solution 371
   14.3.2 Leaky-Box Equations 372
   14.3.3 Illustrative Example:
      Galactic Cosmic Ray Diffusion 1 373
   14.3.4 Momentum Problem 375
  14.4 Solutions Without Momentum Diffusion 378

15. Interplanetary Transport of Cosmic Ray Particles 383
- 15.1 Solar Flare Events 383
  - 15.1.1 Observations 383
  - 15.1.2 Comparison with Quasilinear Theory 387
- 15.2 Solar Modulation of Galactic Cosmic Rays 388

16. Acceleration of Cosmic Ray Particles at Shock Waves 391
- 16.1 Astrophysical Shock Waves 391
- 16.2 Magnetohydrodynamic Shock Discontinuities 392
  - 16.2.1 Stationary Discontinuities 395
  - 16.2.2 Shock Waves 397
- 16.3 Alfvén Wave Transmission Through a Parallel Fast Shock ... 402
  - 16.3.1 Basic Equations 403
  - 16.3.2 Reflection and Transmission Coefficients 405
  - 16.3.3 Only Forward Moving Waves Upstream 405
  - 16.3.4 Only Backward Moving Waves Downstream 409
  - 16.3.5 Alternative Notation 410
  - 16.3.6 Shock Effect on Wavenumbers 410
  - 16.3.7 Precursor Streaming Instability 413
- 16.4 Cosmic Ray Transport and Acceleration Parameters 415
  - 16.4.1 Upstream Cosmic Ray Transport Equation 416
  - 16.4.2 Downstream Cosmic Ray Transport Equation 417
  - 16.4.3 Scattering Center Compression Ratio 418
  - 16.4.4 Importance of Downstream Momentum Diffusion 419
  - 16.4.5 Interlude 420
- 16.5 Solution of the Transport Equations 421
  - 16.5.1 Basic Equations 422
  - 16.5.2 Upstream Solution 424
  - 16.5.3 Downstream Solution 424
- 16.6 Qualitative Results 428
  - 16.6.1 Zero Momentum Diffusion 429
  - 16.6.2 Finite Momentum Diffusion 431
- 16.7 Maximum Particle Energy 433

17. Galactic Cosmic Rays 435
- 17.1 Galactic Cosmic Ray Diffusion 2 435
- 17.2 Cosmic Ray Propagation in the Leaky-Box Approximation ... 436
  - 17.2.1 Secondary/Primary Ratio 437
  - 17.2.2 Secondary Cosmic Ray Clocks 438
- 17.3 Cosmic Ray Propagation Including Interstellar Acceleration .. 439
  - 17.3.1 Mathematical Solution 441
  - 17.3.2 Sources of Cosmic Ray Primaries 443
  - 17.3.3 Steady-State Primary Particle Spectrum Below R_{max} 445

17.3.4 Influence of Source Spectral Index Dispersion
          on the Steady-State Primary Particle Spectrum
          Below R_{max} 450
  17.3.5 Steady-State Primary Particle Spectrum Above R_{max} 452
 17.4 Time-Dependent Calculation 455
  17.4.1 General Values $\chi \neq 0$ 455
  17.4.2 Special Case $\chi = 0$ 457
 17.5 Secondary/Primary Ratio Including Interstellar Acceleration . 458

18. Formation of Cosmic Ray Momentum Spectra 461
 18.1 Fundamental Physical Picture 461
 18.2 Some Exact Solutions for $q = 2$ 464
  18.2.1 Shock Acceleration of Relativistic Electrons
          Undergoing Radiation Losses 465
  18.2.2 Shock Acceleration of Relativistic Electrons
          Undergoing Ionization and Coulomb Losses 467
 18.3 Bessel Function Solutions 467
 18.4 Interlude 468

19. Summary and Outlook 469
 19.1 Cosmic Ray Sources 470
 19.2 Cosmic Ray Isotropy 471
 19.3 Cosmic Ray Isotopic and Elemental Composition 472
 19.4 Electron/Nucleon Ratio 472
 19.5 Formation of Particle Momentum Spectra 474
 19.6 Conclusion 474

Appendix

A. Radiation Transport 475

B. Conductivity Tensor in a Hot Magnetized Plasma 479
 B.1 The General Case 481
 B.2 Isotropic Distribution Functions 482

C. Calculation of the Integral (10.3.10) 485
 C.1 The Integral J for Imaginary Values of y 486
 C.2 The Integral J for Non-imaginary Values of y 487
 C.3 A Second Form of the Integral J 490
 C.4 The Integral J for Non-relativistic Temperatures 491

References 495

Illustration Credits 511

Index 513

1. Introduction

1.1 The Start of Cosmic Ray Research

At six o'clock on the morning of August 7, 1912, the Austrian physicist Victor Hess and two companions climbed into a balloon gondola for the last of a series of seven launches. The flight, which had started at Aussig on the river Elbe, was under the command of Captain W. Hoffory. The meteorological observer was W. Wolf and Hess listed himself as "observer for atmospheric electricity". Over the next three or four hours the balloon rose to an altitude above 5 km, and by noon the group was landing at Pieskow, some 50 km from Berlin. During the six hours of flight Hess had carefully recorded the readings of three electroscopes he used to measure the intensity of radiation and had noted a rise in the radiation level as the balloon rose in altitude.

In the *Physikalische Zeitschrift* of November 1st that year, Hess wrote, "The results of these observations seem best explained by a radiation of great penetrating power entering our atmosphere from above ..." This event, as described by Harwit (1981 [215]), was the beginning of cosmic ray astronomy. Twentyfour years later Hess shared the Nobel prize in physics for his discovery.

1.2 The Scope of Cosmic Ray Research

The main problem of astrophysics is the analysis of the electromagnetic emission from astronomical objects. To a major extent, the early evolution of this discipline was connected with the progress in atomic physics without which the interpretation of optical, ultraviolet, near-infrared, and often of radio spectra would have been impossible. More recently, the opening of new observational windows, going hand-in-hand with the ability of launching satellite observations, involving X-rays, γ-rays, far-infrared, submillimeter and radio waves, has revolutionized the discipline by revealing the existence of new and exciting phenomena. Due to the importance of plasma physics in the understanding of many of these phenomena, the discipline of plasma astrophysics has emerged. The development of modern astrophysics relies to a large extent on our understanding of plasma physics. Of particular interest are the dynamics of energetic charged particles, their interaction processes with electromagnetic fields, their collective processes, their spontaneous and coherent radiation processes and the role of nuclear interactions with ordinary matter.

A problem arises because many astrophysicists are unfamiliar with theoretical plasma physics and nuclear physics, while many plasma and nuclear physicists are unfamiliar with the astronomical observations. For the former to a major extent this stems from the training and education of astrophysics students which still is oriented on the traditional astrophysical topics such as stellar structure, stellar evolution and radiation transfer in stellar atmospheres. In stellar cores and envelopes the matter densities are large enough that hydrodynamics and magnetohydrodynamics are appropriate descriptions of the collision-dominated gases in these objects, and sophisticated analytical and numerical methods have been developed and applied with remarkable success. During the past years we have witnessed, however, that these methods have also been applied to more diffuse objects, such as the interstellar and intergalactic medium, which by volume occupy most of the universe. But for these objects, often ionized and interwoven with electromagnetic fields and a high-energy particle component, because of the low gas densities a hydrodynamical description is not appropriate, and a full kinetic theory for all constituents has to be developed. In view of this, there is a clear need for a monograph on cosmic ray plasma astrophysics which, at a technical level, develops the basic principles of plasma physics with emphasis on the different conditions in cosmic plasmas as compared to terrestrial laboratory and fusion plasmas. Of particular interest are the dynamics of energetic charged particles, their coupling to thermal matter and large-scale magnetic fields, because of their important role in generating observable radiation. These will be the central topics of this book.

Historically, radiation processes have been divided into thermal and nonthermal processes. Nonthermal processes (e.g. synchrotron radiation, curvature radiation, nonthermal bremsstrahlung, inverse Compton scattering, plasma radiation) are caused by interactions of energetic particles with ambient cosmic matter, magnetic and photon fields. The frequency distribution of nonthermal radiation is determined by the energy distribution of the radiating particles. Thermal radiation is emission from an agitated plasma of given temperature and degree of ionization. The temperature of a cosmic plasma is calculated from the balance of heating and cooling processes assuming its chemical composition. One important heating process is due to energetic charged nucleons: a fraction of the energy (about 20 eV) lost by a nucleon in ionizing the plasma is available as heat. The degree of ionization results from the balance of recombination and ionization rates where the latter is mainly due to soft X-radiation and energetic nucleons. Since the degree of ionization and the temperature determine the plasma emissivity, knowledge is required of the flux of energetic nucleons in cosmic plasmas. Today there is observational evidence that the presence or absence of energetic particles in cosmic objects influences the optical line emission.

For the understanding of both thermal and nonthermal radiation processes, knowledge of the energy distribution function of energetic particles is a key problem. This is closely related to the question: how and where are charged particles accelerated to the highest energies. It is the purpose of this monograph to summarize research on the origin and dynamics of cosmic rays. We try to demonstrate that a comprehensive understanding of the behavior of energetic particles in astrophysical objects arises if elementary interaction processes of the cosmic rays are combined. The equilibrium energy distribution of the particles results from the balance of acceleration processes and the relevant energy loss processes. A variety of different observations can be understood by this approach.

1.3 Organization of the Book

As the table of contents indicates, the book is divided into three parts:

Part I: Observations of cosmic rays,
Part II: Theory of cosmic rays,
Part III: Astrophysical applications.

We start in Part I with a definition of what cosmic rays are and in what cosmic environment they exist. In space they interact with ambient photons, matter and magnetic fields whose observational evidence is reviewed in Chap. 2. Chapter 3 summarizes the measurements of cosmic ray particles both by satellites in space and ground-based detectors for the very high energy cosmic rays. We distinguish between cosmic rays of solar origin and galactic origin. Measurements of the elemental and isotopic composition, of the energy spectra and of the anisotropy are discussed. The most remarkable measurements are the high degree of isotropy and the long mean residence time $T \simeq 10^7$ yr of galactic cosmic rays. Since direct measurements of cosmic ray particles are restricted to the near vicinity of the Earth, any information of cosmic radiation in more distant regions of space is based on the detection of cosmic ray interaction products as electromagnetic radiation. In Chaps. 4 and 5 we therefore investigate in detail the various interaction processes of cosmic ray electrons (Chap. 4) and cosmic ray nuclei (Chap. 5) with photon and matter fields. Whereas relativistic electrons produce observable radiation in almost any frequency band, cosmic ray nuclei can be traced by detecting gamma-ray line and continuum radiation. Cosmic objects emitting nonthermal radiation have to be regarded as potential sources of cosmic rays: whether they can power the whole cosmic ray luminosity of a galaxy depends on the source abundance and the efficiency to accelerate particles to high energies. In Chap. 6 we discuss the available radio continuum and γ-ray surveys. Most importantly, they provide valuable information on whether the number densities of cosmic rays measured near Earth are representative for the whole Galaxy. The immediate consequences of the various observations, in particular on the energy content of the Galaxy in cosmic rays and the implied requirements on potential sources of cosmic rays, are discussed in Chap. 7. Part I will provide the reader with an up-to-date review of the observed properties of cosmic radiation in space.

Part II starts from the observational fact of low number densities of cosmic rays in cosmic objects, which requires a kinetic description of the cosmic ray plasma (Chap. 8) because the Coulomb collisional length is much too large. Cosmic rays and the background plasma are coupled by plasma wave turbulence (Chaps. 9 and 10), generated very often by the cosmic rays themselves (Chap. 11). As we make clear in Chap. 8, the fundamental equations to be solved are Maxwell's equations, where the particle distribution functions enter via the charge and current densities and determine the electromagnetic fields, and a collisionless Boltzmann equation for the particle distribution functions in which the electromagnetic fields enter via the Lorentz force. This system of coupled equations, of course, is highly nonlinear, since the particles determine the electromagnetic fields and vice versa, and resembles the fundamental problem of cosmic ray plasma physics. In order to proceed with a solution, two opposite points of view can be taken:

1. the test wave approach (Chaps. 9–11), in which the particle distribution functions are assumed to be given, so that the resulting plasma waves, their properties and their generation can be investigated;
2. the test particle approach (Chaps. 12 and 13), in which the electromagnetic fields are assumed to be given, and the response of the particles can be studied.

A consistent theory should combine both approaches thereby avoiding dramatic contradictions. The test wave approach is used to investigate the properties of the most relevant plasma waves in cold and hot magnetized plasmas and their generation by cosmic-ray-induced instabilities. It is demonstrated that cosmic rays efficiently generate low-frequency magnetohydrodynamic waves. The response of the cosmic rays to these low-frequency plasma waves in the test particle approach is the subject of Chap. 12: we derive the various particle transport equations and discuss their range of validity and applicability. Because of the efficient generation of low-frequency plasma waves, a nearly isotropic cosmic ray particle distribution function is quickly established, in excellent agreement with the observed isotropy of cosmic rays, which evolves according to the fundamental diffusion-convection equation whose diffusion and convection rates are calculated in Chap. 13. The structure of the diffusion-convection equation, the inclusion of energy loss processes, and mathematical solutions, including the limiting case of leaky-box solutions, are studied in Chap. 14. Part II will provide the reader with the necessary tools to understand the current theoretical developments in cosmic ray research.

In Part III several applications of the mathematical solutions of the fundamental diffusion-convection transport equation are presented which demonstrate how the key observations of cosmic radiation can be accounted for. The successful qualitative and quantitative modeling of interplanetary cosmic ray transport serves as an important test for the validity of this transport equation and is described in Chap. 15. A central topic is the acceleration of cosmic rays near shock waves (Chap. 16). After a classification of various types of shock waves we discuss in detail the diffusive acceleration of cosmic rays at parallel fast shock waves and the stochastic acceleration by cyclotron damping of the enhanced Alfvénic turbulence near shocks. Another application is concerned with the propagation of galactic cosmic radiation (Chap. 17) with and without inclusion of interstellar distributed acceleration. Finally in Chap. 18 we demonstrate that for cosmic sources in general the combination of plasma wave acceleration processes with the relevant energy loss processes of the particles has the capability to reproduce the observed particle distributions in solar flares, galactic centers and other objects.

Finally I want to mention what is not covered in the text:

1. experimental techniques of cosmic ray physics,
2. physical processes in magnetospheres of the Earth, other planets and stars.

2. Cosmic Rays as Part of the Universe

In the preceding introductory chapter cosmic radiation has been defined as extraterrestrial charged particle radiation, i.e. it consists of a flux of electrons, positrons and nucleons with kinetic energies greater than 1 keV that bombards the Earth from outside. To understand the origin and dynamics of these particles we have to recall some basic astronomical observations concerning the structure and content of the universe.

2.1 Our Present View of the Structure of the Universe

Naked eye stars are members of a large stellar system–called the *Milky Way* or the *Galaxy* –which has the form of a flat disk with a radius of $\simeq 15$ kpc (1 kpc $= 3.086 \times 10^{21}$ cm) and a thickness of approximately 0.5 kpc (see Fig. 2.1). The disk rotates around its center. The Milky Way contains about 10^{11} stars which exhibit a great variety of properties, from the extremely large red giants to tiny white dwarfs and neutron stars. The Galaxy has not always been the same over its lifetime: new stars are formed in the diffuse interstellar medium between the stars; old stars of earlier generations end their lives sometimes in gigantic explosions as supernovae or in more gentle explosions as planetary nebulae, delivering their burned-out ashes back to the interstellar medium with a new chemical composition as a result of nuclear burning in the stars. With each new star generation the Galaxy becomes a little richer in heavy elements and a little more mass becomes trapped in stellar cinders called white dwarfs and neutron stars.

According to the convention of the International Astronomical Union the solar system lies at a distance of 8.5 kpc away from the center of the disk–called the *galactic center* –nearly in the middle of the galactic disk (Kerr and Lynden-Bell 1986 [259]). With respect to the galactic center the coordinates (longitude l, latitude b) have been introduced. They are the angles of a spherical coordinate system with the Earth as origin; the direction ($l = 0°, b = 0°$) points to the galactic center, at directions $b = +90°$ and $b = -90°$ we look straight up and down out of the disk to the poles, respectively. The longitude l is the angular distance from the center line measured counterclockwise as seen from the north galactic pole, and the latitude b is the angular distance above ($+$) or below ($-$) the plane of the disk. A galactocentric cylin-

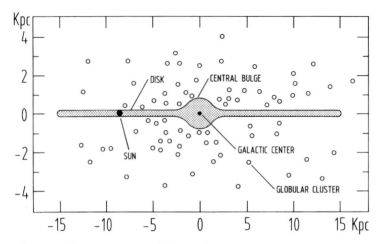

Fig. 2.1. Schematic view of the Milky Way stellar system, viewed edge-on by an observer outside the Galaxy, and its dimensions in units of 1 kpc = 3.086×10^{21} cm. The structure of the Galaxy consists of three components: a thin disk of stars, gas and dust; a central bulge; and a roughly spherical halo of globular star clusters

drical coordinate system (R, ϕ, z) with the galactic center as origin is also useful; in terms of these coordinates the solar system is located at a galactic radius $R_0 = 8.5$ kpc, an azimuthal angle $\phi_0 = 180°$, and a galactic height $z_0 = 0$ kpc. If D denotes the distance of an astronomical object from Earth, the relationship between the two coordinate systems is

$$R = \left[R_0^2 + D^2 \cos^2(b) - 2R_0 D \cos(b)\cos(l)\right]^{1/2}, \quad (2.1.1)$$
$$\phi = \phi_0 + \arcsin\left[D\cos(b)\sin(l)/R\right], \quad (2.1.2)$$
$$z = D\sin(b), \quad (2.1.3)$$

as can be inferred from Fig. 2.2.

(1 A.U. = 1.5×10^{13} cm = 5×10^{-9} kpc) from the center of the Sun, a negligible distance compared to galactic dimensions. The Sun is the closest star to us: light takes about 8 min to travel from the Sun to the Earth. From the second closest star, Proxima Centauri, light takes about 4 yr to reach us. Because of its proximity, the Sun is the best studied astronomical object, so that solar physics has always played a leading role in many branches of astrophysics; as we shall see this is also the case in cosmic ray astrophysics. As illustrated in Fig. 2.3 the Sun blows off a proton–electron gas that streams past the Earth with a mean velocity of 400–500 km s^{-1} and a mean proton and electron density of about 5 particles cm^{-3} which varies with the solar activity (flares and sunspots). This *solar wind* blows a cavity inside the galactic disk which extends to an estimated radius of $\simeq 100$ astronomical units and is referred to as the *interplanetary medium* or *heliosphere*. Besides our own Milky Way, there exist many other stellar systems, the *galaxies*. Galaxies are

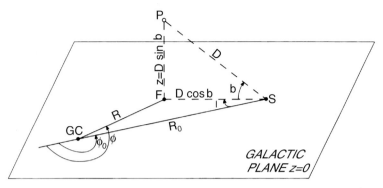

Fig. 2.2. Connection between the heliocentric galactic coordinates (l, b, D) and the galactocentric cylindrical coordinate system (R, ϕ, z). By applying simple trigonometric relations to the triangles Δ (PFS) and Δ (SFGC) (2.1.1)–(2.1.3) follow immediately

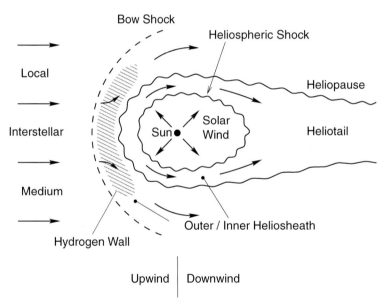

Fig. 2.3. A possible model of the heliosphere. From Fichtner (2000 [167])

not smoothly distributed in space. They occur in *clusters* and *groups*. For example, our own Milky Way is one of the two largest systems in the Local Group of about two dozen galaxies distributed in a volume of about 10^3 kpc in diameter. Apart from small groups like our own Local Group, galaxies are also found in great *clusters*, containing thousands of members. Most cluster galaxies are roundish ellipticals rather than the more common spirals found in the general field. On the basis of their redshift galaxies have been found at distances exceeding 10^6 kpc from us, indicating the tremendous size of the observable universe.

The space between galaxies and between the stars in individual galaxies is not empty. The space between the stars is populated by *interstellar matter*, the total mass of which, in our Milky Way, is about 10% of the stellar mass. The space between the galaxies is filled by *intergalactic matter*. Let us describe the properties of this non-stellar matter.

2.2 Interstellar Medium

The space between the stars has been studied both in our Milky Way and in recent years also in other galaxies. In our Milky Way we can study individual gas clouds in great detail because of their relative proximity which allows good spatial resolution with present-day telescopes. However, because of the location of the solar system within the galactic disk it is difficult to infer the *global* properties of the interstellar medium. The old saying "you can't see the forest for the trees" applies very much to this situation. On the contrary, in the more distant galaxies we can study global properties by viewing, at different angles, galaxies similar to our own but, because of poor resolution, we cannot study the individual "trees" with great detail. The space between the stars of our Milky Way is populated by diffuse matter, photons and cosmic rays. The interstellar matter is made up of *gas* and *dust* with an average mass ratio of 100:1. Interstellar gas is composed mainly of hydrogen (70% of the gas mass), helium (28%) and only a 2% contribution from heavier elements. According to its stellar history the chemical composition of the interstellar medium varies smoothly throughout the Milky Way (Güsten and Megger 1982 [205]). Hydrogen appears in different states: HI, H_2 and HII hydrogen.

The neutral *atomic HI hydrogen* is either clumped in HI clouds of temperature $T \simeq 50$ K and density $n \simeq 30$ cm^{-3}, or distributed diffusely as intercloud medium of temperature $T \simeq 10^4$ K and density $n \simeq 0.1$ cm^{-3}. Its large-scale, i.e. averaged over the azimuthal coordinate ϕ and radial sizes of 0.5 kpc, density distribution can be represented by (Burton 1988 [84])

$$n_{\mathrm{HI}}(R,z) = n_{\mathrm{HI}}^0(R) \exp\left[-\frac{z^2}{2h_{\mathrm{HI}}^2(R)}\right] \text{ cm}^{-3}, \qquad (2.2.1)$$

where the dispersion $h_{\mathrm{HI}}(R)$ is equal to 0.12 kpc for $R \leq R_0$ and increases at a rate $dh_{\mathrm{HI}}/dR = 0.04$ with increasing R. The radial distribution $n_{\mathrm{HI}}^0(R)$

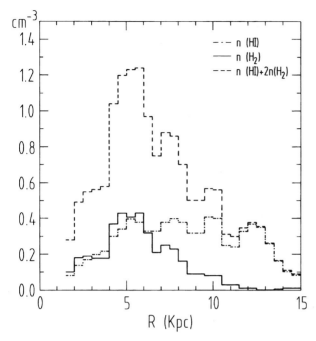

Fig. 2.4. Large-scale radial density distribution of atomic (HI) and molecular (H$_2$) hydrogen in the Galaxy. From Gordon and Burton (1976 [195])

is displayed in Fig. 2.4, leaving out the galactic center region ($R < 1.5$ kpc) where the simple representation (2.2.1) does not hold. Between $R = 4$ and 13 kpc the distribution is nearly constant with $n_{\rm HI}^0 \simeq 0.3$ H atoms cm^{-3}. *Molecular hydrogen* H$_2$ can exist only in dark cool clouds where it is protected against the ionizing stellar ultraviolet radiation. Roughly 40% of the *mass* of the interstellar hydrogen appears in this form. A tracer for the spatial distribution of H$_2$ molecules is the $\lambda = 2.6$ mm ($J = 1 \to 0$) emission line of carbon monoxide CO, since collisions between the CO and H$_2$ molecules in the clouds are responsible for the excitation of the carbon monoxide. The H$_2$ large-scale density distribution can be represented by

$$n_{\rm H_2}(R, z) = n_{\rm H_2}^0(R) \exp\left[-\frac{z^2}{2h_{\rm H_2}^2}\right] \quad {\rm cm}^{-3}, \qquad (2.2.2)$$

with $h_{\rm H_2} = 0.05 \pm 0.01$ kpc and where $n_{\rm H_2}^0(R)$ is shown in Fig. 2.4. The radial distribution $n_{\rm H_2}^0(R)$ exhibits a pronounced maximum between 4 and 6 kpc and has density values between 0.15 and 0.45 H$_2$ molecules cm^{-3} in the inner cycle ($R < R_0$) of the Milky Way whereas in the outer regions ($R > 11$ kpc) the molecular hydrogen density is very small.

In the vicinity of young, bright O and B stars the interstellar gas is ionized by intense ultraviolet radiation from these stars. These regions are called *HII*

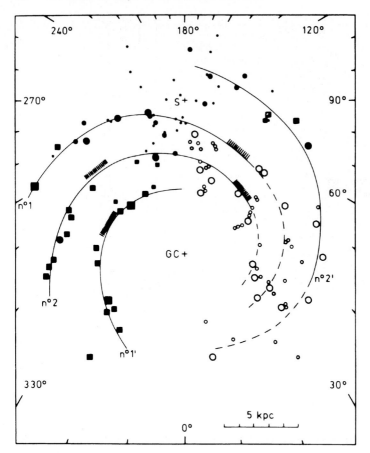

Fig. 2.5. Large-scale distribution of giant HII regions in the galactic plane tentatively represented by a four-arm spiral pattern (1 Sagittarius-Carina arm, 2 Scutum-Crux arm, 1′ Norma arm, 2′ Perseus arm). From Downes and Güsten (1982 [133])

regions. Their radial distribution is similar to that of molecular hydrogen, but mass-wise their contribution is negligible. HII regions are a good tracer of the spiral structure in the Milky Way. In Fig. 2.5 we show the location of the HII regions in the galactic plane in the solar vicinity. One notices the emergence of different spiral arms in our Galaxy. In Fig. 2.6 we show a beautiful example of spiral structure in the galaxy NGC 4622 where one clearly recognizes different spiral arms.

Observations of the soft X-ray background and interstellar ultraviolet O VI absorption lines (York 1982 [571]) have revealed the existence of a further component of the interstellar gas: a dilute (density $n \simeq 10^{-3}$ cm^{-3}), hot (T $\simeq 10^5$–10^6 K) plasma, in its temperature and composition properties similar to the solar corona. This *coronal gas component* may occupy up to 80% of the *volume* of the interstellar gas. According to our current understanding

Fig. 2.6. Optical photograph of the external galaxy NGC 4622

the coronal component results from intersecting supernova remnants. Outgoing supernova shock waves sweep up interstellar material leaving behind a dilute hot gas. The basic theoretical argument goes as follows: suppose that the coronal gas in the interior of a spherical supernova remnant with radius r occupies an average volume $V = 4\pi r^3/3$ and survives for a time τ. Assume supernova explosions occur randomly in the Milky Way with a rate S of supernova explosions per unit volume per unit time. Let f_1 be the probability that some randomly chosen point is inside a supernova remnant at some instant t_1, and let f_2 be the same probability at instant t_2, a short time later. Clearly, the probability f_2 equals the probability f_1 that the point was in a remnant at time t_1 multiplied by the probability $[1-((t_2-t_1)/\tau)]$ that the remnant has survived to time t_2, plus the probability $(1-f_1)$ that the point was not in a remnant at time t_1 multiplied by the probability $SV(t_2-t_1)$ that the remnant has been created at that point during the elapsed time interval $(t_2 - t_1)$:

$$f_2 = f_1 \left[1 - ((t_2 - t_1)/\tau)\right] + \left[(1 - f_1) SV(t_2 - t_1)\right] \ . \tag{2.2.3}$$

In a statistically steady state, f_2 must equal f_1. Setting $f_2 = f_1 = f$ in (2.2.3) one finds for the fraction of the Milky Way volume occupied by the interiors of supernova remnants $f = Q/(1+Q)$ with $Q = SV\tau$.

For a supernova rate of 1 per 30 years in our Galaxy of volume $\pi(15)^2 0.5$ kpc^3 = 350 kpc^3 we obtain $S \simeq 10^{-4}$ kpc^{-3}year^{-1}. The volume and the survival time of a supernova remnant are very sensitive to the density of the pre-existing interstellar medium into which the supernova blast wave evolves. If the supernova explodes in the intercloud medium values of $r = 0.08$ kpc and $\tau = 10^7$ yr can be attained implying $Q = 2$ so that $f = 0.67$, and indeed a large volume fraction of the interstellar medium can be filled by this component.

Now in the real interstellar medium things are much more complicated, as supernova explosions may not occur randomly but correlated, and the background medium is not uniform but highly structured. So the final interstellar medium structure depends sensitively on the assumed initial conditions. Two typical end configurations have been suggested on the basis of detailed numerical simulations:

(i) a matter disk interwoven with a network of hot dilute *tunnels* (Cox and Smith 1974 [108]),
(ii) interstellar matter clumped in clouds embedded in a bath of hot, dilute coronal plasma (Ostriker and McKee 1977 [379]).

In the latter configuration the warm HI intercloud medium exists as a transition region between the clouds and the coronal medium with a volume filling factor $f \simeq 0.2$–0.3.

Having described the basic theoretical reasoning for the coronal phase let us return to its observational properties. Figure 2.7 shows the observed column density of five-times ionized oxygen O^{5+} in the direction of 72 galactic stars of different distance from the Sun as inferred from the ultraviolet O^{5+}

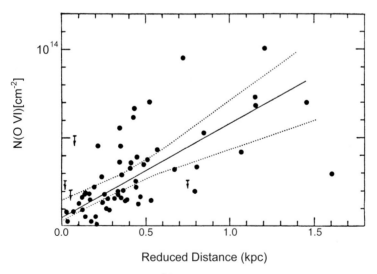

Fig. 2.7. Observed oxygen O^{5+} column density towards 72 galactic stars as a function of 'reduced' distances to the stars. The true distances towards stars away from the galactic plane were foreshortened to compensate for an overall trend toward lower O^{5+} densities at high z. From Jenkins (1978 [242])

absorption line measurements. One notices the correlation of this column density with distance indicating an interstellar origin of the hot absorbing material, which rules out a circumstellar origin. In the latter case the column density is determined by individual stellar properties and should be more or less the same independent of the distance from the star.

Figure 2.8 shows the expected number densities of differently charged species of a plasma of cosmic element abundance as a function of the plasma temperature. By measuring the N^{4+}/O^{5+} ratio $\simeq 0.1$ and other ion ratios in the directions to the stars βCen, αOri, HD28497, μCol it has been possible to restrict the temperature of the intervening gas to the range $(2-3) \times 10^5$ K (Hartquist and Snijders 1982 [213]). Soft X-ray observations (McCammon et al. 1983 [334]) also confirm the existence of this coronal gas, although quantitative measures of its distribution are difficult to obtain due to the photoelectric absorption of the soft X-rays by neutral interstellar gas.

Ultraviolet absorption line spectra have also been obtained from stars in our neighboring galaxy, the Magellanic Clouds, and they indicate the presence of hot $(T \simeq (0.7-1) \times 10^5$ K) gas at large galactic heights, which is responsible for the observed absorption. Figure 2.9 shows the inferred ion density distribution of C^{3+} and Si^{3+} as a function of galactic height which has been deduced by assuming the same rotation curve for gas at different

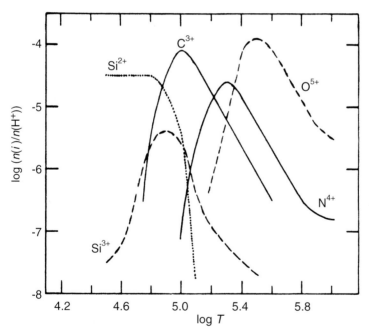

Fig. 2.8. Calculated fractional abundances of different ions in a plasma of cosmical element abundances as a function of the plasma temperature. From Hartquist and Snijders (1982 [213])

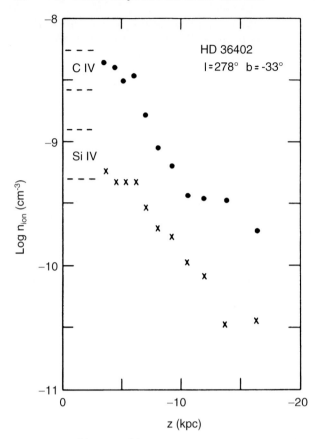

Fig. 2.9. C^{3+} and Si^{3+} densities toward the star HD 36402 located in the Magellanic Clouds as a function of z (distance below the galactic plane). The derived curves assume that the observed velocities in the line profiles represent the radial component of the co-rotating gas in the galactic halo. Due to velocity blending and saturation near $v = 0$ km s^{-1}, the densities between 0 and -3 kpc are only approximate. At very large z the co-rotation assumption is probably invalid. The presence of gas flows along this particular line of sight would of course invalidate the result. Instrumental smearing was ignored in converting the observed C^{3+} and Si^{3+} profiles into the density distribution illustrated. From Savage and DeBoer (1981 [446])

galactic heights (i.e. co-rotation). Adopting cosmic element abundances a total gas density of $n \simeq 3 \times 10^{-5}$ cm^{-3} at a height of 5 kpc can be deduced, yielding a gas pressure of $P = k_B n T \simeq 4 \times 10^{-16}$ dyn cm^{-2} which is large enough to confine the observed in-falling HI clouds at high galactic latitudes.

In Table 2.1 we summarize the various components of the interstellar gas with their main features. One can see that in the galactic plane ($z = 0$) there is approximate pressure equilibrium between them.

Table 2.1. Phases of interstellar matter (after Downes and Güsten (1982 [133]))

State	Gas phase	Number density (cm^{-3})	Kinetic temp. (K)	Volume filling (%)	Mass fract. (%)	Radio tracers
Molecular	Giant molecular clouds	10^3	10	≤ 2	40	CO, NH$_3$, H$_2$CO
	Dark clouds	$10^2 - 10^3$	10			
Atomic	HI clouds	30	50–100		40	HI
	Intercloud	0.1–1	$10^3 - 10^4$	50	20	
Ionized	HII regions	$1 - 10^5$	10^4	≤ 2	≤ 1	Recomb. Lines Free–Free
	Coronal gas	$10^{-4} - 10^{-2}$	$10^4 - 10^6$	20–80	≤ 1	UV-absorp. lines soft X-rays

Interstellar dust is an important constituent of the interstellar medium since it absorbs ultraviolet and optical photons and re-emits the radiation at infrared wavelengths. The dust is generally well-mixed with the total interstellar gas, and the dust density correlates well with the total gas density. In order to explain the observed average extinction curve and the scattering of optical light by dust particles, the dust has to be a mixture of grains of different sizes and chemical composition (Mathis et al. 1977 [330]). According to grain models, dust consists of a mixture of core-mantle grain and bare grains which are believed to be formed in the atmospheres of cool stars and ejected into the interstellar medium by radiation pressure (Sedlmayr 1985 [480]).

2.3 Cosmic Photon Fields

The space between the stars is further populated by cosmic photons. As a function of frequency the interstellar radiation field at any position r in the Milky Way can be well represented by the sum of diluted blackbody distributions (referred to as a *graybody distribution*) plus the isotropic microwave background radiation field. The graybody distribution is characterized by two parameters, the photon energy density $w(r)$ at position r and the temperature T, so that the number density of photons per unit volume and unit photon energy interval is

$$n_G(\epsilon, \boldsymbol{r}) = \frac{15w(\boldsymbol{r})}{\pi^4 (k_B T)^4} \frac{\epsilon^2}{\exp\left[\epsilon/(k_B T)\right] - 1} \quad \text{photons eV}^{-1} \text{ cm}^{-3}. \quad (2.3.1)$$

The identity

$$w(\boldsymbol{r}) = \int_0^\infty d\epsilon \, \epsilon \, n_G(\epsilon, \boldsymbol{r}) \quad \text{eV cm}^{-3} \quad (2.3.2)$$

is readily verified. The graybody distribution (2.3.1) has the same energy dependence as Planck's blackbody distribution but compared to the latter the photon number densities are smaller by the dilution factor $w(\boldsymbol{r})/w_B$ where

$$w_B = \frac{\pi^2 (k_B T)^4}{15(\hbar c)^3} \quad \text{eV cm}^{-3} \quad (2.3.3)$$

is the energy density of the blackbody distribution. $\hbar$ and c denote Planck's constant and the speed of light, respectively. Equation (2.3.3) is the Stefan-

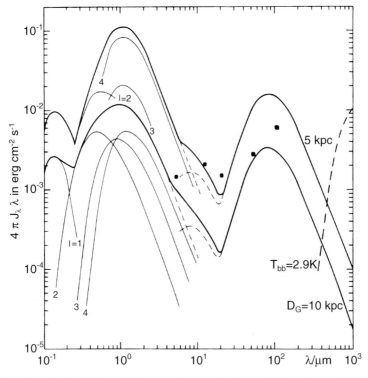

Fig. 2.10. Interstellar radiation field at galactocentric distance R_0 and $0.5 R_0$. (Note that here the old solar system distance $R_0 = 10$ kpc has been used.) **Curves** labeled $l = 1, 2, 3, 4$ relate to the contributions from four stellar components. For wavelengths ≥ 5 μm the radiation field is dominated by re-emission from dust grains. The **filled squares** represent the local interstellar radiation field from the IRAS data and from observations by Matsumoto et al. (1988 [332]) at 5 μm. From Cox and Mezger (1989 [109])

Table 2.2. Electromagnetic radiation field in the vicinity of the solar system

i	T_i/K	$W_i/\text{eV cm}^{-3}$	Comment
1	20 000	0.09	spectral type B
2	5 000	0.3	spectral type G–K
3	20	0.4	infrared
4	2.7	0.25	microwaves

Boltzmann law. Table 2.2 gives the parameters T_i and w_i of the various components in the vicinity of the solar system. The optical, infrared and microwave photon fields dominate the local interstellar radiation field whereas the energy densities of radio, ultraviolet and X-ray photons are negligibly small. Of course, in other regions of the Galaxy the values of these parameters are different due, for example, to a higher number density of photon emitting stars, different dust densities or additional radiation processes. Modeling of the interstellar radiation field over the Galaxy has been performed by Mathis et al. (1984 [331]) and Bloemen (1985 [63]). As an example we show in Fig 2.10 the interstellar radiation field in the galactic plane at the two galactocentric distances $R = R_0$ and $R = R_0/2$ as obtained by Cox and Mezger (1989 [109]). The agreement with recent measurements of the local radiation field is reasonably good.

2.4 Cosmic Magnetic Fields

There are a number of techniques to detect magnetic fields in astronomical objects. The basic information is provided by measuring the polarization of optical, infrared and radio emission (Beck 1986 [33], Sofue et al. 1986 [502]). However, the measurements are indirect. In the case of optical polarization it is believed to be the alignment of dust grains in interstellar magnetic fields that gives an observable effect. At radio frequencies, linearly polarized waves are generated by the nonthermal synchrotron radiation process. In both the optical and radio regimes, information is obtained only on the magnetic field component $B_\perp$, perpendicular to the line of sight. Information on the parallel magnetic field component $B_\parallel$ is provided by the Faraday effect (discussed in more detail in Sect. 9.2.8). The most direct method of measuring magnetic field strengths is to use the Zeeman splitting of spectral lines. In the radio frequency range this effect has been observed for the neutral hydrogen $\lambda = 21$ cm hyperfine structure line and the molecular OH and H_2O lines.

There have been a number of studies of the optical polarization of stars. The analysis by Ellis and Axon (1978 [151]) has given us information about the magnetic field structure in our local galactic neighborhood. They find evidence for an ordered field along the plane of the Galaxy pointing in the direction $l = 45°$, at least within a circle of 0.5 kpc. Beyond this distance the field seems to point towards $l = 60°$. This local field configuration may be due to a local bubble or a single loop of a helical field. Because of the absorption of optical light by dust particles, optical polarization studies allow us to probe only the local field. The polarized Galactic radio emission is dominated by loops and spurs in the solar galactic neighborhood (Spoelstra 1972 [507]); the general distribution is consistent with a local magnetic field in the direction $l \simeq 60°$, as suggested by optical polarization studies.

Measurements of the Faraday effect from extragalactic sources of polarized radio emission have given us a good understanding of the large-scale magnetic field of the Galaxy (Simard-Normandin et al. 1981 [488], MacLeod et al. 1988 [312]). The plane of polarization of a linearly polarized radio wave, as provided by the synchrotron process (see Sect. 4.1), is rotated when the wave passes through a plasma with a uniform magnetic field. Positive (counterclockwise) Faraday rotation occurs if the magnetic field is directed towards the observer; negative Faraday rotation occurs if the field points away from the observer. The rotation angle $\Delta \chi = RM \lambda^2$ increases with the line-of-sight integral

$$RM = \int_0^\infty n_e B_\parallel dl \quad \text{G cm}^{-2} \tag{2.4.1}$$

(where n_e is the thermal electron density along the line of sight), and with the wavelength of observation λ^2. The quantity (2.4.1) is called the *rotation measure*. An accurate determination of RM requires observations at least at three wavelengths because the observed direction of the polarization "vectors" is ambiguous by $\pm m\pi$, where m denotes any integer. One of the interesting results from RM studies is the occurrence of magnetic field direction reversals on scales of a few degrees.

In the case of pulsars (see Sect. 7.2) the RM measurements can be combined with measurements of the *dispersion measure* (DM) from the different time delays of the pulsar signal at different frequencies (see Sect. 9.2.1.1)

$$DM = \int_0^\infty n_e dl \quad \text{cm}^{-2} . \tag{2.4.2}$$

Obviously, the ratio $RM/DM = \langle B_\parallel \rangle$ determines the average parallel magnetic field component along different directions in the Galaxy without further assumptions. Studies by Lyne and Smith (1989 [309]) for 185 pulsars indicate an average field strength of about 2–3 µG (1 µG = 10^{-6} G).

Measurements of the Zeeman effect have now been made in HI clouds (Verschuur 1987 [549]) and molecular clouds employing the splitting of the OH (Crutcher et al. 1987 [110]) and H_2O maser (Fiebig and Güsten 1989 [168]) lines. The measured field strengths in the clouds vary proportional to the cloud's gas density from 10–100 µG in dark clouds up to 5×10^4 µG in high-density ($n \simeq 10^{10}$ cm^{-3}) cloud layers.

2.4 Cosmic Magnetic Fields

All the data gathered so far indicate that the magnetic fields in the Galaxy are along the spiral arms, i.e. azimuthal. The recent analysis by Vallee (1988 [548]) shows that any deviations of the pitch angle of the field from the spiral arms are small, probably less than 6° (see Fig. 2.11). Vallee also deduces a field reversal in the Sagittarius arm. On the other hand, the magnetic field in the center of our Galaxy is poloidal. Polarization measurements at $\lambda = 2.8$ cm and $\lambda = 0.9$ cm (Reich 1989 [425]) show that in the galactic center arc the field lines run exactly perpendicular to the galactic plane, i.e. parallel with galactic height z. This may be a hint of magnetic field generation in the centers of galaxies.

Magnetic fields have also been detected in external galaxies, both by optical and radio polarization studies. In external spiral galaxies the magnetic field in the galactic plane seems to follow the spiral arms, indicating a similar situation as in our own Galaxy. In Fig. 2.12 we show, as an example, the measured magnetic field structure of the galaxy NGC 6946 which we view face-on, i.e. from the poles. Clearly the field lines follow the individual spiral arms.

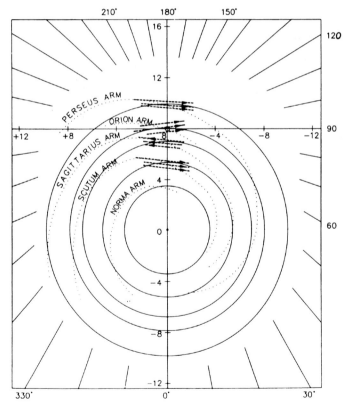

Fig. 2.11. The model of the magnetic field of the Galaxy by Vallee (1988 [548])

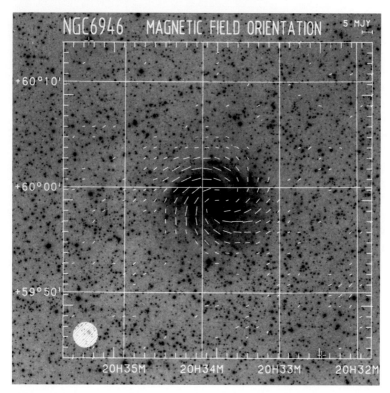

Fig. 2.12. The magnetic field structure in the external face-on spiral galaxy NGC 6946 derived from radio polarization measurements at $\lambda = 6$ cm wavelengths by Harnett et al. (1989 [210]). The lengths of the vectors are proportional to the flux density of the linearly polarized emission

Recently magnetic fields have been observed in the halos, i.e. at large distances out of the stellar disk ($z > 1$ kpc), of external galaxies. Hummel et al. (1988 [232]) have reported the detection of polarized radio emission from the galaxy NGC 4631, which we view from the side, referred to as *edge-on*. Polarized emission has been detected at galactic heights of $z \simeq 6$ kpc from the galactic plane of NGC 4631, and Fig. 2.13 shows the derived magnetic field structure. Apparently over a large halo region the field structure is poloidal, i.e. parallel to z.

From the experimental evidence we can conclude that the space between the stars in galaxies is filled by large-scale magnetic fields.

With respect to cosmic *electric fields* it is concluded that no large-scale steady electric fields can exist. The high electrical conductivity of non-stellar cosmic media would immediately short-circuit any steady electric field. However, transient electric fields may well be generated during non-stationary cosmic induction processes, as, for example, in solar and stellar flares and in pulsars.

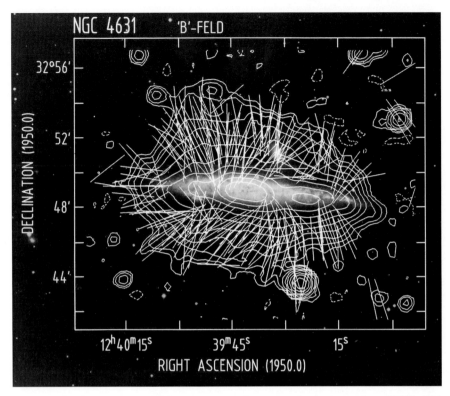

Fig. 2.13. The magnetic field structure in the external edge-on galaxy NGC 4631 derived from radio polarization measurements at $\lambda = 20$ cm wavelength by Hummel et al. (1988 [232]), assuming negligible Faraday rotation as indicated by the rotation measures of nearby extragalactic background radio sources

2.5 Intergalactic Medium

Our information on intergalactic space is much scarcer. From the detection of diffuse extragalactic X-ray background radiation we have evidence for the existence of a hot ($T \simeq 10^7$–10^8 K) dilute *intergalactic plasma*, such as that found near hot stars in the Galaxy. Its existence has important effects on the evolution of galaxies and the universe. A mean density of 3×10^{-6} atoms cm^{-3} would contribute enough mass to close the universe with the Hubble constant $H_0 = 50$ km s^{-1} Mpc^{-1} (Field 1980 [169]). However, quantitative estimates of the density of intergalactic matter from the measured X-ray intensities are very model-dependent and controversial. The only safe conclusion is that intergalactic space is not empty.

Perhaps our best knowledge of the properties of intergalactic matter comes from studies of the diffuse radio and X-ray emission from clusters of galaxies. Clusters of galaxies act as gravitational potential wells in which intergalactic matter clumps. By combining observations of the diffuse radio and hard

X-ray emission from the Coma cluster of galaxies the value of the average magnetic field strength has been restricted to be between 0.04 µG and 12 µG (Schlickeiser et al. 1987 [475]). Similar analyses for six other clusters of galaxies indicate a lower limit for the magnetic field strength of 0.1 µG (Rephaeli 1988 [430]).

2.6 Interlude

In this chapter we have described the astronomical surroundings in which cosmic ray particles exist. Before they can be detected by experimental devices near Earth, in their journey from their birthplaces they have to traverse the interplanetary, the interstellar and perhaps the intergalactic medium. During their travels they undergo interactions with the different constituents of these plasmas, and we will spend a great deal of this book studying these interactions and their consequences. But before we discuss this topic, let us summarize the observational status of cosmic rays.

3. Direct Observations of Cosmic Rays

With the technical capability of flying balloons high in the Earth atmosphere and launching satellites into extraterrestrial space carrying light-weight solid-state particle detectors, which was achieved after World War II, the original ground-based cosmic ray airshower experiments were supplemented by in situ measurements of the cosmic ray flux in the interplanetary space. Today, detectors on board spacecraft routinely provide spatial and temporal information on the particle energy spectra throughout the heliosphere.

The ground-based observations have the advantage of long exposure time and practically no limit on the size of the detectors but their results have to be corrected for the influence of the Earth's atmosphere (~ 1000 g cm^{-2} column depth of matter) through which the cosmic rays have to penetrate to reach the Earth ground. In the atmosphere the in-falling primary cosmic rays undergo inelastic collisions with the atmospheric atoms and molecules producing secondary particles which, again, are subject to further interactions, so that a whole "shower" of particles reaches the ground (Figs. 3.1 and 3.2).

Only with the help of numerical model calculations can one analyze the shower measurements to investigate the properties of the in-falling cosmic rays, so that the quality of the cosmic ray measurements from these methods depends on our understanding of the interaction processes in the atmosphere.

On the other hand, these disturbing atmospheric effects disappear if a cosmic ray experiment is flown outside the atmosphere but here, because of the limited size and weight of the detectors and the limited observing time of the experiment ($\leq$ several weeks in the case of space shuttle experiments), weak cosmic ray intensities cannot be measured. These instrumental limitations may improve with time as more sophisticated platforms (e.g. the space station) become available in the future. Today our information on the flux of cosmic rays below energies of $\sim 10^{14}$ eV/nucleon stems mainly from satellite and balloon experiments, whereas at higher energies $\geq 10^{14}$ eV/nucleon, because of the rather weak intensities, information on cosmic rays is provided solely by ground-based detectors. In this chapter we summarize the results from these direct particle detectors.

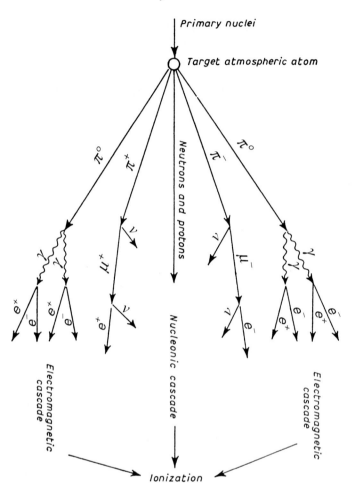

Fig. 3.1. Development of an extensive airshower in the Earth atmosphere. From Weekes (1972 [562])

3.1 The Cosmic Ray Landscape

Before entering a more detailed discussion of the various cosmic ray components, let us take a bird's eye view of the cosmic ray landscape as sketched in Fig. 3.3 (Lund 1986 [308]).

Most of this landscape is still hidden from our view by heavy clouds, but there are some general physical arguments that limit the ultimate extent of the region.

Up- and downwards in nuclear charge, nuclear stability sets two boundaries. Towards the high energy side the boundary is set by the universal 2.7 K background radiation (Sect. 2.3) which destroys heavy cosmic ray nuclei with energies in excess of $\sim 10^{19}$ eV/nucleon by photonuclear collisions

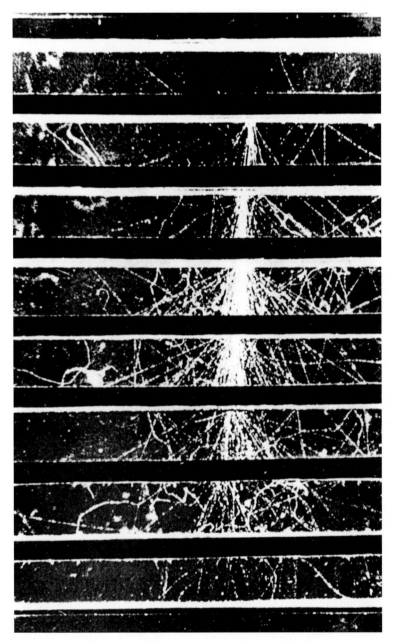

Fig. 3.2. Cloud chamber photograph of cascade shower through several thicknesses of material from an incident proton of energy of the order 4 GeV. Reproduced from Hopper (1964 [231])

28   3. Direct Cosmic Ray Observations

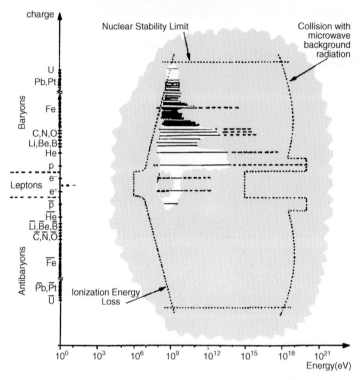

Fig. 3.3. An outline of the cosmic ray landscape after Lund (1986 [308]). Only a limited area is presently observable. The **dotted lines** indicate that total particle spectra are known, but the detailed composition is uncertain

$A + \gamma \to (A-1) + n$ (Greisen 1966 [202], Zatsepin and Kuzmin 1966 [573]) and ultrahigh-energy cosmic ray protons with energies larger than $\sim 10^{20}$ eV by photo-pion production $p + \gamma \to p + \pi^0$ (Sect. 5.2.2). Towards the low energy side a boundary exists due to ionization losses (Sect. 5.3.8) which increase rapidly with decreasing particle energy. These energy boundaries are not sharply defined since they depend on the amount of photons and matter, respectively, the particles have to penetrate on their way from their sources to us.

Lund's diagram in Fig. 3.3 has been drawn as if the Universe were symmetric with respect to matter and antimatter. Cosmic rays may ultimately prove that this symmetry exists because they are the only probe of matter of non-solar origin available to us. But so far this has not happened. The present-day 95% confidence upper limit for the ratio of antinuclei to nuclei with $Z \geq 2$ from cosmic ray observations (rigidity range 1.6–16 GV/c) is $< 8.1 \times 10^{-6}$ (Ormes et al. 1997 [377]) which is consistent with a purely secondary production of antimatter in inelastic baryon-baryon collisions $b + b \to \bar{b} + X$. If the Universe is matter-dominated everywhere, then the lower half of the diagram, with the exception of the antiprotons and positrons, can be removed. In that case we have, over a limited energy range, already explored the full extent of the cosmic ray charge scale. For two recent measurements of the cosmic ray antiproton flux and its interpretation we refer the reader to Hof et al. (1996 [228]), Boezio et al. (1997 [69]) and Simon et al. (1998 [489]). The observed antiproton fluxes in the energy range from 0.62 to 19 GeV are consistent with a purely secondary origin from inelastic nuclear interactions of positively charged cosmic ray nuclei with the atoms and molecules of the interstellar medium (Simon et al. 1998 [489]).

Having now an idea of the contours of the landscape we must ask what causes the clouds in our picture. Besides the instrumental limitations in combination with the weak fluxes of cosmic rays already mentioned, it is mainly the influence of the Sun, the other planets and the interplanetary medium. The interferences on the measurements of the galactic and extragalactic cosmic radiation are two-fold:

1. The outstreaming solar wind (Sect. 2.1) seriously disturbs the determination of particle fluxes below kinetic energies of about 500 MeV/nucleon for nuclei and below 5 GeV for electrons. This phenomenon is referred to as *solar modulation*.
2. The Sun itself and some of the planets produce cosmic radiation which has to be distinguished from the galactic and extragalactic component. The study of these *solar cosmic rays* is itself an interesting area of investigation since in situ observations of these particles provide much more detailed information than the limited studies of extrasolar cosmic rays (see Chap. 15).

Before we describe the observations of extrasolar cosmic rays we will discuss first solar modulation and solar cosmic rays.

3.2 Solar Modulation

The outer atmosphere of the Sun, the solar corona, is in continuous hydrodynamical expansion, producing a flow of plasma out through the interplanetary medium that is known as the solar wind (e.g. Parker 1963 [388]). Current estimates conclude that this expansion continues for at least 100–160 A.U. from the Sun. A magnetic field is frozen into this plasma, and as a dynamically passive partner in the expansion, is dragged radially outward from the Sun. The field remains attached to the rotating Sun while undergoing this expansion, so that the resulting large-scale field pattern is that of an Archimedes spiral (Parker 1958 [387]). Superimposed on this large-scale field are numerous small-scale field irregularities generated by the turbulence and instabilities in the solar corona and interplanetary medium (Fisk 1974 [170]). Figure 3.4 shows measurements of the solar wind velocity and magnetic field strength

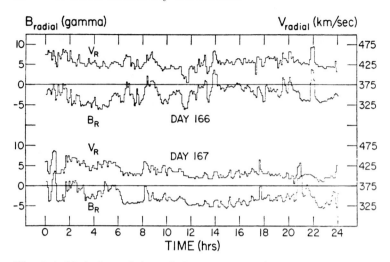

Fig. 3.4. Variations of the radial components of the interplanetary magnetic field, B_R, and of the solar wind velocity, V_R. The very good detailed correlation of the non-sinusoidal fluctuations is characteristic for Alfvén waves moving away from the Sun along B_R. From Belcher et al. (1969 [40])

that have led to the identification of the small-scale irregularities as Alfvén plasma waves.

The charged extrasolar cosmic ray particles are strongly influenced by this magnetic field as they penetrate from the outside into the heliosphere. Cosmic rays with rigidity R ($\equiv$ total momentum per unit charge) $\lesssim 10$ GV have Larmor radii smaller than the characteristic dimension of the sector structure anywhere in the interplanetary medium, and move along the large-scale interplanetary magnetic field. During their movement they are scattered by the small-scale magnetic field irregularities. Since the irregularities, along with the large-scale magnetic field, are moving essentially outward with the solar wind, this scattering process tends to sweep cosmic rays out of the solar system. Then the density of extrasolar cosmic rays that we observe at Earth is lower than the density present in the local interstellar medium, and the observed spectra are altered from their interstellar forms. The spectra are *modulated* as these particles pass through the interplanetary medium. This concept of solar modulation of galactic cosmic radiation was developed originally by Parker (1965 [389]) to explain the observed anticorrelation of the cosmic ray intensity with solar activity as represented in Fig. 3.5.

Roughly speaking, the cosmic ray intensity in the inner solar system is lowest when the solar activity, as measured by the sunspot number, is highest. This anticorrelation was reported first by Forbush (1954 [174]), based on small intensity variations in ion chambers, and has later been fully established after the launch of satellite cosmic ray detectors (for a review see Simpson 1989 [492]). Interplanetary conditions apparently change with the 11-year so-

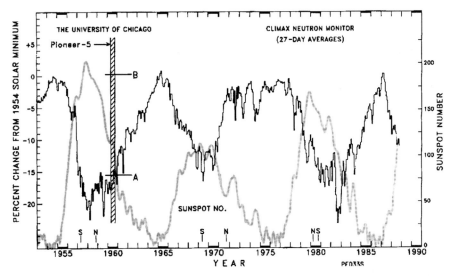

Fig. 3.5. Changes in the nucleonic component integral flux above ~ 3 GV (**solid line**). The solar activity is represented by the sunspot numbers (**dashed line**). From Simpson (1989 [492])

lar cycle and produce corresponding changes in the modulation. While the average solar wind speed, though showing large short-term fluctuations, is relatively constant over the solar cycle it is believed that the magnetic irregularities, responsible for the scattering of the particles, are strongly controlled by solar activity (Fisk 1974 [170]), although the detailed physical reasons are still unknown.

Figures 3.6 and 3.7 show the measured electron and ion energy spectra for the consecutive years 1968 to 1972 and from 1969 to 1972, respectively. The energy spectrum of relativistic electrons in the local interstellar medium can be independently estimated from their nonthermal synchrotron radio emission (Sect. 4.1) in the local interstellar magnetic field in the frequency range 0.4–400 MHz. From the measured frequency spectrum of this radio background Goldstein et al. (1970 [192]) and Cummings et al. (1973 [114]) have determined the interstellar electron spectrum at energies $\gtrsim 200$ MeV. In Fig. 3.8 we show a comparison of the so-obtained interstellar and the measured modulated electron energy spectrum in 1968.

By comparing the interstellar spectrum with the measured spectrum at Earth at every year, we obtain a measure of the total modulation experienced by the electrons. Figure 3.9 shows the derived rigidity dependence of the radial mean free path ($\lambda \propto \kappa/\beta$) resulting from the pitch-angle scattering of the electrons by the interplanetary magnetic irregularities needed to produce the modulated electron spectra. The main modulation parameters, derived from electron spectra, can then be used to demodulate the proton and α-particle spectra at Earth to derive the local interstellar spectra of these cosmic rays. Figure 3.10 shows the results for the 1968 measurements. A consistency check for this procedure is provided by using these interstellar spectra inferred from the 1968 data and calculating the expected modulated spectra in the years 1969–1972 with the respective modulation parameters from Fig. 3.9. Figure 3.7 indicates that the overall agreement between the calculated curves and the data is good above about 40 MeV/nucleon.

In general, modulation theory reproduces the shape of the spectra and the year-to-year changes remarkably well. However, on the basis of more detailed data sets, Garcia-Munoz et al. (1986 [181]) have discovered an underlying dependence of solar modulation on the sign of the cosmic ray particle charge over the solar cycle with relative year-to-year variations up to 60%, and thus weak compared to the order of magnitude modulation of the incoming interstellar cosmic rays. Understanding the basic physical processes that control and influence the penetration of cosmic rays from outside into the heliosphere and its first description by a simple diffusion model (Parker 1965 [389]) was a milestone of cosmic ray astrophysics and has led to the recognition that similar processes must take place on much larger scales in interstellar space.

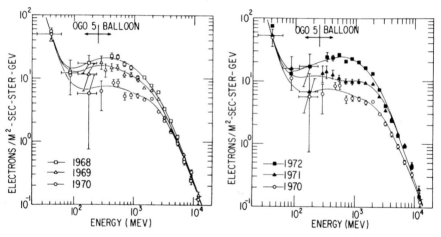

Fig. 3.6. Measured primary electron energy spectra for 1968, 1969, 1970, 1971 and 1972 from the University of Chicago electron experiment on board of OGO 5 and from balloon apparatus. The **curves** represent fits obtained by solutions of the transport equation describing solar modulation. From Fulks (1975 [179])

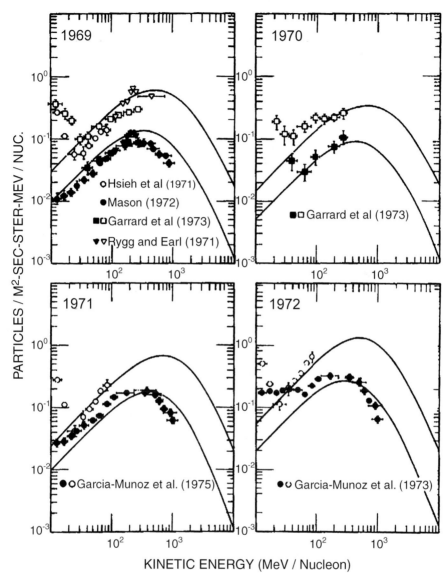

Fig. 3.7. Comparison of measured and calculated proton and α-particle modulated spectra. **Open points** represent proton measurements, **filled points** represent α-particle measurements. From Fulks (1975 [179])

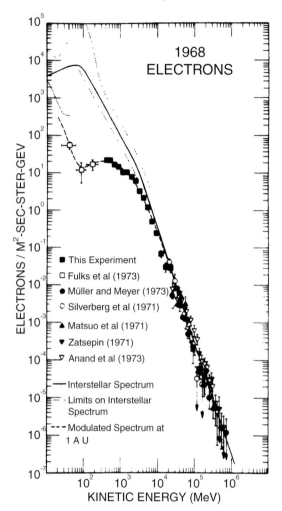

Fig. 3.8. Interstellar and modulated electron energy spectra. The **points** represent 1968 measurements except past 10 GeV, where data from other years are included. The **curves** show the nominal interstellar electron spectrum from Cummings et al. (1973 [114]) and the near-Earth modulated electron spectrum obtained by numerical solution of the transport equation describing solar modulation. From Fulks (1975 [179])

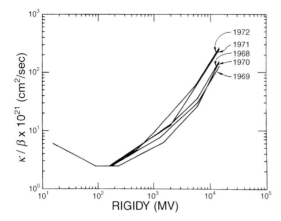

Fig. 3.9. The rigidity dependence of the radial diffusion coefficient κ divided by the particle velocity β needed to produce the modulated electron spectra curves shown in Fig. 3.6. From Fulks (1975 [179])

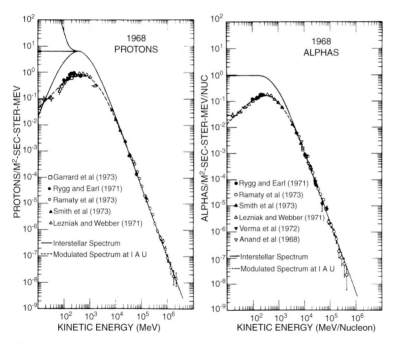

Fig. 3.10. Interstellar and modulated proton left and α-particle right energy spectra. The **points** represent 1968 measurements except past 1 GeV, where other years have been included. The **curves** show the derived interstellar spectrum and the corresponding near-Earth modulated spectrum. From Fulks (1975 [179])

3.3 Solar and Heliospheric Cosmic Rays

Evidence for a variety of particle acceleration processes is found throughout the heliosphere. The regions where acceleration of charged particles occurs, as well as the various particle populations resulting from these processes, are shown schematically in Fig. 3.11.

In many cases the acceleration takes place in association with shock waves (e.g. interplanetary traveling shocks, co-rotating forward-reverse shocks, bow shocks, possibly coronal shocks, solar wind termination shock) emphasizing their importance for particle acceleration. A beautiful example that interplanetary shock waves accelerate charged particles is shown in Fig. 3.12, clearly indicating that order of magnitude enhancements in the cosmic ray particle fluxes occur at the location of the shock wave.

An additional discovered cosmic ray particle component has to be mentioned. The first in situ observations of the cometary plasma environment (see special issues in the year 1987 of Science 232, p. 353–385, Nature 321, p. 259–366, and Astronomy and Astrophysics Vol. 187) resulted in the detection of both very high levels of magnetohydrodynamic turbulence, and large fluxes of energetic particles at the comets P/Giacobini-Zinner and P/Halley. The review by Terasawa and Scholer (1989 [542]) summarizes the highlights of the observations and their interpretation.

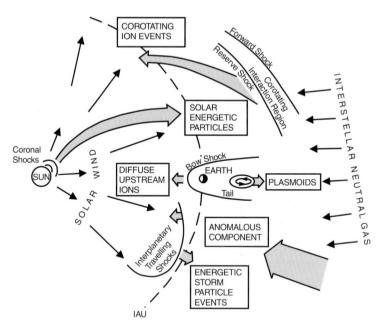

Fig. 3.11. Schematic representation of heliospheric acceleration regions and processes leading to suprathermal and low energy ion populations (listed in **boxes**) observed at the orbit of Earth. After Glöckler (1984 [188])

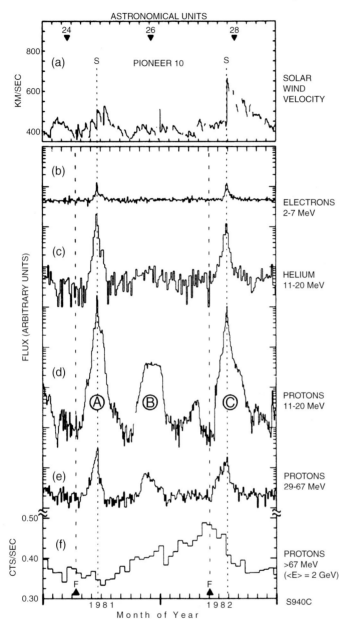

Fig. 3.12. Measurements from Pioneer 10 of the 2-year period 1981–1982. (a) The proton velocity (daily averages) measured by the solar wind plasma instrument; (b)–(f) count rates from the particle detectors. The averaging periods for (b)–(f) are, respectively, 2, 4, 3, and 15 days. **Solid triangles** indicate periods of major solar activity; S indicates the time of strong plasma shock passage. From Pyle et al. (1984 [413])

Here we concentrate on flare-associated events and the anomalous cosmic ray component. The solar associations of the other types of events listed in Table 3.1 are not so well established and some of them may represent a superposition of flare events.

Table 3.1. Nonthermal heliospheric particle populations. After Glöckler (1984 [188])

Population type	Energy range (keV/amu)	Dur.	Source plasma	Acceleration Region	Acceleration Mechanism
1. solar energetic particles (SEP)	$\lesssim 10^2$ to $>10^4$	days	$\sim 2\times 10^6$ K corona and cold material	solar corona	diffus. shock / stoch. Fermi
2. a) energetic storm particles (ESP)	$\lesssim 10$ to $\sim 2\times 10^3$	hrs-days	solar wind? SEP ?	propagating inter-planetary shock	diffus. shock shock drift turb. heating?
b) shock spikes		min-hrs			shock drift
3. diffusive upstream ions	$\lesssim 10$ to $\sim 2\times 10^2$	hrs	solar wind?	bow shock	diffus. shock
4. distant tail plasmoids	$\lesssim 3$ to $\sim 2\times 10^2$	10 s of min.	solar wind ionosphere?	distant tail	neutral sheet? turb. heating?
5. co-rotating ion events	$\lesssim 5\times 10^2$ to $\sim 10^4$	days*	solar wind?	F–R shocks (>2 AU)	diffus. shock turb. heating?
6. anomalous component	$\lesssim 10^3$ to $\geq 5\times 10^4$	yrs*	interstellar gas?	outer heliosphere, heliospheric boundary, or Galaxy	?

(* Observed near Earth primarily around solar minimum.)

3.3.1 Solar Flare Cosmic Ray Particles

Solar energetic particles are commonly accelerated in association with solar flare eruptions. It is widely believed that the onset of the flare is triggered by the sudden release of stored magnetic energy, probably due to reconnection of magnetic field lines, but the details of the flare process are an intense field of research (Stix 1989 [526]). Some of the energetic particles accelerated in these fields escape from the Sun and can be detected by spacecraft in the interplanetary medium. Others propagate down into the solar atmosphere where they produce X-rays and γ-rays which also can be recorded by spacecraft detectors. The recently discovered common 153-day periodicity in the occurrence time of soft and hard X-rays, γ-rays, optically Hα emission, radio and microwave emission and charged particle acceleration in solar flares (Rieger et al. 1984 [433], Dröge et al. 1990 [140]) proves the close physical connection of these phenomena.

Large solar flares are the most energetic natural particle accelerators in the solar system, occasionally accelerating ions to energies of several GeV (Fig. 3.13) and electrons to energies of several 100 MeV. The survey of interplanetary electron spectra from 0.1 to 100 MeV measured with the ICEE-3 satellite (Moses et al. 1989 [368]) has shown that the observed particle events can be divided into two classes, impulsive and long-duration. The long-duration flares, as referred to the > 1 h duration of their soft X-ray thermal emission, are well correlated with coronal and interplanetary shock waves and exhibit a power law behavior of the electron distribution function in rigidity, $N(R) \propto R^{-\gamma}$, $\gamma = $ const., see Fig. 3.14.

Table 3.1 lists the typical energy ranges, durations and other features for each accelerated population.

They also have rather small electron/proton ratios e (4–19 MeV)/p (9–23 MeV) ranging from 0.001 to 0.08 (Cane et al. 1986 [91]). In contrast, the impulsive flares have large electron/proton ratios covering the range from 0.02 to greater than 10, are almost never associated with interplanetary shock waves but correlate well with intense meter wavelength type III bursts with associated type V continuum (for a review of the radio emission from solar flares see Dulk 1985 [146]). Their electron spectra deviate significantly from a power law in rigidity (see Fig. 3.15) and exhibit a flattening below several MeV/c.

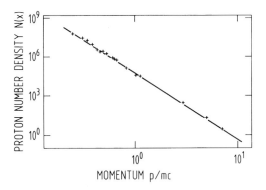

Fig. 3.13. High energy proton spectrum at Feb. 16, 1984 (data from Debrunner et al. 1988 [118]). From Steinacker and Schlickeiser (1989 [521])

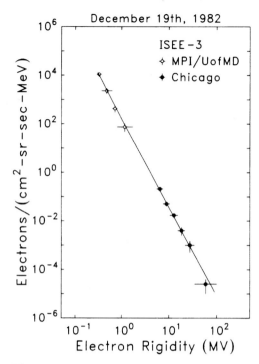

Fig. 3.14. Electron spectrum of the long-duration flares from December 19, 1982. From Dröge et al. (1989 [138])

The existence of two physically distinct classes of flares was originally established by Pallavicini et al. (1977 [384]) on the basis of their optical and soft X-ray emission properties. The impulsive flares have smaller volumes, occur at low coronal heights ($< 10^4$ km) and are associated with coronal mass ejections, whereas the long-duration flares occupy larger volumes and are located at large coronal heights ($> 5 \times 10^4$ km).

The study of ion spectra in different flares has been summarized by McGuire and von Rosenvinge (1984 [335]). Of course, the best studied flares are those with high proton and ion fluxes which, according to the work of Cane et al. (1986 [91]), are long-duration flares. For those it has been found that in the energy range from 0.8 to > 400 MeV for protons and 1.4 to 80 MeV/nucleon for α-particles; the representation as an exponential in rigidity and as a Bessel function in momentum per nucleon (p), $N(p) \propto p\, K_2(Bp^{1/2})$, B = const., best fits the observations. The spectra are seldom power laws. Typically they become steeper at higher momenta. Figure 3.16 shows three examples of proton and α-particle spectra.

Figure 3.16 also indicates that the p/α-ratio varies remarkably from flare to flare. For long-duration flares Lin et al. (1982 [304]) have established a tight correlation between the energy spectral indices of the non-

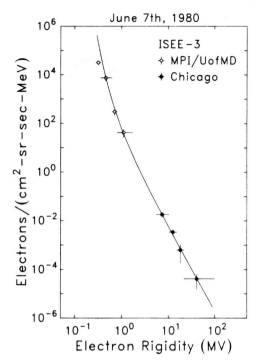

Fig. 3.15. Electron spectrum of the impulsive flares from June 7, 1980 (3:12 UT). From Dröge et al. (1989 [138])

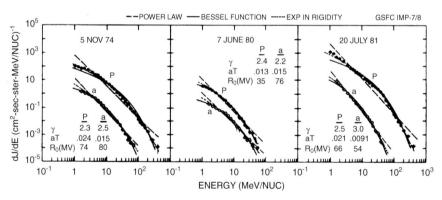

Fig. 3.16. Proton and α-particle spectra for three events. From McGuire and von Rosenvinge (1984 [335])

relativistic electrons below 100 keV (γ_e) and of the protons above 10 MeV ($\gamma_p = -\,\mathrm{d}\ln N(E)/\,\mathrm{d}\ln(E)$), see Fig. 3.17.

Ion spectra in impulsive flares have not been studied very systematically. However, rather unusual ion spectra have been reported for the prompt solar event from November 22, 1977 in the energy range from 20 to > 200 MeV/nucleon which we reproduce in Fig. 3.18.

More comprehensive measurements exist at particle energies below ~ 10 MeV/nucleon referring to the elemental and isotopic composition and the charge state of *solar energetic particles* (SEPs). Although it is now well es-

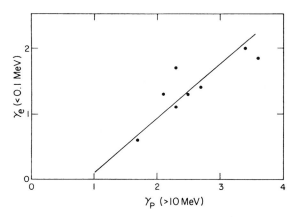

Fig. 3.17. Electron spectral index below 100 keV (γ_e) versus proton spectral index above 10 MeV (γ_p) for different flares. From Lin et al. (1982 [304])

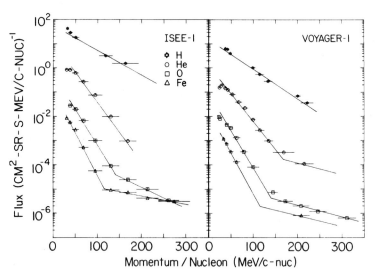

Fig. 3.18. Event-averaged differential spectra for the November 22, 1977 event. From Beeck et al. (1987 [36])

tablished that the relative elemental SEP abundances vary strongly from one event to the next, the composition averaged over many SEP-events shows a systematic deviation from compositions found in other source materials in our solar system (Glöckler 1984 [188]).

Figure 3.19 shows the ratio for the indicated elements of the SEP composition (normalized to oxygen) to the local galactic (LG) abundances similarly normalized as a function of the first ionization potential of that element. It is evident that SEPs with a low ($\lesssim 8$ eV) ionization potential I are overabundant by factors 3 to 5 with respect to the LG composition, while those with $I > 10$ eV have comparable abundances. Comparison with the average composition of the few elements (He, O, Ne, Si, Fe, Ar) which have been measured in the solar wind, on the other hand, shows no systematic differences implying that SEPs at least at these low energies are accelerated in the solar corona where the solar wind originates. The notable influence of the first ionization potential suggests that the lower coronal medium of the Sun is the birth place of SEPs since only in this region are the temperature and radiation fields such as to apply to first ionization potentials. Geiss and Bochsler (1984 [183]) examined ionization times in a highly non-equilibrium solar plasma which is irradiated by the solar UV spectrum, with a temperature 10^4 K and electron density of 10^{10} cm^{-3}. They find that elements with first ionization potential I below 7 eV are ionized essentially instantaneously (< 1 s), while the group N, O, Ne and Ar requires about 50 s to ionize, with He requiring a few hundred seconds to ionize. This fact, very suggestively, separates low and high first ionization potential elements as in Fig. 3.19. These calculations suggest that some charged/neutral separation process could operate in this environment, depleting the flow of high first ionization potential ions to the corona and thus into the reservoir for the solar wind and the solar energetic

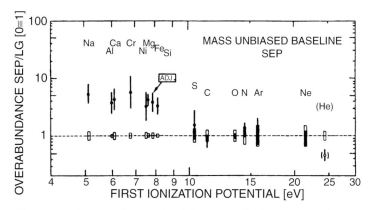

Fig. 3.19. Ratios of normalized (to oxygen) abundances of indicated elements in SEPs to the normalized (to oxygen) abundances of the respective elements in the local galactic (mostly photospheric) composition versus the first ionization potential of the element. From Meyer (1981 [347])

particles. It might also indicate that a "2-plateau" shape is the proper one for the first ionization potential enhancement ordering visible in Fig. 3.19 (Meyer 1985 [348]).

The isotopic composition of SEPs has also been measured. Most remarkable is the occurrence of ^{3}He-rich events. These events represent one of the most striking composition anomalies with ratios of the neighboring isotopes ^{3}He/^{4}He of order unity compared to ratios $\lesssim 10^{-3}$ in most flares, see Fig. 3.21. Figure 3.20 shows the measured ^{3}He/^{4}He-variation as a function of kinetic energy for several flares. Reames et al. (1985 [422]) have found that virtually all solar ^{3}He-rich events observed at kinetic energies $\gtrsim 1.3$ MeV/nucleon with the ISEE-3 spacecraft are associated with impulsive electron events, although many electron events were not accompanied by detectable ^{3}He increases. The ^{3}He-rich events are also associated with enhancements of heavy nuclei, especially iron. The isotopic abundances of

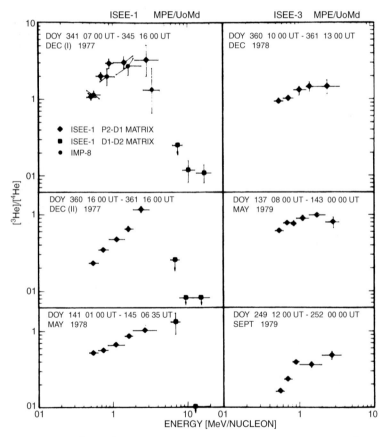

Fig. 3.20. ^{3}He/^{4}He ratio versus kinetic energy for selected ^{3}He-rich periods 1977–1979. From Möbius et al. (1982 [362])

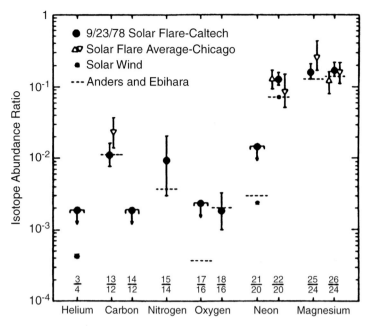

Fig. 3.21. Comparison of SEP-derived coronal isotopic abundances with the solar wind and solar system abundances from Anders and Ebihara. From Mewaldt and Stone (1989 [344])

heavy SEP elements agree well with the solar system abundances as Fig. 3.21 indicates. Compared with the solar wind abundances there is a factor of 2 enhancement in the ^{22}Ne/^{20}Ne ratio.

Another interesting measurement concerns the charge state of SEPs since various electromagnetic processes (plasma heating, acceleration, propagation) which produce SEPs depend on the mass-to-charge ratio of the ions. Measurements of ten SEP events (Fan et al. 1984 [156]), indicate besides the surprisingly large abundance of He$^+$ (He$^+$/He^{++} ∼ 0.1), that

(a) the mean charge states for C, O, and Fe are about 5.8, 7.1 and 12.5, respectively;
(b) several ionization states are present, especially for Fe (e.g. $Q \sim 10$ to 15);
(c) no measurable abundances of low charge-state heavy ions (i.e. C^{+1} − $^{+3}$, O^{+1} − $^{+4}$, Fe^{+1} − $^{+5}$) are observed.

These charge states are consistent with values for the corresponding ions in the solar wind and in the ∼(1–2)×10^6 K corona. This provides further evidence that the acceleration takes place in the corona and that the bulk of the accelerated ions come from coronal material.

3.3.2 The Anomalous Cosmic Ray Component

During the 1972–1978 solar minimum period observations of cosmic rays with energies below ~ 50 MeV/nucleon revealed anomalous enhancements in the low-energy spectra of the elements O, N, He, and Ne, relative to those of other elements such as B and C (Fig. 3.22). Later Voyager observations established similar enhancements in the spectra of H, C, and Ar, and they indicated a continuously increasing intensity with radial distance in the heliosphere out to beyond > 40 A.U. (Cummings and Stone 1988 [112], 1996 [113]; Webber 1989 [559]). This so-called "anomalous" cosmic ray (ACR) component is now generally believed to represent interstellar neutral particles that have drifted into the heliosphere, become ionized by the solar wind or UV radiation, and then accelerated to energies > 10 MeV/nucleon, probably at the solar wind termination shock (Fisk et al. 1974 [172], Pesses et al. 1981 [396]). The presence of the ACR is one good reason to expect the existence of such a shock, which is not required from the plasma physics of the interaction between the solar wind and the interstellar fields and plasmas. A unique prediction of this model, for which there is indirect evidence, is that the ACR component should be singly ionized and one has a population of particles in which each species has a different rigidity at the same energy/nucleon.

The measured energy spectra of the ACR shown in Fig. 3.22 then carry important information to test acceleration theories of cosmic ray particles at shock waves and their transport through the heliosphere. Note that the energy spectra of the ACR are some of the most unusual particle spectra ever measured. The spectrum of each species is different, exhibiting a different energy for the location of the peak in the differential spectrum, for example. Moreover, the elemental abundances of the ACR can be used to measure the neutral composition of the local interstellar medium, yielding essential information about the galactic evolution in the solar neighborhood over the time interval since the formation of the solar nebula.

3.4 Extrasolar Cosmic Rays

Three properties of the cosmic radiation are measurable to any degree of accuracy–at least in principle: the composition, the energy spectrum for each species, and the directional distribution for any species at any energy, and all at any time.

3.4.1 Elemental Composition

The cosmic radiation arriving at the solar heliosphere is composed of $\sim 98\%$ nucleons, stripped of all their orbital electrons, and $\sim 2\%$ electrons and positrons. In the energy range 10^8–10^{10} eV/nucleon where it has its highest intensity, the nuclear component consists of $\sim 87\%$ hydrogen, $\sim 12\%$

3.4 Extrasolar Cosmic Rays   47

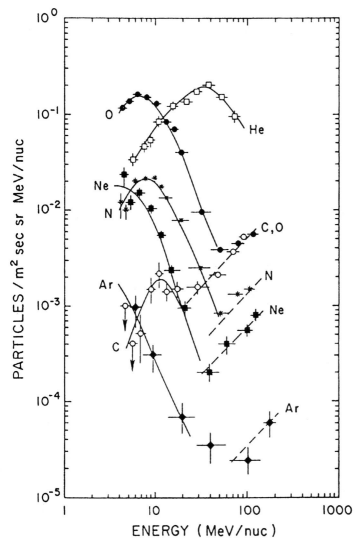

Fig. 3.22. Spectra of low energy He, C, N, O, Ne and Ar nuclei. Data are average spectra from Voyager 1 and Voyager 2 from 1985/274 to 1986/254 at an average radial distance $\sim$ 25 A.U. Total observed spectra are shown for all nuclei except He where only the anomalous component is shown. For all nuclei the anomalous component part is outlined with a **solid line**. From Webber (1989 [559])

helium and $\sim 1\%$ for all heavier nuclei from carbon to the actinides. Figure 3.23 shows the present status of the relative elemental abundances for the elements between hydrogen and nickel as measured at energies between 70 and 280 MeV/nucleon together with two compilations of the solar system abundances, all relative to carbon (C=100).

The solar system abundance is representative for a sample of matter produced in the interiors of stars, and these elements are also seen in optical lines in stellar spectra. As Fig. 3.23 exhibits, the cosmic ray abundances are very different from the solar system abundances. If the carbon abundances in both samples are comparable we note that

(a) hydrogen and helium are under-abundant in cosmic rays as compared to stellar material, and
(b) nuclei of the (Li, Be, B)-group and the sub-Fe-group (Sc, Ti, V, Cr, Mn) are overabundant by several orders of magnitude.

If we had normalized the samples at hydrogen we would have to conclude that the abundances of heavy elements ($Z > 2$) in the cosmic radiation lie significantly above the stellar abundances, and that the abundance variations from element to element in the cosmic radiation are much smaller than in stellar matter (see Fig. 2 of Lund 1986 [308]). The lack of strong abundance variations in the arriving cosmic rays finds a natural explanation in the spallation of heavier cosmic ray nuclei during their propagation from their sources to the solar system. This process distributes nuclear fragments over all chemical elements, and for those which have a low source abundance this contamination can be very severe, effectively precluding a study of the source abundance of many elements. However, the observed abundance of precisely these elements (such as Li, Be, B and the sub-Fe group) can be used to determine the amount of matter through which the cosmic ray particles have had to pass during their journey to the solar system. This amount of matter can then be used with the known spallation cross-section to correct for the effect of cosmic ray spallation in the interstellar medium and to infer the true source cosmic ray elemental composition.[1] Although the mechanisms that accelerate the cosmic ray particles to their enormous energies are not yet understood, there is no doubt that the place of their birth lies in the interior of stars, the sites where, as F. Hoyle was the first to suggest, all elements, with the exception of the very lightest ones, originate. Hence, cosmic ray elements, which have a large abundance also in stellar material, are referred to as *primary cosmic rays* since they can be produced in stellar sources probably by processes similar to solar flares, whereas elements with low stellar abundances, like Li, Be, B, and the sub-Fe-group, are referred to as *secondary cosmic rays*

---

[1] This situation is completely different from the case of SEP composition: there the lack of observed spallation products such as ^{2}H, of rare nuclei such as Li, Be, and B, and of ionization energy loss effects makes it possible to show that SEPs observed in interplanetary space have traversed less than 0.03 g cm^{-2} of material.

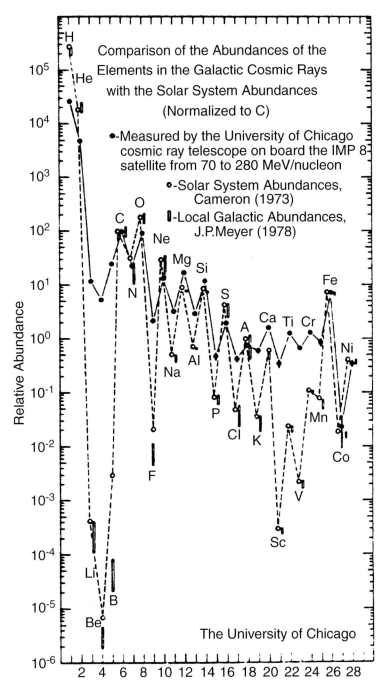

Fig. 3.23. The elemental abundance of the cosmic rays compared with two compilations of the solar system abundances. From Meyer (1980 [346])

since they result mainly from fragmentation of the primaries. With respect to this classification the nuclei of the Fe-group are the purest primary particles since all species heavier than iron and nickel are extremely rare in cosmic rays, whereas cosmic ray ions of lower charge can be both of primary and secondary origin due to the fragmentation of Fe-nuclei and their daughter particles. The information contained in the elemental abundance distribution is therefore two-fold. The spallation products provide insight into the question of cosmic ray propagation and containment. The abundance of source nuclei, after correcting for the effect of propagation, are used for diagnosing the nature of the sources. With respect to the secondary cosmic rays, by using the relevant fragmentation cross-sections as $p + C^{12} \rightarrow Li^7 + \chi$ (where χ stands for anything else) from laboratory measurements (for a collection of fragmentation

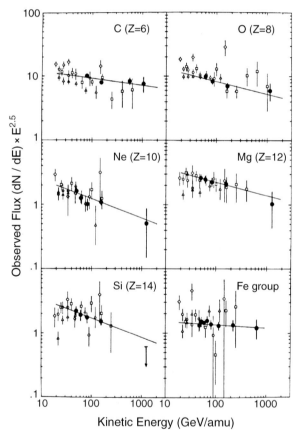

Fig. 3.24. The differential energy spectra of C, O, Ne, Mg, Si, and Fe-group ($Z = 25, 26, 27$) nuclei. The spectra are multiplied by $E^{2.5}$ to emphasize spectral differences. Power law fits to the data represented as **filled circles** are given. From Grunsfeld et al. (1988 [203])

cross-sections see Silberberg and Tsao 1973 [486]) it has been inferred that at non-relativistic energies the primary cosmic rays have to penetrate a total column density of matter

$$X = \int_0^\infty dl\, n(\boldsymbol{r}) \simeq n_0 \tau v \simeq (6-9) \quad \text{g cm}^{-2}\,, \qquad (3.4.1)$$

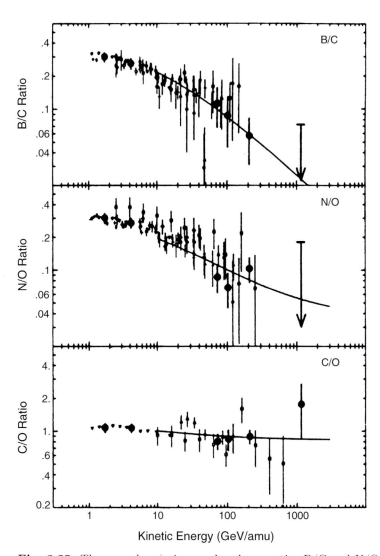

Fig. 3.25. The secondary/primary abundance ratios B/C and N/O and the primary/primary abundance ratio C/O as a function of kinetic energy per nucleon. From Swordy et al. (1990 [535])

Table 3.2. Spectral indices above 50 GeV/nucleon

Element	Charge	Value	Reference
Proton	1	2.80 ±0.04	Asakimori et al. 1998 [18]
Helium	2	2.68 ±0.06	Asakimori et al. 1998 [18]
Boron	5	3.20 ±0.20	Swordy et al. 1990 [535]
Carbon	6	2.60 ±0.17	Grunsfeld et al. 1988 [203]
Nitrogen	7	3.25 ±0.20	Swordy et al. 1990 [535]
Oxygen	8	2.65 ±0.08	Grunsfeld et al. 1988 [203]
Neon	10	2.86 ±0.10	Grunsfeld et al. 1988 [203]
Magnesium	12	2.77 ±0.08	Grunsfeld et al. 1988 [203]
Silicon	14	2.97 ±0.12	Grunsfeld et al. 1988 [203]
Fe-group	25–27	2.55 ±0.09	Grunsfeld et al. 1988 [203]

where n is the interstellar gas density. If the gas density is approximated as uniform, (3.4.1) reduces to $X \simeq n_0 \tau v$, where v is the cosmic ray velocity and τ the mean residence time of the primary cosmic ray particle in the Galaxy.

The composition of cosmic rays has also been measured as a function of energy. In Figs. 3.11 and 3.12 we have shown already the energy spectra of protons and α-particles to energies up to 10^{12} eV/nucleon. In Fig. 3.24 we show the energy spectra of the primary nuclei C, O, Ne, Mg, Si, and the Fe-group at the same energy/nucleon, whereas in Fig. 3.25 the secondary/primary abundance ratios B/C, N/O, and the primary/primary ratio C/O are displayed as a function of kinetic energy/nucleon. One notices immediately that all spectra are well represented by simple power law distributions in kinetic energy per nucleon ($dN/dE \propto E^{-\Gamma}$) over the whole energy range, but that the power law spectral index Γ varies significantly from element to element. Table 3.2 shows the spectral indices observed. The Fe-group energy spectrum is the flattest, indicating a steady increase in the relative abundance of iron in cosmic rays with increasing energy. The measured decrease of the abundance ratio of the secondary nuclei to their primary pregenitors as B/C and N/O in Fig. 3.25 implies that particles with higher energy traverse less interstellar matter than those of lower energy, and the measured ratio implies a variation of the total column density (3.1) as a function of rigidity as (Swordy et al. 1990 [535])

$$X(R) = 6.9 \ (R/[20 \ \mathrm{GV/nucl}])^{-0.6} \ \mathrm{g \ cm}^{-2} \ \mathrm{for} \ R > 20 \ \mathrm{GV/nucl}. \quad (3.4.2)$$

Several explanations of this effect are possible (see Chap. 17), the simplest being that the cosmic ray residence time τ decreases with increasing rigidity.

Using the inferred energy rigidity dependence for the traversed column density of the cosmic rays, the measured primary energy spectra (Fig. 3.24) have been corrected for the effect of propagation, and the calculated cosmic ray source abundances at 100 GeV/nucleon and 1 TeV/nucleon have been compared with the local galactic abundances. Figure 3.26 shows the result. The data are normalized to Fe and are plotted versus the first ionization potential of the various species. The abundances at both energies follow the pattern of depletion of elements with high first ionization potential, which is well established both at lower energies (see Fig. 3.26) and for solar flare particles (see Sect. 3.3.1 and Fig. 3.19). Also, most source abundances do not seem to change noticeably with energy, with the exception of silicon whose depletion at 1 TeV/nucleon is peculiar. Moreover, the flatness of the observed energy spectrum of iron is likely to be a consequence of the competition

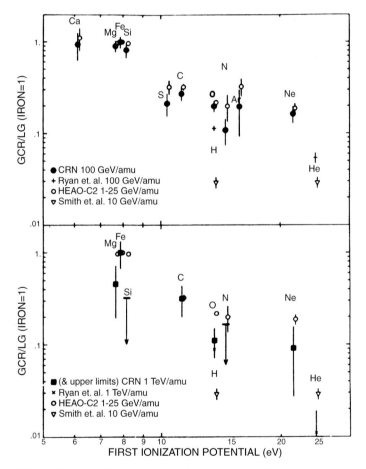

Fig. 3.26. Galactic cosmic ray abundances relative to the local galactic abundance (from Meyer 1985 [348]) plotted versus the first ionization potential. In the **top panel**, the data at 100 GeV/nucleon (**filled circles**) are compared with data at lower energies. In the **lower panel**, the data at 1 TeV/nucleon (**filled squares**) are compared with data at lower energies. From Grunsfeld et al. (1990 [204])

between spallation and propagation, and not inherent to a peculiar source abundance. From these studies we conclude:

1. The source composition of galactic cosmic rays is similar to that of the solar system.
2. The injection process and probably also the acceleration process is similar to that operating in solar flares. Elements with values of the first ionization potential less than about 9 eV are preferentially accelerated.

These findings emphasize again the importance of solar flare particle studies in clarifying the processes responsible for the origin of galactic cosmic rays.

In a way, the finding that cosmic rays probably originate in similar astrophysical sites as the solar corona has been a disappointment for workers in the field since the original aim has been to detect signposts of specific explosive nucleosynthesis processes–the r-process or the s-process–in the cosmic ray element abundances to prove that supernovae do indeed produce and ac-

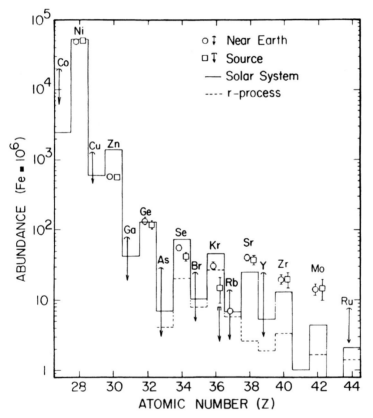

Fig. 3.27. Relative abundances observed with the HEAO-3 satellite experiment (**circles**) extrapolated to the cosmic ray source (**squares**), and in the solar system (**solid histogram**). From Israel (1983 [234])

celerate cosmic radiation (Ginzburg and Syrovatskii 1964 [185], Shapiro and Silberberg 1970 [481]). In that respect the abundances of ultraheavy ($Z > 30$) elements beyond the iron group are important (Israel 1983 [234]). Beyond the iron peak the element abundance decreases very rapidly by about four orders of magnitude. Summarizing the data obtained, Israel (1983 [234]) and Binns (1988 [51]) have stated that there is no strong evidence for r-process dominance in the cosmic ray source.

For $30 < Z < 60$, where individual element abundances have been resolved (see Fig. 3.27), the cosmic ray source composition looks like that of the solar system, with a similar first ionization potential pattern as established before, or possibly some s-process enhancement. In the $76 \leq Z \leq 83$

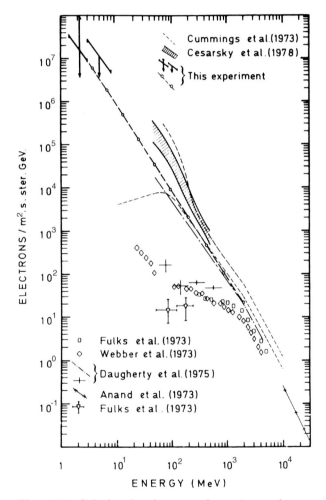

Fig. 3.28. Calculated and measured cosmic ray electron spectrum below 10 GeV. From Mandrou et al. (1980 [314])

interval the data suggest a vague ^{78}Pt-peak, but the conclusion is not yet certain because individual elements have not been resolved, and the solar system s-process (mainly ^{82}Pb) abundance is uncertain. The abundance of actinides, which are results only of the r-process, is not greatly enhanced in the cosmic rays. The first detection of actinides by Fowler et al. (1967 [176]) had been taken as strong evidence for a supernova origin of cosmic rays, but later ultraheavy data do not support this conclusion. They certainly do not contradict a supernova source, especially as the energy source (see the discussion in Chap. 7), but they do not *prove* such a source since they do not display a composition which is obviously synthesized in supernovae. The ultraheavy data are fully consistent with elemental results for cosmic rays of $Z < 30$ which are similar to the composition of solar energetic particles as noted above.

With respect to relativistic electrons in cosmic radiation we have already shown their energy spectrum in Fig. 3.8. As noted there, the true interstellar spectrum is very uncertain due to the influence of solar modulation, and its variation has been derived from studies of the galactic radio synchrotron radiation background (Cummings et al. 1973 [114]) and of the diffuse galactic gamma rays. Figure 3.28 summarizes the various estimates. Mandrou et al. (1980 [314]) have concluded that an electron power law $dN/dE \propto E^{-2}$ in the range from 10^{-3} to 2 GeV fits the observations best. At higher energies ($E > 10$ GeV) the observations (see Fig. 3.29) of Prince (1979 [408]), Tang

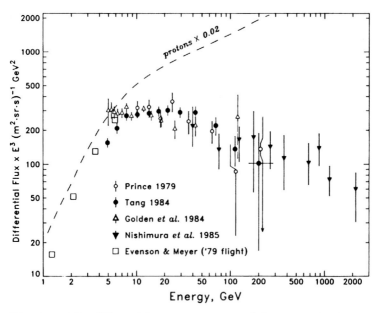

Fig. 3.29. The differential energy spectrum of electrons. The flux is multiplied by $E^{3.0}$. The spectrum of protons is shown for comparison. From Müller (1989 [369])

(1984 [540]) and Nishimura et al. (1990 [375]) indicate a steepening of the spectrum between 10 and 30 GeV to $dN/dE \propto E^{-3.2 \pm 0.2}$ above 30 GeV.

Also of interest is the positron/electron ratio $f^+ \equiv e^+/(e^+ + e^-)$ which is shown in Fig. 3.30. Whereas the measurements below 2 GeV may be influenced again by solar modulation, the Müller and Tang (1987 [370]) data at higher energies indicate that the positron fraction increases above 10 GeV.

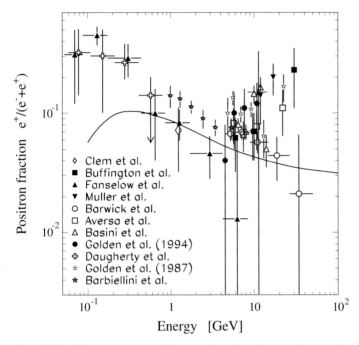

Fig. 3.30. Measurements of the positron fraction in the cosmic rays as a function of the energy. Private communication Menn (1999 [341])

3.4.2 Isotopic Composition

In addition to the elemental composition, sophisticated cosmic ray experiments have elucidated the 'fine structure' of the abundance distributions–the isotopic abundances. Both nucleosynthesis and spallation involve nuclear processes rather than atomic processes, so that the determination of individual nuclei, that is isotopes, is crucial for the understanding of the origin, acceleration and propagation of cosmic rays. Two excellent reviews by Meyer (1978 [345]) and Mewaldt (1989 [343]) have been written on this subject on which most of the following is based.

Our knowledge of the isotopic composition of cosmic rays is still very limited. Only the Ne, Mg, and Si isotopes among the primary cosmic rays have had their source abundance determined to an accuracy of ∼ 30% or better; among the secondary cosmic rays the radioactive clocks as ^{10}Be, ^{26}Al and others have been at the center of interest. We consider each in turn.

3.4.2.1 Cosmic Ray Clocks and Secondary Nuclei. From the abundances of secondary nuclei (Sect. 3.4.1) it is known that typical cosmic ray primary nuclei below a few GeV/nucleon have traversed an average of 6 to 9 g/cm^2 of material during their lifetime. This pathlength (which actually is the mean of a pathlength distribution) is known to decrease with increasing energy (3.4.2). Studies of secondary nuclei in different parts of the periodic table can reveal whether their parents have shared a common history.

The rare isotopes ^{2}H and ^{3}He in cosmic rays are believed to be of secondary origin, produced mainly by the breakup of primary cosmic ray ^{4}He. Figure 3.31 shows selected measurements of the ^{2}H/^{4}He and ^{3}He/^{4}He ratios obtained during solar minimum in comparison with a leaky-box calculation that fits measurements of heavier cosmic ray secondary/primary ratios such as B/C. Apart from the balloon measurement of ^{2}H at ∼80 to 150 MeV/nucleon by Webber and Yushak (1983 [560]) the measurements are generally consistent with a purely secondary origin of these isotopes.

Particularly interesting among the isotopes that are produced as secondaries by the fragmentation of heavier cosmic rays are those which are radioactive, with half-lives suitable for measuring the average residence time of cosmic ray nuclei in the Galaxy. Examples are: ^{10}Be ($t_{1/2} = 1.6 \times 10^6$ yr), ^{14}C ($t_{1/2} = 5730$ yr), ^{26}Al ($t_{1/2} = 9 \times 10^5$ yr), ^{36}Cl ($t_{1/2} = 3 \times 10^5$ yr) and ^{54}Mn ($t_{1/2}$ estimated to be ∼ 2×10^6 yr). Since these nuclei would not be expected to be present in any significant amount in cosmic ray source material, their relative abundance in cosmic rays is a calculable function of the amount of material traversed by heavier nuclei and the mean lifetime since acceleration. Figure 3.32a displays the mass distribution for the element Be as observed in the cosmic rays at energies between 30 and 150 MeV/nucleon and shows the near absence of the isotope ^{10}Be. The same detector, when exposed to a flux of Be produced by spallation of a nitrogen beam from an accelerator, responds as shown in Fig. 3.32b. A distribution of this general form could be expected in the cosmic rays had all ^{10}Be survived or, in other words, if the cosmic ray residence time in the Galaxy is much shorter than the ^{10}Be half-life $t_{1/2} = 1.6 \times 10^6$ yr. The near absence of the isotope ^{10}Be implies that the mean cosmic ray residence time is in excess of 10^7 yr, so that most of the ^{10}Be-isotopes decayed during their propagation in the Galaxy. Using the mean cosmic ray residence time $\tau = 15\ (+7, -4) \times 10^6$ yrs measured by Garcia-Munoz et al. (1977 [180]) in (3.4.1) in the limit of an uniform interstellar medium, the measured penetrated column density of matter implies (with 1 g of cosmic matter corresponding to 6×10^{23} hydrogen atoms) that the cosmic ray particles have propagated in a medium of average density less than $0.4 - 0.6$ hydrogen atoms/cm^3. This is considerably less than the average density of matter in the galactic plane of $\gtrsim 1$ atoms/cm^3, see Sect. 2.2. Evidently

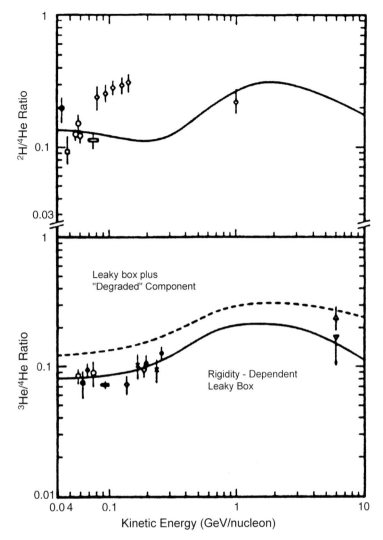

Fig. 3.31. Selected solar minimum satellite and balloon data for the ^{2}H/^{4}He and ^{3}He/^{4}He ratios in comparison with a standard leaky-box propagation model. From Mewaldt (1989 [343])

this involves the cosmic rays spending $\lesssim 1/3$ of their lifetime in the galactic matter disk and $\gtrsim 2/3$ in low-density regions of the interstellar medium; this can either be the hot coronal phase of the interstellar medium and/or a low-density region around the matter disk–often referred to as the galactic halo. The latter hypothesis is supported by the radio measurement evidence of detecting significant amounts of synchrotron radiation far away from the stellar disk both in our Galaxy and external edge-on Galaxies (Sect. 2.4) which,

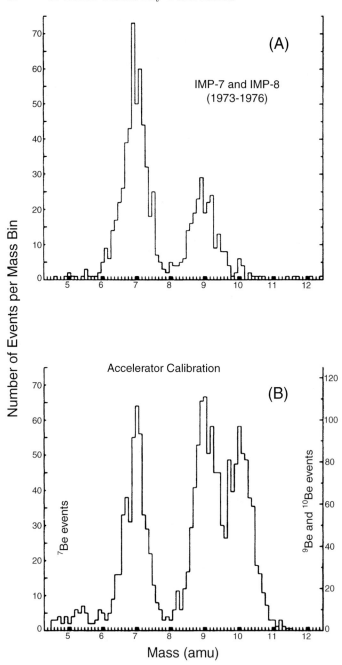

Fig. 3.32. Comparison of the Be mass histogram obtained: **(A)** with flight data from the IMP 7/8 satellite experiments, and **(B)** with data from the calibration of the IMP backup telescope at the Bevalac accelerator of the Lawrence Berkeley Laboratory, using individual Be isotopic beams. From Simpson and Garcia-Munoz (1988 [493])

besides the presence of magnetic fields, requires the existence of cosmic ray electrons at these distances from the plane.

Figure 3.33 compares measurements of two cosmic ray clocks, ^{10}Be and ^{26}Al at different energies with calculations parameterized by the average density of material in the propagation region and a mean pathlength of ~ 6 g/cm^2 appropriate for non-relativistic particles. As the figure indicates the various experiments yield roughly the same value for the mean density of the propagation region although there is considerable energy dependence re-

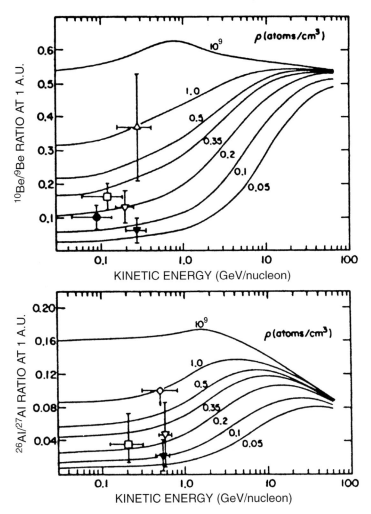

Fig. 3.33. Calculations of two isotope ratios, parameterized by the density of the propagation region, are compared with measurements. The **curves** labeled 10^9 show the expected result if there were no radioactive decay. From Mewaldt (1989 [343])

sulting from time-dilation effects in the radioactive decay and the energy dependence of the production cross-sections.

In principle, measurements of cosmic ray clocks with different half-lives can measure the distribution of the material traversed by the cosmic rays. For example, if much of this material immediately surrounds cosmic ray sources we would expect a relatively greater production of radioactive secondaries in the distant past, so that the ratio of long-lived species to short-lived species should be enhanced compared to the case of uniform distribution of the traversed matter. At the other extreme, the well-known radioactive isotope ^{14}C would be expected to have a measurable abundance in cosmic rays that would allow a comparison of the density of material traversed over the past $\sim 10^4$ yr with the average density.

3.4.2.2 Isotopic Composition of Cosmic Ray Source Material. As with cosmic ray elements, the accuracy with which isotope source abundances can be determined depends on the magnitude of the fragmentation contribution to the measured abundance. There are 25 to 30 isotopes with $Z < 30$ where this secondary contribution is $< 50\%$ and propagation corrections are relatively minor, including several elements where two or more isotopes can be studied (e.g., Ne, Mg, Si, Fe, and Ni). In other interesting cases (e.g. ^{13}C, ^{14}N, ^{18}O) propagation corrections limit the source abundance accuracy.

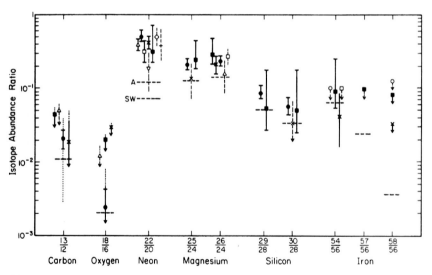

Fig. 3.34. Comparison of selected cosmic ray source measurements with solar system measurements for various isotope abundance ratios. The solar system values (**horizontal dashed lines**), including neon-A (A), are from Cameron (1981 [88]), except for solar wind neon (SW). Cosmic ray measurements with unresolved mass resolution are shown with **dashed vertical error bars**, and should be given less weight. **Dotted extensions** to the C and O error bars indicate propagation uncertainties. From Mewaldt (1983 [343])

Figure 3.34 summarizes selected determinations of ten cosmic ray source isotopic ratios relative to the solar system abundances of Cameron (1981 [88]). A distinction has been made between 'resolved' and 'unresolved' isotope measurements, based on whether the mass resolution achieved in a particular experiment is high enough to avoid statistical uncertainties.

Table 3.3 presents a summary of the isotopic composition of cosmic ray source material, based on a weighted mean of the results included in Fig. 3.34. The results in Table 3.3 have been broken down into three categories in an effort to separate those confirmed characteristics that cosmic ray origin theories should address, from other less-established results and limits. The major features are the excess of neutron-rich Ne and Mg isotopes. The reported excess of neutron-rich Si isotopes is a noteworthy 'probable' result. From these differences in isotopic composition between cosmic ray source material and solar system material it has been concluded that the nucleosynthesis of these two samples of matter has differed, a conclusion that has stimulated theoretical work as to how such differences might have occurred. This includes the possible influence of interstellar re-acceleration of injected primary cosmic rays as well as special type of stars as cosmic ray particle injectors. However, we think the conclusion of different nucleosynthesis has been premature and too far-reaching given the fact that the solar flare particle isotopic composition exhibits similar anomalies and excesses as in the

Table 3.3. Isotopic composition of cosmic ray source material. After Mewaldt (1989 [343])

Established results		Probable/possible require confirmation		Other limits	
Isotope ratio	Cosmic ray source / Solar system	Isotope ratio	Cosmic ray source / Solar system	Isotope ratio	Cosmic ray source / Solar system
^{22}Ne/^{20}Ne	3.3 ± 0.45^a or 5.5 ± 0.75	^{29}Si/^{28}Si	1.5 ± 0.3	^{13}C/^{12}C	1.55 ± 1.25
^{25}Mg/^{24}Mg	1.6 ± 0.25	^{30}Si/^{28}Si	1.4 ± 0.35	^{18}O/^{16}O	≤ 4
^{26}Mg/^{24}Mg	1.5 ± 0.20			^{34}S/^{32}S	≤ 3
^{24}N/^{16}O	0.25 ± 0.10			^{54}Fe/^{56}Fe	1.15 ± 0.5
^{12}C/^{16}O	~ 2			^{57}Fe/^{56}Fe	≤ 4
				^{57}Fe/^{56}Fe	≤ 10
				^{60}Ni/^{58}Ni	1.9 ± 1.3

a) Depending upon whether neon-A or solar wind neon is used as a solar system standard

^{3}He/^{4}He and ^{22}Ne/^{20}Ne ratios (Sect. 3.3.1), so that an explanation must be found within the concept of accelerating stellar matter. There surely exist selective acceleration mechanisms that modify the isotopic composition of the accelerated seed particles.

3.4.3 Anisotropy

The study of anisotropy in the arrival directions of cosmic rays is clearly of great interest to locate their possible sources. With the peculiar position of our solar system with respect to the galactic disk (Sect. 2.1) one would expect an anisotropy towards the direction ($l \simeq 0$, $b \simeq 0$) if the cosmic ray sources are galactic objects. Experimental data on anisotropy predominantly come from ground-based shower detectors. There are a number of problems in interpreting the data on anisotropy. At the highest energies, data are statistically limited and subject to fluctuations. Watson (1991 [555]) notes that above the total cosmic ray energy 10^{18} eV up to July 1990 with all stations worldwide $N = 23\,555$ particle events have been recorded, implying a counting error $N^{1/2} = 153 = 0.7\%$, so that a convincing anisotropy has to be established at a level larger than 1%. Apparently this has not been achieved so far, see below. Another experimental problem is the bias of each detector due to the non-uniform acceptance of cosmic ray arrival directions. For instance, detectors located in the northern hemisphere do not see a region around the south magnetic pole. This hole in acceptance, if not properly accounted for, can produce biases in event distributions (Sokolsky 1979 [503]). Lastly, since one typically searches for anisotropy as a function of energy, biases in determining energy may cause problems.

3.4.3.1 Harmonic Analysis. For a detector operating approximately uniformly with respect to sidereal time the zenith angle dependent shower detection and direction reconstruction efficiency is a strong function of declination but not of the right ascension (R.A.). Therefore one usually searches for anisotropy in R.A. only within a given declination band. This is done by measuring the counting rate as a function of sidereal time (R.A.) and performing a harmonic analysis, e.g. fitting the data by

$$R(t) = A_0 + A_1 \sin\left(\frac{2\pi t}{24} + \phi_1\right) + A_2 \sin\left(\frac{2\pi t}{12} + \phi_2\right), \quad (3.4.3)$$

where A_0, A_1, and A_2 are the amplitudes of the zeroth, first and second harmonics, respectively, and ϕ_1 and ϕ_2 are the phases of the first and second harmonics.

In Fig. 3.35 we show the amplitude A_1/A_0 and the phase ϕ_1 of the first harmonic. One notes that (a) at energies less than 10^{14} eV the anisotropy is small ($\sim 0.07\%$) and has constant phase ($\sim 3^h$ R.A.), (b) the anisotropy starts to increase as the energy rises above $\sim 10^{15}$ eV, and (c) the phases of the relatively large amplitude anisotropies (several percent) above 10^{17}eV

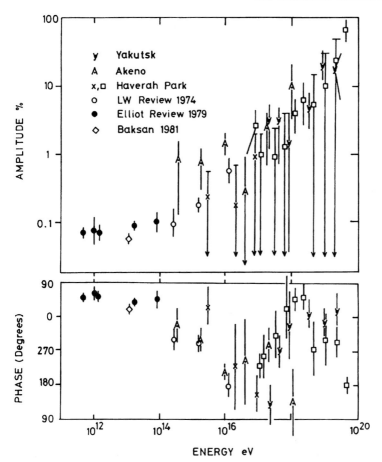

Fig. 3.35. A summary of data on the phase and the amplitude of the first harmonic of cosmic ray anisotropy in right ascension. Error bars correspond to ±1 standard deviation and upper limits are at the 95% confidence level. From Watson (1984 [554])

vary rapidly with energy but are consistent between the various experiments. Contrary to its visual appearance these data do not convincingly establish a significant anisotropy beyond 10^{17}eV. If these data are analyzed as a function of galactic latitude the gradient of a linear least squares fit to the latitude distribution, shown in Fig. 3.36, is consistent with zero implying isotropy up to the highest energies. The data show no evidence of anisotropy at any energy, regardless of the energy calibration model chosen.

3.4.3.2 Sources of Neutral Primaries–The Cyg X-3 Story. In early 1983 Samorski and Stamm (1983 [444], 1984 [445]) analyzed their airshower data operating at energies above 10^{15} eV for directional anisotropies. They found six potential directions where significant excesses of showers (chance

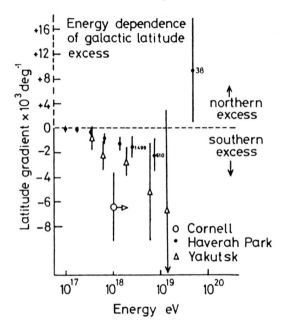

Fig. 3.36. The gradient of a linear least squares fit to the galactic latitude distribution of cosmic rays above 10^{17} eV. From Watson (1984 [554])

probability less than 10^{-4}) occurred. Among them was the binary X-ray source Cyg X-3. Figure 3.37 shows the number of detected extensive airshowers as a function of R.A. in the declination band of Cyg X-3. One notices the peak at the position of this source: over the time of observations from 1976 March 18 until 1980 January 7 the station recorded 33 showers from that direction whereas the average off-source background was 14.4 ± 0.4. The authors estimated the probability for random excess of the 18.6 ± 5.8 excess showers to be 1.8×10^{-5}.

From infrared to X-ray observations Cyg X-3 is known to be a binary system with an orbital period of 4.8 h. Samorski and Stamm therefore folded (modulo 4.8 h) their shower data and the resulting, periodogram (Fig. 3.38) showed a strong peak in phase bin 4. Assuming Poissonian fluctuations the probability that 13 events might occur in any of the 10 phase bins at random out of a background of average value 1.44 $(= 14.4/10)$ is 5×10^{-8}. This finding made their claim that Cyg X-3 is responsible for the shower excess very compelling. The Haverah Park group (Lloyd-Evans et al. 1983 [306]), using portions of, and supplements to, their large ground array, also observed a signal about the same phase bin in the time interval 1982–1984, supporting the Samorski and Stamm detection. However, when making phase plots, corrections for the slight deceleration of the orbital motion of the source and the orbital Doppler shifts have to be applied by using astronomical ephemeris

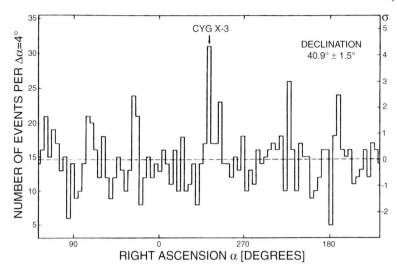

Fig. 3.37. Number of detected extensive airshowers in the declination band $40.9° \pm 1.5°$ as a function of right ascension with the University of Kiel shower array. The **dashed line** represents the average number of showers per bin over the total band. From Samorski and Stamm (1983 [444])

calculations, and care must be taken in comparing the results of different experiments that consistent ephemerides are used. It is evident that modulating 33 events taken over nearly four years, corresponding to 3.5×10^4 h, with a 4.8 h period that is increasing slightly, is non-trivial and that the period change with time has to be known very accurately. Furthermore, the 33 events will group somehow over the 7300 periods observed in the four years cover time of the Kiel array; therefore there is considerable debate in the literature on the statistical significance of the Samorski and Stamm and similar detections. Also, with the at least ten times more sensitive HEGRA- and CASA-arrays it has not been possible to establish Cyg X-3 as a point source in the time from July 1988. Although Cyg X-3 is highly time-variable at radio wavelength, these negative detections may mean that either a similar variability occurs at energies of 10^{15} eV, or that the Samorski and Stamm claim was a false effect. We have to wait for further measurements for a final assessment.

If Cyg X-3 is confirmed as being responsible for shower excesses this will indeed be a major step forward in the search for the sources of high-energy cosmic rays. Due to the influence of the interstellar magnetic fields that will deflect all charged particles, the emitted particles have to be neutrals: photons, neutrons, neutrinos or even more exotic particles. Neutrinos are ruled out because their small interaction cross-section would require excessively large fluxes to account for the observed rates; neutrons are extremely unlikely because with a minimum distance of Cyg X-3 from the solar system of

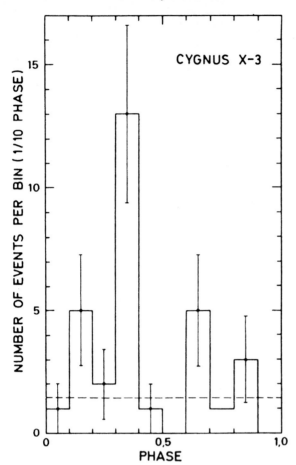

Fig. 3.38. Phase histogram of arrival times of extensive airshowers from the Cyg X-3 direction. The **dashed line** corresponds to the average off-source rate of 1.44 ± 0.04 showers per bin. From Samorski and Stamm (1983 [444])

12 kpc (Dickey 1983 [129]) the neutrons would have to survive a large number of lifetimes. This leaves γ-ray photons or hitherto unobserved long-lived neutral particles as candidates.[2]

A peculiarity of the Samorski and Stamm result is that the array, which was also fitted with low energy muon detectors, did not see the significant decrease in the number of muons in showers coming from the direction of Cyg X-3 which would be expected if these showers were indeed initiated by γ-rays. Two other experiments equipped with muon detectors, the Akeno array (Kifune et al 1986 [263]) and the Cygnus array (Goodman 1987 [193]),

---

[2] The reader can imagine that the latter prospect renewed the interest of many high-energy particle physicists in cosmic ray physics.

give conflicting results. The Akeno experiment sees a phase enhancement from Cyg X-3 only when muon-poor showers are selected, while the Cygnus array sees no significant decrease in muon number from showers contributing to the phase enhancement. The muon deficiency implies either that nuclear physics changes considerably at energies of 10^{15} eV with a larger muon production by high energy photons, different from what is known at energies accessible in terrestrial laboratories, a further exciting alternative for high energy physicists, or it emphasizes again the detection of a new long-lived neutral particle. A successful confirmation of the Samorski and Stamm discovery would have far-reaching consequences not only for cosmic rays but also for high-energy physics.

3.4.4 Time History from Cosmogenic Nuclei

The variations of cosmic ray intensity on time scales longer than the modern era of direct observations can be inferred from the study of cosmogenic nuclei which are defined as nuclides formed by nuclear interactions of galactic and solar cosmic rays with extraterrestrial (meteorites, moon, interplanetary dust) or terrestrial (atmosphere, lithosphere) matter. The nuclides, produced in these reactions range from short-lived radioactive species to stable isotopes with some remaining in various geological reservoirs today, serve as a link to the cosmic ray activity in the past. The studies of extrasolar cosmic ray flux variations compare activities of radioactive nuclides with different half-lives such as ^{10}Be and ^{14}C in meteoritic and lunar samples. Ratios of measured activities to those predicted by various models or to concentrations of stable cosmogenic nuclides allow us to estimate the variations in the average fluxes of cosmic rays over the mean lives of various radionuclides. It has been established (Schaeffer 1975 [449], Reedy et al. 1983 [423]) that:

1. within a factor of two the flux of extrasolar cosmic rays has been constant over the past 10^9 yr;
2. the cosmic ray intensity averaged over the past 400 yr is within 10% of that averaged over the past 4×10^5 yr;
3. the cosmic ray intensity averaged over the past 4×10^5 yr is 50% higher than that averaged over the past 10^9 yr.

The first two findings establish galactic cosmic rays as a steady phenomenon within a factor 2 over time scales much longer than the average residence time of cosmic rays in the Galaxy.

The temporal variation of cosmic rays over the last 1.5×10^5 yr has been studied using the ^{10}Be-abundance in polar ice (Konstantinov et al. 1990 [268]). The measurements suggest that a supernova explosion which took place at a distance of ~ 50 pc from the solar system about 3.5×10^4 yr ago has been the cause for the observed ^{10}Be production rate enhancement (Kocharov 1991 [266], 1992 [267]).

3.4.5 Ultrahigh Energy Cosmic Rays

In the context of cosmic ray anisotropy and cosmic ray time history we have already described some results of ground-based observations of cosmic rays with energies above 10^{14} eV. We will concentrate here on the measurements of the energy spectrum and the composition of these ultrahigh energy particles following the summary by Sokolsky (1979 [503]). Studies of the composition of cosmic rays in this energy range are indirect by measuring the muon content of the airshowers, since their muon multiplicity depends on the atomic number of the primary particle, and by analyzing the lateral shower distribution on the ground. A common problem of both methods is that such high primary energies are not available in terrestrial accelerators, so that the interpretation relies on the theoretical extrapolation of nuclear physics interactions to these high energies. It is clear that the same observations made in this way can be interpreted quite differently and that many controversial results exist on the composition ranging from a dominantly protonic composition to a dominantly Fe composition of ultrahigh energy cosmic rays.

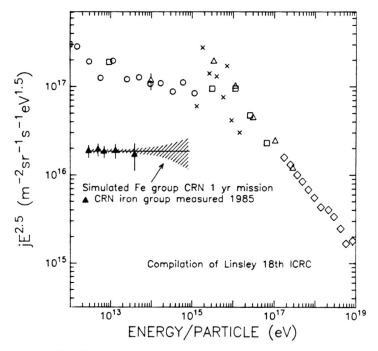

Fig. 3.39. The all-particle energy spectrum as compiled by Linsley (1983 [305]) and the spectra of the Fe-group ($Z = 25, 26, 27$) as measured by the space shuttle experiment of Grunsfeld et al. (1988 [203]). In the absence of absolute fluxes the normalization of the space shuttle data are somewhat arbitrary. From Müller (1989 [369])

Here we restrict our discussion to the energy spectrum as a function of the total energy of the primary particle. Figure 3.39 shows the all-particle energy spectrum as compiled by Linsley (1983 [305]) in comparison with the O and Fe-group spectra measured in space by Grunsfeld et al. (1988 [203]). The spectrum extends to beyond 10^{20} eV. Between 10^{12} and 10^{19} eV the intensity is known to at least $\pm$ 20% and the most prominent feature, the 'knee' at 5×10^{15} eV is well established. It should be mentioned that it is actually much harder to measure the all-particle spectrum at 10^{15} eV than it is at 10^{18} eV for example, since showers produced by 10^{15} eV primaries are rather small and past their maximum development, even at the highest altitude laboratory, so that fluctuations are a severe problem (Watson 1984 [554]). This may explain some of the scatter in the data points in Fig. 3.39 at this energy. Above 10^{16} eV the spectrum J appears to be quite featureless and can be well represented by

$$J(E) = (1.31 \pm 0.14) \times 10^{27} [E(\mathrm{eV})]^{-s} \quad \mathrm{m}^{-2} \quad \mathrm{s}^{-1} \quad \mathrm{sr}^{-1} \quad \mathrm{eV}^{-1}, \quad (3.4.4)$$

with $s = 3.14^{+0.06}_{-0.05}$ (Watson 1991 [555]).

Of particular interest are cosmic ray particles with total energy larger than 10^{19} eV. Soon after the discovery in 1965 of the universal microwave background radiation Greisen (1966 [202]) and Zatsepin and Kuzmin (1966 [573]) pointed out that protons of this energy would interact with this microwave background through photo-pion reactions (discussed in Sect. 5.2) and lose energy on a length scale of about $\simeq$ 10 Mpc which is a relatively short distance in cosmological terms. It was therefore anticipated that the ultrahigh energy cosmic ray spectrum would show a cutoff at these energies. However, airshower arrays in the US and in Japan (Bird et al. 1995 [54], Takeda et al. 1998 [539]) have detected seven particles at energies clearly above 10^{20} eV with no indication of the Greisen-Zatsepin-Kuzmin cutoff. Within conventional explanations of cosmic ray acceleration this means that the potential sources have to be relatively nearby within 50 Mpc. But the detection of these ultrahigh energy particles has also stimulated a great deal of speculation about possible new physics. Surely, more data are required, although the fluxes at these energies are only of order one per square kilometer per century.

4. Interactions of Cosmic Ray Electrons

Because of the influence of non-uniform cosmic magnetic fields it is impossible to observe directly the intensity of cosmic rays in regions of the Universe other than the neighborhood of our solar system. However, by measuring the electromagnetic radiation that results from interactions of cosmic rays with other constituents of the Universe we can infer the cosmic ray intensity in other regions of space. Since neutrino astronomy is still in its infancy this is the only method existing today. This chapter is devoted to a discussion of the radiation and interaction processes of cosmic ray electrons. In the next chapter we will discuss the radiation and interaction processes of cosmic ray nucleons. The purposes of these two chapters are two-fold. First, we want to investigate the influence of these interaction processes on the dynamics of the cosmic rays. As the particles interact and produce observable electromagnetic radiation they continuously lose energy and that should affect their propagation and their original energy spectrum. Second, we want to establish the relation between the observable electromagnetic radiation spectra and properties of the radiating particles. Armed with these relations we are then in a position to glean information on cosmic rays in other regions of space using the observed radiation from radio to γ-ray frequencies (Chap. 6).

Because of limited space we cannot derive the relevant interaction cross-sections and photon emissivities ab initio from first principles of electrodynamics and quantum mechanics. There are excellent monographs and articles on this topic to which we shall refer the interested reader at the appropriate place. Here our main interest is to establish the link between astrophysical radiation processes and cosmic radiation.

Due to the content and structure of the Galaxy and the Universe (Chap. 2) the following interaction processes of relativistic electrons may occur:

(a) synchrotron radiation in cosmic magnetic fields,
(b) inverse Compton scattering of ambient photon gases,
(c) triplet pair production in ambient photon gases,
(d) nonthermal electron bremsstrahlung in ambient matter fields like the interstellar medium,
(e) ionization and excitation of atoms and molecules in ambient matter fields as well as Coulomb interactions with ionized plasmas.

We discuss each process in turn.

4.1 Synchrotron Radiation

Synchrotron radiation is the radiation of a relativistic particle of charge Ze, mass m, and energy $E = \gamma mc^2$ in a magnetic field of strength B that is uniform on scales much larger than the gyroradius of the particle. For a detailed account of the theory of synchrotron radiation the reader is referred to the two books of Pacholczyk (1970 [381], 1977 [382]). We consider the synchrotron emission of relativistic electrons (mass m), negatrons e^- and positrons e^+, at frequencies ν which are large compared to the plasma frequency ν_p and the non-relativistic gyrofrequency $\nu_0 = eB/(2\pi mc)$ of the thermal electrons in the magnetic field, i.e.

$$\nu \gg \nu_p = 8980 \, (n_e/1 \text{ cm}^{-3})^{1/2} \text{ Hz} \tag{4.1.1}$$

and

$$\nu \gg \nu_0 = 2.8 \times 10^6 \, (B/1 \text{ G}) \text{ Hz}, \tag{4.1.2}$$

where n_e is the thermal electron density. To avoid quantum effects (for a discussion of those see Erber 1966 [153], Brainerd and Petrosian 1987 [74]) we demand the product $\gamma B \ll 4.4 \times 10^{13}$ G.

4.1.1 Synchrotron Power

We start from the spontaneously emitted spectral synchrotron power $P_\parallel$ and $P_\perp$ in directions parallel and perpendicular to the projection of the magnetic field on the plane of the sky from a single relativistic electron

$$P_\parallel(\nu,\theta,\gamma) = \pi\sqrt{3}r_0 mc\nu_0 \sin\theta \left[F(x/\sin\theta) - G(x/\sin\theta) \right]$$
$$\text{erg s}^{-1} \text{ Hz}^{-1}, \tag{4.1.3a}$$

$$P_\perp(\nu,\theta,\gamma) = \pi\sqrt{3}r_0 mc\nu_0 \sin\theta \left[F(x/\sin\theta) + G(x/\sin\theta) \right]$$
$$\text{erg s}^{-1} \text{ Hz}^{-1}, \tag{4.1.3b}$$

with

$$x = \frac{2\nu}{3\nu_0\gamma^2}\left[1 + \left(\frac{\gamma\nu_p}{\nu}\right)^2\right]^{3/2}, \tag{4.1.4}$$

where $r_0 = e^2/(mc^2) = 2.82 \times 10^{-13}$ cm is the classical electron radius and θ is the inclination angle of the magnetic field $\boldsymbol{B}$ with respect to the line of sight to the observer. The two dimensionless functions

$$F(t) \equiv t \int_t^\infty K_{5/3}(z) \, dz; \quad G(t) \equiv tK_{2/3}(t), \tag{4.1.5}$$

determine the spectral behavior of synchrotron radiation and are calculated from two modified Bessel functions of the second kind. They are shown in

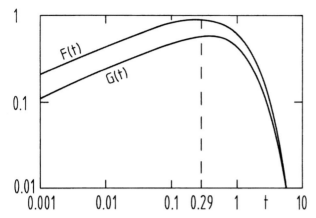

Fig. 4.1. The spectral functions F and G in logarithmic scale. From Pfleiderer et al. (1973 [397])

Fig. 4.1. As one can see a single relativistic electron produces a broad continuum of synchrotron photons.

Their maxima occur at $t = 0.29$ for F and $t = 0.40$ for G, and for small and large arguments they vary as

$$2G(t \ll 1) \simeq F(t \ll 1) \simeq \frac{4\pi}{\sqrt{3}\Gamma(1/3)}(t/2)^{1/3} , \qquad (4.1.6\text{a})$$

$$G(t \gg 1) \simeq F(t \gg 1) \simeq \sqrt{\pi t/2} \, \exp(-t) , \qquad (4.1.6\text{b})$$

where $\Gamma(z)$ denotes the gamma function. Wallis (1959 [552]) has found a very convenient approximation for the function F that is sufficiently accurate for most astronomical applications,

$$F_\text{W}(t) \simeq 1.78 \, t^{0.3} \, \exp(-t) . \qquad (4.1.6\text{c})$$

In practical computations a useful approximation to the spectral function $F(t)$ is given by

$$F_\text{SO}(t) \simeq 1.25 \, t^{1/3} \, \exp(-t) \left[648 + t^2\right]^{1/12} . \qquad (4.1.6\text{d})$$

The total (i.e. summed over both polarization modes) spontaneously emitted spectral synchrotron power of a single relativistic electron is the sum of $P_\perp$ and $P_\parallel$ yielding

$$P(\nu, \theta, \gamma) = 2\pi\sqrt{3} \, r_0 \, mc \, \nu_0 \sin\theta \, F(x/\sin\theta) \text{ erg s}^{-1} \text{ Hz}^{-1} . \qquad (4.1.7)$$

In the vacuum approximation ($\nu_\text{p} \propto n_\text{e} = 0$), where the influence of the background plasma is neglected and which is fulfilled at frequencies above $\nu > \gamma\nu_\text{p}$, (4.1.4) and (4.1.7) simplify to $x = \nu/\nu_\text{c}$ and

$$P_v(\nu,\theta,\gamma) = 2\pi\sqrt{3}\, r_0\, mc\, \nu_0 \sin\theta\, F\left(\frac{\nu}{\nu_c \sin\theta}\right) \text{ erg s}^{-1}\text{ Hz}^{-1}, \quad (4.1.8)$$

where

$$\nu_c = \frac{3}{2}\nu_0\,\gamma^2 = 4.2\times 10^6\,(B/1\text{ G})\,\gamma^2 \text{ Hz} \quad (4.1.9)$$

is referred to as the *characteristic synchrotron radiation frequency*, since it determines the frequency $\nu_{\max} = 0.29\nu_c \sin\theta$ where a single relativistic electron emits most of its synchrotron radiation.

Despite generating a continuum of synchrotron photons the spectral power (4.1.8) has been approximated as (Felten and Morrison 1966 [161])

$$P_{v,\text{ monochromatic}}(\nu,\theta,\gamma) \simeq P_0 \delta(\nu - \nu_c);$$

$$P_0 = \int_0^\infty d\nu\, P_v(\nu,\theta,\gamma)$$

$$= \frac{2}{3}\, c\, r_0^2\,(B\sin\theta)^2 \gamma^2 \text{ erg s}^{-1}. \quad (4.1.10)$$

This 'monochromatic approximation' is often used to calculate synchrotron spectra in a first crude approach.

4.1.2 Emission and Transfer of Synchrotron Radiation

For a given distribution function of relativistic electrons, i.e. the differential number density per unit energy and volume interval $N(\gamma)$, which are assumed to be directionally isotropic, the emitted synchrotron radiation intensity from a considered source can be calculated for a prescribed source geometry from the emission and absorption coefficients of the four Stokes parameters I, Q, U, V and their transport through the source. For ease of exposition we consider a simple slab source of size L containing an ordered magnetic field that is embedded in a thermal background plasma of density n_e (Fig. 4.2).

At relativistic energies of the radiating particles the emission and absorption of elliptical polarization is negligible and the equations of radiation transfer reduce to (Pacholczyk 1977 [382], Chap. 3)

$$\frac{dI}{dr} = \epsilon_I - \kappa I - qQ \quad (4.1.11a)$$

$$\frac{dQ}{dr} = \epsilon_Q - qI - \kappa Q - fU \quad (4.1.11b)$$

$$\frac{dU}{dr} = -\kappa U + fQ \quad (4.1.11c)$$

with the emission coefficients

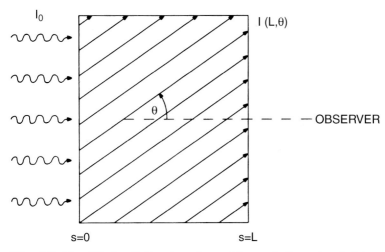

Fig. 4.2. Simplified radiation transfer problem: a 1-dimensional slab source of size L contains relativistic electrons and a uniform magnetic field that is inclined by the angle θ to the line of sight to the observer. The slab is illuminated from behind by unpolarized radiation of intensity I_0

$$\epsilon_I(\nu,\theta) = \frac{1}{4\pi}\int_1^\infty d\gamma\, N(\gamma)P(\nu,\theta,\gamma)$$
$$\text{erg cm}^{-3}\text{ ster}^{-1}\text{ s}^{-1}\text{ Hz}^{-1}, \qquad (4.1.12a)$$

$$\epsilon_Q(\nu,\theta) = \frac{1}{4\pi}\int_1^\infty d\gamma\, N(\gamma)P(\nu,\theta,\gamma)\frac{G(x/\sin\theta)}{F(x/\sin\theta)}$$
$$\text{erg cm}^{-3}\text{ ster}^{-1}\text{ s}^{-1}\text{ Hz}^{-1}, \qquad (4.1.12b)$$

the absorption coefficients

$$\kappa(\nu,\theta) = -\frac{c^2}{8\pi mc^2\nu^2}\int_1^\infty d\gamma\gamma^2 P(\nu,\theta,\gamma)\frac{d}{d\gamma}\left[\gamma^{-2}N(\gamma)\right]\text{ cm}^{-1}, \quad (4.1.13a)$$

$$q(\nu,\theta) = -\frac{c^2}{8\pi mc^2\nu^2}\int_1^\infty d\gamma\gamma^2\left[P_\perp(\nu,\theta,\gamma)\right.$$
$$\left. - P_\parallel(\nu,\theta,\gamma)\right]\frac{d}{d\gamma}\left[\gamma^{-2}N(\gamma)\right]\text{ cm}^{-1}, \qquad (4.1.13b)$$

and the Faraday rotation coefficient

$$f = \frac{2\pi\nu_p^2\nu_0\cos\theta}{c\,\nu^2}$$
$$= 4.73\times 10^4 \frac{n_e}{1\text{ cm}^{-3}}\frac{B}{1\text{ G}}\cos\theta\left(\frac{\nu}{\text{Hz}}\right)^{-2}\text{ cm}^{-1}. \qquad (4.1.14)$$

The degree of linear polarization Π_L and the polarization position angle χ are given by

$$\Pi_{\rm L}(\nu,\theta) = \sqrt{Q^2(\nu,\theta) + U^2(\nu,\theta)}/I(\nu,\theta) \,, \qquad (4.1.15)$$

and

$$\chi(\nu,\theta) = \frac{1}{2} \arctan\left[U(\nu,\theta)/Q(\nu,\theta)\right] \,, \qquad (4.1.16)$$

respectively. Equations (4.1.11)–(4.1.16) indicate that the resulting synchrotron radiation spectrum from a considered source is determined both by radiation transport parameters (as the Faraday rotation coefficient) and by the energy distribution function $N(\gamma)$ of the radiating relativistic electrons that determines the emission and absorption coefficients. The task is to find clues on $N(\gamma)$ from the measured intensity and polarization variation with frequency, and so to disentangle spectral variations due to radiation transport parameters from those due to the energy distribution function of the radiating particles.

Note that the absorption and the Faraday rotation coefficients vary $\propto \nu^{-2}$ so that their effects will predominantly show up at small frequencies. At large enough frequencies the equations of radiative transfer then simplify enormously. We consider such a situation first.

4.1.3 Synchrotron Radiation from Cosmic Ray Electrons in the Interstellar Medium

At frequencies $\nu > 50$ MHz one may neglect the influence of the ionized gas in the interstellar medium on the synchrotron radiation process (Ramaty 1974 [415]), so that the vacuum approximation ($\nu_{\rm p} = 0$) to the synchrotron powers is appropriate, and any synchrotron self-absorption process ($\kappa = q = 0$) on the scale of the Galaxy. The equations of radiation transfer (4.1.11) then readily solve to line-of-sight integrals over the spatial ($\boldsymbol{r} = (l,b,r)$) variation of the electron distribution function and the magnetic field strength (Cioffi and Jones 1980 [101]), where l and b denote galactic longitude and latitude and r the coordinate along the line of sight,

$$I(\nu,\theta,l,b) = \int_0^\infty {\rm d}r\ \epsilon_I(\nu,\theta,l,b,r)$$
$$\text{erg cm}^{-2}\ \text{ster}^{-1}\ \text{s}^{-1}\ \text{Hz}^{-1}\,, \qquad (4.1.17a)$$

$$Q(\nu,\theta,l,b) = \int_0^\infty {\rm d}r\ \epsilon_Q(\nu,\theta,l,b,r) \cos\left[\int_0^r {\rm d}\xi\ f(\xi)\right]$$
$$\text{erg cm}^{-2}\ \text{ster}^{-1}\ \text{s}^{-1}\ \text{Hz}^{-1}\,, \qquad (4.1.17b)$$

$$U(\nu,\theta,l,b) = \int_0^\infty {\rm d}r\ \epsilon_Q(\nu,\theta,l,b,r) \sin\left[\int_0^r {\rm d}\xi\ f(\xi)\right]$$
$$\text{erg cm}^{-2}\ \text{ster}^{-1}\ \text{s}^{-1}\ \text{Hz}^{-1}\,. \qquad (4.1.17c)$$

The energy spectrum of relativistic electrons in the interstellar medium of our Galaxy can be well represented as a power law (Sect. 3.4.1) over large energy intervals,

$$N(\mathbf{r},\gamma) = \begin{cases} N_0(\mathbf{r})\gamma^{-s} & \text{for } \gamma_1 < \gamma < \gamma_2 \\ 0 & \text{otherwise} \end{cases} \quad \text{cm}^{-3} \quad (4.1.18)$$

with $s > 1$, in which case the two emission coefficients (4.1.12) reduce to

$$\epsilon_I(\nu,\theta,l,b) = N_0(\mathbf{r})\, r_0\, mc\, 2^{(5-s)/2}\, 3^{(s-1)/2}$$
$$\times [\nu_0 \sin\theta]^{(s+1)/2}\, \nu^{(1-s)/2}$$
$$\times g\left[\frac{\nu}{\nu_c(\gamma_2)\sin\theta}, \frac{\nu}{\nu_c(\gamma_1)\sin\theta}, s\right]$$
$$\text{erg cm}^{-3}\,\text{ster}^{-1}\,\text{s}^{-1}\,\text{Hz}^{-1}, \quad (4.1.19a)$$

$$\epsilon_Q(\nu,\theta,l,b) = \frac{s+1}{s+(7/3)}\,\epsilon_I(\nu,\theta,l,b)$$
$$\text{erg cm}^{-3}\,\text{ster}^{-1}\,\text{s}^{-1}\,\text{Hz}^{-1}, \quad (4.1.19b)$$

where the function

$$g\left[\frac{\nu}{\nu_c(\gamma_2)\sin\theta}, \frac{\nu}{\nu_c(\gamma_1)\sin\theta}, s\right] \equiv \int_{(\nu/\nu_c(\gamma_2)\sin\theta)}^{(\nu/\nu_c(\gamma_1)\sin\theta)} du\, u^{(s-3)/2}\, F(u) \quad (4.1.19c)$$

describes the influence of the sharp cutoffs in the energy spectrum (4.1.18) at γ_1 and γ_2. For $\nu_c(\gamma_1)\sin\theta \ll \nu \ll \nu_c(\gamma_2)\sin\theta$ the function g becomes independent of frequency,

$$g[0,\infty,s] = g_0(s)$$
$$= 2^{(s-3)/2}\,\frac{s+(7/3)}{s+1}\,\Gamma\left[\frac{(3s-1)}{12}\right]\Gamma\left[\frac{(3s+7)}{12}\right]. \quad (4.1.19d)$$

The two emission coefficients (4.1.19a)–(4.1.19b) in this case reduce to

$$\epsilon_I(\nu,\theta,l,b) = c_1(s)\, m^{(3-s)/4}\, c\, r_0^{(s+5)/4}$$
$$\times N_0(\mathbf{r})\, B_\perp^{(s+1)/2}(\mathbf{r})\, \nu^{(1-s)/2}$$
$$\text{erg cm}^{-3}\,\text{ster}^{-1}\,\text{s}^{-1}\,\text{Hz}^{-1}, \quad (4.1.20a)$$

$$\epsilon_Q(\nu,\theta,l,b) = \frac{s+1}{s+(7/3)}\,\epsilon_I(\nu,\theta,l,b)$$
$$\text{erg cm}^{-3}\,\text{ster}^{-1}\,\text{s}^{-1}\,\text{Hz}^{-1}, \quad (4.1.20b)$$

with $B_\perp \equiv B\sin\theta$ and

$$c_1(s) \equiv \frac{4}{3}\left(\frac{2\pi}{3}\right)^{(s+1)/2}\frac{s+(7/3)}{s+1}\,\Gamma\left[\frac{3s+7}{12}\right]\Gamma\left[\frac{3s-1}{12}\right], \quad (4.1.20c)$$

Using (4.1.20) in (4.1.17) the total synchrotron radiation intensity varies as

$$I(\nu,\theta,l,b) = c_1(s)\, m^{(3-s)/4}\, c\, r_0^{(s+5)/4}\, \nu^{(1-s)/2}$$
$$\times \int_0^\infty dr\, N_0(\mathbf{r})\, B_\perp^{(s+1)/2}(\mathbf{r})$$
$$\text{erg cm}^{-2}\,\text{ster}^{-1}\,\text{s}^{-1}\,\text{Hz}^{-1}, \quad (4.1.21a)$$

which is a power law in frequency ($I \propto \nu^{-\alpha}$) with the spectral index $\alpha = (s-1)/2$ and the proportionality factor is given by the line of sight over the relativistic electron number density $N_0(\mathbf{r})$ times some power (given by s) of the magnetic field strength perpendicular to the line of sight. If we know the latter we can infer the variation of the cosmic ray electron density from the measured synchrotron radiation intensity distribution.

In an emission region with uniform magnetic field and distribution of thermal and relativistic electrons the degree of linear polarization (4.1.15) is frequency independent and given by

$$\Pi_\mathrm{L}(\nu,\theta,l,b) = \frac{s+1}{s+(7/3)} = \frac{\alpha+1}{\alpha+(5/3)}, \qquad (4.1.21\mathrm{b})$$

which can attain large values as $\Pi_\mathrm{L}(\alpha=0.5) = 0.69$, $\Pi_\mathrm{L}(\alpha=1.0) = 0.75$.

4.1.4 Synchrotron Energy Loss of Relativistic Electrons

Integrating equation (4.1.8) over all frequencies ν we obtain the energy loss of a single relativistic electron due to synchrotron radiation (compare with (4.1.10))

$$-\frac{dE}{dt} = -mc^2 \frac{d\gamma}{dt} = 2\pi\sqrt{3}\, r_0\, mc\, \nu_0\, \nu_c\, \sin^2\theta \int_0^\infty du\, F(u)$$

$$= P_0 = \frac{c\sigma_\mathrm{T}}{4\pi} (B\sin\theta)^2 \gamma^2 \text{ erg s}^{-1}, \qquad (4.1.22)$$

where we introduce the Thomson cross-section $\sigma_\mathrm{T} = 8\pi r_0^2/3 = 6.65 \times 10^{-25}$ cm^2. This energy loss rate has been calculated with the vacuum approximation (4.1.8) of the total synchrotron power; a more general calculation (Schlickeiser and Crusius 1988 [464]) using the total synchrotron power (4.1.7) shows that this approximation is valid for relativistic electron Lorentz factors larger than the Razin-Lorentz factor

$$\gamma_\mathrm{R} \equiv \frac{2\nu_\mathrm{P}}{3\nu_0} = 2.1 \times 10^{-3}\, (n_e/1 \text{ cm}^{-3})^{1/2}\, (B/1 \text{ G})^{-1}, \qquad (4.1.23)$$

whereas below γ_R the energy loss is reduced compared to the vacuum case by the exponential factor $\exp(-\gamma_\mathrm{R}/\gamma \sin\theta)$. Figure 4.3 shows the synchrotron energy loss rate of a single electron for different inclination angles.

4.2 Inverse Compton Scattering

The inverse Compton effect of relativistic electrons traversing a photon gas was originally discussed as an energy loss process of cosmic ray electrons by Follin (1947 [173]) and Feenberg and Primakoff (1948 [158]). Savedoff

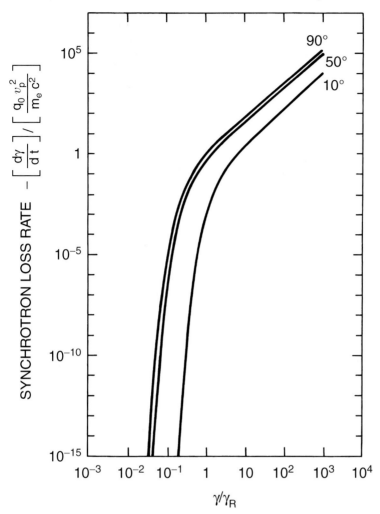

Fig. 4.3. Synchrotron energy loss rate of a single relativistic electron in a plasma for different magnetic field inclination angles. From Schlickeiser and Crusius (1988 [464])

(1959 [447]) and Felten and Morrison (1963 [160]) have pointed out that this interaction may be an important source of cosmic high-energy photons. In this interaction the target photon receives part of the kinetic energy of the relativistic electron and is scattered to higher frequencies.

4.2.1 Inverse Compton Power

Assuming that target photons and relativistic electrons are spatially isotropic distributed–for anisotropic problems see Jones et al. (1974 [251])–the differential cross-section for this process is given by the Klein-Nishina formula (e.g. Jauch and Rohrlich 1955 [239]) averaged over the initial photon polarization states and summed over the final photon polarization states (Ginzburg and Syrovatskii 1964 [185], Jones 1968 [250], Blumenthal and Gould 1970 [68]):

$$\sigma(E_\gamma, \epsilon, \gamma) = \frac{3\sigma_T}{4\epsilon\gamma^2} \, G(q, \Gamma_e) \; \text{cm}^2 \; \text{eV}^{-1} \, , \quad (4.2.1)$$

with

$$G(q, \Gamma_e) \equiv \left[2q \ln q + (1+2q)(1-q) + \frac{(\Gamma_e q)^2 (1-q)}{2\,(1+\Gamma_e q)} \right] \, , \quad (4.2.2)$$

$$\Gamma_e = 4\epsilon\gamma/(mc^2) \quad q = E_\gamma / \left[\Gamma_e (\gamma mc^2 - E_\gamma) \right] \, , \quad (4.2.3)$$

where γ is the initial Lorentz factor of the relativistic electron, ϵ and E_Γ are the photon energies before and after the scattering, respectively. This cross-section defines the probability that an electron of energy γmc^2 generates a photon of energy E_γ by scattering a target photon of energy ϵ. From the kinematics of the scattering process the range of q is restricted to values of (Blumenthal and Gould 1970 [68])

$$0 \simeq \frac{1}{4\gamma^2} \leq q \leq 1 \, . \quad (4.2.4)$$

The inverse Compton power of photons scattered spontaneously by a single relativistic electron is obtained by folding the Klein-Nishina cross-section (4.2.1) with the differential number density of target photons at position $\boldsymbol{r}$, i.e. the number of photons per unit volume and energy element $n(\epsilon, \boldsymbol{r})$, as

$$P(E_\gamma, \gamma, \boldsymbol{r}) = c \, E_\gamma \int_0^\infty \mathrm{d}\epsilon \; n(\epsilon, \boldsymbol{r}) \, \sigma(E_\gamma, \epsilon, \gamma) \; \text{erg s}^{-1} \, \text{eV}^{-1} \, . \quad (4.2.5)$$

The spectral distribution of the inverse Compton scattering power depends sensitively on the value of the Compton parameter Γ_e. This is most clearly seen if we consider for ease of exposition the scattering off a monochromatic target photon distribution $n(\epsilon, \boldsymbol{r}) = n_0(\boldsymbol{r})\delta(\epsilon - \epsilon_0)$. In this case (4.2.5) reduces to

$$P(E_\gamma, \gamma, \boldsymbol{r}) = \frac{3 \, \sigma_T \, c \, E_\gamma \, n_0(\boldsymbol{r})}{4 \, \epsilon_0 \, \gamma^2} \, G(q, \Gamma) \, ,$$

with $\Gamma = \Gamma_e(\epsilon = \epsilon_0)$. In Fig. 4.4 we have calculated the function $G(q, \Gamma)$ for three different values of the parameter Γ and an electron Lorentz factor $\gamma = 10^6$, dependent on the normalized scattered photon energy $S/S_{\max}$, where $S = E_\gamma/E$ and $S_{\max} = \Gamma/(\Gamma + 1)$.

In the Thomson limit ($\Gamma \ll 1$) the resulting scattered photon distribution is a broad continuum ranging from $E_\gamma = \epsilon_0$ to $E_{\gamma, \max} = 4\epsilon_0 \gamma^2$ and a mean scattered photon energy of

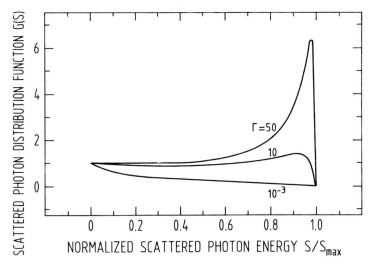

Fig. 4.4. Inverse Compton scattered photon distribution function $G(S)$ for three values of the Compton parameter Γ and an electron Lorentz factor $\gamma = 10^6$ as a function of the normalized scattered photon energy $S/S_{\max}$ where $S = E_\gamma/E$ and $S_{\max} = \Gamma/(\Gamma + 1)$. From Schlickeiser (1989a [457])

$$\langle E_\gamma \rangle \equiv \frac{\int_0^\infty dE_\gamma E_\gamma P(E_\gamma, \gamma, \boldsymbol{r})}{\int_0^\infty dE_\gamma P(E_\gamma, \gamma, \boldsymbol{r})}$$

$$= 4\epsilon_0 \gamma^2 \frac{\int_0^1 dq\, q\, [1 + q - 2q^2 + 2q \ln q]}{\int_0^1 dq\, [1 + q - 2q^2 + 2q \ln q]} = \frac{4}{3}\epsilon_0 \gamma^2 \,. \quad (4.2.6)$$

This broad continuum distribution of the inverse Compton scattering power in the Thomson limit has been approximated similar to (4.1.10) by Felten and Morrison (1966 [161]) as a 'monochromatic approximation'

$$P_{\text{monochromatic}}(E_\gamma, \gamma, \boldsymbol{r}) \simeq P_c \delta(E_\gamma - \langle E_\gamma \rangle)\ \text{erg s}^{-1}\ \text{eV}^{-1}\,, \quad (4.2.7a)$$

with

$$P_c = \int_0^\infty dE_\gamma P(E_\gamma, \gamma, \boldsymbol{r}) = \frac{4}{3} c\, \sigma_T\, w(\boldsymbol{r})\, \gamma^2\,, \quad (4.2.7b)$$

and $\langle E_\gamma \rangle$ given in (4.2.6). However, in the extreme Klein-Nishina limit ($\Gamma \gg 1$) a different dependence results. Here the function $G(S)$ is equal to unity for practically all γ-ray energies except for a sharp and pronounced maximum $G_{\max} = (\Gamma + 1)/8$ at $S_0 = (\Gamma - 1)/(\Gamma + 1)$ close to the maximum energy $S_{\max} = \Gamma/(\Gamma + 1)$. Instead of a broad continuum of scattered photons a very sharp and narrow distribution at almost the relativistic electron energy results.

The inverse Compton scattering emission and absorption coefficients are calculated by expressions analogous to equations (4.1.12a) and (4.1.13a) (e.g. Schlickeiser 1982a [453]) with the additional constraint (4.2.4) that in order to scatter a photon of energy ϵ into one with energy E_γ the minimum electron Lorentz factor has to be

$$\gamma_{\min} = \frac{E_\gamma}{2mc^2}\left[1 + \left(1 + \frac{m^2 c^4}{\epsilon E_\gamma}\right)^{1/2}\right]. \quad (4.2.8)$$

4.2.2 Inverse Compton Scattering of Target Photons in the Interstellar Medium by Cosmic Ray Electrons

At high scattered photon energies inverse Compton self-absorption is usually negligible (Schlickeiser 1982a [453]) and the intensity of inverse Compton scattering photons is obtained by integrating over the relativistic electron distribution

$$I_c(E_\gamma, l, b) = \frac{3c\sigma_T E_\gamma}{16\pi} \int_0^\infty dr \int_0^\infty d\epsilon\, \epsilon^{-1}\, n(\mathbf{r}, \epsilon)$$

$$\times \int_{\gamma_{\min}}^\infty d\gamma\, \gamma^{-2} N(\mathbf{r}, \gamma) G(q, \Gamma_e)$$

$$\mathrm{erg\ cm^{-2}\ s^{-1}\ ster^{-1}\ eV^{-1}}. \quad (4.2.9)$$

For the important case of the power law distribution function (4.1.18) and a separable target photon distribution $n(\mathbf{r}, \epsilon) = w(\mathbf{r})\, n(\epsilon)$ we obtain

$$I_c(E_\gamma, l, b) = \frac{E_\gamma}{4\pi} \int_0^\infty dr\, S(E_\gamma, \mathbf{r}), \quad (4.2.10)$$

where we introduce the differential omnidirectional source function $S(E_\gamma, \mathbf{r})$, which gives the number of inverse Compton scattered photons per volume, time and energy element at position $\mathbf{r}$

$$S(E_\gamma, \mathbf{r}) = 3 \times 2^{s-2}\ c\, \sigma_T\, N_0(\mathbf{r})\, w(\mathbf{r})\, E_\gamma^{-(s+1)/2}$$

$$\times \int_0^\infty d\epsilon\, \epsilon^{(s-1)/2}\, n(\epsilon)\, M[\eta(\epsilon, E_\gamma), s]$$

$$\mathrm{photons\ cm^{-3}\ s^{-1}\ eV^{-1}}, \quad (4.2.11)$$

with the integral (Blumenthal and Gould 1970 [68])

$$M[\eta, s] \equiv \int_0^1 dq\, q^{(s-1)/2}$$

$$\times \frac{2q \ln q + 1 + q - 2q^2 + 2\eta q(1-q)}{[1 + [\eta q/(1+\eta q)]^{1/2}]^{s+2}\,(1+\eta q)^{(s+3)/2}}, \quad (4.2.12)$$

and the dimensionless parameter

$$\eta = \epsilon E_\gamma/(mc^2)^2 \,, \tag{4.2.13}$$

which determines the domain of scattering. Expression (4.2.11) is exact and demonstrates that under these assumptions the source function $S(E_\gamma, \mathbf{r})$ also is a separable function in energy and position. The quantity $w(\mathbf{r})$ is referred to as the target photon energy density if $\int_0^\infty d\epsilon\, \epsilon n(\epsilon) = 1$.

In order to reduce the general result (4.2.11) to simpler expressions, asymptotic forms have been discussed in the literature:

(i) Thomson limit: $\eta \ll 1$,
(ii) near Klein-Nishina limit: $\eta \leq 1$,
(iii) extreme Klein-Nishina limit: $\eta \gg 1$.

For values of η smaller than unity (cases (i) and (ii)) the integrand in (4.2.12) can be expanded by keeping the lowest order terms in $(q\eta)^{1/2}$, so that the q-integration can be performed. By keeping terms up to fourth-order Schlickeiser (1979 [451]) has obtained

$$M^{(4)}[\eta, s] = M^{(0)}[s] \left[1 - \frac{(s^2 + 6s + 16)(s+1)(s+3)^2(s+5)}{(s^2 + 4s + 11)(s+4)^2(s+6)} \eta^{1/2} \right.$$
$$+ \frac{(s^4 + 12s^3 + 62s^2 + 164s + 209)(s+1)(s+3)}{2(s^2 + 4s + 11)(s+5)(s+7)} \eta$$
$$- \frac{(s^4 + 14s^3 + 84s^2 + 296s + 576)}{6(s^2 + 4s + 11)(s+4)(s+6)^2(s+8)}$$
$$\left. \times (s+1)(s+2)(s+3)^2(s+5)\, \eta^{3/2} \right] \tag{4.2.14}$$

with

$$M^0[s] = \frac{4(s^2 + 4s + 11)}{(s+1)(s+5)(s+3)^2} \tag{4.2.15}$$

being, to zeroth order, independent of η. For large values $\eta \gg 1$ (case (iii)) Blumenthal and Gould (1970 [68]) have derived

$$M[\eta \gg 1, s] \to 2^{-(s+1)}\, \eta^{-(s+1)/2} \left[\ln(\eta) + C(s) \right], \tag{4.2.16}$$

where $C(s)$ is a parameter of order unity. These asymptotic expressions agree well with the numerical evaluation of the function $M[\eta, s]$ as a function of η for various values of the relativistic electron spectral index shown in Fig. 4.5. For small values of η the function M approaches the value $M^{(0)}[s]$ which is independent of η, whereas at large values of η the drastic power law decrease according to equation (4.2.16) sets in. At large values of η the efficiency of inverse Compton scattering is strongly reduced, which is commonly referred to as "Klein-Nishina cutoff". With the zeroth order term $M^{(0)}[s]$ one obtains the well-known approximation of (4.2.11) in the Thomson limit

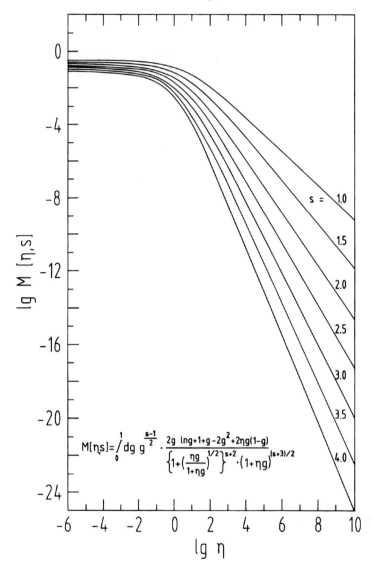

Fig. 4.5. Variation of the function $M[\eta, s]$ as a function of η for different values of the spectral index of the power law distribution of the relativistic electrons

$$S_T(E_\gamma, r) = \frac{3 \times 2^s \left(s^2 + 4s + 11\right)}{(s+1)(s+5)(s+3)^2}$$
$$\times c\, \sigma_T\, N_0(r)\, w(r)\, E_\gamma^{-(s+1)/2}$$
$$\times \int_0^\infty d\epsilon\, \epsilon^{(s-1)/2} n(\epsilon) \text{ photons cm}^{-3}\, \text{s}^{-1}\, \text{eV}^{-1}\,, \quad (4.2.17)$$

a power law in E_γ with the spectral index $(s+1)/2$.

The question of validity of the Thomson limit in the context of calculating inverse Compton scattering source functions is answered in Fig. 4.6. It displays the energy of the scattered photon $E_\gamma{}^0$ up to which the Thomson limit

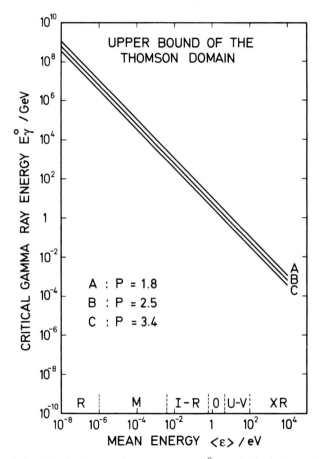

Fig. 4.6. Scattered photon energy $E_\gamma{}^0$, at which the Compton parameter η becomes unity, which is a reasonable upper bound for using the Thomson limit as a function of the mean energy $\langle\epsilon\rangle$ of the target photon distribution for different values of the electron spectral index s. The values of $\langle\epsilon\rangle$ cover the spectral range from radio (R), microwave (M), infrared (I-R), optical (O), ultraviolet (U-V) and X-ray (XR) wavelengths. From Schlickeiser (1979 [451])

may be used as a function of the mean energy $<\epsilon>$ of the target photons for different values of the relativistic electron power law spectral index s. The critical energy E_γ^0 has been defined by the condition that the first-order correction term in (4.2.14) $M^{(1)}[\eta(\epsilon, E_\gamma^0), s] = 0.5$. It can be seen that in order to calculate γ-ray source functions with $E_\gamma \geq 100$ MeV the Thomson limit of the Klein-Nishina cross-section is not appropriate if target photons with mean energies greater than 1 eV, i.e. optical, ultraviolet and X-ray photons, are involved. In this case we have to use the general result (4.2.11). On the other hand for the calculation of X-ray source functions in this example the Thomson limit is highly appropriate.

With the extreme Klein-Nishina limit approximation (4.2.16) the source function (4.2.11) reduces to

$$S_{\rm KN}(E_\gamma, \boldsymbol{r}) = \frac{3}{8} \, c \, \sigma_{\rm T} \, N_0(\boldsymbol{r}) \, w(\boldsymbol{r}) \, (mc^2)^{1+s} \, E_\gamma^{-(s+1)}$$
$$\times \int_0^\infty {\rm d}\epsilon \, \epsilon^{-1} \, n(\epsilon) \left[\ln\left(\frac{\epsilon E_\gamma}{m^2 c^4}\right) + C(s) \right]$$
$$\text{photons cm}^{-3} \, \text{s}^{-1} \, \text{eV}^{-1} \,, \qquad (4.2.18)$$

a power law with the spectral index $(s+1)$, which is much steeper than the Thomson limit power law (4.2.17). Due to the onset of the Klein-Nishina cutoff the scattered photon source function steepens considerably beyond the energy E_γ^0.

4.2.3 Inverse Compton Energy Loss of Relativistic Electrons

Integrating the inverse Compton power (4.2.5) over all scattered photon energies E_γ we obtain the energy loss of a single relativistic electron due to inverse Compton scattering

$$-\frac{{\rm d}E}{{\rm d}t} = -mc^2 \frac{{\rm d}\gamma}{{\rm d}t}$$
$$= \int_0^{E_{\gamma,\,{\rm max}}} {\rm d}E_\gamma P(E_\gamma, \gamma, \boldsymbol{r})$$
$$= \frac{3\,\sigma_{\rm T}\,c\,m^2\,c^4}{4} \times \int_0^\infty {\rm d}\epsilon \, \epsilon^{-1} \, n(\epsilon, \boldsymbol{r})$$
$$\times \int_0^1 {\rm d}q \frac{\Gamma_{\rm e}^2 q}{(1+\Gamma_{\rm e} q)^3} \, G(q, \Gamma_{\rm e}) \, \text{erg s}^{-1} \,, \qquad (4.2.19)$$

where we used (4.2.1) and (4.2.4).

In the Thomson limit, $\Gamma_{\rm e} \ll 1$, the second integral in (4.2.19) gives to lowest order in $\Gamma_{\rm e}$

$$\int_0^1 {\rm d}q \frac{\Gamma_{\rm e}^2 q}{(1+\Gamma_{\rm e} q)^3} \, G(q, \Gamma_{\rm e})$$

$$\simeq \Gamma_{\rm e}^2 \int_0^1 dq \; q[2q \ln q \; + \; (1+2q)(1-q)]$$

$$= \frac{1}{9}\Gamma_{\rm e}^2$$

so that

$$-\left(\frac{dE}{dt}\right)_{\rm T} \simeq \frac{4}{3} \; c \; \sigma_{\rm T} \; w(\boldsymbol{r}) \; \gamma^2 \; {\rm erg \; s^{-1}} \; , \qquad (4.2.20)$$

where we introduced the photon energy density $w(\boldsymbol{r}) \equiv \int_0^\infty d\epsilon\, n(\epsilon, \boldsymbol{r})$. Comparing (4.2.20) with (4.1.22) we note that in the Thomson limit the inverse Compton scattering losses exhibit the same γ^2-dependence as the synchrotron radiation losses.

In the extreme Klein-Nishina limit, $\Gamma_{\rm e} \gg 1$, the evaluation of the second integral in (4.2.19) gives, to leading order in $\Gamma_{\rm e}$ (Jones 1965 [249], Blumenthal and Gould 1970 [68]),

$$-\left(\frac{dE}{dt}\right)_{\rm KN} \simeq \frac{3}{8} \; c \; \sigma_{\rm T} \; m^2 c^4 \int_0^\infty d\epsilon\, \epsilon^{-1} \; n(\epsilon, \boldsymbol{r}) \left[\ln \Gamma_{\rm e} - \frac{11}{6}\right]. \qquad (4.2.21)$$

In the extreme Klein-Nishina limit the energy loss rate increases only logarithmically with E (or γ) while in the Thomson limit the loss rate increases with γ^2. However, in the extreme Klein-Nishina limit the energy loss rate does not have the same meaning as in the Thomson limit, where in each Compton collision the relativistic electron loses a small fraction of its energy. In the extreme Klein-Nishina limit, i.e. at very high energies, the relativistic electron loses its energy in discrete amounts which are a sizeable fraction of its initial energy.

4.3 Triplet Pair Production

Pair production in photon-electron collisions

$$e_0 + \gamma \to e + e^+ + e^-, \qquad (4.3.1)$$

is usually referred to as triplet pair production (TPP), and can be important when ultrarelativistic particles scatter photons in the extreme Klein-Nishina limit. TPP not only acts as a mechanism for electron energy-loss, but also provides a source of electron-positron pairs that can initiate an electromagnetic pair cascade.

The differential cross-section for TPP was derived by Votruba (1948 [551]), Mork (1967 [366]) and Haug (1975 [219]). Jarp and Mork (1973 [240]), Mastichiadis et al. (1986 [326]), and Mastichiadis (1991 [324]) calculated energy spectra of secondary positrons formed in TPP by numerically integrating the TPP differential cross-section over the energies and angles of the primary and secondary electrons and over the outgoing angles of the secondary positron. The total TPP cross-section requires a further integration over the energy of the produced positron, and was numerically evaluated by Haug (1981 [220]), who gives an analytical fit to the cross-section accurate to better than 0.3%. Here, we follow the treatment of Dermer and Schlickeiser (1991 [124]) who have derived a simple analytical formula for the TPP electron energy loss rate, and compare this expression with the numerical results obtained by Mastichiadis (1991 [324]).

The magnitude of the cross-section for TPP depends solely on the value of the invariant photon energy $k' \equiv \epsilon'/(m_e c^2)$ in the incident electron rest frame. For TPP to occur, $k' \geq 4$. TPP is a third-order electromagnetic process, and the value of its cross-section well above threshold is given to order-of-magnitude accuracy by $\sigma_{\text{TPP}} \sim \alpha_f^3 \lambda_c^2$ (Gould 1979 [199]); where $\alpha_f = 1/137.037$ is the fine-structure constant and $\lambda_c = r_0/\alpha_f = 3.862 \times 10^{-11}$ cm is the electron Compton wavelength. The peak value of the cross-section for ordinary inverse Compton scattering (Sect. 4.2) is $\propto \alpha_f^2 \lambda_c^2$, but decreases as $\ln(2k')/(k')$ in the Klein-Nishina regime ($k' > 1$) due to the effects of electron recoil (see Sect. 4.2). By contrast, σ_{TPP} remains logarithmically constant. A direct comparison of the two cross-sections shows that $\sigma_{\text{TPP}} > \sigma_{\text{KN}}$ when $k' > 300$. The effects of the two processes on the energy distribution of the primary electrons depend, however, on the respective energy loss rates which, in turn, depend on the energy transferred to the electron-positron pair in TPP and the energy of the scattered photon on inverse Compton scattering.

Here we argue that the electron-positron pairs are formed, on average, at rest in the frame in which the total x-momentum of the photon-electron system is zero, where the x axis is defined by the direction of the electrons' motion. This assumption is suggested by collider experiments on hadron-hadron interactions in the scaling regime (e.g. Perl 1974 [394]). When the total center-of-momentum energy greatly exceeds the rest-mass energy of the colliding particles, secondary particles are generally well-described by a scaling distribution with zero net momentum in the center-of-momentum of the collision. The simplest such distribution is a δ-function in the zero x-momentum frame. The accuracy of this assumption can, of course, only be tested by comparing the analytic results directly with numerical integrations of the exact TPP cross-section.

Suppose an electron with energy $E = \gamma m_e c^2$ and momentum $P_x = p m_e c$ in the x direction collides with a photon with energy $k m_e c^2$ in the frame K. If μ denotes the cosine of the angle between the directions of the electron and the photon, then $k' = k(\gamma - p\mu)$. In a frame K', moving with velocity βc with respect to frame K along the x direction, the total x-momentum of the photon/electron system vanishes by definition. The x-component of the electron's momentum is simply $\Gamma(p - \beta\gamma)$, and $\Gamma = (1 - \beta^2)^{-1/2}$. The x-component of the photon's momentum in K' is $k'\mu'$, where $k' = \Gamma k(1-\beta\mu)$ and $\mu' = (\mu - \beta)/(1 - \beta\mu)$. If the total x-momentum of the photon/electron system equals zero, then $p - \beta\gamma = -k\mu + \beta k$, implying $\beta = (p + k\mu)/(\gamma + k)$. Thus the Lorentz factor of the transformation

$$\Gamma = \frac{\gamma + k}{[1 + k^2(1 - \mu^2) + 2k(\gamma - p\mu)]^{1/2}}$$
$$\simeq \frac{\gamma}{[2k(\gamma - p\mu)]^{1/2}} = (2k')^{-1/2}\gamma , \qquad (4.3.2)$$

where the approximate equality holds when $k \ll 1 \ll \gamma$, and since $k(\gamma - p\mu) = k' > 4$ that is, for soft photons and relativistic electrons.

We assume that the Lorentz factor of an electron or positron formed in TPP equals $\Gamma(\mu)$. If triplet pairs are formed in head-on collisions, $\mu = -1$, and (4.3.2) shows that $\Gamma \simeq \gamma^{1/2}/(2k^{1/2})$. In glancing collisions, $\mu \to \mu_{\mathrm{thr}} \equiv (k\gamma - 4)/(kp) \simeq 1 - (k\gamma/4)^{-1}$, and triplet pairs are made with energies $\Gamma \simeq \gamma/3$. Thus head-on collisions produce much lower-energy pairs than those formed in glancing collisions, in qualitative agreement with the findings of Mastichiadis et al. (1986 [326]). The average energy lost per TPP interaction by an electron traveling through an isotropic soft photon field, which we denote by $\Delta\Gamma$, can be found by averaging (4.3.2) over the angles of the collision. Changing variables from μ to k' and replacing $\Gamma(\mu)$ by the final expression in (4.3.2) we obtain

$$\Delta\Gamma = 2\frac{\int_{-1}^{\mu_{\mathrm{thr}}} d\mu \Gamma(\mu)(1 - p\mu/\gamma)}{\int_{-1}^{\mu_{\mathrm{thr}}} d\mu(1 - p\mu/\gamma)} \simeq \frac{\sqrt{8}}{3}\left(\frac{\gamma}{k}\right)^{1/2} . \qquad (4.3.3)$$

From (4.3.3) we see that the average energy of triplet pairs formed in frame K is, in general, considerably less than the energy of the primary electron. In Klein-Nishina inverse Compton interactions, in contrast, the scattered photon carries away a large fraction of the energy of the primary electron. Thus we expect the TPP energy loss rate to dominate the Klein-Nishina energy loss rate only when $k' \gg 300$. The energy loss rate of an electron through TPP in an isotropic soft photon field is given by

$$-\left(\frac{d\gamma}{dt}\right)_{\mathrm{TPP}} \simeq 2c\, n_{\mathrm{ph}}^0 \times \frac{1}{2}\int_{-1}^{\mu_{\mathrm{thr}}} d\mu \sigma_{\mathrm{TPP}}(k')(1 - p\mu/\gamma)\Gamma(\mu) ,$$

where n_{ph}^0 denotes the number density of monochromatic soft photons with energy k. Following the procedure used to evaluate (4.3.3), we find

$$-\left(\frac{d\gamma}{dt}\right)_{\mathrm{TPP}} \simeq \frac{2^{-1/2}\alpha_f c\sigma_T n_{\mathrm{ph}}^0}{k^2\gamma}\int_4^{2\gamma k} dk' k'^{1/2} f(k')$$
$$\simeq \frac{4}{3}c\, \sigma_T\, n_{\mathrm{ph}}^0 \alpha_f \left(\frac{\gamma}{k}\right)^{1/2} . \qquad (4.3.4)$$

Here we have written $\sigma_{\mathrm{TPP}}(k') \equiv \alpha_f \sigma_T f(k')$, where σ_T is the Thomson cross-section. By comparison with Haug's (1981 [220]) expression for the TPP cross-section, we find that the function $f(k') = 0.86, 1.32, 1.81$, and 2.70 when $k' = 100, 300, 10^3$, and 10^4, respectively. The simplest approximation consists in letting $f(k') = 1$, giving the final result in (4.3.4).

In Fig. 4.7 we plot the approximate TPP energy loss rate given by (4.3.4) against the exact rate obtained by numerically integrating the full TPP cross-section (Mastichiadis 1991 [324]). In this figure the photon energy is $k = 10^{-3}$. As can be seen, approximation (4.3.4) is accurate to within a factor 2 over 4 orders of magnitude in the variable $k\gamma$, and accurate to better than a factor of 3 over 5.5 orders of magnitude. At low values of $k\gamma$, approximation (4.3.4) overestimates the TPP energy loss rate, which is probably due to an overestimation of the cross-section by using the approximation $f(k') = 1$ in (4.3.4). By the same reasoning, we should expect an underestimation of the energy loss rate at large values of $k\gamma (\gg 10^4)$, but we see instead that expression (4.3.4) overestimates the energy loss rate. This implies that our simple argument leading to (4.3.2) is only approximately correct, and that triplet pairs are actually produced with somewhat smaller average energies than estimated. Equation (4.3.4) is, however, sufficiently accurate and simple to characterize regimes where the TPP process is important.

By replacing n_{ph}^0 with the differential photon number density $n_{ph}(k)dk$ and integrating over k, one can determine the TPP energy loss rate in an arbitrary isotropic photon distribution. For the graybody distribution (2.3.1) we obtain

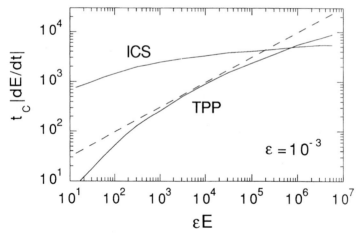

Fig. 4.7. Comparison of electron energy loss rates through inverse Compton scattering (ICS) and triplet pair production (TPP). The electron energy in units of the electron rest mass is denoted by E, and the photons are assumed to be isotropically distributed with energy $\epsilon = h\nu/(m_e c^2) = 10^{-3}$. The **solid curves** are from the work of Mastichiadis (1991 [324]), and the **dashed line** is the approximate analytic result for the TPP energy loss rate given by (4.3.4). From Dermer and Schlickeiser (1991 [124])

$$-\left\langle\frac{d\gamma}{dt}\right\rangle_{\text{TPP}} = \frac{8\zeta(5/2)\alpha_f^3 c\rho}{3\pi^{1/2}\bar{\lambda}_c}\gamma^{1/2}\theta^{5/2}, \qquad (4.3.5)$$

where $\zeta(x)$ denotes the zeta function, $\zeta(5/2) = 1.341$, $\theta = k_B T/(m_e c^2)$ is the dimensionless photon temperature, and the brackets refer to a thermal averaging.

Also shown in Fig. 4.7 is the inverse Compton scattering energy loss rate, given in the Klein-Nishina regime $k' \gg 1$ the logarithmic dependence (4.2.21). Averaging (4.2.21) over the graybody distribution (2.3.1) gives

$$-\left\langle\frac{d\gamma}{dt}\right\rangle_{\text{KN}} = \frac{\pi\alpha_f^2 c\rho}{6\bar{\lambda}_c}\theta^2\ln(a'\theta\gamma), \qquad (4.3.6)$$

where $a' = 0.55$.

A direct comparison of (4.2.21) and (4.3.4) shows that the TPP and KN energy loss rates are equal when $\gamma = \gamma_t = 2.0 \times 10^5 k^{-1}$. The numerical results of Mastichiadis (1991 [324]) show that $\gamma_t = 8 \times 10^5 k^{-1}$. This difference is less significant than may appear, since the energy loss rates are slowly varying in the regime $k\gamma > 10^4$. For $\gamma < \gamma_t$ inverse Compton losses dominate, while for $\gamma > \gamma_t$ triplet pair production losses are more important. The thermal-averaged rates (4.3.5) and (4.3.6) are equal when $k\theta = 1.7 \times 10^5$. This shows that the thermal-averaging does not introduce any significant change in the energy loss rate behavior, as expected since the monochromatic energy loss rates are slowly varying quantities.

4.4 Nonthermal Relativistic Electron Bremsstrahlung

If a relativistic electron (negatron or positron) of energy $E = \gamma mc^2$ is deflected in the electrostatic potential of an atom, ion or molecule, it emits a γ-ray photon of typical energy $E_\gamma \simeq E/2$.

4.4.1 Nonthermal Bremsstrahlung Power

Bethe and Heitler (1934 [48]) have calculated the differential cross-section for this interaction to be

$$\sigma(E_\gamma, E) = \frac{3}{8\pi}\alpha\,\sigma_T\left[\left[1+\left(1-\frac{E_\gamma}{E}\right)^2\right]\phi_1\right.$$
$$\left.-\frac{2}{3}\left(1-\frac{E_\gamma}{E}\right)\phi_2\right]\text{cm}^2\,\text{eV}^{-1}, \qquad (4.4.1)$$

where $\alpha = 1/137.037$ is the fine structure constant and σ_T the Thomson cross-section. $\phi_1(E_\gamma, E)$ and $\phi_2(E_\gamma, E)$ are energy-dependent scattering functions which depend on the properties of the scattering system. When the scattering system is an unshielded charge (Ze), then $\phi_1 = \phi_2 = Z^2 \phi_u$ where

$$\phi_u = 4 \left[\ln \left[\frac{2E}{mc^2} \left(\frac{E - E_\gamma}{E_\gamma} \right) \right] - \frac{1}{2} \right]. \qquad (4.4.2)$$

Gould (1969 [196]) has considered the astrophysically important cases of nonthermal electron bremsstrahlung scattering in the field of hydrogen and helium atoms. Given in Table 4.1 are the scattering functions ϕ_1 and ϕ_2 when the target is a hydrogen atom or a helium atom as a function of the parameter

$$\Delta = \frac{E_\gamma mc^2}{4\alpha E(E - E_\gamma)}. \qquad (4.4.3)$$

For $\Delta > 2$ one should use (4.4.2). If the scattering system is a hydrogen molecule H_2 the cross-section $\sigma_{H_2}(E_\gamma, E)$ is given by $2\sigma_{HI}(E_\gamma, E)$, where the error of this approximation is less than 3% (Gould 1969 [196]).

The nonthermal electron bremsstrahlung power emitted spontaneously by a single relativistic electron in a medium with j different species (atoms, ions and electrons) of corresponding densities n_j is

$$P_B(E_\gamma, E) = c\, E_\gamma \sum_j n_j\, \sigma_j(E_\gamma, E) \text{ erg s}^{-1} \text{ eV}^{-1}, \qquad (4.4.4)$$

from which the nonthermal electron bremsstrahlung omnidirectional source function $S_B(E_\gamma, \boldsymbol{r})$ and the nonthermal electron bremsstrahlung emission coefficient $\epsilon_B(E_\gamma, \boldsymbol{r})$ at position $\boldsymbol{r}$ follow as

Table 4.1. Scattering functions ϕ_1 and ϕ_2 for the hydrogen atom and helium atom. After Blumenthal and Gould (1970 [68])

Δ	H		He	
	ϕ_1	ϕ_2	ϕ_1	ϕ_2
0	45.79	44.46	134.60	131.40
0.01	45.43	44.38	133.85	130.51
0.02	45.09	44.24	133.11	130.33
0.05	44.11	43.65	130.86	129.26
0.1	42.64	42.49	127.17	126.76
0.2	40.16	40.19	120.35	120.80
0.5	34.97	34.93	104.60	105.21
1	29.97	29.78	89.94	89.46
2	24.73	24.34	74.19	73.03
5	18.09	17.28	54.26	51.84
10	13.65	12.41	40.94	37.24

$$S_B(E_\gamma, r) = 4\pi\, \epsilon_B(E_\gamma, r)/E_\gamma$$

$$= \frac{1}{E_\gamma} \int_{E_L}^{\infty} d\gamma\, N(\gamma, r)\, P_B(E_\gamma, E)$$

$$= c \sum_j \int_{E_L}^{\infty} d\gamma\, N(\gamma, r)\, n_j(r)\, \sigma_j(E_\gamma, \gamma)$$

$$\text{photons cm}^{-3}\, \text{s}^{-1}\, \text{eV}^{-1}, \tag{4.4.5}$$

where $E_L = \max(E_\gamma, E_1)$ where $E_1 = \gamma_1 m c^2$ denotes a low-energy cutoff in the relativistic electron distribution (see (4.1.18)).

4.4.2 Nonthermal Bremsstrahlung of Cosmic Ray Electrons in the Interstellar Medium

In the local interstellar medium we may neglect the ionized gas component which has a low density (see Sect. 2.2) and take into account that

(i) the helium/total hydrogen number ratio in the interstellar medium is 0.1, and
(ii) the ratio of the scattering functions $\phi_1^{He}/\phi_1^H \simeq \phi_2^{He}/\phi_2^H \simeq 3.0$ for all values of the parameter $\Delta < 2$.

We then find

$$\sum_j n_j(r)\, \sigma_j(E_\gamma, E) \simeq [n_{HI}(r) + 2 n_{H2}(r)]\, \sigma_{HI}(E_\gamma, E)$$

$$+ n_{He}(r) \sigma_{He}(E_\gamma, E) = 1.3\, [n_{HI}(r) + 2 n_{H2}(r)]\, \sigma_{HI}(E_\gamma, E). \tag{4.4.6}$$

For the separable single power law distribution (4.1.18) for the cosmic ray electrons a simple expression for the nonthermal electron bremsstrahlung source function (4.4.5) results in the strong-shielding ($\Delta \ll 1$) limit which is appropriate for relativistic electron energies $\gamma \gtrsim 15/Z$ (Stecker 1977 [519])

$$S_B(E_\gamma, r) = \frac{3.9\, \alpha\, \phi_{HI}^{s-s}\, \sigma_T}{8\pi}\, N_0(r)\, [n_{HI}(r) + n_{H2}(r)]$$

$$\times \frac{f_B(s)}{s-1}\, E_\gamma^{-s}\, \text{photons cm}^{-3}\, \text{s}^{-1}\, \text{eV}^{-1}, \tag{4.4.7}$$

with

$$f_B(s) \equiv 1 - \frac{2(s-2)}{3s(s+1)}, \tag{4.4.8}$$

and where $\phi_{HI}^{s-s} \simeq \phi_{1,HI}(\Delta = 0) \simeq \phi_{2,HI}(\Delta = 0) \simeq 45$. Here we assume that the low-energy cutoff γ_1 in the relativistic electron distribution (4.1.18) is below $15/Z$. As can be seen power law distributions for the radiating relativistic electrons generate power law bremsstrahlung photon number distributions with the same power law spectral index s.

In the case of a two-fold power law dependence of the differential electron density

$$N(\gamma, \boldsymbol{r}) = \begin{cases} N_1(\boldsymbol{r})\gamma^{-s1} & \text{for } \gamma \leq \gamma_{\text{b}} \\ N_2(\boldsymbol{r})\gamma^{-s2} & \text{for } \gamma \geq \gamma_{\text{b}} \end{cases} \quad \text{cm}^{-3} \qquad (4.4.9)$$

one obtains

$$S_{\text{B}}(E_\gamma \leq E_{\text{b}}, \boldsymbol{r}) = \frac{3.9 \, \alpha \, \phi_{\text{HI}}^{s-s} \, \sigma_{\text{T}}}{8\pi} N_1(\boldsymbol{r}) \left[n_{\text{HI}}(\boldsymbol{r}) + 2n_{\text{H2}}(\boldsymbol{r})\right]$$

$$\times \left[\frac{f_{\text{B}}(s1) E_\gamma^{-s1}}{s1 - 1} - \frac{4}{3 E_\gamma} \left(\frac{E_{\text{b}}^{-(s1-1)}}{s1 - 1} - \frac{\kappa E_{\text{b}}^{-(s2-1)}}{s2 - 1} \right) \right.$$

$$+ \frac{4}{3} \left(\frac{E_{\text{b}}^{-s1}}{s1} - \frac{\kappa E_{\text{b}}^{-s2}}{s2} \right)$$

$$\left. - E_\gamma \left(\frac{E_{\text{b}}^{-(s1+1)}}{s1+1} - \frac{\kappa E_{\text{b}}^{-(s2+1)}}{s2+1} \right) \right]$$

$$\text{cm}^{-3} \, \text{s}^{-1} \, \text{eV}^{-1}, \qquad (4.4.10)$$

and

$$S_{\text{B}}(E_\gamma \geq E_{\text{b}}, \boldsymbol{r}) = \frac{3.9 \, \alpha \, \phi_{\text{HI}}^{s-s} \, \sigma_{\text{T}} \, f_{\text{B}}(s2) \, E_\gamma^{-s2}}{8\pi \quad s2 - 1}$$

$$\times N_2(\boldsymbol{r}) \left[n_{\text{HI}}(\boldsymbol{r}) + 2n_{\text{H2}}(\boldsymbol{r})\right]$$

$$\text{cm}^{-3} \, \text{s}^{-1} \, \text{eV}^{-1}, \qquad (4.4.11)$$

where $E_{\text{b}} = \gamma_{\text{b}} mc^2$ and $\kappa = N_1(\boldsymbol{r})/N_2(\boldsymbol{r}) = \gamma_{\text{b}}^{s1-s2} = \text{const.}$

In the case of general or moderate shielding ($\Delta \simeq 1$) the integral in (4.4.5) cannot be evaluated exactly in terms of simple functions. However, a simple approximate formula may be obtained by bringing the slowly varying scattering functions outside the integrand and setting it equal to some typical value $\phi_1(\Delta) \simeq \phi_2(\Delta) \simeq \phi(\bar{\Delta})$ (Blumenthal and Gould 1970 [68]) with

$$\bar{\Delta} \simeq \begin{cases} mc^2/8\alpha E_\gamma & \text{if } E_\gamma \gtrsim E_1 \\ (1/4\alpha E_1) \left[E_\gamma/(E_1 - E_\gamma)\right] & \text{if } E_\gamma \lesssim E_1 \end{cases} \qquad (4.4.12)$$

The remaining γ-integral in (4.4.5) then reduces to

$$\int_{E_{\text{L}}}^\infty d\gamma N(\gamma, \boldsymbol{r}) \sigma_j(E_\gamma, \gamma) \simeq \phi(\bar{\Delta})$$

$$\times \int_{E_{\text{L}}}^\infty d\gamma N(\gamma, \boldsymbol{r}) \left[\frac{4}{3}\gamma^2 m^2 c^4 - \frac{4}{3}\gamma mc^2 E_\gamma + E_\gamma^2 \right], \qquad (4.4.13)$$

which can be performed once the relativistic electron distribution function $N(\gamma, \boldsymbol{r})$ is specified.

From the respective source functions the nonthermal bremsstrahlung photon intensity is obtained as the line-of-sight integral

$$I_{\rm B}(E_\gamma, l, b) = \frac{E_\gamma}{4\pi} \int_0^\infty dr\ S_{\rm B}(E_\gamma, \boldsymbol{r})$$

$$\text{erg cm}^{-2}\ \text{s}^{-1}\ \text{ster}^{-1}\ \text{eV}^{-1}\ . \qquad (4.4.14)$$

In γ-ray astronomy, measurements are often given in terms of the differential photon number flux from a direction (l, b) which is related to the photon intensity as

$$\frac{{\rm d}N_\gamma(E_\gamma, l, b)}{{\rm d}t\ {\rm d}E_\gamma\ {\rm d}\Omega} \equiv I_{\rm B}(E_\gamma, l, b)/E_\gamma$$

$$\text{photons cm}^{-2}\ \text{s}^{-1}\ \text{ster}^{-1}\ \text{eV}^{-1}\ . \qquad (4.4.15)$$

4.4.3 Bremsstrahlung Energy Loss of Relativistic Electrons

Integrating the nonthermal electron bremsstrahlung power (4.4.4) over all photon energies E_γ we obtain the energy loss of a single relativistic electron due to bremsstrahlung

$$\begin{aligned}
-\left(\frac{{\rm d}E}{{\rm d}t}\right)_{\rm B} &= -mc^2 \left(\frac{{\rm d}\gamma}{{\rm d}t}\right)_{\rm B} \\
&= \int_0^E {\rm d}E_\gamma P_{\rm B}(E_\gamma, E) \\
&= c \sum_j n_j \int_0^E {\rm d}E_\gamma E_\gamma \sigma_j(E_\gamma, E)\ \text{erg s}^{-1}\ , \quad (4.4.16)
\end{aligned}$$

where the respective targets define the scattering functions ϕ_i^j (i=1,2) of $\sigma_j(E_\gamma, E)$. Explicit expressions for (4.4.16) can be found only in the strong-shielding ($\Delta \ll 1$) and weak-shielding ($\Delta \gg 1$) limits (Blumenthal and Gould 1970 [68]). In the case of weak-shielding one finds for an overall neutral plasma

$$-\left(\frac{{\rm d}\gamma}{{\rm d}t}\right)_{\rm B} = \frac{3\alpha\ c\ \sigma_{\rm T}}{2\pi}\ \gamma \sum_Z n_Z\ Z(Z+1)$$

$$\times \left[\ln\gamma + \ln 2 - \frac{1}{3}\right]\ \text{s}^{-1}$$

(weak-shielding or completely ionized) . (4.4.17)

Equation (4.4.17) is exact for all energies in a completely ionized medium and appropriate for $\gamma \lesssim 15/Z$ in dominantly neutral gases. In the strong-shielding limit $\gamma \gtrsim 15/Z$ the scattering functions for the neutral species of the interstellar medium are constant, see Table 4.1, and practically equal $\phi_{1,j} \simeq \phi_{2,j} \simeq \phi_j^{s-s}$ so that

$$-\left(\frac{{\rm d}\gamma}{{\rm d}t}\right)_{\rm B} = \frac{3\alpha\ c\ \sigma_{\rm T}}{8\pi}\ \gamma \sum_j n_j\ \phi_j^{s-s}$$

$$\simeq \frac{3.9\alpha\, c\, \phi_{\mathrm{HI}}^{s-s}\, \sigma_{\mathrm{T}}}{8\pi}\, \gamma\, [n_{\mathrm{HI}}(\boldsymbol{r}) + 2n_{\mathrm{H2}}(\boldsymbol{r})]\ \mathrm{s}^{-1}$$
(strong shielding) . (4.4.18)

In the intermediate range $\Delta \simeq 1$ the E_γ-integration in (4.4.16) has to be done numerically. However, in the range $\Delta \simeq 1$ the scattering functions ϕ_1 and ϕ_2 are nearly equal and slowly varying functions of E_γ. Hence in an approximation similar to (4.4.12) and (4.4.13) we may bring the scattering function outside the integral and set it equal to its value at the characteristic energy $\bar{E}_\gamma = E/2$ which, according to (4.4.3), corresponds to $\Delta = (4\alpha\gamma)^{-1}$. In this case one obtains

$$-\left(\frac{d\gamma}{dt}\right)_{\mathrm{B}} \simeq \frac{3.9\alpha\, c\, \sigma_{\mathrm{T}}}{8\pi}\, \gamma\, \phi_{1,\mathrm{HI}}\left(\Delta = (4\alpha\gamma)^{-1}\right)$$
$$\times\ [n_{\mathrm{HI}}(\boldsymbol{r}) + 2n_{\mathrm{H2}}(\boldsymbol{r})]\ \mathrm{s}^{-1}$$
(general case) . (4.4.19)

In all cases (4.4.17)–(4.4.19) the bremsstrahlung energy loss rate is essentially linearly proportional to the relativistic electron Lorentz factor γ, in contrast to the synchrotron (4.1.22) and inverse Compton losses (4.2.20) where a quadratic dependence has been found. Note also that similar to the Compton losses in the extreme Klein-Nishina limit, in each bremsstrahlung event the relativistic electron loses a significant fraction of its energy.

4.5 Ionization and Coulomb Interactions

The energy loss of relativistic electrons and positrons traversing both atomic matter by ionizing neutral atoms and molecules as well as fully ionized plasmas has been carefully discussed by Gould (1975 [198]). For high-energy particles traversing atomic matter, energy loss occurs because of ionization and excitation to bound atomic levels. In a plasma the energy loss takes place through the scattering of individual plasma electrons where kinetic energy is transferred and also by the excitations of large-scale systematic or collective motions of many plasma electrons. For the plasma problem, quantum mechanics comes in through the effects of electron spin on individual electron-electron and positron-electron scattering and from the influence of the quantization of the plasma oscillations. The minimum excitation energy is $h\nu_{\mathrm{p}}$. Quantum mechanics enters the atomic or neutral matter problem through the effects of atomic binding. As in bremsstrahlung losses, the energy loss rates here are different in ionized and neutral matter. They are also slightly different for relativistic electrons and positrons but these differences are negligibly small in the low-density cosmic plasmas.

In a plasma the energy loss rate is

$$-\frac{d\gamma}{dt} = \frac{3}{4} c \, \sigma_T \, n_e \left[\ln \gamma + 2 \ln \frac{mc^2}{h\nu_p} \right]$$

$$= \frac{3}{4} c \, \sigma_T \, n_e \left[74.3 + \ln \left(\frac{\gamma}{n_e/1 \text{ cm}^{-3}} \right) \right] \text{ s}^{-1}. \quad (4.5.1)$$

Because of the large first term in the brackets in (4.5.1) the logarithmic dependence on γ/n_e is negligibly small, and the energy loss rate is essentially independent of γ.

In neutral atomic matter the energy loss rate is

$$-\frac{d\gamma}{dt} = \frac{9}{4} c \, \sigma_T \sum_j n_j \, Z_j \left[\ln \gamma + \frac{2}{3} \ln \frac{mc^2}{\langle \Delta E \rangle_j} \right], \quad (4.5.2)$$

where Z_j is the number of atomic electrons associated with the species j and $\langle \Delta E \rangle_j$ is the average electronic excitation energy, equal to 15.0 eV for hydrogen and 41.5 eV for helium (Dalgarno 1962 [115]). For the interstellar medium consisting of hydrogen and helium with a total number abundance ratio of 10:1 we obtain

$$-\frac{d\gamma}{dt} = 2.7 \, c \, \sigma_T \left[6.85 + \ln \gamma \right] \left[n_{\text{HI}}(r) + 2 n_{\text{H2}}(r) \right] \text{ s}^{-1}. \quad (4.5.3)$$

Again the logarithmic energy dependence is negligibly small and the loss rate is essentially independent of γ.

As a consequence of the ionization and excitation processes of the relativistic electrons the ambient cosmic matter gas becomes heated and ionized. The gas emits thermal radiation the amount of which is determined by the temperature and the degree of ionization of the gas. The degree of ionization results from the balance of recombination and ionization rates and through the latter by the relativistic electrons. A fraction of the kinetic energy lost by the relativistic electrons is available as a heating source of the plasma and thus determines the gas temperature. So there is an indirect link between the thermal emission of cosmic plasmas and the relativistic electrons of the cosmic radiation. We will come back to this point after we have discussed the ionization interactions of cosmic ray nucleons (Sect. 5.3.8) which also significantly ionize and heat cosmic gases.

4.6 Total Energy Loss Rate

Combining the energy loss rates due to radiation ((4.1.22) and (4.2.20)), bremsstrahlung ((4.4.17) and (4.4.18)) and ionization and Coulomb ((4.5.1) and (4.5.2)) losses and introducing the degree of ionization, $x = n_e/n_{\text{HI}}$, we obtain for the total energy loss rate of relativistic electrons

$$-\left(\frac{d\gamma}{dt}\right)_{\text{total}} = 36.18 \, c \, \sigma_T \, n_H \left[1 + 0.146 \ln\gamma + 1.54x(n_{HI}/n_H)\right.$$
$$+ (\gamma/709)\left[1 + 0.137(x \, n_{HI}/n_H)(\ln\gamma + 0.36)\right]$$
$$+ 2.685 \times 10^{-9}\gamma^2$$
$$\left.\times \left[\frac{(B_\perp/10^{-6} \, \text{G})^2}{n_H} + 26.84\frac{(w/1 \, \text{eV cm}^{-3})}{n_H}\right]\right] \, \text{s}^{-1} \, , \quad (4.6.1)$$

where $n_H \equiv n_{HI} + 2n_{H2}$ is measured in particles per cm^3.

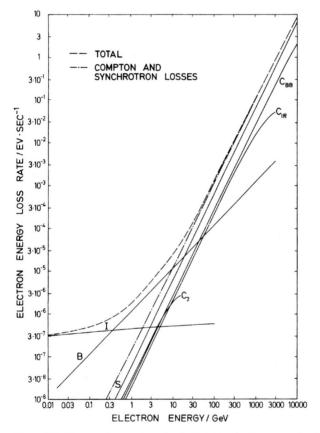

Fig. 4.8. Energy loss rates of relativistic electrons in the local interstellar medium taking a neutral gas density $n_H = 1.14$ cm^{-3}, degree of ionization $x = 0$, magnetic field strength $B_\perp = 4 \times 10^{-6}$ G and the photon energy densities from Table 2.2 (I: ionization, B: bremsstrahlung, C: inverse Compton scattering, S: synchrotron radiation). Index 2 of C_2 refers to component 2 of Table 2.2. From Schlickeiser (1981 [452])

In order to understand the relative importance of the different energy loss processes in the local interstellar medium we calculate the respective loss rates by using the values of the parameters as measured near the solar system. Figure 4.8 shows the result. As one can see, ionization losses dominate for electron energies smaller than 360 MeV. In the range 360 MeV to 8 GeV relativistic electrons lose their energy predominantly by bremsstrahlung, whereas at higher energies Compton and synchrotron losses dominate.

A discussion of the energy loss processes in other regions of the Universe requires knowledge of the parameter values there. Gould (1975 [198]) has considered relativistic electron loss rates in the intergalactic medium. As one can see from Fig. 4.9, due to the small matter densities and magnetic field strength in this environment, inverse Compton losses with the universal 2.7 K microwave background radiation dominate already for energies greater than 50 MeV.

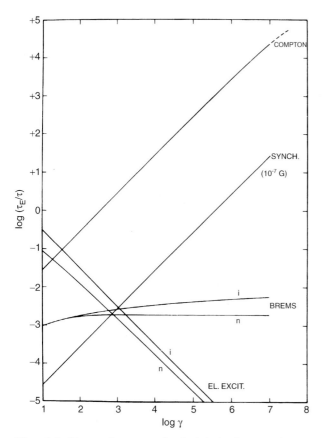

Fig. 4.9. Energy loss rate of relativistic electrons in intergalactic space. Shown is the ratio τ_E/τ where $\tau_E = \frac{1}{3}H_0^{-1}$ (Hubble constant $H_0 = 50$ km s^{-1} Mpc^{-1}) is the cosmic expansion time and $1/\tau = (dE/dt)/E$. From Gould (1975 [198])

4.7 Continuum Radiation Processes of Relativistic Electrons

Cosmic relativistic electrons produce a nonthermal continuum by synchrotron radiation in cosmic magnetic fields, inverse Compton radiation by scattering ambient photon gases and nonthermal electron bremsstrahlung. Table 4.2 summarizes the main features and predictions of these radiation processes under the assumption of optically thin emission (i.e. no absorption processes), if the spatially isotropic differential number density of relativistic electrons is given by (4.1.18), i.e. a single power law, which in the interstellar medium of our Galaxy appears appropriate.

The interstellar magnetic field strength is of the order $B_\perp \simeq 4 \times 10^{-6}$ G, so that the synchrotron radiation of cosmic ray electrons in the Galaxy produces diffuse continuum radio emission in the MHz/GHz frequency range. In objects with larger magnetic fields, synchrotron radiation by relativistic electrons may be responsible for continuum infrared, optical, ultraviolet and even X-ray emission. By measuring the predicted linear degree of polarization of the

Table 4.2. Electron continuum radiation processes

Process	Energy flux $\propto$	Remarks
Synchrotron radiation	$\nu^{(1-p)/2} \int_0^\infty dl I(\boldsymbol{r})$ $[B_\perp(\boldsymbol{r})]^{(p+1)/2}$	polarization $\langle \nu \rangle$ (MHz) $\cong 16$ $B_\perp$ (µG) E^2 (GeV)
Inverse Compton scattering	$\nu^{(1-p)/2} \int_0^\infty dl I(\boldsymbol{r}) \, n_{\rm ph}(\boldsymbol{r})$	for $\varepsilon E \ll (mc^2)^2$ "Thomson limit" $\langle E_\gamma \rangle \cong \varepsilon E^2/(mc^2)^2$
	$\nu^{-p} \int_0^\infty dl I(\boldsymbol{r}) \, n_{\rm ph}(\boldsymbol{r})$	for $\varepsilon E \gg (mc^2)^2$ "Klein-Nishina" cutoff $\langle E_\gamma \rangle \cong E$
Nonthermal bremsstrahlung	$\nu^{1-p} \int_0^\infty dl n_{\rm gas}(\boldsymbol{r}) I(\boldsymbol{r})$	for $E > 10$ MeV, $\langle E_\gamma \rangle \cong 0.5 \, E$

$n_{\rm ph}(\varepsilon, \boldsymbol{r})$: target photon distribution; $B_\perp(\boldsymbol{r})$: magnetic field strength; isotropic relativistic electron distribution assumed: $I(E, \boldsymbol{r}) = I(\boldsymbol{r}) E^{-p}$

synchrotron radiation (4.1.21b) this effect has, indeed, been confirmed for many sources.

The mean energy of a bremsstrahlung photon is typically half of the radiating relativistic electrons, so that this radiation may be detected in the γ-ray range.

In inverse Compton interactions of relativistic electrons we have to consider two cases:

(i) in the Thomson limit the typical energy of the scattered photon is $\langle E_\gamma \rangle \simeq \frac{4}{3}\epsilon\gamma^2$ (see (4.2.6)) which (depending on ϵ and γ) can generate photons from radio to γ-ray frequencies. At infrared, optical, ultraviolet and soft X-ray wavelengths these inverse Compton scattering photons are very often unobservable because of other much stronger radiation fields such as dust emission, emission of stars, thermal plasma emission, etc. So the inverse Compton scattering radiation products in this case are best looked for in the radio, hard X-ray and γ-ray wavelength bands.

(ii) in the extreme Klein-Nishina limit, the scattered photon energy equals the radiating relativistic electron energy, so that this radiation is detectable again in the γ-ray regime. In the extreme Klein-Nishina limit, recoil effects in the interaction become important, lowering the cross-section value, which results in the so-called Klein-Nishina cutoff in the inverse Compton radiation spectrum. The detection of such a continuous steepening of the γ-ray energy spectrum with increasing energy would reveal the inverse Compton origin of this radiation.

To summarize: for studying the distribution of relativistic electrons in other parts of the Galaxy and the Universe the measurements of the radio, hard X-ray and γ-ray diffuse backgrounds are important because we expect the radiation products from the interactions of relativistic electrons to be observable in these frequency bands. In Chap. 6 we will summarize the radio, hard X-ray and γ-ray observations and their implications.

5. Interactions of Cosmic Ray Nuclei

In Chap. 4 we have seen that the cross-sections for point-like electromagnetic interactions of cosmic ray electrons (synchrotron radiation, inverse Compton scattering, nonthermal bremsstrahlung) involve the Thomson cross-section being proportional to the square of the electron radius of the radiating particle. Therefore, if we consider here a heavy cosmic ray nucleus of charge Ze and mass $M = Am_\mathrm{p}$ instead of a cosmic ray electron, the corresponding cross-sections involve the square of the nucleon radius $R_0 = (Ze)^2/(Am_\mathrm{p}c^2) = (Z^2/A)(m_\mathrm{e}/m_\mathrm{p})r_0$ yielding the Thomson cross-section for nuclei

$$\sigma_\mathrm{N} = (Z^4/A^2)(m_\mathrm{e}/m_\mathrm{p})^2 \sigma_\mathrm{T} = 1.97 \times 10^{-31} \frac{Z^4}{A^2} \ \mathrm{cm}^2 \ .$$

As a consequence the cross-sections of electromagnetic interactions of cosmic ray nucleus are much smaller than those of the cosmic ray electrons of the same Lorentz factor, and can be neglected in practically all astrophysical applications as production processes for cosmic photons and as energy loss processes for cosmic ray nuclei.

The remaining interaction processes can be divided into two groups:

(i) interactions with cosmic photons,
(ii) interactions with cosmic matter fields.

Before we consider each in turn, we discuss the kinematics of inelastic collisions.

5.1 Relativistic Kinematics of Inelastic Collisions

The interactions that we consider involve energies at least of the order of magnitude of the rest masses of the particles involved. Therefore we must use results of the special theory of relativity to describe their kinematics (e.g. Hagedorn 1973 [207]).

Special relativity makes use of a four-vector algebra. All quantities consisting of a set of four numbers that transform under a Lorentz transformation exactly as the component of the line element $\mathrm{d}s = (c\,\mathrm{d}t, \mathrm{d}x, \mathrm{d}y, \mathrm{d}z)$ are called four-vectors. A Lorentz transformation leaves the square of the line element

$$ds^2 = c^2 dt^2 - dx^2 - dy^2 - dz^2 \tag{5.1.1}$$

invariant, i.e. its value is the same in each inertial frame of reference. This requirement of invariance is the consequence of the experimental result that the velocity of light $c = 2.99793 \times 10^{10}$ cm/s is the same in all inertial frames. Consequently, if four-vectors transform like ds, then the scalar product of the four-vector $P = (P_t, P_x, P_y, P_z)$ with itself,

$$P^2 = P_t^2 - P_x^2 - P_y^2 - P_z^2$$

is an invariant; then if P and Q are four-vectors,

$$(P+Q)^2 = P^2 + 2PQ + Q^2 = P'^2 + 2P'Q' + Q'^2 \,.$$

Hence

$$PQ = P_t Q_t - P_x Q_x - P_y Q_y - P_z Q_z = \text{invariant} \,.$$

An important four-vector is the energy-momentum vector

$$P = \left(\frac{E}{c}, \boldsymbol{p}\right), \quad P^2 = m^2 c^2 \,. \tag{5.1.2}$$

5.1.1 Threshold Energy

Suppose we consider in a certain Lorentz system—we call it the laboratory system—the collision of two particles a and b with four-momenta $P_a = (\epsilon_a/c, \boldsymbol{p}_a)$ and $P_b = (\epsilon_b/c, \boldsymbol{p}_b)$ and with masses m_a and m_b leading to the creation of particles $c, d, \ldots$

$$a + b \to c + d + \ldots \,. \tag{5.1.3}$$

To create new particles in this reaction the minimum energy of the incoming particles a and b has to be equivalent to just enough available energy in the center-of-momentum system (CMS) to produce the rest mass of all outgoing particles $c, d, \ldots$. The CMS is characterized by $\boldsymbol{p}'_a + \boldsymbol{p}'_b = 0$, i.e. the sum of the momentum vectors of the incoming particles is zero. (We denote CMS quantities with a prime.) From the invariance of the four-momentum we obtain for the total energy E' in the CMS

$$\begin{aligned} E'^2 \equiv (E'_a + E'_b)^2 &= c^2 (P'_a + P'_b)^2 = c^2 (P_a + P_b)^2 \\ &= m_a^2 c^4 + m_b^2 c^4 + 2\epsilon_a \epsilon_b - 2\boldsymbol{p}_a \boldsymbol{p}_b c^2 \,. \end{aligned} \tag{5.1.4}$$

The threshold energy for the creation of particles $c, d, \ldots$ in reaction (5.1.3) is

$$E'_{\text{th}} = m_a c^2 + m_b c^2 + \Delta m c^2 = m_c c^2 + m_d c^2 + \ldots \,, \tag{5.1.5}$$

where Δm is the mass difference between incoming and outgoing particles. Combining (5.1.4) and (5.1.5) we obtain the relation

$$\epsilon_a \epsilon_b - \boldsymbol{p}_a \boldsymbol{p}_b c^2 = m_a m_b c^4 + \Delta m c^4 \left(m_a + m_b + \frac{\Delta m}{2} \right). \tag{5.1.6}$$

When both incoming particles have non-zero masses (5.1.6) reduces to

$$\gamma_a \gamma_b - \sqrt{(\gamma_a^2 - 1)(\gamma_b^2 - 1)} \cos\theta = 1 + \Delta m \left(\frac{1}{m_a} + \frac{1}{m_b} + \frac{\Delta m}{2 m_a m_b} \right), \tag{5.1.7}$$

where we introduced the lab-system Lorentz factors of the particles and where θ denotes the collision angle of the reaction in the lab-system.

When particle b is a photon ($m_b = 0$), (5.1.6) reduces to

$$\epsilon_b \left(\gamma_a - \sqrt{\gamma_a^2 - 1} \cos\theta \right) = \Delta m c^2 \left(1 + \frac{\Delta m}{2 m_a} \right). \tag{5.1.8}$$

5.1.2 The Energy of One Particle Seen from the Rest System of Another

Another useful calculation concerns the energy of particle b in reaction (5.1.3) seen from the rest system of particle a. We call that energy E_{ba}. Using the invariance of $P_a P_b$ we find

$$P_a P_b = \frac{\epsilon_a \epsilon_b}{c^2} - \boldsymbol{p}_a \boldsymbol{p}_b = m_a E_{ba}. \tag{5.1.9}$$

When both incoming particles have non-zero masses (5.1.9) reduces to

$$E_{ba} = m_b c^2 \left[\gamma_a \gamma_b - \sqrt{(\gamma_a^2 - 1)(\gamma_b^2 - 1)} \cos\theta \right], \tag{5.1.10}$$

whereas when particle b is a photon we obtain

$$E_{ba} = \epsilon_b \left[\gamma_a - \sqrt{\gamma_a^2 - 1} \cos\theta \right]. \tag{5.1.11}$$

5.2 Interactions Between Cosmic Ray Nuclei and Cosmic Photons

Three basic interactions have important astrophysical consequences for relativistic nuclei:

1. pair production (particularly of $e^+ e^-$ pairs) in the field of the nucleus $N + \gamma \rightarrow N + e^+ + e^-$,
2. photo-production of hadrons mostly pions $p + \gamma \rightarrow p + \pi' s$,

3. photo-disintegration of the nucleus $A + \gamma \to (A-1) + N$, where $N =$ n, p.

As will be shown in Sect. 5.2.2, when process (2) becomes important, only protons need to be considered.

In the rest system of the cosmic ray nucleus we find according to (5.1.6) for $\boldsymbol{p}_a = 0$, $\epsilon_a = m_a c^2$, and $m_b = 0$ for the photon threshold energy

$$\epsilon'_{\text{th}} = \Delta m c^2 \left[1 + \frac{\Delta m}{2 m_a} \right]. \tag{5.2.1}$$

Process (1) ($\Delta m = 2 m_e$) then occurs for photon energies above

$$\epsilon'_{\text{th},e^+e^-} = 2 m_e c^2 \left(1 + \frac{m_e}{A m_p} \right) \simeq 2 m_e c^2, \tag{5.2.2}$$

and process (2) at

$$\epsilon'_{\text{th},K\pi} = K m_\pi c^2 \left[1 + \frac{K m_\pi}{2 m_p} \right], \tag{5.2.3}$$

for the production of K pions so that for the production of a single ($K = 1$) pion the rest system threshold energy is $\epsilon'_{\text{th},\pi} = 145$ MeV. Process (3) is particularly important in most cases between 15 and 25 MeV where the giant dipole resonance has its peak (Puget et al. 1976 [411]).

If the target photons are distributed isotropically in space and have mean energy $\langle \epsilon \rangle$ in our observing laboratory frame of reference, the minimum threshold Lorentz factor of the relativistic ($\gamma_a \gg 1$) cosmic ray nucleus according to (5.1.8) occurs for a head-on collision ($\cos\theta = -1$) and is

$$\gamma_{\min} = \frac{\Delta m c^2}{2 \langle \epsilon \rangle} \left[1 + \frac{\Delta m}{2 m_a} \right]. \tag{5.2.4}$$

In intergalactic space (see Sect. 2.3) the most prominent radiation field is the microwave background radiation with $\langle \epsilon \rangle = 7 \times 10^{-4}$ eV, so that processes (1) and (2) occur only for ultrahigh energy cosmic ray nuclei,

$$\gamma_{\min}^{(1)} \simeq m_e c^2 / \langle \epsilon \rangle = 7 \times 10^8, \tag{5.2.5}$$

and

$$\gamma_{\min}^{(2)} \simeq \frac{K m_\pi c^2}{2 \langle \epsilon \rangle} \left(1 + \frac{K m_\pi}{2 m_p} \right) \simeq K \times 10^{11}, \tag{5.2.6}$$

respectively. The contribution from collisions with extragalactic optical and infrared photons are negligible.

5.2.1 Photo-Pair Production

Pair production in the field of the nucleus in an isotropic radiation field has been treated in detail in its astrophysical context by Blumenthal (1970 [67]), Berezinsky and Grigoreva (1988 [42]) and Chodorowski et al. (1992 [100]). The energy loss rate for a cosmic ray nucleus of charge Ze is

$$-\frac{dE}{dt} = \frac{3c\alpha\sigma_T}{8\pi} Z^2 (mc^2)^2 \int_2^\infty d\xi\, n\left(\frac{\xi mc^2}{2\gamma}\right) \frac{\phi(\xi)}{\xi^2}, \qquad (5.2.7)$$

where m is the electron mass, $\sigma_T = 6.65 \times 10^{-25}$ cm^2 is the electron Thomson cross-section, $\phi(\xi)$ is an integral over the pair production cross-section and $n(\epsilon)$ is the number density distribution of the target photons. The asymptotic formula for $\xi \gg 1$,

$$\phi(\xi \gg 1) \to \xi\left[-86.07 + 50.95 \ln\xi - 14.45(\ln\xi)^2 + 2.667(\ln\xi)^3\right] \qquad (5.2.8)$$

works reasonably well for all values of $\xi \geq 2$ in (5.2.7) when the target photon distribution $n(\epsilon)$ is such that for all contributing photons $\gamma\epsilon \gg mc^2$. With the graybody distribution function (2.3.1) we obtain for cosmic ray protons

$$-\frac{dE}{dt}\bigg|_{e^\pm} = \frac{45\, c\, w(\mathbf{r})\, \alpha\sigma_T (mc^2)^2}{8\pi^5\, (kT)^2} f\left(\frac{mc^2}{2\gamma kT}\right), \qquad (5.2.9)$$

where

$$f(\nu) \equiv \nu^2 \int_2^\infty d\xi\, \phi(\xi)(e^{\nu\xi} - 1)^{-1}. \qquad (5.2.10)$$

$f(\nu)$ calculated for the asymptotic expansion (5.2.8) for $\phi(\xi)$ is shown in Fig. 5.1. It is clear from this figure that the threshold for pair production is at $\nu \sim 1$ corresponding to $\gamma \sim (mc^2/2k_BT)$ in agreement with the estimate (5.2.5) since for a graybody distribution $\langle\epsilon\rangle = 2.7 k_B T$.

For the special case of the microwave background blackbody radiation field one may express the photon energy density in terms of the blackbody temperature according to the Stefan–Boltzmann law, $w = aT^4$ where $\sigma = 7.56 \times 10^{-15}$ erg cm^{-3} K^{-4}. As an example we obtain the energy loss rate for protons at threshold

$$-\left(\frac{dE}{dt}\right)_{e^\pm} = 2.4 \times 10^5 \frac{w}{\text{erg cm}^{-3}} \left(\frac{k_BT}{\text{eV}}\right)^{-2} \text{ eV s}^{-1}. \qquad (5.2.11)$$

5.2.2 Photo-Hadron Production

The photo-hadron production process is dominated by photo-pion production, e.g. for protons at threshold the channels p $+\gamma \to \pi^0+$ p and p $+\gamma \to \pi^++$ n dominate and at higher energies by multi-pion production.

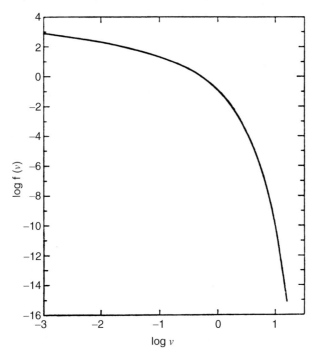

Fig. 5.1. Graph of $\log_{10} f(\nu)$ versus $\log_{10}\nu$, where $f(\nu)$ is defined by (5.2.10). From Blumenthal (1970 [67])

Baryon production and K-meson production can be neglected in most astrophysical applications. Pion production has been treated in detail by Stecker (1968 [514]), Sikora et al. (1987 [484]), Mannheim and Biermann (1989 [316]) and Begelman et al. (1990 [38]).

Photo-pion production by nuclei of mass number A obeys the simple Glauber rule $\sigma_A \simeq A^{2/3}\sigma_p$. However, in an astrophysical environment nuclei cannot be accelerated to energies above pion threshold, since they are destroyed before by photo-disintegration (see Sect. 5.2.3) that has a much lower threshold of about 10 MeV in the nucleus rest frame compared to pion production which requires at least 145 MeV.

For a cosmic ray proton of Lorentz factor γ_p, traversing an isotropic photon field of number density $n(\epsilon, \boldsymbol{r})$, one obtains the energy loss rate

$$-\frac{d\gamma}{dt} = \frac{c}{2\gamma_p}\int_{\epsilon'_{\rm th}/(2\gamma_p)}^{\infty} d\epsilon\, n(\epsilon, \boldsymbol{r})\epsilon^{-2}\int_{\epsilon'_{\rm th}}^{2\gamma_p\epsilon} d\epsilon'\epsilon'\sigma(\epsilon')K_p(\epsilon')\,, \qquad (5.2.12)$$

where $\sigma(\epsilon')$ and $K_p(\epsilon')$ are the total photo-hadron production cross-section and inelasticity, respectively, as a function of the photon energy in the proton rest frame. The proton rest frame threshold energy $\epsilon'_{\rm th}$ is given by (5.2.3) for the respective reaction. Crucial for further evaluation is the knowledge of both the photo-hadron inelasticity and the cross-section as a function of energy ϵ'.

There exist a vast number of phenomenological particle physics models for the latter two whose free parameters are adjusted by available accelerator studies of individual reactions. Stecker (1968 [514]) has used an interpolation of the available experimental photo-meson production studies to infer the energy dependence of $\sigma(\epsilon')$ and $K_p(\epsilon')$. Equation (5.2.12) was then numerically integrated to calculate the energy loss rate of protons for photo-hadron production in the microwave background radiation field and to compare this with the energy loss rate for pair production and redshift expansion. Figure 5.2 shows the energy loss time scales $\tau \equiv \gamma/(|d\gamma/dt|)$ and the associated attenuation lengths $l = c\tau$ for cosmic ray protons from pair production and photo-meson production in comparison with the redshift expansion losses. It can be seen that these two loss processes are only significant at total cosmic ray energies above 10^{18} eV in this radiation field.

To discuss photo-pion production in more general radiation fields Sikora et al. (1987 [484]) have used approximate formulae to the inelasticity and the cross-section by Greisen (1966 [202]) and Maximon (1966 [333]), whereas Mannheim and Biermann (1989 [316]) and Begelman et al. (1990 [38]) used the compilation of data by Genzel, et al. (1973 [184]). For interactions at high CMS-energies they used interpolation formulae suggested by specific assumptions about the gluon content of the photon structure function. These authors have calculated attenuation lengths of high-energy nucleons in nonthermal radiation fields with a power law energy distribution with low- and high-energy cutoffs. They concluded that in sources with strong X-ray photon fields, like active galactic nuclei, protons with Lorentz factors larger than $\sim 10^6$ are cooled very rapidly by pair production and photo-pion production.

The resulting energy loss rate in a thermal radiation field is given by (Mannheim and Schlickeiser 1994 [317])

$$-\left(\frac{dE}{dt}\right)^{p\gamma}_{\pi} = 1.8 \times 10^{10} \frac{w}{\mathrm{erg\ cm^{-3}}} \left(\frac{k_B T}{\mathrm{eV}}\right)^{-2} \mathrm{eV\ s^{-1}}, \qquad (5.2.13)$$

at threshold. Comparing with the loss rate (5.2.11) we find that photo-hadron production, albeit having a cross-section of only $\alpha_f \sigma_{pp}$, can be regarded as an energy loss mechanism as important as photo-pair production with cross-section $\alpha_f \sigma_T$, where α_f is the fine-structure constant. This is because the former process has an inelasticity of $1/4$ at threshold, whereas the latter only reaches an inelasticity of order $2m_e/m_p$. For flat target photon distributions, such as the thermal distribution, photo-hadron production can be the dominant cooling mechanism, although photo-pair production has a factor $1/145$ smaller threshold energy.

112   5. Interactions of Cosmic Ray Nuclei

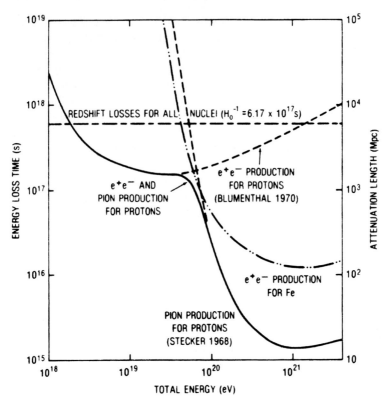

Fig. 5.2. The energy loss time and attenuation length of cosmic ray protons from pair production, photo-pion production in the microwave background radiation field of 2.7 K in comparison with the redshift expansion loss time. Also shown is the attenuation length for ^{56}Fe from pair production losses. From Puget et al. (1976 [411])

5.2.3 Photo-Disintegration

The photo-disintegration rate for a nucleus of mass A with the subsequent release of i nucleons in an isotropic photon field of number density $n(\epsilon, \boldsymbol{r})$ is given by the expression (Stecker 1969 [515])

$$R_{A,i} = \frac{1}{2} \gamma_A^{-2} \int_0^\infty d\epsilon n(\epsilon, \boldsymbol{r}) \epsilon^{-2} \int_0^{2\gamma_A \epsilon} d\epsilon' \epsilon' \sigma_{A,i}(\epsilon') \; s^{-1} \, , \qquad (5.2.14)$$

where γ_A is the Lorentz factor of the cosmic ray nucleus and ϵ the photon energy measured in the rest frame of the nucleus. For some nuclei, for example helium, the dominant photo-disintegration process is proton emission rather than neutron emission. Puget et al. (1976 [411]) established that a Gaussian approximation to the cross-section is useful both as an adequate fit to the cross-section data and as an expedient in performing the numerical integration in (5.2.14). These authors calculated the photo-disintegration rates for

5.2 Interactions Between Cosmic Ray Nuclei and Cosmic Photons 113

nuclei up to $A = 56$ in a radiation field characteristic for the intergalactic medium being dominated by the 2.7 K microwave background radiation and the infrared radiation background. Since only upper limits existed in 1976 for the latter they chose two power law representations which can be considered as giving reasonable lower (LIR) and upper (HIR) limits on the intergalactic infrared flux.

Figure 5.3 shows the comparison of the energy loss time scales $\tau \equiv (dE/dt)^{-1} E$ for ^{56}Fe due to the effect of redshift expansion in the Hubble flow, photo-disintegration and pair production losses. The photo-disintegra-

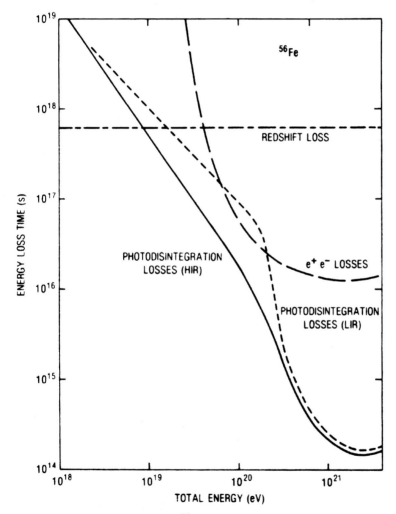

Fig. 5.3. Energy loss time for ^{56}Fe shown for redshift expansion, pair production and photo-disintegration losses in the HIR and LIR cases. From Puget et al. (1976 [411])

tion time scale is shown for the two representations LIR and HIR of the intergalactic infrared background. It can be seen that for ^{56}Fe nuclei the energy loss rates from pair production and photo-disintegration are comparable in the energy range near 10^{20} eV, whereas below 10^{19} eV the photo-disintegration time scale becomes longer than the expansion losses. Puget et al. (1976 [411]) have also pointed out that for photo-disintegration the average fractional energy loss rate

$$E^{-1}(\mathrm{d}E/\mathrm{d}t) = A^{-1}(\mathrm{d}A/\mathrm{d}t) , \qquad (5.2.15)$$

because the nucleon emission is isotropic in the rest system of the nucleus. The Lorentz factor, which is equivalent to energy per nucleon, is conserved for photo-disintegration as opposed to the pair production and photo-pion production processes which involve the creation of new particles which carry off energy.

Another noteworthy difference concerns the (A/Z) dependence of the loss rates. The energy loss time for pair production calculated from (5.2.7) varies as $\tau_{pp} \propto AZ^{-2}$, whereas the energy loss time from photo-disintegration implied by (5.2.15) is almost independent of A. Thus the importance of pair production as an energy loss process relative to photo-disintegration losses decreases with decreasing Z when we consider cosmic ray nuclei lighter than iron.

5.3 Interactions Between Cosmic Ray Nuclei and Matter

Equally as important as particle-photon collisions are inelastic reactions of cosmic ray nuclei with atoms and molecules of interstellar and intergalactic matter. In inelastic p–p, p–α, and α–p collisions mainly charged and neutral pions are produced. The charged pions $\pi^{\pm}$ decay into muons and neutrinos, and the muons decay into electrons. These decays are regarded as the most important secondary source of cosmic electrons, positrons and neutrinos. Neutral pions π^{0} decay after a mean lifetime of 9×10^{-17} s into two high-energy γ-rays. Assuming that the decaying particles are distributed isotropically in their rest frame we can relate the energy spectrum of the decay products to the energy spectrum of the decaying particles (see (5.3.1)).[1]

Also of interest, in particular for the origin of nuclear γ-ray lines, is the excitation of nuclei by collisions between fast cosmic ray particles and nuclei of the interstellar medium. A discussion of these interactions is given after the study of pion production.

Secondary neutrons and protons can also be produced in reactions not involving pions. Such processes have importance in high-energy processes in

---

[1] For two-body and three-body decays these relations are derived in Sect. 7.6 of Hagedorn (1973 [207])

5.3.1 Gamma-Ray, Electron, Positron and Neutrino Source Functions

For the $\pi^0 \to 2\gamma$ decay the omnidirectional (i.e. integrated over the whole solid angle) differential γ-ray source function $q_{\pi^0}(E_\gamma; \boldsymbol{r})$ at the position $\boldsymbol{r} = (l, b, r)$ in space is related to the omnidirectional differential neutral pion source function

$$q_{\pi^0}(E_\gamma; \boldsymbol{r}) = 2 \int_{E_\gamma + [(m_\pi c^2)^2/(4E_\gamma)]}^{\infty} dE_\pi Q_{\pi^0}(E_\pi; \boldsymbol{r}) \left[E_\pi^2 - m_\pi^2 c^4\right]^{-1/2}$$

photons cm^{-3} s^{-1} eV^{-1} . (5.3.1)

The decay of the charged pions is more complicated and has been treated in detail by Scanlon and Milford (1965 [448]). Let us first consider the decay $\pi^\pm \to \mu^\pm + \nu_\mu(\bar{\nu}_\mu)$. The Lorentz factor γ'_μ of the muon in the rest frame of the pion follows readily from the conservation of four momentum in the 2-body decay $P_\pi = P_\mu + P_\nu$ implying with $m_\nu = 0$ that $P_\pi P_\mu = [m_\pi^2 + m_\mu^2]c^2/2$. In the rest frame of the pion $P_\pi P_\mu = m_\pi m_\mu c^2 \gamma'_\mu$ implying

$$\gamma'_\mu = \frac{[m_\pi + m_\mu]^2}{2 m_\pi m_\mu} . \qquad (5.3.2)$$

With $m_\pi c^2 = 140$ MeV and $m_\mu c^2 = 106$ MeV we obtain $\gamma'_\mu = 1.04$. This low value implies that, as long as the velocity of the pion in the laboratory frame is not exceedingly small (i.e. $\gamma_\pi > 1.04$), the muon can be treated as essentially at rest in the rest frame of the pion. Therefore, in the laboratory frame, the muon moves with almost the same speed as the pion, so that per unit Lorentz factor, the muon and pion source functions are equal

$$q_\mu(\gamma_\mu) = q_\pi(\gamma_\pi) . \qquad (5.3.3)$$

The decay $\mu^\pm \to e^\pm + \nu_e(\bar{\nu}_e) + \bar{\nu}_\mu(\nu_\mu)$ is a 3-body process. The relation of the secondary electron and positron production spectrum to the muon source function has been calculated by Scanlon and Milford (1965 [448]), Perola et al. (1967 [395]) and Ramaty (1974 [415]); the corresponding relation of the neutrino production spectra to the muon source function has been given by Zatsepin and Kuzmin (1962 [572]) and Marscher et al. (1980 [323]). For example, for secondary electrons and positrons Ramaty (1974 [415]) obtains

$$q_e(\gamma_e) = \int_1^{\gamma'_e \text{ (max)}} d\gamma'_e \frac{1}{2} \frac{P(\gamma'_e)}{\sqrt{\gamma'^2_e - 1}} \int_{\gamma_\mu^-}^{\gamma_\mu^+} d\gamma_\mu \frac{q_\mu(\gamma_\mu)}{\sqrt{\gamma_\mu^2 - 1}}$$

$$= \int_{\gamma_1}^{\gamma_\mu} d\gamma_\mu \frac{q_\mu(\gamma_\mu)}{\sqrt{\gamma_\mu^2 - 1}} \int_{\gamma'_1}^{\gamma'_\mu} d\gamma_e \frac{P(\gamma'_e)}{2\sqrt{\gamma'^2_e - 1}} , \qquad (5.3.4)$$

where

$$\gamma_\mu^\pm = \gamma_e \gamma'_e \pm \sqrt{\gamma_e^2 - 1}\sqrt{\gamma_e'^2 - 1}$$

$$\gamma'_1 = \gamma_\mu \gamma_e - \sqrt{\gamma_\mu^2 - 1}\sqrt{\gamma_e^2 - 1}$$

$$\gamma'_\mu = \begin{cases} \gamma'_e \,(\text{max}) & \text{for } \gamma_e > \gamma'_e \,(\text{max}) \\ \gamma'_e \,(\text{max}) & \text{for } \gamma'_e \,(\text{max}) < \gamma_\mu \gamma_e + \sqrt{\gamma_\mu^2 - 1}\sqrt{\gamma_e^2 - 1} \\ \gamma_\mu \gamma_e + \sqrt{\gamma_\mu^2 - 1}\sqrt{\gamma_e^2 - 1} & \text{for } \gamma'_e \,(\text{max}) > \gamma_\mu \gamma_e + \sqrt{\gamma_\mu^2 - 1}\sqrt{\gamma_e^2 - 1} \end{cases},$$

$$\gamma_1 = \begin{cases} 1 & \text{for } \gamma_e < \gamma'_e \,(\text{max}) \\ \gamma'_e \,(\text{max})\,\gamma_e - \sqrt{\gamma_e'^2 \,(\text{max}) - 1}\sqrt{\gamma_e^2 - 1} & \text{for } \gamma_e > \gamma'_e \,(\text{max}) \end{cases},$$

$$\gamma_\mu = \gamma'_e \,(\text{max})\,\gamma_e + \sqrt{\gamma_e'^2 - 1}\sqrt{\gamma_e^2 - 1},$$

and with $\gamma'_e(\text{max}) = 104$ and the electron distribution in the muon's rest frame

$$P(\gamma'_e) = 2\gamma_e'^2 \left[3 - \frac{2\gamma'_e}{\gamma'_e \,(\text{max})}\right] / \gamma_e'^3 \,(\text{max}).$$

5.3.2 Pion Production Spectra

To obtain the pion source spectra required in (5.3.1) and (5.3.3) we use the same formalism as in Chap. 4 and start with the respective $(\pi^0, \pi^\pm)$ power of a single relativistic nucleon of total energy $E_N = \gamma_N mc^2$ and mass m

$$P(E_\pi, E_N) = cE_\pi \sum_j n_j \sum_k \sigma_{jk}(E_\pi, E_N) H[E_N - E_{k,\text{th}}] \text{ erg s}^{-1} \text{ eV}^{-1}, \quad (5.3.5)$$

where $\sigma_{jk}(E_\pi, E_N)$ is the differential cross-section for the interaction process k of a primary cosmic ray nucleus of total energy E_N with atoms and molecules of species j and density n_j. H denotes the Heaviside step-function ($H(x) = 1$ for $x \geq 0$, and $H(x) = 0$ for $x < 0$) and $E_{\text{th},k}$ the threshold energy of the interaction process k. Table 5.1 lists the important high-energy nuclear interactions leading to pion production in p–p and p–α collisions. Most experimental data refer to the total inclusive cross-sections

$$\xi\sigma_{\text{pp}}^{\pi^0}(T_\text{p}) = \sum_k \int_0^{E_{\pi\text{max}}} dE_\pi \sigma_k(E_\pi, E_\text{p}) \text{ cm}^2 \quad (5.3.6)$$

for the reaction p+p→ π^0 + X, where X stands for anything else, and $\xi\sigma_{\text{p}\alpha}^{\pi^0}(T_\text{p})$ for the reaction p + α → π^0 + X, respectively, with multiplicities ξ included. Cross-sections for the interaction of higher metalicity cosmic ray nuclei are less well-studied. Figures 5.4 and 5.5 show the measured inclusive cross-sections for π^0 and π^+ production in p–p and p–α collisions as a function of the proton kinetic energy $T_\text{p} = E_\text{p} - m_\text{p}c^2$.

5.3 Interactions Between Cosmic Ray Nuclei and Matter

Table 5.1. Principal interactions leading to pion production (a and b are positive integers)

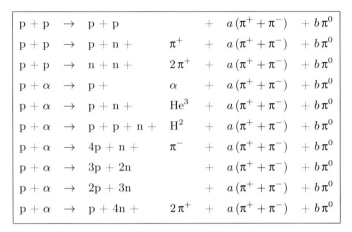

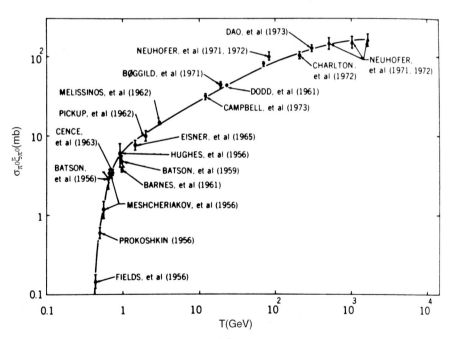

Fig. 5.4. Inclusive cross-sections for π^0 production in p–p interactions. From Stecker (1973 [517])

According to Stecker (1973 [517]) the accelerator data on the total cross-section σ times multiplicity ξ for neutral pion production in p–p interactions for energies up to ~ 1500 GeV shown in Fig. 5.4 can be well approximated by the broken power law in kinetic energy $T = E_p - m_p c^2 = m_p c^2 (\gamma_p - 1)$

$$\xi \sigma_{pp}^{\pi^0}(T) = \begin{cases} 6.13 \times 10^{-26} (\gamma_p - 1)^{7.64} \text{ cm}^2 & \text{for } 1.43 \leq \gamma_p \leq 1.75 \\ 8.12 \times 10^{-27} (\gamma_p - 1)^{0.53} \text{ cm}^2 & \text{for } \gamma_p \geq 1.75 \end{cases}. \quad (5.3.7)$$

For charged pions the mean multiplicity has an energy dependence $\xi \simeq 2 E_p^{1/4}$ (GeV) where E_p is in GeV. Laboratory measurements of the mean pion energy in p–p and p–α interactions indicate that a constant fraction, about 30%, of the incident kinetic energy of protons goes to pion energy. This is in agreement with Fermi's theory of pion production, in which a thermal equilibrium is assumed in the resulting pion cloud and in which the mean Lorentz factor of pions in the laboratory frame (where one of the protons is initially at rest) is given by

$$\bar{\gamma}_\pi \simeq \gamma_p^{3/4}, \quad (5.3.8)$$

where γ_p denotes the Lorentz factor of the incident proton. In terms of total energy (5.3.8) reads $\bar{E}_\pi \simeq 0.14 E_p^{3/4}$, where both energies are in GeV.

In order to utilize the experimentally measured inclusive cross-sections as given in Fig. 5.5 and 5.6 we approximate the differential cross-section by

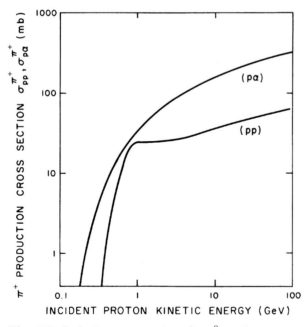

Fig. 5.5. Inclusive cross-sections for π^0 production in p–p and p–α interactions. From Cheung (1972 [99])

a δ-function centered at the mean pion energy (5.3.8) that is

$$\sum_k \sigma_{k,\mathrm{pp,\,p\alpha}}^{\pi^0,\pi^\pm}(E_{\pi^0,\pi^\pm}, E_\mathrm{p}) \simeq \xi\sigma_{\mathrm{pp,\,p\alpha}}^{\pi^0,\pi^\pm}(E_\mathrm{p})\delta(E_{\pi^0,\pi^\pm} - \bar{E}_\pi)$$

$$= \frac{\xi\sigma_{\mathrm{pp,\,p\alpha}}^{\pi^0,\pi^\pm}(E_\mathrm{p})}{m_{\pi^0,\pi^\pm}c^2}\delta(\gamma_{\pi^0,\pi^\pm} - \bar{\gamma}_\pi). \quad (5.3.9)$$

This δ-function approximation and the use of the same mean pion energy (5.3.8) in p–p and p–α collisions do not introduce appreciable error except in interactions of nucleons with energies greater than 10^4 GeV (Gould and Burbidge 1965 [200], Cheung 1972 [99]).[2] For the pion power (5.3.5) we then obtain, with (5.3.8)

$$P(E_\pi, E_\mathrm{N}) = c\gamma_\pi\ 1.30\ [n_\mathrm{HI}(\boldsymbol{r}) + 2n_{\mathrm{H}_2}(\boldsymbol{r})]\xi\sigma_\mathrm{pp}^\pi(E_\mathrm{p})$$

$$\times\ \delta(\gamma_\pi - \gamma_\mathrm{N}^{3/4})H[E_\mathrm{N} - E_\mathrm{th}]\ \mathrm{erg\ s^{-1}\ eV^{-1}}, \quad (5.3.10)$$

where the factor 1.30 accounts for the known chemical composition of the interstellar medium. $E_\mathrm{th} = \gamma_\mathrm{th}m_\mathrm{p}c^2 = 1.22$ GeV is the nucleon total energy threshold.

In analogy to (4.4.5)–mutatis mutandis–for a differential number density distribution of cosmic ray protons $N(\gamma_\mathrm{p}, \boldsymbol{r})$ at position $\boldsymbol{r} = (l, b, r)$ in the universe we obtain for the three pion source functions

$$Q_{\pi^0,\pi^\pm}(E_\pi; \boldsymbol{r}) = \frac{1.26}{E_\pi}\int_1^\infty d\gamma_\mathrm{p} N(\gamma_\mathrm{p}, \boldsymbol{r})P_{\pi^0,\pi^\pm}(E_\pi, E_\mathrm{p})$$

$$= \frac{1.64\ c}{m_\pi c^2}[n_\mathrm{HI}(\boldsymbol{r}) + 2n_{\mathrm{H}_2}(\boldsymbol{r})]\int_{\gamma_\mathrm{th}}^\infty d\gamma_\mathrm{p} N(\gamma_\mathrm{p}; \boldsymbol{r})$$

$$\times\ \xi\sigma_\mathrm{pp}^{\pi^0,\pi^\pm}(E_\mathrm{p})\delta(\gamma_\pi - \gamma_\mathrm{p}^{3/4})\ \mathrm{pions\ cm^{-3}\ s^{-1}\ eV^{-1}}, \quad (5.3.11)$$

where the factor 1.26 accounts for the contribution of α–p, α–α collisions and collisions of higher metalicity cosmic rays, and is derived using the ratio of the respective inclusive cross-section to the p–p inclusive cross-section and the known elemental composition of the cosmic rays.

For the astrophysically important case (see Sect. 3.4.1) of the power law distribution (4.1.18) of cosmic ray protons and α-particles with $\gamma_2 = \infty$ and $\gamma_1 < \gamma_\mathrm{th} = 1.30$ the source function (5.3.11) reduces to

$$Q_{\pi^0,\pi^\pm}(\gamma_\pi; \boldsymbol{r}) = \frac{2.2\ c}{m_\pi c^2}N_0(\boldsymbol{r})[n_\mathrm{HI}(\boldsymbol{r}) + 2n_{\mathrm{H}_2}(\boldsymbol{r})]\gamma_\pi^{-(4s-1)/3}$$

$$\times\ \xi\sigma_\mathrm{pp}^{\pi^0}(m_\mathrm{p}c^2(\gamma_\pi^{4/3} - 1)) = \frac{c}{m_\pi c^2}N_0(\boldsymbol{r})[n_\mathrm{HI}(\boldsymbol{r}) + n_{\mathrm{H}_2}(\boldsymbol{r})]$$

$$\times\ \begin{cases} 1.35\times 10^{-25}\gamma_\pi^{-(4s-1)/3}(\gamma_\pi^{4/3} - 1)^{7.64} & \text{for } 1.31 < \gamma_\pi \leq 1.75 \\ 1.79\times 10^{-26}\gamma_\pi^{-(4s-1)/3}(\gamma_\pi^{4/3} - 1)^{0.53} & \text{for } \gamma_\pi > 1.75 \end{cases}, (5.3.12)$$

---

[2] For a more sophisticated treatment on the basis of more recent and advanced models of pion production the reader is referred to Dermer (1986 [121]) and Mori (1997 [365]).

where we have used (5.3.7). At relativistic pion energies ($\gamma_\pi \gg 1$) the pion source functions (5.3.12) are power laws

$$Q_{\pi^0,\pi^\pm}(\gamma_\pi; \boldsymbol{r}) = K_\pi(\boldsymbol{r})\gamma_\pi^{-s_\pi}, \qquad (5.3.13)$$

with $s_\pi = \frac{4}{3}s - 1.04$ and $K_\pi(\boldsymbol{r})$ independent of γ_π.

Figure 5.6 shows the resulting π^+ and π^- source functions in interstellar space as calculated by Ramaty (1974 [415]) with a more general expression for the differential pion production cross-section and two representations of the local cosmic ray proton spectrum. The power law behavior at relativistic pion energies is again clearly visible.

Equation (5.3.12) can be used readily in (5.3.1), (5.3.3) and (5.3.4) to infer the differential γ-ray, secondary electron, positron and neutrino source functions. We consider each in turn.

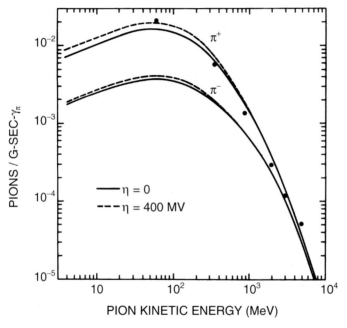

Fig. 5.6. Pion production spectrum in interstellar space giving the number of pions per s, gram and unit pion Lorentz factor. The **solid line** is calculated with the cosmic ray spectrum as observed near Earth whereas the **dashed line** is calculated with the demodulated cosmic ray spectrum. From Ramaty (1974 [415])

5.3.3 π^0 Decay γ-Rays

Irrespective of the pion source function Q_{π^0} the resulting differential γ-ray source function $q_{\pi^0}(E_\gamma)$ as calculated from (5.3.1) is symmetric around

$\frac{1}{2}m_{\pi^0}c^2 = 70$ MeV in a double-logarithmic plot. Moreover, because of the power law behavior of the pion source function at relativistic pion energies (5.3.13) we find at small and large γ-ray energies $E_\gamma \ll m_{\pi^0}c^2$ and $E_\gamma \gg m_{\pi^0}c^2$, respectively, the power law dependence with the same power law spectral index

$$q_{\pi^0}(E_\gamma \ll m_{\pi^0}c^2; \boldsymbol{r}) = \frac{2K_\pi}{s_\pi}(m_{\pi^0}c^2)^{-s_\pi} E_\gamma^{s_\pi}, \tag{5.3.14}$$

and

$$q_{\pi^0}(E_\gamma \gg m_{\pi^0}c^2; \boldsymbol{r}) = \frac{2K_\pi}{s_\pi}(m_{\pi^0}c^2)^{s_\pi} E_\gamma^{-s_\pi}. \tag{5.3.15}$$

The differential source function (5.3.1) and the associated integral source function

$$qi_{\pi^0}(> E_\gamma; \boldsymbol{r}) \equiv \int_{E_\gamma}^\infty \mathrm{d}E'_\gamma\, q_{\pi^0}(E'_\gamma; \boldsymbol{r}) \tag{5.3.16}$$

have been calculated numerically for all γ-ray energies by several authors (Stecker 1970 [516], Cavallo and Gould 1971 [95], Stephens and Badhwar 1981 [523], Dermer 1986 [121]). Their results shown in Fig. 5.7 agree remarkably well, with differences stemming mainly from using different representations of the local interstellar cosmic ray proton energy spectrum. Stecker (1975 [518]) has noted that, using the upper limit to the demodulated cosmic ray energy spectrum given by Comstock et al. (1972 [104]), an upper limit to the local integral γ-ray source function of

$$qi_{\pi^0,l}(E_\gamma \geq 100 \text{ MeV}) \leq (1.51 \pm 0.23) \times 10^{-25} \text{ s}^{-1}, \tag{5.3.17}$$

is obtained, with the error bracket reflecting the experimental error in the accelerator data on the inclusive cross-section.

In analogy to (4.4.14) we obtain the differential photon number flux of π^0-decay γ-rays from a direction (l,b) as the line-of-sight integral of the source function (5.3.1)

$$\frac{\mathrm{d}N_\gamma(E_\gamma; l, b)}{\mathrm{d}t\, \mathrm{d}E_\gamma\, \mathrm{d}\Omega} = \frac{1}{4\pi}\int_0^\infty \mathrm{d}r\, q_{\pi^0}(E_\gamma; \boldsymbol{r}) \text{ photons cm}^{-2} \text{ s}^{-1} \text{ ster}^{-1} \text{ eV}^{-1}, \tag{5.3.18}$$

and for the integral photon number flux

$$\frac{\mathrm{d}N_\gamma(\geq E_\gamma; l, b)}{\mathrm{d}t\, \mathrm{d}\Omega} = \frac{1}{4\pi}\int_0^\infty \mathrm{d}r\, qi_{\pi^0}(\geq E_\gamma; \boldsymbol{r}) \text{ photons cm}^{-2} \text{ s}^{-1} \text{ ster}^{-1}. \tag{5.3.19}$$

Additionally, if we assume that the shape of the cosmic ray nucleus energy spectrum does not vary with position $\boldsymbol{r}$,

$$N(\gamma; \boldsymbol{r}) = \frac{N_0(\boldsymbol{r})}{N_{0,l}}\gamma^{-s}, \tag{5.3.20}$$

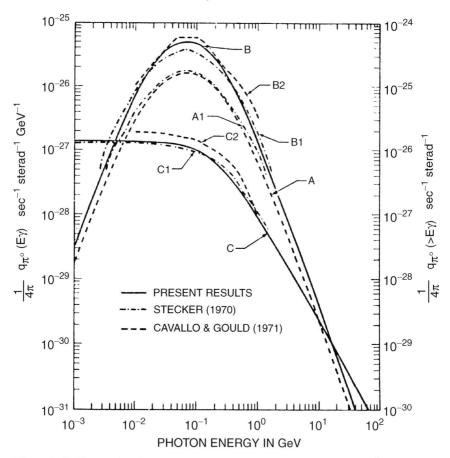

Fig. 5.7. Differential and integral source functions of γ-rays from the $\pi^0 \to 2\gamma$ decay calculated by various authors with different representations of the local interstellar cosmic ray proton spectrum. After Badhwar and Stephens (1977 [24])

where the subscript l in (5.3.17) and (5.3.20) denotes the position of the solar system in our Galaxy, we can use the source functions calculated for cosmic ray parameter values near the solar system in (5.3.19) to derive for the integral number flux

$$\frac{dN_\gamma(\geq E_\gamma; l, b)}{dt\, d\Omega} = \frac{qi_{\pi^0, l}(\geq E_\gamma)}{4\pi} \int_0^\infty dr\, \frac{N_0(l, b, r)}{N_{0, l}}$$
$$\times\, [n_{HI}(l, b, r) + 2n_{H_2}(l, b, r)]\ \text{photons cm}^{-2}\ \text{s}^{-1}\ \text{ster}^{-1}\,. \quad (5.3.21)$$

The study of γ-rays from the π^0-decay therefore allows one to investigate the large-scale distribution of cosmic ray nuclei if the interstellar gas distribution is known. This fact has stimulated the interest in the development of γ-ray astronomy for a long time. Stecker (1973 [517]) has shown that the majority of γ-rays from the π^0-decay are produced by cosmic ray nuclei with energies between 1 and 30 GeV.

5.3.4 Secondary Electrons and Positrons

Using (5.3.11) in (5.3.3) and (5.3.4) and performing the indicated integrations yields the secondary electron and positron source functions. Figure 5.8 shows the resulting spectra for electrons and positrons as obtained by Ramaty (1974 [415]).

Secondary electrons can also be produced by the knock-on process whereby ambient atomic and free electrons of the interstellar matter achieve relativistic energies when they collide with cosmic ray protons and nuclei (Abraham et al. 1966 [2]; Baring 1991 [29]). According to Abraham et al. (1966 [2]) the source function of knock-on electrons is

$$Q_{\rm kn}(\gamma_{\rm e}; \boldsymbol{r}) = 4.9 \rho(\boldsymbol{r}) (\gamma_{\rm e} - 1)^{-2.76} \text{ cm}^{-3} \text{ s}^{-1} \gamma_{\rm e}^{-1} \,, \qquad (5.3.22)$$

where $\gamma_{\rm e}$ denotes the Lorentz factor of the electrons and $\rho(\boldsymbol{r})$ the density of the target material in g cm^{-3}. This source function dominates the electron

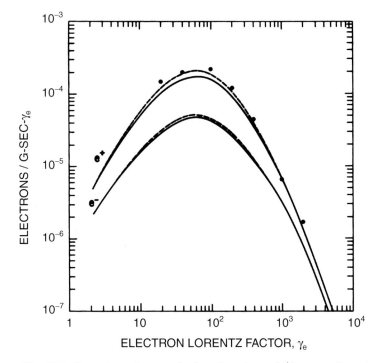

Fig. 5.8. Secondary electron (e$^-$) and positron (e$^+$) source functions per s, gram and unit Lorentz factor. The **solid line** is calculated with the cosmic ray spectrum as observed near Earth whereas the **dashed line** is calculated with the demodulated cosmic ray spectrum. From Ramaty (1974 [415])

source function from the π^--decay only at energies less than $\gamma_e \leq 20$. Other sources of secondary electrons and positrons, such as neutrons and radioactive positron emitters, are negligible at energies greater than a few MeV (Ramaty 1974 [415]).

The added source functions from the pion decay and the knock-on electrons calculated for a cosmic ray nucleon spectrum as measured near the solar system are shown in Fig. 5.9. Here q_+ and q are the source functions of positrons (e^+) and secondary electrons and positrons ($e^+ + e^-$), respectively. The turnup in the total electron and positron spectrum below about 50 MeV is due to the knock-on process. Above 5 GeV the ratio e^+/e^- reproduces the measured $\mu^+/\mu^- = 1.28$ ratio measured at sea level for high energy muons in the Earth atmosphere, which also are of secondary origin from cosmic ray showers.

Marscher and Brown (1978 [322]) have deduced for the positron content at production

$$Q_+(\gamma)/Q(\gamma) \simeq 0.675 - 0.126 \exp[-1710/\gamma] , \qquad (5.3.23)$$

which agrees well with the results shown in Fig. 5.9. According to Ramaty and Westergaard (1976 [418]) the differential secondary electron and positron production rate at energies above 100 GeV can be well approximated by

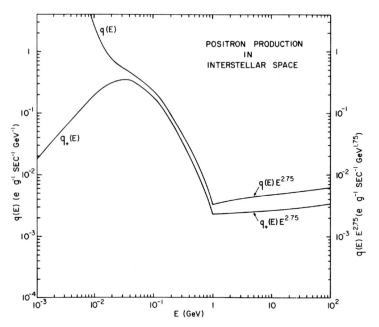

Fig. 5.9. The source functions of positrons, $q_+(E)$, and secondary electrons and positrons, $q(E)$, from charged pion decay and the knock-on process. Above 1 GeV these functions are multiplied by $E^{2.75}$. From Ramaty and Westergaard (1976 [418])

$$Q(E > 100 \text{ GeV}) = 6.4 \times 10^{-3} (E/\text{GeV})^{-2.75} \text{ g}^{-1} \text{ s}^{-1} \text{ GeV}^{-1}, \quad (5.3.24)$$

where the exponent 2.75 reproduces the measured cosmic ray proton spectral index at high energies (Sect. 3.4.1).

Another useful approximation to the numerically calculated secondary electron and positron spectra which is accurate to within 5% has been given by Marscher and Brown (1978 [322]) for electron and positron energies between 17 MeV and 100 GeV:

$$\begin{aligned} Q(\gamma) &= Q_+(\gamma) + Q_-(\gamma) \\ &= [n_{\text{HI}}(\boldsymbol{r}) + 2n_{\text{H}_2}(\boldsymbol{r})] K(\gamma) f(\gamma) \text{ cm}^{-3} \text{ s}^{-1} \gamma^{-1} \end{aligned} \quad (5.3.25)$$

with

$$f(\gamma) = \begin{cases} 2.85 \times 10^{-28} \gamma^{-0.06} \left[1 + (\gamma/292)^{5.52}\right]^{-1/3} & \text{for } 33 < \gamma \le 977 \\ 4.19 \times 10^{-22} \gamma^{-2.44} & \text{for } 977 < \gamma < 2 \times 10^5 \end{cases}, \quad (5.3.26)$$

and

$$K(\gamma) = 0.935 + 0.05 \log_{10} \gamma . \quad (5.3.27)$$

Although based on a different representation of the local cosmic ray nuclei energy spectrum this analytical approximation agrees well with the results of Ramaty and Westergaard (1976 [418]) shown in Fig. 5.9.

5.3.5 Secondary Neutrinos

The neutrino source function from the pion decay and the subsequent muon decay with cosmic ray nucleus spectra characteristic for the local interstellar medium have been calculated by Berezinsky and Zatsepin (1977 [43]), Silberberg and Shapiro (1977 [485]), Margolis et al. (1978 [320]) and Stecker (1979 [520]). Figure 5.10 shows the results for the ν_μ component in comparison with the γ-ray source function from the π^0-decay. Since each charged pion decay results in one ν_μ, one $\bar{\nu}_\mu$, and one $\nu_e(\bar{\nu}_e)$, all of roughly the same energy, the total number of neutrinos produced (although of four different types) is a factor of 3 higher than that shown for muon neutrinos. In this regard it should be kept in mind that the experimental cross-section ratios $\sigma_{\nu_e N}/\sigma_{\nu_\mu N} = 1.26 \pm 0.23$, $\sigma_{\bar{\nu}_e N}/\sigma_{\bar{\nu}_\mu N} = 1.32 \pm 0.32$, $\sigma_{\bar{\nu}_\mu N}/\sigma_{\nu_\mu N} = 0.40 \pm 0.12$, $\sigma_{\bar{\nu}_e N}/\sigma_{\nu_e N} = 0.38 \pm 0.02$ (Stecker 1979 [520]) apply.

The inclusion of the γ-ray production spectrum in Fig. 5.10 is convenient in the context of estimating the flux of diffuse secondary neutrinos from the Galaxy. Since the flux of diffuse galactic γ-rays has been measured (see Chap. 6), the predicted neutrino flux follows readily by using the relation between the source functions of neutrinos and γ-rays shown in Fig. 5.10. For further details see Stecker (1979 [520]).

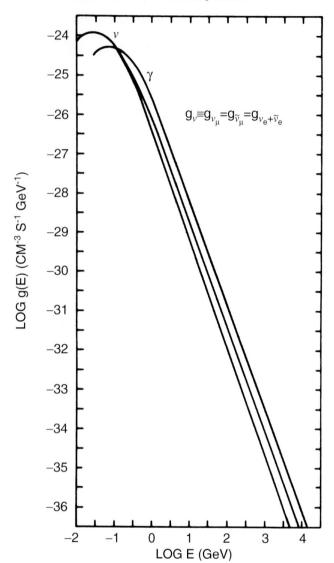

Fig. 5.10. Differential source function of neutrinos and γ-rays from the decay of pions produced by interactions of cosmic rays in our galactic neighborhood per hydrogen atom. The **γ-ray curve** and the **upper neutrino curve** are calculated for cosmic rays having a spectral index of 2.67 between 10 and 3×10^6 GeV; the **lower neutrino curve** is for a cosmic ray spectrum with index 2.75. The spread in the curve is indicative of the uncertainty in such calculations. From Stecker (1979 [520])

Treating the neutrino production cross-section similarly to (5.3.9) Stecker (1979 [520]) has shown that for the power law pion spectrum (5.3.13) at large neutrino energies $E_\nu \gg m_\pi c^2$ the neutrino source function is also of the power law form $Q_\nu(E_\nu) \propto E_\nu^{-s_\nu}$ with the same value for the spectral index $s_\nu = s_\pi$, as is clearly visible in Fig. 5.10.

At very high cosmic ray nucleus energies secondary neutrinos are also generated by photo-hadron production of pions (Sect. 5.2.2). Figure 5.11 shows the predicted diffuse ν_μ and $\bar\nu_\mu$ fluxes from the nucleon-nucleon and photo-hadron interactions (the corresponding flux of electron-neutrinos and antineutrinos is lower by a factor of 2). The hatched region marked p–p is for galactic ν-production coming from the galactic center region defined by galactic longitude $340° \le l \le 40°$ and latitudes $-10° \le b \le 10°$ assuming the same cosmic ray spectrum in that region as in our local galactic neighborhood. The curve marked PCL(UL) is an upper limit on the neutrino flux from primordial cosmic ray interactions related to the extragalactic γ-ray background.

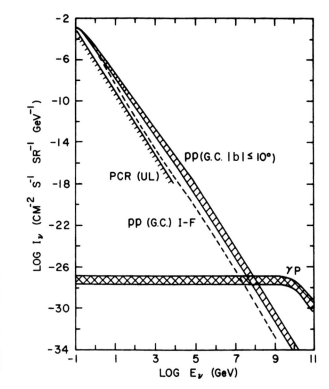

Fig. 5.11. Predicted diffuse background muon neutrino and antineutrino differential fluxes. From Stecker (1979 [520])

5.3.6 Energy Loss by Pion Production

Integrating the pion power (5.3.10) over all pion energies E_π we obtain the energy loss of a single relativistic nucleus of mass m due to pion production

$$-\left(\frac{dE_N}{dt}\right)_\pi = -mc^2\left(\frac{d\gamma_N}{dt}\right)_\pi = 3\int_0^{E_{\pi\max}} dE_\pi P(E_\pi, E_N)$$
$$= 4.9\, c\, m_\pi c^2 \left[n_{\rm HI}(r) + 2n_{\rm H_2}(r)\right] H\left[\gamma_N - 1.30\right] \sigma_{\rm pp}^\pi(\gamma_N)\gamma_N\,,$$
(5.3.28)

where the factor 3 accounts for the production of π^0-, π^+- and π^--mesons, respectively. After a strong energy dependence near the non-relativistic threshold Lorentz factor the use of the p–p pion production cross-section (5.3.7) at relativistic energies and a multiplicity law $\xi = \gamma_N^{1/4}$ gives

$$-\left(\frac{d\gamma_N}{dt}\right)_\pi (1 \ll \gamma_N \leq 3000) = 1.4 \times 10^{-16}\left[n_{\rm HI}(r) + 2n_{\rm H_2}(r)\right]$$
$$\times A^{-0.47}\gamma_N^{1.28}\ {\rm s}^{-1}\,.$$
(5.3.29)

5.3.7 Excitation of Nuclei

Excited nuclei produced by collisions between fast particles and nuclei undergo nuclear transition to their ground states whereby γ-rays may be emitted. We consider only those nuclei that are appreciably abundant in stellar and interstellar matter, i.e. the nuclei C, N, O, Ne, Mg, Si, and S. The most important γ-ray lines from the excited states of these nuclei and their physical properties are listed in Table 5.2.

Two types of collisions occur:

(i) collision of fast cosmic ray nuclei with p and α-particles of the stellar or interstellar target gas at rest,
(ii) collision of fast p and α-particles of the cosmic radiation with heavier target nuclei at rest.

Summing over all inelastic cross-sections and the appropriate lifetimes of the excited states for the various nuclei, the emergent nuclear γ-ray line intensities have been calculated in detail for different energy spectra of the energetic cosmic ray nuclei in astrophysical sites as solar and stellar flares, galactic centers and supernova envelopes. Sophisticated computations have been performed by Ramaty et al. (1979 [419]). The resulting γ-ray intensities exhibit a great wealth of spectral structure, ranging from very narrow to broad features, depending on the composition and energy spectrum of the energetic particles, and the composition and state of the ambient medium. As an illustrative example in Fig. 5.12 we show the calculated γ-spectrum for energetic particles with a power law energy spectrum and an ambient medium both having solar composition.

Table 5.2. Most prominent reactions leading to nuclear γ-ray line emission. After Meneguzzi and Reeves (1975 [340])

Reaction	E_γ (MeV)
^{12}C (p,p′) ^{12}C	4.44
^{12}C (p,2p) ^{11}B	4.44
^{14}N (p,p′) ^{14}N	2.31
^{14}N (p,p′) ^{14}N	3.94
^{16}O (p,p′α) ^{12}C	4.44
^{16}O (p,pn) ^{15}O	5.18
^{16}O (p,2p) ^{15}N	5.27
^{16}O (p,p′) ^{16}O	6.13
^{16}O (p,p′) ^{16}O	7.12
^{20}Ne (p,p′) ^{20}Ne	1.63
^{24}Mg (p,p′) ^{24}Mg	1.37
^{28}Si (p,p′) ^{28}Si	1.78
^{56}Fe (p,p′) ^{56}Fe	0.845
^{56}Fe (p,p′) ^{56}Fe	1.24

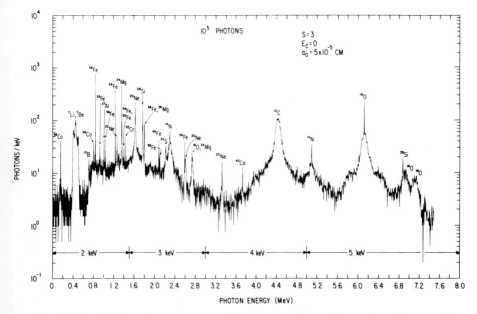

Fig. 5.12. Calculated γ-ray spectrum from energetic particles and ambient medium having solar composition. The energetic particles obey a power law spectrum in kinetic energy with spectral index s above a low-energy cutoff E_c, and the contribution from lines from interstellar grains of characteristic radius a_0 is included. From Ramaty et al. (1979 [419])

With the emergent intensity being very sensitive to the assumed state of the target material and to the spectrum of incoming cosmic ray nuclei it is clear that successful spectroscopy of these lines will provide many detailed clues on the ongoing astrophysical processes. The contribution as an energy loss process of cosmic ray nuclei, however, is negligibly small as compared to the Coulomb and ionization losses which we are going to discuss next.

5.3.8 Coulomb and Ionization Interactions

5.3.8.1 Interactions in Fully Ionized Plasma.

The rate of energy loss of a fast test ion of mass $M = Am_p$, charge Ze and velocity $v = \beta c$ in a completely ionized thermal plasma consisting of s species of respective mass m_s, charge $Z_s e$, density n_s and temperature T_s, density n_e has been calculated by Butler and Buckingham (1962 [85]), Sivukhin (1965 [494]) and Gould (1972 [197]) as

$$-\left(\frac{dE}{dt}\right) = \frac{4\pi Z^2 e^4 \ln\lambda}{v} \sum_s \frac{Z_s^2 n_s}{m_s} W_s\left(\frac{\beta}{\beta_s}\right)$$

$$= \frac{3\, c\, \sigma_T\, \ln\lambda (m_e c^2) Z^2}{2\beta} \sum_s \frac{m_e}{m_s} Z_s^2 n_s\, W_s\left(\frac{\beta}{\beta_s}\right) \text{ eV s}^{-1}, \qquad (5.3.30)$$

where

$$\beta_s \equiv \left(\frac{2kT_s}{m_s c^2}\right)^{1/2}, \qquad (5.3.31)$$

and

$$W_s(x) \equiv \frac{2}{\pi^{1/2}} \left[\int_0^x dy \exp(-y^2) - \left(1 + \frac{m_s}{M}\right) x \exp(-x^2)\right]. \qquad (5.3.32)$$

The Coulomb logarithm appearing in (5.3.30) $\ln\lambda$ is equal to $\simeq 20$ for a wide range of plasma densities and temperatures (see Sivukhin 1965 [494]). Because of the small electron/proton mass ratio $m_e/m_p = 1/1836$ it is easy to see that for all values of the metalicity of the plasma the Coulomb collisions are dominated by scattering off the thermal electrons, i.e.

$$\sum_s \frac{m_e}{m_s} Z_s^2 n_s W_s \simeq n_e W_e, \qquad (5.3.33)$$

so that (5.3.30) reduces to

$$-\left(\frac{dE}{dt}\right) \simeq \frac{30\, c\, \sigma_T (m_e c^2) Z^2}{\beta} n_e\, W_e\left(\frac{\beta}{\beta_e}\right) \text{ eV s}^{-1}, \qquad (5.3.34)$$

where numerically

$$\beta_e = 0.026(T_e/2 \times 10^6 \text{ K})^{1/2} \, . \qquad (5.3.35)$$

For small and large arguments we may approximate (5.3.32) as

$$W_e \left(\frac{\beta}{\beta_e} \right) \simeq \begin{cases} \frac{2}{\pi^{1/2}} \frac{\beta}{\beta_e} \left[-\frac{m_e}{M} + \left(\frac{2}{3} + \frac{m_e}{M} \right) \left(\frac{\beta}{\beta_e} \right)^2 \right] & \text{for } \beta \ll \beta_e \\ 1 & \text{for } \beta \gg \beta_e \end{cases} , \qquad (5.3.36)$$

For values of $\beta \leq \beta_c \, (A, T_e)$ where

$$\beta_c \, (A, T_e) \equiv [1.5 \, m_e/(Am_p)]^{1/2} \, \beta_e = 0.0286 \, A^{-1/2} \, \beta_e \qquad (5.3.37)$$

we find $W_e < 0$, indicating that at low particle velocities Coulomb collisions accelerate $((dE/dt) > 0)$ the test ion up to velocity $\beta_c(A, T_e)$. In the intermediate range $\beta_c(A, T_e) \leq \beta \leq \beta_e$ we obtain $W_e \simeq (4/3)\pi^{-1/2}(\beta/\beta_e)^3$ according to (5.3.36a) and, beyond $\beta \geq \beta_e$, W_e quickly approaches the asymptotic value $W_e \simeq 1$. The interpolation formula

$$W_e \, (\beta \geq \beta_c \, (A, T_e)) = \frac{\beta^3}{x_m^3 + \beta^3} \, , \qquad (5.3.38)$$

with

$$x_m \equiv \left(\frac{3\pi^{1/2}}{4} \right)^{1/3} \beta_e = 1.10 \, \beta_e = 0.0286 \, (T_e/2 \times 10^6 \text{ K})^{1/2} \qquad (5.3.39)$$

being independent of A, provides an excellent fit to the exact variation over the whole range $\beta \geq \beta_c$ as can be seen from Fig. 5.13.

Using the interpolation formula (5.3.38) in (5.3.34) we obtain

$$-\left(\frac{dE}{dt} \right) \simeq 3.1 \times 10^{-7} \, Z^2 \, n_e \, \frac{\beta^2}{x_m^3 + \beta^3} \text{ eV s}^{-1} \qquad (5.3.40)$$

and for the associated total momentum loss rate

$$-\left(\frac{dp_{\text{tot}}}{dt} \right) = 3.1 \times 10^{-7} \, Z^2 \, n_e \, \frac{\beta}{x_m^3 + \beta^3} \text{ eV c}^{-1} \text{ s}^{-1} \, . \qquad (5.3.41)$$

Below the Coulomb barrier $p_{\text{tot,M}} = 0.87\beta_e = 0.0227(T_e/2 \times 10^6 K)^{1/2}Mc$ the total momentum loss rate varies linearly proportional to p_{tot}, whereas beyond the barrier it decreases $\propto p_{\text{tot}}^{-2}$.

5.3.8.2 Interactions in Neutral Matter. The energy loss of particles traversing neutral matter was first calculated by Bohr using classical theory. In quantum mechanics the problem has been investigated by several authors including Moeller, Bethe, Williams and Bloch (for references see the monograph of Heitler (1954 [226])) in a satisfactory way using various atomic

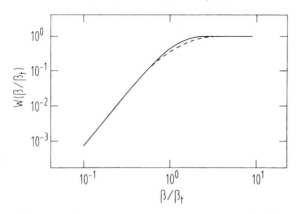

Fig. 5.13. Coulomb energy loss function $W_e(\beta/\beta_e)$ defined in (5.3.32) for protons ($A = 1$) as a function of β/β_e for values $\beta \geq \beta_c$. The **full curve** represents the exact variation from (5.3.32) whereas the **dashed curve** has been calculated with the interpolation formula (5.3.38). From Schlickeiser and Steinacker (1989 [470])

models. We follow the treatment of Heitler (1954 [226]); a detailed summary of the results can also be found in Hayakawa (1969 [221], his Sect. 2.2).

The kinetic energy ($T = E - Mc^2$) loss rate of an ion traversing neutral matter consisting of s sorts of atoms and molecules with concentrations n_s and charges z_s is (Heitler 1954 [226], Chap. 37; Evans 1955 [155], Chap.18; Barkas and Berger 1964 [30], Northcliffe 1963 [376])

$$-\left(\frac{dT}{dt}\right)(\beta \geq \beta_0) = \frac{3\,c\sigma_T\,Z^2\,(m_e c^2)}{4\beta} \sum_s n_s [B_s + B'(\alpha Z/\beta)], \quad (5.3.42)$$

with

$$B_s = \left[\ln\left(\frac{2m_e c^2 \beta^2 Q_{max}}{I_s^2(1-\beta^2)}\right) - 2\beta^2 - \frac{2C_s}{z_s} - \delta_s\right] \text{eV s}^{-1}. \quad (5.3.43)$$

The largest possible energy transfer from the incident particle to an initially stationary atomic electron follows from the conservation laws for momentum and energy and is

$$Q_{max} = \frac{T + 2Mc^2}{1 + [(M + m_e)^2 c^2/(2m_e T)]} \simeq \frac{2m_e c^2 \beta^2 \gamma^2}{1 + [2m_e T/(m^2 c^2)]} \text{ eV}, \quad (5.3.44)$$

since $M \gg m_e$. The function

$$B'(x) = 2[\Psi(0) - \Re\Psi(\imath x)], \quad (5.3.45)$$

where $\Psi(x)$ denotes the Digamma function, is important for slow particles or particles with large atomic number so that the product $\alpha Z/\beta = Z/(137\beta)$ is no longer small compared to unity. It is straightforward to show that for small values of $x \ll 1$, $B'(x \ll 1) \to 0$, whereas in the opposite case, $B'(x \gg 1) \to -2\ln x$. I_s denotes the geometrical mean of all ionization and excitation potentials of the absorbing atom or molecule s and is defined by

$$z_s \ln I_s \equiv \sum_{n,l} f_{n,l}^s A_{n,l}^s , \qquad (5.3.46)$$

where $f_{n,l}^s$ is the sum of the oscillator strengths for all optical transitions of the electron in the n, l shell of the atom s and is close to unity, while $A_{n,l}^s$ is the mean excitation energy of the n, l shell and can be put equal to the ionization potential with sufficient accuracy. Empirical values for I_s can be determined for each value of z_s. For most elements I_s in eV is roughly $13 z_s$; for the very light elements hydrogen and helium, which are important for astronomical applications, the experimental values are $I_H = 19$ eV and $I_{He} = 44$ eV, respectively. The shell-correction term (C_s/z_s) and the density correction term δ_s of Sternheimer (1952 [524]), resembling the effect of the polarization of the medium on the ionization loss, lead to a reduction of the logarithmic $\ln \gamma$ increase of the ionization rate at relativistic energies to a constant value.

Equation (5.3.42) can be used for particle velocities large compared to the characteristic velocity of the medium's electrons. In the case of atomic hydrogen this characteristic velocity is determined by the orbital velocity of the electrons $\beta_0 = 1.4 e^2 (\hbar c)^{-1} = 0.01$ corresponding to a kinetic energy of

$$T_0 = 49 A \text{ keV} . \qquad (5.3.47)$$

From the asymptotic behavior of the function B' one deduces immediately that $B_s + B' \simeq B_s$ for $\beta > \max[\beta_0, Z/137]$, whereas in the case $\beta_0 < \beta < Z/137$ we obtain Bohr's original classical formula

$$B_s + B' \simeq 2 \ln \frac{2c\hbar m_e c^2 \beta^3}{I_s Z e^2} . \qquad (5.3.48)$$

Below the kinetic energy T_0 the ionization rate changes. This has been clearly established by laboratory measurements of the stopping cross-section per atom as collected by Whaling (1958 [564]) and shown in Fig. 5.14. For ions of velocity large compared to β_0 there is excellent agreement between formula (5.3.42) and the experimental values. However, for ions of lower velocity there is, according to Whaling (1958 [564]), no regularity in the energy dependence of the stopping cross-section for all absorbing atoms, and each curve at low velocities in Fig. 5.14 has been simply drawn by eye to fit the experimental points in a reasonably smooth manner. Hayakawa (1969 [221]) discusses some relevant physical processes at these low ion velocities. Ginzburg and Syrovatskii (1964 [185]) give a useful ionization loss rate for slow ion velocities for the case that the ion atomic number Z lies between $Z/4 \leq z_s \leq 4Z$,

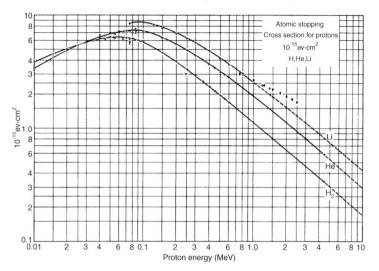

Fig. 5.14. Laboratory measurements of the stopping cross-section per atom of incident protons in hydrogen, helium and lithium. The stopping cross-section per atom is defined by $-1/n_s \beta c \times \mathrm{d}T/\mathrm{d}t$. Values for hydrogen are one-half of the molecular stopping cross-section. From Whaling (1958 [564])

$$-\left(\frac{\mathrm{d}T}{\mathrm{d}t}\right)(\beta \leq \beta_0) = 2.34 \times 10^{-23} c^2 \beta^2 \sum_s n_s (Z + z_s) \text{ eV s}^{-1}, \quad (5.3.49)$$

which agrees well with the experimental data in Fig. 5.14. So again, as in the case of Coulomb losses, the ionization losses have a maximum at T_0 given in (5.3.47) below which the loss rate varies proportional to $\propto T$ while it varies $\propto T^{-1/2}$ at non-relativistic energies above T_0 according to (5.3.42).

Using the asymptotic behavior of $Q_{\max}$ from (5.3.44)

$$Q_{\max}(T \leq 918 A^2 m_p c^2) \simeq 2 m_e (c \beta \gamma)^2$$

and

$$Q_{\max}(T \geq 918 A^2 m_p c^2) \simeq T$$

as well as the values of the ionization potentials of hydrogen and helium, taking into account the chemical composition of the interstellar medium, we obtain for (5.3.42)

$$-\left(\frac{\mathrm{d}T}{\mathrm{d}t}\right) = \frac{3}{2} \sigma_\mathrm{T}\, c\, Z^2 (m_e c^2) \sum_s n_s$$
$$\times \begin{cases} \ln \frac{2 m_e c^2}{I_s} + \ln \beta + \beta^4/2 & \text{for } T_0 \leq T \leq 918 A^2 m_p c^2 \\ \ln \frac{2 m_e c^2}{I_s} + 3.411 + 0.5 \ln A & \text{for } T \geq 918 A^2 m_p c^2 \end{cases}$$

$$\simeq 3.56 \times 10^{-13} \frac{m_e c^2}{\beta} Z^2 \left[n_{HI} + 2n_{H_2}\right]$$
$$\times \begin{cases} 1 + 0.0185 \ln \beta & \text{for } T_0 \leq T \leq 918 A^2 m_p c^2 \\ 1.315 \left[1 + 0.035 \ln A\right] & \text{for } T \geq 918 A^2 m_p c^2 \end{cases} \text{eV s}^{-1}.$$
(5.3.50)

At very low energies, $T \leq T_0$, (5.3.49) has to be used.

For cosmic ray protons a useful interpolation formula for all energies below $E \leq 918 m_p c^2$ is

$$-\left(\frac{dE}{dt}\right)_{\text{proton}} = 1.82 \times 10^{-7} \left[n_{HI} + 2n_{H_2}\right]$$
$$\times (1 + 0.0185 \ln \beta H[\beta - \beta_0]) \frac{2\beta^2}{\beta_0^3 + 2\beta^3} \text{ eV s}^{-1}, \quad (5.3.51)$$

which attains its maximum at $\beta_0 = 0.01$ corresponding to the kinetic energy T_0 given in (5.3.47). The function H in (5.3.51) denotes the Heaviside step function.

The ionization loss for a higher metalicity ion of charge Z can be expressed in terms of the ionization loss of protons, (5.3.51), as (Brown and Moak 1972 [77])

$$\left(\frac{dE_N}{dt}\right)_{\text{ion}} = Z_{\text{eff}}^2 \left(\frac{dE}{dt}\right)_{\text{proton}} \quad (5.3.52a)$$

at the same kinetic energy per nucleon with the effective charge

$$Z_{\text{eff}} = Z \left[1 - 1.034 \exp\left(-137\beta Z^{-0.688}\right)\right]. \quad (5.3.52b)$$

where β is the ion's velocity in units of c. The effective charge at small energies is less than Z since complete stripping of the heaviest ions occurs only at energies above 100 MeV/nuc.

If the primary particle is an electron or a positron, a modification ought to be made in (5.3.42) reflecting the fact that, for large energy, transfer exchange and spin effect play some role. This leads to (4.5.2) discussed before.

5.3.9 Catastrophic Losses from Fragmentation and Radioactive Decay

In inelastic proton-nucleus and α-nucleus collisions with atoms and molecules of the interstellar and intergalactic target gas cosmic ray nuclei with charge greater than 1 can fragment in reactions $A + (p, \alpha) \to (A - k) +$ anything, with $k = 1, 2, 3 \ldots, (A-1)$. The individual fragmentation cross-sections have been tabulated by Silberberg and Tsao (1973 [486], 1990 [487]). Letaw et al.

(1983 [302]) have derived a very useful empirical formula for the total inelastic cross-section of protons on nuclei with $A > 1$. At high energies ($T \geq 2$ GeV) the formula reproduces experimental data to within reported errors ($\leq 2\%$) whereas at lower energies the maximum error is 20% at 40 MeV. According to Letaw et al. (1983 [302]) (1 mb = 10^{-27} cm^2) at high energies

$$\sigma_{\text{tot}}(T \geq 2 \text{ GeV}) = 45\ A^{0.7}\left[1 + 0.016\sin(5.3 - 2.63\ln A)\right] \text{ mb} \quad (5.3.53)$$

is independent of energy and increases with increasing nucleus mass number A.

At energies below 2 GeV the total inelastic cross-section is not independent of energy. In general terms, it decreases to a minimum ($\sim 15\%$ below the high energy value) at 200 MeV. It then sharply increases to a maximum at ~ 20 MeV (60% above the high energy value). Below this energy resonance effects become dominant and the cross-section fluctuates rapidly. In the low-energy range the approximation is

$$\sigma_{\text{tot}}(T \leq 2 \text{ GeV}) = \sigma_{\text{tot}}(T \geq 2 \text{ GeV}) \left[1 - 0.62\exp(-T/200)\sin\left(10.9 \times T^{-0.28}\right)\right] \text{ mb}, \quad (5.3.54)$$

where T is in MeV. Fragmentation loss times for cosmic ray nuclei follow immediately from the total inelastic cross-sections as

$$\tau_f(T) = \left[\sum_s n_s c\beta\sigma_{\text{tot}}(T)\right]^{-1} = \left[1.3\left(n_{\text{HI}} + 2n_{\text{H}_2}\right)c\beta\sigma_{\text{tot}}(T)\right]^{-1} \text{ s}. \quad (5.3.55)$$

At relativistic energies we obtain an energy-independent fragmentation lifetime of

$$\tau_f(T > 2 \text{ GeV}) \simeq 2 \times 10^7\ A^{-0.7}\left[n_{\text{HI}} + 2n_{\text{H}_2}\right]^{-1} \text{ yr}. \quad (5.3.56)$$

The fragmentation losses are catastrophic losses, i.e. they do not conserve the total particle number of the considered nuclei. For this reason we have to describe the loss process by a loss time rather than an energy loss rate (dT/dt) which is appropriate for continuous losses that redistribute the particles in energy space but conserve the total number of particles. For a more thorough discussion of this point see Sect. 14.1.

To estimate the yield of fragmentation products from inelastic nucleon-proton interactions, individual partial fragmentation cross-sections, are needed. Table 5.3 shows the compilation by Silberberg and Tsao (1973 [486]) at relativistic energies where the partial cross-sections become independent of energy.

5.3 Interactions Between Cosmic Ray Nuclei and Matter

Table 5.3. Partial cross-sections in millibarns for collisions of various "heavy nuclei" with hydrogen at kinetic energies T ≥ 2.3 GeV. After Silberberg and Tsao (1973 [486])

Product	Target							
	^{11}B	^{12}C	^{14}N	^{16}O	^{20}Ne	^{24}Mg	^{28}Si	^{56}Fe
^{6}Li	14	*15**	11	*14*	13	13	13	30
^{7}Li	20	*13*	13	*14*	13	13	13	20
^{7}Be	7	10	12	11	10	*10*	10	8.5
^{9}Be	5.5	6	3	*3.7*	3	3	3	5
^{10}Be	*14*	3.5	1.9	*1.0*	1.9	1.9	1.9	4
^{10}B	26	16	14	*12*	11	10	9	9
^{11}B	...	*51*	25	*25*	18	15	12	9
^{12}C	...	...	26	24	18	13	10	7
^{13}C	...	...	25	20	14	10	8	5
^{14}N	...	...	...	26	18	13	10	6
^{15}N	...	...	...	*50*	23	17	13	6
^{16}O	...	...	...	...	24	18	13	6
^{17}O	...	...	...	...	25	19	14	6
^{18}O	...	...	...	...	23	16	*12*	6
^{19}F	...	...	...	...	45	19	14	7
^{20,21,22}Ne	...	...	...	...	...	*69*	55	21
^{23}Na	...	...	...	...	...	51	23	9
^{24,25,26}Mg	...	...	...	...	...	...	77	25
^{27}Al	...	...	...	...	...	...	52	10
σ_i	195 †	205	235	260	315	355	400	676

* Italicized values refer to cross-sections based primarily on experimental information
† σ_i is the total inelastic cross-section

In case of unstable particles (such as, for example, Be^{10}, Al^{26}) the total catastrophic loss time τ_c is determined by

$$\tau_c^{-1} = \tau_f^{-1} + \left(\gamma T^0_{\text{decay}}\right)^{-1}, \quad (5.3.57)$$

where T^0_{decay} is the radioactive half-life time at rest.

5.3.10 Total Energy Loss Rate from Interactions with Matter

Combining the energy loss rates due to pion production (5.3.28/29), Coulomb (5.3.40) and ionization (5.3.51) interactions and introducing the degree of ionization, $x = n_e/n_{HI}$, we obtain for the total energy loss rate of cosmic ray protons (A = 1) in interactions with matter

$$-\left(\frac{dT}{dt}\right)_{\text{proton}}(\beta) = 1.82 \times 10^{-7} [n_{HI} + 2n_{H_2}]$$
$$\times \left[[1 + (0.0185 \ln \beta H[\beta - 0.01])] \frac{2\beta^2}{10^{-6} + 2\beta^3} + 1.69 \frac{xn_{HI}}{n_{HI} + 2n_{H_2}} \right.$$
$$\times H\left[\beta - 7.4 \times 10^{-4}(T_e/2 \times 10^6 K)^{1/2}\right] \frac{\beta^2}{\beta^3 + 2.34 \times 10^{-5}(T_e/2 \times 10^6 K)^{3/2}}$$
$$\left. + 0.72\gamma^{1.28} H[\gamma - 1.3] \right] \text{ eV s}^{-1} . \tag{5.3.58}$$

In order to understand the relative importance of the different energy loss processes in the local interstellar medium we have calculated the individual loss rates by using parameters as measured near the solar system in the local galactic neighborhood. Fig. 5.15 shows the result. Whereas Coulomb losses play only a minor role, ionization losses dominate the total energy loss rate below the kinetic energy T_k = 450 MeV, corresponding to a velocity β_k = 0.74, whereas pion production losses become dominant above T_k. It

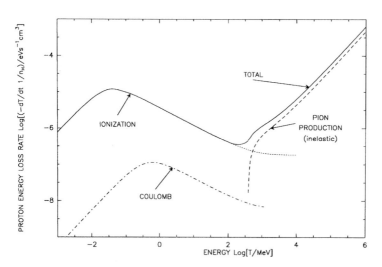

Fig. 5.15. Total energy loss rate of cosmic ray protons in the local interstellar medium by pion production, Coulomb and ionization interactions taking a neutral gas density $n_{HI} + 2n_{H_2} = 1.14$ cm^{-3} and a thermal electron distribution of density $n_e = 0.01$ cm^{-3} and temperature $T_e = 3\times10^5$ K

should be noted that the value of T_k is independent of the value used for the interstellar gas density, since both ionization and pion production scale with the gas density, and are only determined by the ratio of the respective interaction cross-sections. For the associated total loss time scale we find

$$\tau_{\text{proton}} \equiv T/[\,|dT/dt|\,] \simeq \frac{1}{[n_{\text{HI}} + 2n_{\text{H}_2}]}$$
$$\times \begin{cases} 45 \text{ yr} & \text{for } T \leq T_0 = 45 \text{ keV} \\ 3.8 \times 10^7 \, (T/T_k)^{1.5} \text{ yr} & \text{for } T_0 \leq T \leq T_k = 550 \text{ MeV} \\ 10^8 \, (\gamma/10)^{-0.28} \text{ yr} & \text{for } T \geq T_k \end{cases}, \quad (5.3.59)$$

which is comparable to the catastrophic fragmentation time scale (5.3.56) at relativistic energies.

The total energy loss rate for higher metalicity ions ($Z \geq 2$) in matter follows readily from (5.3.58) using (5.3.52) and (5.3.34). We turn now to the observational consequences of cosmic ray nucleus interactions.

5.3.11 Ionization and Heating Rate of Interstellar Matter by Cosmic Rays

As a result of ionization interactions of neutral interstellar matter with cosmic ray nuclei (Sect. 5.3.8.2) the former becomes ionized. According to Heitler (1954 [226]) the average energy loss is also, with fairly good accuracy, representative of the average primary ionization of the particle. The fractional number of cases in which a collision with an atom results in ionization rather than excitation to a discrete level (discussed in Sect. 5.3.7) is almost independent of the energy of the primary, and the same is true for the average energy transferred to the ionized electron. Thus the number of primary ion pairs formed per time interval is very nearly proportional to the average loss by collision given in (5.3.42). The number of primary ion pairs can be inferred from the fact that about one ion pair is formed when the primary loses an amount of 32 eV. This figure is practically independent of the nature of the particle and its energy (at least if the latter is not excessively small). It should be noted that the secondary particles produce further ion pairs and the total ionization differs from the primary ionization (see factor η_1 below).

The primary ionization rate per unit volume per unit time interval at galactic location $\boldsymbol{r}$ caused by cosmic ray protons of differential number density $N(\boldsymbol{r},T)$ is then obtained by integrating the average ionization loss rate of cosmic ray protons from (5.3.51) multiplied with $N(\boldsymbol{r},T)$ over all kinetic energies and dividing by 32 eV, yielding

$$f_{\text{CR}}(\boldsymbol{r}) = \int_{13.6 \text{ eV}}^{\infty} dT \left|\frac{dT}{dt}(\boldsymbol{r},T)\right| N(\boldsymbol{r},T)/(32 \text{ eV}) \text{ cm}^{-3} \text{ s}^{-1}. \quad (5.3.60)$$

The total cosmic ray ionization rate is then calculated from (5.3.60) as

$$\xi_{\rm CR}(r) = \eta_1\eta_2\eta_3 f_{\rm CR}(r)/[n_{\rm HI}(r) + 2n_{\rm H_2}(r)]\ {\rm s}^{-1}\,, \tag{5.3.61}$$

where according to Takayanagi (1973 [538]) $\eta_1 = 5/3$ accounts for secondary ionization from electrons produced in the ionization by the primary cosmic ray particle (Dalgarno and Griffing 1958 [116]), $\eta_2 \simeq 1.17$ accounts for a 10% helium abundance by number density in the interstellar medium, and $\eta_3 \simeq 1.43$ accounts for the contribution of heavy cosmic ray nuclei and cosmic ray electrons and positrons. Strictly, (5.3.60) and (5.3.61) apply only for the ionization of neutral HI gas. However, it may also be used for estimating the ionization rate in a completely ionized HII gas (Spitzer 1948 [505], Ginzburg 1969 [186]).

The heating rate $\Gamma_{\rm CR}(r)$ of matter by cosmic rays is obtained by multiplying the total ionization rate (5.3.61) with the matter density and the energy $\Delta Q \simeq 20$ eV that is finally deposited as heat from each ionization process (Spitzer and Scott 1969 [506], Goldsmith and Langer 1978 [190]). With (5.3.60) we then obtain

$$\begin{aligned}\Gamma_{\rm CR}(r) &= \Delta Q \xi_{\rm CR}(r)\left[n_{\rm HI}(r) + 2n_{\rm H_2}(r)\right] \\ &= (5/8)\eta_1\eta_2\eta_3 \int_{13.6\ {\rm eV}}^{\infty} {\rm d}T \left|\frac{{\rm d}T}{{\rm d}t}(r,T)\right| N(r,T)\ {\rm erg\ cm^{-3}\ s^{-1}}\,.\end{aligned} \tag{5.3.62}$$

The evaluation of (5.3.61) and (5.3.62) in the Galaxy requires knowledge of the interstellar cosmic ray proton energy spectrum down to energies of 13.6 eV, in a range where it is not accessible to direct observations because of the solar modulation.

5.3.12 Continuum Radiation Processes of Relativistic Nuclei

Cosmic ray nuclei in distant parts of the universe may be detected by their radiation products: photons and neutrinos. Neutrino astronomy has just started to provide detectable fluxes with the few neutrinos measured at the explosion of SN 1987A in the Large Magellanic Cloud with the Kamiokande neutrino detector.

Electromagnetic radiation from cosmic ray nuclei lies in the γ-ray band and results from excitation of target nuclei by subrelativistic nuclei (Sect. 5.3.7) and from neutral pion production by 1–30 GeV nuclei (Sect. 5.3.3) with subsequent decay into two γ-rays. The radiation spectrum of the latter process reveals a remarkable symmetry around $m_{\pi^0}c^2/2 \simeq 70$ MeV, and detection of this characteristic spectrum would be a strong argument in favor of a cosmic ray nucleus origin of this radiation. Unfortunately, as we have discussed in Sect. 5.3.4, nuclei also produce secondary charged pions which decay into electrons and positrons, the bremsstrahlung of which is also emitted in the MeV–GeV photon energy range, and may mask the π^0-bump at 70 MeV. So non-detection of the π^0-bump at 70 MeV does not naturally argue against a nucleonic origin of these γ-rays, but indicates the presence of many secondary electrons and positrons which were also generated by cosmic ray nuclei in their interaction with matter (see Schlickeiser (1982c [455]) for further discussion of this point). Hence, as nonthermal synchrotron radiation is indicative of the presence of cosmic ray electrons, γ-rays from π^0-decay (as well as from bremsstrahlung radiation of secondary electrons and positrons from charged pion decay) are indicative of the presence of cosmic ray nuclei. In the following chapters we will therefore summarize the main results from radio and γ-ray astronomy with respect to this question.

6. Indirect Observations of Cosmic Rays

The measured frequency spectrum of electromagnetic radiation from a cosmic object provides clues on the presence and the properties of the radiating cosmic ray particles once the radiation process and the source geometry are specified. As discussed in Chaps. 4 and 5 for cosmic ray electrons synchrotron radiation in the radio wavelength band and for cosmic ray nuclei emission at γ-ray frequencies are most decisive. In the following we describe the results from radio and γ-ray astronomy concerning the origin of cosmic rays.

6.1 Clues from Radio Astronomy

Synchrotron radiation of relativistic electrons in ambient magnetic fields is not the only continuum radiation process operating at radio frequencies ($\nu \lesssim 10^{11}$ Hz). A competing process is radiation due to the deflection of a charge in the Coulomb field of another charge in a plasma, commonly referred to as *free–free emission* or *thermal bremsstrahlung*.

6.1.1 Free–Free Emission

Calculations of the emission coefficient $\epsilon_{\rm ff}(\boldsymbol{r},\nu)$ and the absorption coefficient $\kappa_{\rm ff}(\boldsymbol{r},\nu)$ for this process from a plasma of temperature T, metallicity Z and thermal electron and ion densities $n_{\rm e}(\boldsymbol{r})$ and $n_{\rm i}(\boldsymbol{r})$, respectively, are reviewed in Oster (1961 [378]) and Bekefi (1966 [39], Chap. 3), and can be well approximated by (e.g. Rybicki and Lightman 1979 [441], Chap. 5)

$$\epsilon_{\rm ff}(\boldsymbol{r},\nu) \simeq 6.8 \times 10^{-38} Z^2 n_{\rm e}(\boldsymbol{r}) n_{\rm i}(\boldsymbol{r}) T^{-1/2}$$
$$\times (\nu/{\rm GHz})^{-0.1} \exp\left(-\frac{h\nu}{k_{\rm B}T}\right) \; {\rm erg\; cm^{-3}\; s^{-1}\; Hz^{-1}} \quad (6.1.1)$$

and

$$\kappa_{\rm ff}(\boldsymbol{r},\nu) \simeq 2.665 \times 10^{-20} Z^2 (T/K)^{-1.35} (n_{\rm e}(\boldsymbol{r})/{\rm cm}^{-3})$$
$$\times (n_{\rm i}(\boldsymbol{r})/{\rm cm}^{-3})(\nu/{\rm GHz})^{-2.1} \; {\rm cm}^{-1}\,. \quad (6.1.2)$$

Equation (6.1.2) implies a free–free optically depth (see A.1 for definition) (Mezger and Henderson 1967 [350])

$$\tau_{\mathrm{ff}} \equiv \int_0^\infty dr \kappa_{\mathrm{ff}} \simeq 8.235 \times 10^{-2} (T/K)^{-1.35} (\nu/\mathrm{GHz})^{-2.1} \left(\frac{EM}{\mathrm{pc\ cm}^{-6}}\right), \tag{6.1.3}$$

where the emission measure $EM = \int_0^\infty dr n_i n_e$ has dimensions pc cm^{-6}. It is convenient to introduce the turnover frequency ν_t, where the optical depth $\tau_{\mathrm{ff}}(\nu_t) = 1$ is unity,

$$\nu_t \simeq 0.3[(T/K)^{-1.35} EM]^{1/2}\ \mathrm{GHz}\,. \tag{6.1.4}$$

For frequencies less than ν_t free–free emission is optically thick (see A.7) whereas at larger frequencies free–free emission is optically thin.

If free–free emission is the only relevant radiation process in a given source the solution (A.4) of the equation of radiation transfer in the optically thin (A.6) case with zero foreground and background emission yields

$$I_{\mathrm{ff}}(\nu > \nu_t, l, b) = \int_0^\infty dr \epsilon_{\mathrm{ff}}(\nu, l, b, r)$$

$$\propto EM\, Z^2 T^{-1/2} \nu^{-0.1} \exp\left(-\frac{h\nu}{k_B T}\right), \tag{6.1.5}$$

a flat intensity distribution with respect to frequency up to photon energies comparable with $k_B T$, whereas in the optically thick case the intensity distribution approaches the Rayleigh-Jeans approximation of Planck's law

$$I_{\mathrm{ff}}(\nu < \nu_t, l, b) = \epsilon_{\mathrm{ff}}/\kappa_{\mathrm{ff}} = \frac{2 k_B T}{c^2} \nu^2\,. \tag{6.1.6}$$

This intensity spectrum can be distinguished clearly from the usually steeper power laws (4.1.21a) $\propto \nu^{-\alpha}$ provided by optically thin synchrotron radiation. Moreover free–free emission is not linearly polarized.

Figure 6.1 shows a collection of radio-frequency spectra from selected sources in the sky. The thermal bremsstrahlung nature of the radio emission from the Orion nebula and the synchrotron origin of the radio emission from the supernova remnant Cas A follow immediately. However, the spectra from the Seyfert galaxy NGC 1275 and the two pulsars indicate that more complex spectra with more than one component and radiation process can occur.

If, in the Galaxy, the cosmic ray electrons, the galactic magnetic field, and the ionized component of the interstellar medium are well mixed, the total received intensity at the position of the solar system is determined by the sum of the respective emission and absorption coefficients for synchrotron radiation and free–free emission according to (A.2). The measured diffuse radio spectrum from the anticenter ($l \sim 180°$) direction of our Galaxy is displayed in Fig. 6.2, and can be well understood by optically thin synchrotron emission of cosmic ray electrons in the galactic magnetic field at frequencies above 10 MHz, whereas at lower frequencies the synchrotron emission is modified by free–free absorption, i.e.

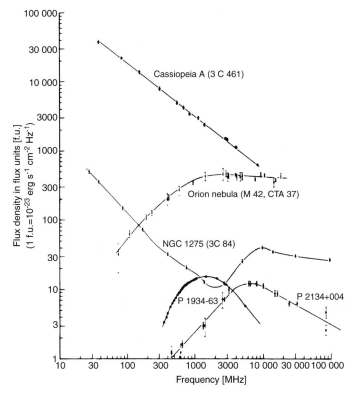

Fig. 6.1. Radio-frequency spectra of sources exhibiting the power law spectrum of optically thin synchrotron radiation (Cassiopeia A), the flat spectrum of free–free emission with low-frequency self-absorption (Orion Nebula), a high-frequency free–free emission component superposed on a low-frequency optically thin synchrotron radiation component (NGC 1275), and low-frequency absorption processes in the two pulsars P 1934−63 and P 2134+004. From Lang (1974 [283])

$$I_{\text{anticenter}} \propto \epsilon_{\text{synchrotron}}/\kappa_{\text{ff}} \propto \nu^{2.1-\alpha} \ . \tag{6.1.7}$$

As mentioned above in Sect. 3.2 this anticenter radio spectrum has been used to infer the local interstellar cosmic ray electron spectrum at energies > 200 MeV to account for the influence of solar modulation.

6.1.2 Radio Continuum Surveys

If properly corrected for free–free emission the measured radio continuum emission can be used to trace cosmic ray electrons in the universe at locations different from our local interstellar neighborhood. It is reasonable to expect that the diffuse distributed emission stems from cosmic ray electrons emitting in the ambient galactic or extragalactic magnetic fields, whereas compact

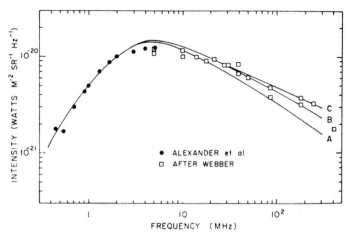

Fig. 6.2. Radio background spectrum from the galactic anticenter measured by Alexander et al. (1970 [11]) and Webber (1968 [558]). **Curves A, B, C** are the expected intensity distributions from synchrotron radiation of cosmic ray electrons with a power law spectral index 1.8 below 1, 2, and 5 GeV for A, B, and C, and 2.5 above in the galactic magnetic field of 5 μG in a two-component interstellar medium model of length 4 kpc consisting of cold clouds of 0.05 cm^{-3} and 60 K and an intercloud medium of electron density 0.03 cm^{-3} and 4000 K, providing the free–free absorption of radio waves below 10 MHz, as calculated by Goldstein et al. (1970 [192]). Similar calculations by Cummings et al. (1973 [114]) have led to the nominal electron spectrum shown in Fig. 3.8. From Goldstein et al. (1970 [192])

sources of synchrotron radiation may reveal the birth places of these cosmic ray particles.

6.1.2.1 Total Intensity Surveys. The first all-sky total intensity survey made by Dröge and Priester (1956 [135]) combined their own northern hemisphere 200 MHz observations with those of the southern sky by Allen and Gum (1950 [12]). Since then only three independent attempts to prepare all-sky maps of the cosmic radio emission have been made (Salter and Brown 1988 [443]):

Landecker and Wielebinski (1970 [282]) combined data between 85 and 178 MHz at various resolution to produce a 150 MHz picture of $\sim 3°$ resolution. At lower frequency, Cane (1978 [90]) used the northern hemisphere survey of Milogradev-Turin and Smith (1973 [360]) together with the southern sky observations to obtain a 30 MHz map of about 9° resolution. Haslam et al. (1982 [217]) combined data taken at a single frequency (408 MHz), standardized at a single resolution (0.85°), from three of the world's largest radio telescopes. This survey is shown in Fig. 6.3.

The global properties of the cosmic radio emission are clearly visible. The galactic center serves as an approximate center of symmetry for the entire sky, with the low-latitude intensity dropping away steeply on each side up to $l \sim 60°$ and $l \sim 280°$. Such a behavior is to be expected for the emission

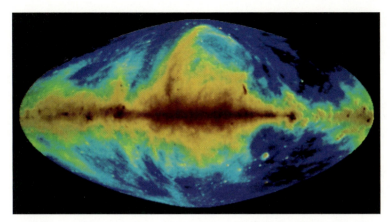

Fig. 6.3. The all-sky radio continuum emission at 408 MHz smoothed to an angular resolution of $2° \times 2°$. From Haslam et al. (1982 [217])

from the galactic disk seen from the special location of the solar system in our Galaxy. Substantial emission is present over the whole sky, but its distribution is far from uniform. Large-scale prominences rising from low latitudes and extending nearly to the poles in both hemispheres, distorting the uniformity of the high-latitude emission. These are known as the loops and spurs. The most prominent is the North Polar Spur starting at $l \sim 30°$. One also notes external galaxies such as the Magellanic Clouds, the Andromeda nebula, Cen A as sources in this survey. The observed distribution of the cosmic radio radiation is often described in terms of a superposition of a number of components: the galactic center region; the galactic disk centered on the galactic center; a population of discrete galactic sources; relatively local features comprising the loops and spurs; extragalactic point sources; a possible galactic halo symmetric about the galactic disk; and an isotropic contribution due to unresolved extragalactic systems. At comparable angular resolution the northern sky is also completely covered at 820 MHz (Berkhuijsen 1972 [44]) and at 1.4 GHz (Reich and Reich 1986 [426]). There is ongoing work to cover the southern sky both at 1.4 and 2.7 GHz. These higher frequency surveys reproduce all nonthermal features seen at lower frequencies but are superior in outlining thermal bremsstrahlung features.

6.1.2.2 Surveys of Linear Polarization of the Galactic Emission. Polarization of the galactic continuum emission was first clearly demonstrated by Westerhout et al. (1962 [563]). This detection was the final evidence to confirm the synchrotron radiation theory for the galactic nonthermal emission. In succeeding years, polarization surveys were made between 240 and 2695 MHz (Berkhuijsen 1975 [45], Brouw and Spoelstra 1976 [76], Spoelstra 1984 [508], Junkes et al. 1987 [253]) for the northern sky plane, and by Mathewson and Milne (1965 [328]) at 408 MHz for the southern sky. Since free–free emission is unpolarized this provides an unambiguous detection of celestial synchrotron radiation and, if the synchrotron radiation origin is accepted, estimates of the uniformness and the direction of the galactic magnetic field and/or of the influence of Faraday depolarization effects in different parts of the galactic plane (see Sect. 4.1.2) are possible. Figure 6.4 shows part of the survey by Junkes et al. (1987 [253]). Quite generally, the polarized intensities are distributed differently from the total intensity picture also shown in the figure, suggesting a rather clumpy and irregular distribution of both the thermal and nonthermal emission regions.

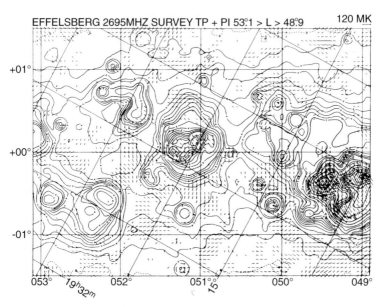

Fig. 6.4. The atlas of contour maps with polarization bars in galactic coordinates. Total intensity contour lines refer to radiation temperature and are labeled in K as relative main beam brightness temperature. The scale of polarization intensity is indicated in the length of bars with the scale indicated in the **upper right corner** of the map. The bars point to the E-field direction, so that the galactic magnetic field direction can be inferred by rotating by 90 degrees assuming no Faraday rotation. The angular resolution is 6'. From Junkes et al. (1987 [253])

6.1.2.3 Implications for Cosmic Ray Electrons. *These surveys provide overwhelming evidence that cosmic ray electrons are present in other parts of the universe: in our Galaxy with number densities comparable to those found from direct observations in our local galactic neighborhood.* This is the result of sophisticated modeling these surveys with reasonable assumptions concerning the distribution and the strength of the galactic magnetic field in the Galaxy, and assuming that the *spectral shape* of the cosmic ray electron distribution does not vary with galactic position, so that, for example, (4.1.20) hold. These calculations have usually been of two types. Either the distributions of cosmic ray electrons and the galactic magnetic field have been prescribed and the resulting radio intensity distribution has been evaluated and compared with measurements in a trial-and-error calculation (e.g. Price 1974 [407], Jäkel et al. 1977 [238]), or the observed profile has been "unfolded" directly under assumed spatial symmetries to give the distribution of synchrotron emissivity, which is the product of cosmic ray electron density times some power of the magnetic field strength (see (4.1.20a)) (Philipps et al. 1981a [398], 1981b [399]; Beuermann et al. 1985 [49]). Besides the clue on the presence of cosmic ray electrons over the whole Galaxy these studies provided important evidence for the relation between nonthermal radio emission and spiral structure as first established by Mills (1959 [356]).

The underlying assumption of these studies, that the shape of the energy distribution of the radiating cosmic ray electrons does not change with galactic position, can be tested by combining total intensity surveys at two wavelengths to deduce any celestial variation of the synchrotron radiation spectral index α. Reich and Reich (1988 [427]) have presented a map of spectral indices of the northern sky based on the surveys at 408 and 1420 MHz with an angular resolution of $2°$ and an absolute spectral index error of ~ 0.1. They find significant variations of the spectral index of the brightness temperature $T_{\rm b} \propto \nu^{-\beta}, \beta = 2 + \alpha$, along the galactic plane and from the plane toward higher galactic latitudes. Most noteworthy is the reported *flattening* of spectral indices with increasing latitude both in the inner and outer Galaxy (Fig. 6.5). Towards the outer Galaxy $\beta(|b| = 0°) = 2.7$–2.8 near the plane and β reduces to 2.5–2.6 at $|b| \simeq 30°$. After correcting for the thermal free–free emission in the plane Reich and Reich (1988 [427]) have found that the nonthermal spectral index $\beta_{\rm nth}$ flattens from a value of 2.85 in the plane $(|\beta| = 0°)$ by $\Delta\beta = 0.35 \pm 0.2$ with increasing latitude. Towards the inner Galaxy the nonthermal spectral index changes from $\beta_{\rm nth} \simeq 3.1$ in the plane to values of $\beta_{\rm nth} \simeq 2.7$ at high latitudes.

External galaxies have been already mentioned in Sect. 2.4. With respect to cosmic ray electrons the situation seems to be comparable to our own Milky Way: *cosmic ray electrons are a common constituent of galaxies*, and exhibit themselves by nonthermal synchrotron radiation. Detailed studies of synchrotron intensity and spectral index variations exist for several individual galaxies, and have become an important diagnostic tool for the nonthermal properties of galaxies.

With respect to the space between galaxies, however, a different picture emerges. The discovery of the universal 2.7 K background radiation has made it clear (Fazio et al. 1966 [157]) that *cosmic ray electrons cannot fill the whole universe uniformly*. The inverse Compton losses with the microwave photons would give a significant X-ray background radiation which is not observed. This establishes cosmic ray electrons as a phenomenon associated with galaxies. Their birth places have to be galactic objects. They can then propagate enormous distances, as the jets and hotspots in active galaxies show, and the

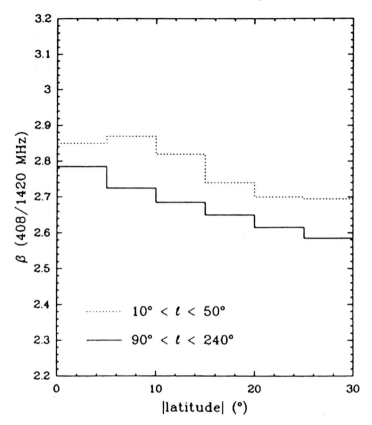

Fig. 6.5. Latitude distributions of the spectral index β of the galactic radio emission at 408 and 1420 MHz ($T_b \propto \nu^{-\beta}$) in the inner and outer parts of the Galaxy. From Bloemen et al. (1988 [65])

huge radio halos around edge-on galaxies and in clusters of galaxies, such as the Coma cluster (Schlickeiser et al. 1987 [475]), indicate. But they cannot permeate the whole universe uniformly. There is no dependence of the presence of cosmic ray electrons on the type of galaxies: nonthermal synchrotron radiation is fairly common in spiral galaxies, elliptical galaxies, dwarf galaxies, active galaxies as Seyferts, BL-Lacs and quasars. Although the synchrotron radiation spectra differ from object to object, they unambiguously trace the presence of cosmic ray electrons (and magnetic fields) in those distant parts of the universe. The measured frequency spectra of synchrotron radiation from cosmic objects provide clues on the energy distribution of the radiating cosmic ray electrons once effects due to the source geometry and radiation transfer have been properly taken into account. Radio spectra from normal galaxies (Klein and Emerson 1981 [264]) and clusters of galaxies (Andernach et al. 1981 [13]) exhibit power law energy spectra with and without breaks, whereas shell-type supernova remnants in our Galaxy and the Magellanic

Clouds show mainly power laws. Figure 6.6 shows the integrated radio flux spectra from 15 spiral galaxies: four of them exhibit a break from one power law to a steeper one, 11 exhibit a single power law in frequency, corresponding to a single power law for the radiating particles according to (4.1.21a). Figure 6.7 shows the distribution of the power law spectral index α ($S_\nu \propto \nu^{-\alpha}$) from a sample of bright spiral galaxies (Gioia et al. 1982 [187]). The distribution has a mean of $\langle \alpha \rangle = 0.74$ with a dispersion of about 30 %, and this seems to be independent of the morphology and/or the evolutionary stage of a galaxy.

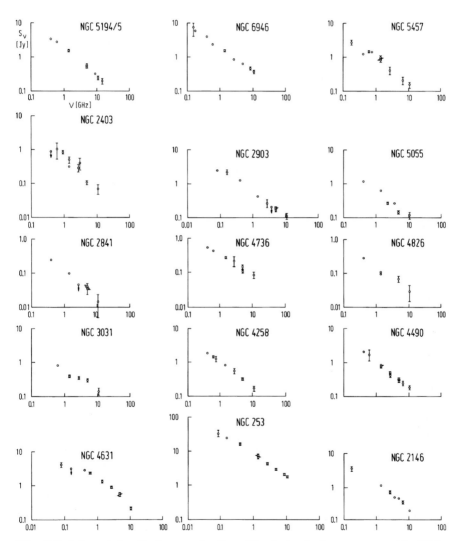

Fig. 6.6. Integrated radio flux spectra from 15 spiral galaxies as measured by Klein and Emerson (1981 [264])

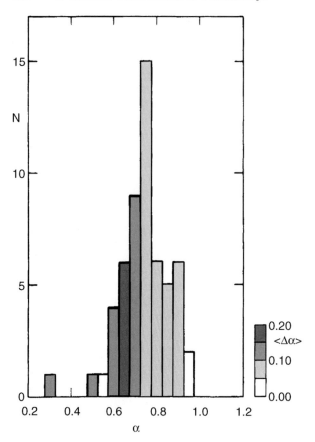

Fig. 6.7. Distribution of radio spectral indices of bright spiral galaxies between 408 MHz and 10.7 GHz. From Gioia et al. 1982 [187]

Figure 6.8 shows a similar distribution of the radio power law spectral indices from shell-type supernova remnants in our Galaxy (Clark and Caswell 1976 [102], Milne 1979 [359]) and in the Magellanic Clouds (Mathewson et al. 1983 [329], Mills et al. 1984 [357]). This distribution has a mean of approximately $\langle \alpha \rangle = 0.5$ with a standard deviation of $\Delta \alpha = 0.15$. As Fig. 6.8 indicates the spectral indices are independent of the evolutionary stage of the individual supernova remnant. It has been argued (Dröge et al. 1987 [137]) that the difference in the mean values (0.74–0.5 = 0.24) between the samples of shell-type supernova remnants and bright spiral galaxies is understandable in terms of energy-dependent propagation effects of relativistic electrons in the galaxies after they have left their sources (likely supernova remnants, see discussion in Chap. 7).

Active galaxies as BL-Lac objects and quasars exhibit power law radiation spectra with high-frequency cutoffs (Rieke et al. 1982 [434], Ennis et al.

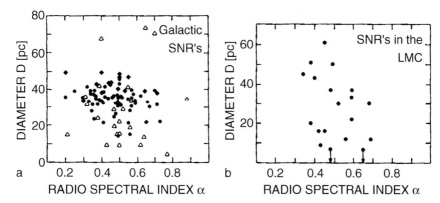

Fig. 6.8. Diameter versus radio spectral index of shell-type supernova remnants in **(a)** our Galaxy and **(b)** the Large Magellanic Cloud as collected by Bogdan et al. (1985 [72]). In **(a)** the **filled circles** and **diamonds** are taken from Clark and Caswell's (1976 [102]) Tables 1 and 2, respectively, and the **open triangles** are from Göbel et al. (1981 [189]). Data in **(b)** are from Mills et al. (1984 [357]). From Dröge et al. (1987 [137])

1982 [152], Cruz-Gonzales and Huchra 1984 [111], Röser and Meisenheimer 1986 [438], Meisenheimer et al. 1989 [337]). Due to their higher magnetic field strength as compared to normal galaxies the synchrotron radiation extends into the mm, infrared, and optical frequency band, in case of Pictor A (Röser 1989 [437]) even to X-ray frequencies. However, there is observational evidence for temporal evolution of relativistic electron spectra in at least some of these sources: if interpreted as optically thin synchrotron radiation, the BL-Lac object AO 0235+164 exhibited a nearly monoenergetic electron spectrum in the initial phases of its outburst in 1975 which, after a few days, developed into an E^{-2}-power law due to synchrotron radiation losses (Rieke et al. 1982 [434]). A similar behavior was found during the 1982 mm outburst of 3C 273 (Robson et al. 1983 [435]). A stationary monoenergetic distribution function of relativistic electrons has been found in our galactic center, including Sgr A* and the extended Bridge and Arc components (Reich et al. 1988 [428], Lesch et al. 1988 [300]).

This short summary of synchrotron radiation measurements indicates that no single unique particle distribution function is present in different objects. Rather, the observations point to a variety of energy spectra although power law energy spectra are dominant. The explanation of this diversity of cosmic relativistic electron spectra, with dominance of power laws, is one important task of cosmic ray origin theories.

6.2 Clues from γ-Ray Astronomy

Continuum and line γ-radiation is produced in interactions of energetic cosmic ray particles with ambient target matter and photon gases. It, therefore, traces the global spatial distribution of these particles, knowing the distribution of the target fields. Much as continuum radio astronomy has been used to infer the global distribution of cosmic ray electrons, a prime motivation for the development of γ-ray astronomy (Morrison 1958 [367]) has been to use γ-radiation to deduce the global distribution of cosmic ray nucleons. However, there is no clear-cut division between cosmic ray electrons radiating only at radio and not at γ-ray frequencies. On the contrary, cosmic ray electrons are responsible for a substantial part of the galactic γ-ray emissivities, and this "electron contamination" limits the conclusions on the global distribution of cosmic ray nucleons.

The celestial distribution of γ-rays has been measured by experiments on board the satellites OSO-3 (Kraushaar et al. 1972 [274]), SAS-2 (Fichtel et al. 1975 [165]), COS-B (Bignami and Hermsen 1983 [53]) and the Compton Gamma Ray Observatory (Hunter et al. 1997 [233]). The first three experiments employed the spark-chamber technique where the incoming γ-ray is converted in an absorber into an electron-positron pair, whose measured properties allow the determination of the direction and energy of the incoming photon. While OSO-3, due to its small size, recorded during its whole mission only 800 γ-rays, the larger experiments on SAS-2 and COS-B collected 8000 and more than 100 000 photons, respectively, during their operation, and thus provided firm statistical bases for the investigation of the γ-ray sky. These two experiments operated in the energy band from 35–2000 MeV (SAS-2) and 70–5000 MeV (COS-B). Besides the spark-chamber experiment EGRET the Compton Observatory carried three additional experiments (BATSE, OSSE and COMPTEL) into space which provided a photon energy coverage from 50 keV to 10 GeV. Since its launch in 1991 until the year 2000 this mission measured cosmic gamma-ray photons.

Here we first discuss the Compton Observatory results on γ-radiation with photon energies above 10 MeV. No line emission occurs in this energy range. As we shall see, these results rule out any metagalactic origin of cosmic ray nucleons with energies below 100 GeV. Gamma radiation below 10 MeV is a mixture of line radiation generated by low-energy cosmic ray nucleons and continuum radiation produced by cosmic ray electrons (Ramaty and Murphy 1987 [417]). The electron generated inverse Compton continuum radiation also excludes any metagalactic origin of these particles, and the detected galactic electron bremsstrahlung radiation indicates the abundance of many semirelativistic cosmic ray electrons in the Galaxy.

6.2.1 γ-Ray Emission Above 10 MeV

The γ-ray sky above 10 MeV, as we know it today, is a composite of the emission from point sources and diffuse emission. Figure 6.9 shows the first complete all-sky survey of high-energy (> 100 MeV) γ-ray emission obtained with the EGRET experiment.

The most prominent feature is the galactic plane, in which interactions between cosmic rays and the thermal gas lead to γ-ray emission by π^0-decay and bremsstrahlung (see Sects. 5.3 and 4.7). On top of this diffuse emission we see point sources all over the sky, part being pulsars, another part distant extragalactic objects (see Sect. 6.3), and also a significant fraction of yet unidentified sources. But we also know that the diffuse emission extends out of the galactic plane up to the poles. There appears to be an isotropic emission component which is presumably extragalactic in origin and may be understood as a blend of unresolved extragalactic sources, but there is also considerable galactic emission. At higher latitudes interactions between cosmic rays and thermal gas still play a role, but inverse-Compton scattering of ambient photons by cosmic ray electrons becomes increasingly important.

Fig. 6.9. EGRET all-sky survey of > 100 MeV γ-ray emission

6.2.1.1 The Locality of Cosmic Rays. There has been a long-standing debate on whether cosmic rays in the GeV energy range are galactic or extragalactic (Brecher and Burbidge 1972 [75]). The electrons have to be galactic since the energy losses by comptonization of the microwave background prevent propagation from one galaxy to another. In the case of protons the situation has been improved significantly by the Compton Observatory results. EGRET observations of the Magellanic Clouds have revealed that the γ-ray flux of the Large Magellanic Cloud is weakly inconsistent (Sreekumar et al. 1992 [510]) and that of the Small Magellanic Cloud is strongly inconsistent

(Sreekumar et al. 1993 [511]) with cosmic ray protons having uniform density in space. *Therefore the bulk of the locally observed protons at GeV energies must be galactic.* Neither our neighboring Galaxy M31 nor the nearby starburst galaxies M82 and NGC 253 have been detected by EGRET (Sreekumar et al. 1994 [512]), in accord with detailed model predictions for the flux of these sources being below the EGRET threshold (Pohl 1994 [401], Paglione et al. 1996 [383]).

6.2.1.2 Confusion and Point Sources.
Any analysis of the galactic diffuse emission can be seriously hampered by unresolved galactic point sources which may have a sky distribution similar to that of gas. Because six objects have already been detected, pulsars are the most likely input from discrete sources. Many authors have addressed this problem on the basis of pulsar emission models (Harding 1981 [209], Bailes and Kniffen 1992 [25], Yadigaroglu and Romani 1995 [568], Sturner and Dermer 1996 [531]) and consistently estimate the contribution of pulsars to the diffuse γ-ray intensity above 100 MeV integrated over the whole sky to be a few per cent.

Estimates for the contributions of other discrete sources are very uncertain due to the lack of a clear identification of γ-ray sources with any known population of galactic objects. It is interesting to see that roughly ten unidentified EGRET sources can be associated with supernova remnants (SNRs) or with OB associations, or with both (SNOBs) (Sturner and Dermer 1995 [530], Esposito et al. 1996 [154], Kaaret and Cottam 1996 [254]). Obviously these sources may also be radio-quiet pulsars or highly dispersed radio pulsars.

Especially SNR are likely sites of particle acceleration in our Galaxy in view of the energetics (see also Chap. 7). We know by their synchrotron emission that SNRs accelerate electrons, in the cases of the Crab nebula (DeJager et al. 1996 [119]) and SN1006 (Koyama 1995 [270], Reynolds 1996 [431]) up to TeV energies. However, there is still no clear observational proof to date that the nuclear component of cosmic rays is produced in SNRs. Models predict that the detection of SNRs in the EGRET range is difficult but not impossible, and that detection at TeV energies appears more feasible (Drury et al. 1994 [143]). Prime candidates for detection are SNR overtaking a molecular cloud like, for example, G78.2+2.1 (Aharonian et al. 1994 [10]), which, however, are often located in crowded regions of the sky. As a result of the limited spatial resolution, confusion is the major problem for the identification of SNRs as emitters of GeV γ-rays. At TeV energies emission from SN1006 and SNR RX J 1713.7–3946, reported by the CANGEROO experiment (Tanimori et al. 1998 [541], Muraishi et al. 2000 [371]), have been interpreted by inverse Compton upscattering of ambient microwave background photons.

6.2.1.3 The γ-Ray Emissivity.
In galactic halos inverse Compton scattering of ambient photons by cosmic ray electrons is an important production process for γ-rays. Due to the large scale-height of far-infrared photons and the universal nature of the cosmic microwave background radiation, the latitude distribution of this emission is much broader than in the case of the

bremsstrahlung and π^0-decay, and may reflect directly the latitude distribution of the radiating cosmic ray electrons (Schlickeiser and Thielheim 1977 [471]). Thus the inverse Compton component can tell us about the life and the propagation history of cosmic ray electrons in the Galaxy. By comparison of the spatial distribution of γ-rays, unrelated to the interstellar gas, to the brightness temperature distribution in radio surveys, the average intensity of inverse Compton emission at high latitudes is determined to be $(5.0 \pm 0.8) \times 10^{-6}$ cm^{-2} s^{-1} sr^{-1} for $E > 100$ MeV with a photon spectral index of -1.85 ± 0.17 (Chen et al. 1996 [98]).

To investigate the γ-ray emission originating from π^0-decay and bremsstrahlung, we need some prior knowledge of the distribution of interstellar gas in the Galaxy. This includes not only HI but also H_2, which is indirectly traced by CO emission lines, and HII, which is traced by Hα and pulsar dispersion measurements. Even in the case of the directly observable atomic hydrogen we obtain only line of sight integrals, albeit with some kinematical information. Any deconvolution of the velocity shifts into distance is hampered by the line broadening of individual gas clouds and by the proper motion of clouds with respect to the main rotation flow. The distance uncertainty will, in general, be around 1 kpc.

Thus, it may be appropriate to use only a few resolution elements and investigate the γ-ray emissivity per H-atom in galactocentric rings, i.e. to assume azimuthal symmetry for the emissivity. This kind of approach has

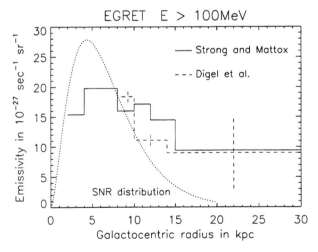

Fig. 6.10. A comparison between the γ-ray emissivity gradient (**solid histogram**) to the distribution of SNR as possible acceleration sites (**dotted line**). The statistical uncertainties of the gradient are typically below 10%. The obvious discrepancy implies that either SNR are not accelerating the bulk of GeV cosmic rays, or diffusive re-acceleration is operative, or galactic cosmic rays are confined on a scale of many kpc's. Note that locally derived emissivities (**dashed histogram**) can differ significantly from the global trend. From Strong and Mattox (1996 [528])

already been successfully applied to the earlier COS-B data. When repeated for the analysis on EGRET data at γ-ray energies above 100 MeV one finds (Strong and Mattox 1996 [528]) that the γ-ray emissivity per H-atom declines with galactocentric radius, confirming earlier COS-B results. In Fig. 6.10 we compare this gradient to the distribution of SNR (Case and Bhattacharya 1996 [93]) as possible acceleration sites for the cosmic rays. Obviously the cosmic ray gradient derived from the γ-ray data is rather weak compared to the SNR distribution, which implies that, if SNR are the sources of galactic cosmic rays, either diffusive re-acceleration is operating, or that the spatial propagation disperses the distribution of cosmic rays from that of their sources by many kpc's with obvious consequences for the halo size in diffusion models, or a galactic variation of the H_2 column density to ^{12}CO intensity ratio (Sodroski et al. 1995 [501]).

It should be noted that the emissivity gradient does not necessarily hold locally. A comparison of the γ-ray emissivity in the Perseus arm at 3 kpc distance in the outer galaxy to the Cepheus and Polaris flare at 250 pc gives a difference of a factor 1.7 ± 0.2 (Digel et al. 1996 [131]), far exceeding the overall gradient (see Fig. 6.10).

6.2.1.4 Spectral Information. The observed γ-ray spectrum of emission related to the gas is difficult to understand. It had been hoped that if the electron contribution is negligibly small, the measured γ-ray energy spectrum would exhibit the characteristic π^0-decay spectrum (see Fig. 5.7) with its symmetry around 70 MeV. Already from the COS-B observations it has become obvious that the observed spectrum does not exhibit the characteristic π^0-decay hump at 70 MeV (see Fig. 6.11).

As noted by Cesarsky and Paul (1981 [96]), with this measurement γ-ray astronomy has failed to prove beyond any doubt the presence of cosmic ray nucleons throughout the galactic disk. The emphasis clearly lies on the words "beyond any doubt" since there is a very good physical argument why pure π^0-decay γ-ray spectra should not occur: it has been pointed out by Schlickeiser (1982c [455]) that the same cosmic ray nucleons that generate the neutral pions, also produce charged pions $\pi^\pm$ which, as we discussed in Sect. 5.3.4, decay via muons into secondary electrons and positrons, and those secondaries will produce nonthermal bremsstrahlung γ-rays, if they are confined long enough in the galactic disk. Apparently, cosmic ray nuclei in our Galaxy use both channels of γ-ray production in the galactic disk: via π^0-production and via secondary electron and positron production and subsequent nonthermal bremsstrahlung. The non-occurrence of pure π^0-decay γ-ray spectrum does not argue against a predominantly nucleonic origin of the diffuse galactic γ-ray emission. In Fig. 6.12 we show the total differential source function of γ-rays per interstellar gas atom by cosmic ray nuclei, which fit the observed COS-B γ-ray spectrum rather well.

For the integral γ-ray source function above 100 MeV Schlickeiser (1982c [455]) estimated $(2.1 \pm 0.2) \times 10^{-25} n_H$ cm^{-3} s^{-1}, if all produced secondary

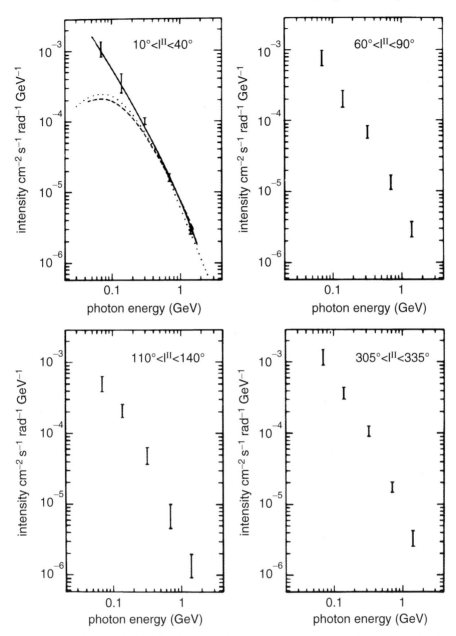

Fig. 6.11. Observed COS-B differential γ-ray energy spectrum from the galactic disk compared to various predictions for the diffuse spectrum. From Paul et al. (1978 [392])

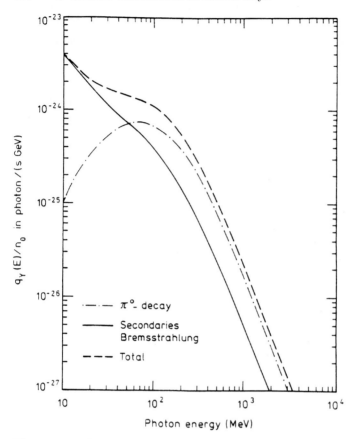

Fig. 6.12. Differential source function of γ-rays per interstellar gas atom from π^0-decay and nonthermal bremsstrahlung from cosmic ray nucleon secondaries. The sum of both contributions is the source function from cosmic ray nucleon interactions. From Schlickeiser (1982c [455])

electrons and positrons radiate γ-rays, and this value agrees well with the emissivity value of 2.5×10^{-25} cm^{-3} s^{-1}, derived from observations of the γ-ray intensity and of the interstellar gas content at large galactic latitudes, which stems from the first few hundreds of parsecs along the line of sight, assuming that the local interstellar neighborhood cosmic ray flux equals that measured near the solar system. So, despite the non-occurrence of the pure π^0-decay γ-ray spectrum, which can be accounted for by including the contribution from the secondary electron and positron bremsstrahlung production channel, the observed diffuse γ-ray spectra before the Compton Observatory era seemed to be compatible with a predominantly nucleonic origin.

However, the recent EGRET measurements of the diffuse galactic γ-ray photon energy spectrum from different directions (Hunter et al. 1997 [233]) deviate significantly from the expected superposition of π^0-decay and bremsstrahlung (see also Fig. 6.14). At energies above 1 GeV there is a clear excess of emission which may be only slightly relaxed by variation of the input proton spectrum to the pion production process (Mori 1997 [365]). This excess can not be explained as superposition of unresolved sources, since pulsars (the only known candidates with an appropriate spectrum) fail to provide the required intensity and latitude distribution (Pohl et al. 1997 [405]). The excess at higher γ-ray energies may also indicate that the average spectrum of high energy electrons is different from what we observe locally. This would be a natural consequence if electrons were not accelerated isotropically throughout our Galaxy, but in discrete objects like SNRs (Cowsik and Lee 1979 [107]).

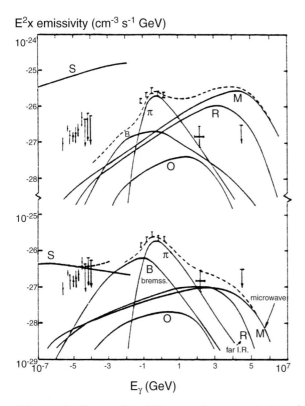

Fig. 6.13. The predicted X-ray and γ-ray emissivity from various production processes: inverse Compton-interactions with the microwave background (**M**), far infrared radiation (**R**), starlight (**O**); synchrotron radiation (**S**); nonthermal bremsstrahlung (**B**); the decay of neutral pions produced in nuclear interactions of cosmic rays (π). The total emissivity is shown as the **dashed curve**. The predictions are compared with the derived emissivity from observations in different energy bands. From Protheroe and Wolfendale (1980 [409])

The production of diffuse γ-rays over a wider energy band than that accessible to the satellite γ-ray detectors has been considered by Protheroe and Wolfendale (1980 [409]). These authors include the modification of the injected source power law electron spectrum by energy loss processes in the interstellar medium. Their predicted X-ray and γ-ray emissivity of photon energies from 100 eV to 10^{15} eV is shown in Fig. 6.13. One notices that at large γ-ray energies eventually the inverse Compton scattering of the microwave and far infrared background radiation becomes dominant, before the Klein-Nishina cutoff (see Sect. 4.2.2) at $E_\gamma \sim 10^{13}$ eV sets in.

6.2.2 Continuum γ-Ray Emission Below 10 MeV

The OSSE and COMPTEL instruments have provided us with the first clear picture of the galactic plane in the light of ≃0.1–10 MeV photons. In this photon band the observed emission is a mixture from continuum radiation processes of cosmic ray electrons and line emission resulting from positron-electron annihilation, radioactive decays and de-excitation of nuclei in the interstellar medium and cosmic ray nuclei. Results in the sense of a clear separation of continuum and line emission at MeV energies are still preliminary.

In the context of cosmic ray origin, the information from the inverse Compton and bremsstrahlung continuum and the cosmic ray de-excitation lines are most conclusive. First and most important, the OSSE and COMPTEL instruments have not detected huge fluxes of inverse Compton scattered microwave background radiation photons, which are to be expected (Fazio et al. 1966 [157]) for any metagalactic origin of cosmic ray electrons. In fact, *this non-detection rules out any metagalactic origin of cosmic ray electrons.*

Moreover, the OSSE and COMPTEL instruments provided evidence that the diffuse galactic continuum emission extends down to photon energies below 100 keV, see Fig. 6.14. The diffuse origin of this radiation has been suggested by correlated SIGMA measurements, to estimate the galactic point source contribution (Purcell et al. 1996 [412]), and the analysis of the GINGA measurements of the galactic ridge emission at much lower energies (Yamasaki et al. 1996 [569], 1997 [570]). Although one still should keep an open mind to alternative explanations, the γ-ray emission in this energy band is most likely electron bremsstrahlung in the interstellar medium, which implies the presence of many low-energy (<10 MeV) cosmic ray electrons in the Galaxy. The power in cosmic ray electrons to produce a given amount of bremsstrahlung is a fixed quantity that depends only on the energy spectrum of the radiating electrons and (weakly) on the ionization state of the interstellar medium. Attributing this power input to injection in cosmic ray electron sources it has been estimated that, integrated over the whole Galaxy, a source power of about ∼4 ×10^{41} erg s^{-1} (Skibo and Ramaty 1993 [495]) or even ∼10^{43} erg s^{-1} (Skibo et al. 1996 [496]) in low-energy (<10 MeV) electrons is required, to maintain these electrons against the severe Coulomb and ionization losses. The increase in the later estimate results from the GINGA result that the

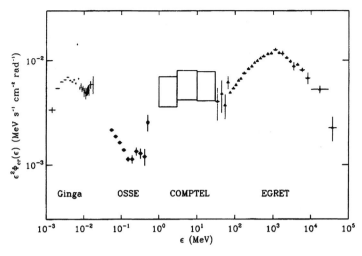

Fig. 6.14. The current best estimate of the diffuse high energy continuum emission from the central radian of the Galaxy (GINGA – Yamasaki et al. 1996 [569]; OSSE – Purcell et al. 1996 [412]; COMPTEL – Strong et al. 1996 [529]; EGRET – Hunter et al. 1997 [233]). The **left vertical axis** is the measured flux multiplied by ϵ^2. From Schlickeiser et al. (1997a [476])

diffuse bremsstrahlung emission may extend down to photon energies ~ 10 keV, well below the OSSE energy band. Taking the first number as a lower limit, this electron power exceeds the power supplied to the nuclear cosmic ray component by at least an order of magnitude.

6.2.3 γ-Ray Line Emission Below 10 MeV from the Orion Region

Intense γ-ray emission in the 3–7 MeV energy band was observed from the Orion molecular cloud complex with the COMPTEL instrument (Bloemen et al. 1994 [60], 1997 [61]).[1] Since the source of emission appears to be spatially extended, models based on line emission from ^{12}C and ^{16}O de-excitations (see Sect. 5.3.7) in a diffuse medium (Ramaty et al. 1996 [420]) seem to be favored. A related issue is the amount of the expected X-ray emission that accompanies the observed gamma ray lines. ROSAT data was used to place an upper limit on the total X-ray emission from the gamma ray emitting region (Dogiel et al. 1997 [132]). The predicted inverse bremsstrahlung and bremsstrahlung from secondary knock-on electrons are not inconsistent with this upper limit. However, the predicted line emission following electron capture onto fast O ions is quite large, and potentially inconsistent with the

---

[1] The recent announcement (Bloemen et al. 1999 [62]), that most of this emission is instrumental, indicates that the following discussion is hypothetical.

ROSAT upper limits (Ramaty et al. 1997 [421]). But this line emission depends sensitively on the spatial distribution of the nuclear γ-ray de-excitation line emission, which is rather uncertain, and on the metallicity and column density of the interstellar medium towards and in Orion, which determines the level of photoelectric absorption.

The observed spectrum suggests that the 3–7 MeV γ-ray emission mostly consists of line emission from ^{12}C and ^{16}O de-excitations. Such line emission can only be produced by accelerated particle interactions. The observed spectrum suggests that the emission mostly consists of broad lines, resulting from collisions of accelerated C- and O-nuclei with ambient H and He, rather than narrow lines due to the collisions of accelerated protons and He-nuclei with ambient C and O (Kozlovsky et al. 1997 [271]). Apart from the observed emission in the 3–7 MeV band, the COMPTEL observations revealed only upper limits at the neighboring γ-ray energy bands. In particular, the upper limit on the 1–3 MeV band sets tight constraints on the accelerated Ne-Fe abundances relative to those of C and O. A further constraint is set by Orion's γ-ray emission at photon energies above 30 MeV (Digel et al. 1995 [130]) which is consistent with pion production and bremsstrahlung due to the irradiation of the molecular clouds by standard Galactic cosmic rays. As such cosmic rays underproduce the observed line emission by at least three orders of magnitude, the gamma ray line production in Orion must be predominantly a low-energy cosmic ray phenomenon, requiring very large low-energy cosmic ray fluxes in Orion. The bulk of the line emission is produced by accelerated particles with energies of at least several tens of MeV/nucleon but not exceeding about 100 MeV/nucleon (Kozlovsky et al. 1997 [271]). The explanations proposed for the suppression of both the Ne-Fe abundance (because of the 1–3 MeV upper limit) and the p-He abundance (because of the limits on the narrow lines), relative to C and O invoked the injection into the particle accelerator of seed particles from a variety of sources: the winds of massive stars in late stages of evolution, the ejecta of supernovae from massive star progenitors, and the acceleration of ions resulting from the breakup of interstellar dust. Apparently these nuclear gamma ray lines allow valuable conclusions to be drawn for the first time on the injection and acceleration processes of low-energy cosmic rays in the interstellar environment.

6.2.4 Interlude

The instruments on board the Compton Observatory have provided important constraints on the origin and the acceleration of cosmic rays. Most importantly, they verified that *the bulk of cosmic ray particles is of galactic origin*: the EGRET observations of the Magellanic Clouds and the OSSE and COMPTEL measurements of the extragalactic diffuse hard X-ray and gamma-ray background radiation exclude any metagalactic origin of cosmic ray nucleons with energies below 100 GeV and cosmic ray electrons, respectively. Cosmic ray nucleons with energies above 100 GeV are not constrained by the Compton Observatory results.

More specific properties of the cosmic ray sources in our Galaxy can be derived from the observations of individual galactic supernova remnants, gradient and spectral measurements of the diffuse galactic continuum radiation and of gamma-ray lines. The measured energy spectrum of the diffuse galactic continuum radiation now available over more than six frequency decades provides new insight into the relative contributions from cosmic ray nucleons and cosmic ray electrons, and thus into the relative acceleration of cosmic ray electrons and nucleons. The interpretation of the strong 3–7 MeV emission from the Orion molecular clouds as cosmic ray nucleon generated de-excitation line emission allows conclusions on the cosmic ray source abundances to be drawn.

6.3 γ-Ray Point Sources

After earlier SAS-2 and COS-B discoveries of the existence of about 25 galactic γ-ray point sources (Swanenburg et al. 1981 [533]), the Compton Observatory has detected many galactic and extragalactic point sources of γ-ray emission. Obviously, the detection of these point sources directly indicates the astrophysical site of intense cosmic ray production.

6.3.1 γ-Ray Observations of Active Galactic Nuclei

Apart from the Magellanic Clouds (Sect. 6.2.1.1) all identified extragalactic sources detected by the experiments on board of the Compton Observatory are active galactic nuclei (AGNs): external galaxies with a bright and variable core.

1. 66 AGNs have been reported as ($> 3\sigma$) sources of > 100 MeV γ-rays (Fichtel et al. 1995 [166], Dermer and Gehrels 1995 [123], Hartman et al. 1999 [212]). Several of them are also seen by the COMPTEL and OSSE instruments (Kurfess 1994 [279]). All these AGNs belong to the *blazar category*, i.e. they exhibit high optical polarization, rapid optical variability, flat-spectrum radio emission from a compact core, and very often apparent superluminal ($V_{\text{app}} > c$) motion components.
2. 24 AGNs have been detected by the OSSE experiment in the energy range 50–150 keV (Kurfess 1994 [279]): 13 of these are classified as Seyfert 1 galaxies, 5 are Seyfert 2s, 2 are the quasars 3C273 and 3C279 and one is the BL-Lac object PKS 2155–304. Five of the sources are also radio galaxies. No Seyfert AGNs or radio galaxies are detected at photon energies above $E_\gamma > 1$ MeV.

Apparently, there exist two classes of γ-ray emitting AGNs (Dermer and Gehrels 1995 [123]):

(a) *γ-ray hard blazars* with power law energy spectra extending to energies ≥ 10 GeV, and in the case of Mrk 421 even to TeV energies (Punch et al. 1992 [410]). The γ-ray emission in these sources most probably is nonthermal radiation from relativistic jets;
(b) *γ-ray soft Seyferts* with energy spectra that cutoff sharply above ~ 100 keV. The γ-ray emission from these sources most probably is hot thermal emission from the inner accretion disk.

Here we concentrate on the first class of γ-ray AGNs and discuss the implications of the observations for the nonthermal radiation and particle acceleration processes.

The most remarkable properties of the γ-ray hard blazars are

1) the high fraction ($> 17\%$) of superluminal sources in the sample: 0234+285, 0528+134, 0836+710, 3C273, 3C279, 1633+382, 2251+158;
2) the dramatic peak of the luminosity spectra (= power per $\lg \nu$) at γ-ray frequencies (see Fig. 6.18) which indicates the importance of nonthermal emission processes in these objects;
3) the short time variability of the γ-ray emission which is observed in most sources (Montigny et al. 1995 [364]). Correcting for the redshift z of the sources the intrinsic variability time scale in the cosmological frame $\delta t_i = \delta t_{obs}/(1+z)$ becomes as short as 0.65 day (0528+134), 0.70 day (1633+382), and 1.3 day (3C279), implying a very compact source from simple light-travel arguments

$$R \leq c\delta t_i \, . \tag{6.3.1}$$

4) the time-correlated monitoring of the sources at lower frequencies (e.g. Reich et al. 1993 [429], Pohl et al. 1995 [406], Lichti et al. 1995 [303]) which very often demonstrates that the γ-ray flare seems to proceed outbursts at lower frequencies, although exceptions to this rule are frequent.

Because of its rapid variability, its high compactness, lack of radio-quiet sources in the sample, and high presence of superluminal motion sources, it is generally agreed (Blandford 1993 [56], Dermer 1993 [122], Dermer and Schlickeiser 1992 [125]) that the γ-ray emission originates in strongly beamed sources, in accord with the relativistic beaming hypothesis and the unified scenario for AGNs that has served well as the baseline model for the central engines of active galactic nuclei (Blandford and Rees 1978 [59], Scheuer and Readhead 1979 [450], Blandford 1990 [55]).

Simplifying many details, the basic features of this model sketched in Figs. 6.15 and 6.16 read as follows:

(i) AGNs are powered by accretion of matter onto a super-massive black hole;
(ii) in the case of radio-loud AGNs, collimated relativistic jets composed of shocks or plasma blobs move with speed $V_{jet} > 0.7c$ outward along the

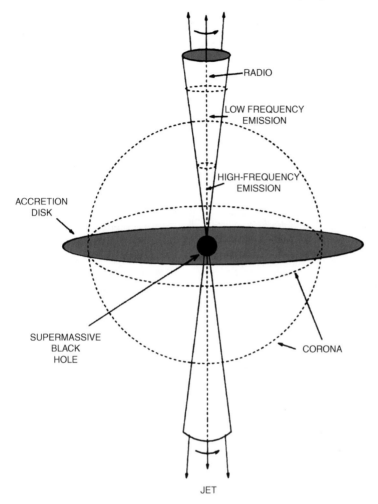

Fig. 6.15. Schematic model of the inner part of an AGN jet system. A supermassive black hole at the center accretes matter in an accretion disk and ejects jet material perpendicular to the disk. From Fegan (1998 [159])

axes of the accretion disk. The orientation angle θ^* of the jet axis with respect to the observer in individual AGNs is random; and

(iii) the line of sight of the observer with respect to the jet axis determines the classification of the particular AGN (for review see Antonucci (1993 [15]), Dermer (1993 [122]), Urry and Padovani (1995 [546])).

The strongest arguments in favor of the conclusion that the high-energy X-ray and γ-ray emission is *beamed radiation from the jet*, are both the absence of intrinsic γ–γ pair absorption features in the observed γ-ray energy

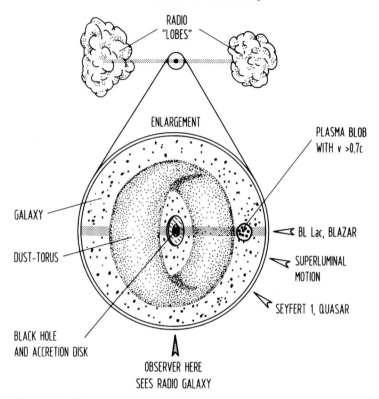

Fig. 6.16. Schematic model of the outer part of an AGN jet system. The disk jet system is surrounded by a dust torus and the central stellar cluster. The various classifications with respect to the viewing angle of the observer are indicated

spectra, and the violation of the Elliot-Shapiro relation in at least four γ-ray AGNs. We discuss both arguments in turn.

6.3.1.1 Intrinsic Compactness. Let us discuss the hypothesis that the high-energy X-ray and γ-ray blazar emission originates as isotropic emission from an uniform spherical (of radius R) source at rest. The optical depth for intrinsic $\gamma + \gamma \to e^+ + e^-$ attenuation then is (Dermer and Schlickeiser 1994 [127])

$$\tau \simeq \sigma_T n_\gamma R = \frac{\sigma_T}{4\pi c \langle \epsilon \rangle} \frac{L}{R}, \qquad (6.3.2)$$

where σ_T denotes the Thomson cross-section. Clearly, the optical depth is determined by the ratio of the intrinsic source luminosity L to the radius R, a quantity referred to as the *source compactness* $l = L/R$. Using a mean photon energy $\langle \epsilon \rangle = 1$ MeV and scaling the intrinsic luminosity in units of the solar luminosity $L_\odot = 2 \times 10^{33}$ erg s^{-1} (6.3.2) becomes

$$\tau = 10^3 \frac{L/L_\odot}{R/\text{ cm}} . \qquad (6.3.3)$$

The measured γ-ray fluxes above 1 MeV combined with the distance derived from the measured cosmological redshifts of the γ-ray blazars imply isotropic intrinsic luminosities of typically $L = 10^{48} L_{48}$ erg s^{-1} (Fichtel et al. 1995 [166]) and using the size constraint (6.3.1) from the observed time variability in (6.3.3) yields values of the optical depth

$$\tau \geq 200 \frac{L_{48}}{\delta t_i / 1 \text{ day}} \tag{6.3.4}$$

much larger than unity. Such large values for the optical depth for pair attenuation are incompatible with the observed straight power law γ-ray energy spectra over four decades of energy, and rule out isotropic emission at rest.

6.3.1.2 Violation of the Elliot-Shapiro Relation.

A second related argument against simple isotropic high-energy emission from spherical accretion is given by applying the argument by Elliot and Shapiro (1974 [150]) originally developed for X-ray observations. For spherical black hole accretion the source luminosity is limited by the Eddington luminosity

$$L \leq L_{\text{Edd}} < 1.3 \times 10^{38} \frac{(M/M_\odot)}{\alpha} \text{ erg s}^{-1} , \tag{6.3.5}$$

where M is the mass of the black hole and $\alpha \leq 1$ is a correction factor (Pohl et al. 1995 [406], Dermer and Gehrels 1995 [123]) that accounts for the deviation of the Klein-Nishina from the Thomson cross-section when calculating the balance between radiation and gravitation forces.

Alternatively, the minimum intrinsic time scale for flux variations is

$$\delta t_{i,\text{ min}} \geq \frac{R_s}{c} = 0.98 \times 10^{-5} \left(\frac{M}{M_\odot} \right) \text{ s} , \tag{6.3.6}$$

since for the generated high-energy photons to escape from the central source the source size has to be larger than the Schwarzschild radius R_s. Eliminating the mass of the black hole from (6.3.5) and (6.3.6) yields ($\lg = \log_{10}$)

$$\lg \delta t_{i,\text{ min}}(s) \geq \lg L \left(\text{erg s}^{-1}\right) - 43.1 + \lg \alpha , \tag{6.3.7}$$

which for $\alpha = 1$ agrees with the Elliot-Shapiro relation. Scaling the observed isotropic unbeamed γ-ray luminosity, again as $L = 10^{48} L_{48}$ erg s^{-1}, we obtain from (6.3.7) that

$$\lg \left[\frac{\delta t_{\text{obs}}/(1+z)}{\text{day}} \right] \geq \lg L_{48} + \lg \alpha . \tag{6.3.8}$$

In Fig. 6.17 we compare the observed variability time scales and γ-ray luminosities with the Elliot-Shapiro relation, i.e. (6.3.8) calculated with $\alpha = 1$. It can be seen that four sources (0528+134, 1633+382, 3C279, 1406-076) violate the Elliot-Shapiro relation since they lie in the shaded area below the dividing line. At least for these four sources isotropic emission from a simple steady spherical accreting source at rest is ruled out.

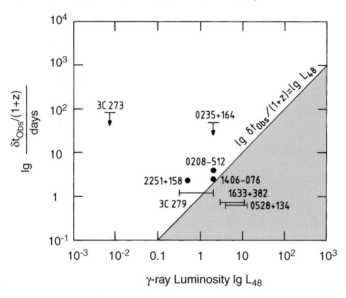

Fig. 6.17. Comparison of the observed γ-ray variability time scales and luminosities with the Elliot-Shapiro relation for several γ-ray blazars

Of course, correcting for deviations from the Thomson cross-section, i.e. using a value of $\lg \alpha = -1.3$ for a mean photon energy of $\langle \epsilon \rangle = 10$ MeV, will move the dividing line to the right so that 3C279 and 1406–076 no longer violate this restriction (as has been pointed out by Pohl et al. (1995 [406]) and Dermer and Gehrels (1995 [123])); but on the other hand most of this correction is counterbalanced by the fact that the actual source size is certainly larger than the Schwarzschild radius, say $\sim 10R_s$, so that (6.3.8) with $\alpha = 1$ still serves as a good approximation.

6.3.1.3 Relativistic Beaming. The discussed absence of photon pair attenuation and the violation of the Elliot-Shapiro relation force us to discard the hypothesis of high-energy X-ray and γ-ray emission originating in isotropic sources at rest, and strongly indicate that this emission originates in a source that moves with large bulk speed with respect to us. The natural candidate for this moving source is the relativistic jet hypothesized to exist in AGNs (see Fig. 6.16) according to the unified scenario.

In the case of emission from the fast moving jet, the theory of special relativity provides the relation between observed (denoted by an asterisk *) and intrinsic physical quantities (Rybicki and Lightman 1979 [441], Begelman et al. 1984 [37]), as

$$\text{the apparent luminosity} \quad L^*_{\text{app}} = \delta^n L, \quad 3 \leq n \leq 4, \quad (6.3.9a)$$
$$\text{the blueshifted frequency} \quad \nu^* = \delta \nu, \quad (6.3.9b)$$
$$\text{the time dilatation} \quad t^* = t/\delta, \quad (6.3.9c)$$

where the Doppler factor

$$\delta = [\Gamma(1 - \beta\cos\theta^*)]^{-1}, \quad \beta = V_{\text{jet}}/c, \quad \Gamma = (1-\beta^2)^{-1/2}, \quad (6.3.9d)$$

is determined by the jet's bulk speed and orientation angle with respect to the observer. The factor n in the Doppler boosting formula (6.3.9a) is model dependent, but for many relevant radiation models lies between 3 and 4. A consequence of the Doppler boosting (6.3.9a) is that observers are particularly sensitive to AGNs whose jet axis points towards them ($\theta^* \to 0$) since the apparent luminosity from these objects is Doppler boosted by a large factor ($\to (2\Gamma)^n$) compared to intrinsically equally bright AGNs with misaligned jet directions.

Since the *intrinsic* luminosity and source size enter the calculation of the compactness in (6.3.2)–(6.3.3) we find with (6.3.9)

$$l = L/R = \delta^{-(1+n)} \frac{L^*_{\text{app}}}{ct^*} \quad (6.3.10)$$

a dramatic reduction by the factor $\delta^{(1+n)}$ in the value of the optical depth (6.3.4) to values smaller than unity provided $\delta \geq 6$. The same reduction factor enters the calculation of the Elliot-Shapiro relation and then implies the non-violation of this relation for the four sources 0528+134, 1633+382, 3C279 and 1406–076. 3C279 is already a well-known superluminal source and the remaining three are prime candidates for superluminal motion. This prediction has been verified by the recent discoveries of superluminal motion components in 0528+134 with apparent speed $\simeq 15c$ (Pohl et al. 1995 [406]) and in 1633+382 with apparent speed $\simeq 6c$ (Barthel et al. 1995 [32]). This discovery also provides conclusive evidence for a close physical connection between γ-ray flaring and the ejection of superluminal blobs in blazars.

6.3.1.4 External Compactness. In this context the decisive role of external compactness for γ-ray production has to be emphasized. It has been pointed out by many authors (Carraminana 1992 [92], Bednarek 1993a [34], Blandford 1993 [56], Dermer and Schlickeiser 1994 [127], Sikora et al. 1994 [483], Blandford and Levinson 1995 [57]) that target photon fields external to the jet plasma like the accretion disk photons and/or the reprocesses disk photons in the emission line clouds provide compactness $\gg 1$ inside the γ-ray photosphere. The optical depth for γ-γ absorption (Dermer and Schlickeiser 1994 [127]) produced in the jet at height z above the central black hole due to interactions

– with accretion disk photons $\tau_A = 6 \times 10^{-6} \dfrac{L_{46} M_8}{\Theta_{-4} z^2(\text{pc})} \dfrac{F(E_\gamma)}{F(E_{\gamma,\text{max}})}$, (6.3.11)

– with reprocessed photons $\tau_s = 0.06 E_\gamma^{0.7}(\text{GeV}) L_{46} \tau_{-2} \ln(z/z_i)$, (6.3.12)

where L_{46} denotes the disk luminosity in units of 10^{46} erg/s, M_8 the black hole mass in units of 10^8 solar masses, τ_{-2} the optical depth for photon scattering in the emission cloud material in units of 0.01, $\Theta_{-4} = 10^4 k_B T_A/(m_e c^2)$ the dimensionless accretion disk temperature in units of 51 eV. The height z_i denotes the base of the relativistic jet, $E_{\gamma,\max} = 3.6/\Theta_{-4}$ GeV is defined by the maximum value of the function

$$F(x) = \left[x^3[\exp\left(\frac{2}{\Theta x}\right) - 1]\right]^{-1} \tag{6.3.13}$$

characterizing the accretion disk's thermal radiation field.

The estimate (6.3.11) yields γ-ray transparency ($\tau_A < 1$) at heights

$$z(\mathrm{pc}) > 2.5 \times 10^{-3} L_{46}^{1/2} M_8^{1/2} \Theta_{-4}^{-1/2}, \tag{6.3.14}$$

which corresponds to $\simeq 500 M_8^{-1/2} R_s$ Schwarzschild radii. At smaller heights the generated γ-ray photons are strongly attenuated. This large size of the γ-ray photosphere is fatal for γ-ray production models close to the central black hole as proposed by Slane and Wagh (1990 [499]) and Bednarek (1993b [35]).

The interaction with the reprocessed photons (6.3.12) provides transparency ($\tau_s < 1$) at all EGRET energies, but would yield strong photon attenuation at TeV energies,

$$\tau_s = 7.5 E_\gamma^{0.7}(\mathrm{TeV}) L_{46} \tau_{-2} \ln(z/z_i). \tag{6.3.15}$$

As a consequence quasars, which have emission line clouds, should not be sources of TeV γ-ray emission. Nearby lineless BL-Lac objects are more likely to be seen at TeV energies.

6.3.2 Nonthermal γ-Ray Emission Processes in the Jets of Active Galactic Nuclei

We briefly discuss relevant γ-ray production processes in relativistic jets under the hypothesis that the jets sometimes are filled with blobs containing relativistic charged particles (e^+, e^-, p?) and thermal particles (p, e^-) which are confined to the jet by suitable magnetic fields of strength B_0. The blobs move with relativistic bulk speed V_{jet}, $\Gamma \gg 1$. The origins of these jets and blobs are still a matter of intensive debate. Models include MHD winds ejected from the surface of the rotating accretion disk (for a review see Blandford 1990 [55], Camenzind 1990 [87]) as well as current-free relativistic e^+– e^- beams (Pelletier and Sol 1992 [393]), and it is hoped that the γ-ray observations will provide an important test for these ideas.

Moreover, we assume that at the start of the γ-ray flare, which is related to the release of a new blob, the relativistic particles in the blob have an initial isotropic power law energy spectrum at height z_i:

$$N(\gamma_i, z_i) = N_0 \gamma_i^{-s}, \quad \text{for } \gamma_1 \leq \gamma_i \leq \gamma_2. \tag{6.3.16}$$

This relativistic particle distribution function may result from "slow" acceleration processes during a "quiescent" AGN phase before the γ-ray flare, and there is no need for rapid particle energization during the γ-ray flare rise time. The total density of relativistic electrons and positrons N_0 in (6.3.16) is assumed to be so small that the inverse Compton optical depth

$$\tau_{\rm IC} = \sigma_{\rm T} N_0 r_{\rm b} = 0.6 \left(\frac{N_0}{10^{11} \text{ cm}^{-3}}\right) \left(\frac{r_{\rm b}}{10^{13} \text{ cm}}\right) \leq 1 \qquad (6.3.17)$$

is smaller than unity; $r_{\rm b}$ denotes the size of the blob.

6.3.2.1 The Role of Energetic Hadrons. We restrict the discussion mainly to the emission processes of relativistic electrons and positrons and do not discuss the emission from relativistic hadrons since hadrons most probably are less important (but see Mannheim 1996 [315], Mastichiadis 1996 [325] for an alternative view) for the following reasons:

1. due to their much smaller masses electrons and positrons radiate more efficient than nuclei of charge Ze and mass $Am_{\rm p}$: for point-like electromagnetic interactions the "hadron Thomson" cross-section

$$\sigma_{\rm H} = \frac{Z^4}{A^2} \left(\frac{m_e}{m_p}\right)^2 \sigma_{\rm T} = 1.97 \times 10^{-31} \frac{Z^4}{A^2} \quad \text{cm}^2 \qquad (6.3.18)$$

   is several orders of magnitude smaller than the Thomson cross-section;
2. as we discussed in Sect. 5.2 inelastic hadron (A) interactions as $A + {\rm p} \to (\pi^\circ, \pi^+, \pi^-)$, $A + \gamma \to A + {\rm e}^+ + {\rm e}^-$, $A + \gamma \to A + (\pi^\circ, \pi^+, \pi^-)$, $A + \gamma \to (A-1) + N$ provide many *secondary* electrons and positrons and, although hadrons serve as sources, the observed radiation is generated in interactions of these secondary electrons and positrons;
3. the photo-pair, pion and photo-pion production processes operate most efficiently in regions of high target photon and matter densities, i.e. close to the central black hole. As shown in Sect. 6.3.1.4, such a thick target will be optically thick for the generated γ-rays, and therefore not in accord with the observed straight power law γ-ray energy spectra and the short variability.

To summarize: hadron acceleration may be important during the quiet phase of γ-ray AGNs as a source of plenty of secondary electrons and positrons, but during the flaring phase the γ-ray radiation is produced by electrons and positrons probably of secondary origin.

6.3.2.2 Nonthermal γ-Ray Production by Relativistic Electrons and Positrons. There are three main radiation production processes by relativistic electrons and positrons that are important in relativistic jets:

1. *the synchrotron self-Compton (SSC) process*: the relativistic electrons and positrons in the jet inverse Compton scatter their own synchrotron radiation to γ-ray energies (Bloom and Marscher 1993 [66], Maraschi et al. 1993 [319]);

2. *inverse Compton scattering of radiation external to the jet*: the relativistic electrons and positrons in the jet inverse Compton scatter
   - the accretion disk photons (Dermer et al. 1992 [128], Dermer and Schlickeiser 1993 [126]). The anisotropy of the incoming target photons in combination with the angular dependence of the relativistic Doppler boosting favors γ-ray emission at observer angles where the source exhibits superluminal motion. This γ-ray production process successfully explains the key observational features as **(i)** the peak of the bolometric luminosity at γ-ray frequencies **(ii)** the γ-ray variability time scale ≤ few days, **(iii)** the softening of the spectra between the hard X-ray and medium γ-ray regime, **(iv)** the high percentage of superluminal sources in the sample, **(v)** the generation of TeV emission for favorably aligned observers;
   - the accretion disk photons reflected from the material through which the jet is passing as, for example, the emission line clouds in quasars (Sikora et al. 1994 [483]);
   - photons from the universal 2.7 K microwave background and optical photons from the central star cluster;
3. *pair annihilation radiation*: the relativistic electrons and positrons in the jet annihilate and including all spectral broadening effects due to the energy distribution of the annihilating particles and the jet bulk motion give rise to broad continuum emission in the MeV γ-ray energy range (Henri et al. 1993 [225], Böttcher and Schlickeiser 1995 [70], Skibo et al. 1997 [497]). If the blobs in the jet also contain thermal electrons the relativistic positrons additionally can annihilate with the thermal electrons.

6.3.2.3 Comparison of Leptonic γ-Ray Production Processes.

The observed EGRET γ-ray energy spectra alone are inconclusive to decide on the issue of the most important γ-ray production process in blazars. Advocates of the individual processes have produced remarkable good fits to the EGRET observations. The best tests for the proposed models will be provided by multifrequency time-correlated monitoring of these sources that now are operating (e.g. Lichti et al. 1995 [303]).

It is important to realize that probably all three basic radiation mechanisms contribute to the observed γ-ray emission during different phases of the flare with individual contributions that vary as a function of time or height z of the blob above the central object. Moreover, each of these three γ-ray production processes has different Doppler boosting factors and antenna characteristics so that estimates of the relative importance depend strongly on the observer's viewing angle. Another complicating factor is the shortness of energy loss times of the radiating particles, implying that the time evolution of the energy spectra of the relativistic electrons and positrons has to be taken into account; not all theoretical calculations include this crucial issue. With all these caveats on the basis of estimates for the total electron loss rates in the various interaction processes (see Dermer and Schlickeiser 1993 [126], 1994 [127] for details), we find that for parameters appropriate to 3C279, the inverse Compton scattering of accretion disk photons dominates the production of γ-rays by

- microwave scattering if $z(\mathrm{pc}) < 360 L_{46}^{1/2}/(\Gamma/10)^2(1+z)^2$,
- Compton scattering of stellar optical photons if
  $z(\mathrm{pc}) < 4.3\,(L_\mathrm{d}/L_\mathrm{opt})^{1/2}/(\Gamma/10)$,
- the synchrotron self-Compton process if $z(\mathrm{pc}) < 0.1\,L_{46}^{1/2}/\Gamma(B_0/1\,\mathrm{G})$,
- Comptonization of scattered radiation if $z(\mathrm{pc}) < 0.06\,M_8^{1/3}\,R_\mathrm{ELC}^{2/3}(\mathrm{pc})/\tau_{-2}^{1/3}$,

where M_8 denotes the central black hole mass in units of 10^8 solar masses, R_ELC the distance in units of pc of the emission line clouds from the central black hole, and τ_{-2} the electron scattering optical depth in emission line clouds in units of 10^{-2}. Note that the lack of an observed soft X-ray turnover due to photoelectric absorption in contemporaneous GINGA observations of 3C279 constrains the value $\tau_{-2} \leq 0.25$.

The above constraints, together with the size estimate (6.3.14) for the γ-ray photosphere, place the location of the γ-ray emission site at distances beyond $z \geq 10^{15}$ cm from the central object, which is comparable to the radial extent of the accretion disk and is determined by the obvious absence of γ-γ pair attenuation in the observed γ-ray spectra.

γ-ray production by inverse Compton scattering of accretion disk photons by outflowing relativistic electrons and positrons in a relativistic jet is an economical process: for favorable viewing conditions of the observer a total energy in relativistic electrons and positrons in the blob of

$$E_\mathrm{tot} \geq 10^{50}(r_\mathrm{b}/10^{15}\,\mathrm{cm})^2(\gamma_1/10)\,\mathrm{erg} \qquad (6.3.19)$$

is sufficient to explain the observed intensities. Combined with the observed γ-ray luminosity $L(> 100\,\mathrm{MeV}) = 10^{48}L_{48}f$ erg s^{-1}, where f is the beaming factor, (6.3.19) implies that the intrinsic relativistic electron and positron loss time

$$t_\mathrm{R} = E_\mathrm{tot}/L(> 100\,\mathrm{MeV}) \geq 10^4 L_{48}^{-1}(f/10^{-2})^{-1}\,\mathrm{s}\,. \qquad (6.3.20)$$

Coupled with the Doppler time-dilation effect, we see from (6.3.20) that very short γ-ray flare durations on the order of a few days can be explained in this model. However, the short radiation loss time (6.3.20) implies that the time evolution of the electron energy spectrum during the γ-ray flare has to be taken into account consistently. Dermer and Schlickeiser (1993 [126]) calculated the time-integrated γ-ray emission from a modified Kardashev (1962 [256]) model by instantaneously injecting the power law electron distribution (6.3.16) at height z_i at the beginning of the flare, calculating its modification with height in the relativistically outflowing blob due to the operation of the various inverse Compton energy loss processes, and finally integrating the

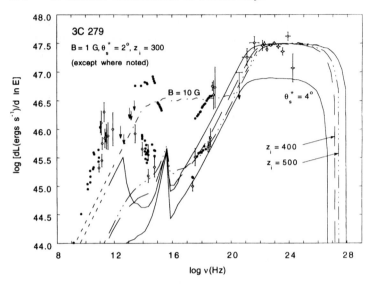

Fig. 6.18. Model results for the time-integrated luminosity spectrum compared to the multiwavelength observations of 3C279. The radio-through-optical synchrotron emission, which is probably made farther out in the radio jet, is not modeled here. z_i denotes the starting height of the blob at the beginning of the flare in units of half of the Schwarzschild radius of the central object. From Dermer and Schlickeiser (1993 [126])

hard X-ray and γ-ray emissivities over height. Comparison with multiwavelength observations of 3C279 is shown in Fig. 6.18. One notices the good fit to the high-energy emission from this source, including the break at MeV energies which results from incomplete relativistic electron cooling. Moreover, the peak of the luminosity spectrum at γ-ray energies is well reproduced by the model.

6.3.3 Interlude

The high-energy X-ray and γ-ray emission from flaring blazars is beamed radiation from the relativistic jet supporting the relativistic beaming hypothesis and the unified scenario for AGNs. Most probably the high-energy emission results from inverse Compton scattering by relativistic electrons and positrons in the jet of radiation originating *external* to the jet plus pair annihilation radiation from the jet. Future positive TeV detections of EGRET AGN sources will be decisive to identify the prominent target photon radiation field. The model of Sikora et al. (1994 [483]), where reprocessed accretion disk photons in the emission line region are comptonized, apparently does not operate in lineless BL Lac objects and implies significant intrinsic γ–γ pair attenuation of the generated TeV photons. In the model of Dermer and Schlickeiser (1993 [126]), where the accretion disk photons are comptonized, TeV emission is possible from nearby BL Lac objects but triplet pair production and the extreme Klein-Nishina cutoff give rise to a significant steepening of the γ-ray energy spectra in this range, whereas in quasars the intrinsic γ–γ attenuation from the reprocessed radiation absorbs the TeV photons. The synchrotron self-Compton models in quasars also suffer from the γ–γ attenuation.

Direct γ-ray production by energetic hadrons is not important for the flaring phase in γ-ray blazars, but the acceleration of energetic hadrons during the quiescent phase of AGNs may be decisive as the source of secondary electrons and positrons through photo-pair and photo-pion production. In any case, the detected γ-ray emission is inconclusive to prove either that AGNs do or do not accelerate cosmic ray hadrons.

6.3.4 Galactic γ-Ray Sources

Six Compton γ-ray point sources have been identified with the known pulsars on the basis of their periodic emission, including the Crab pulsar (NP 0532), the Vela pulsar (PSR 0833–45) and Geminga CG 195+4 that had been enigmatic in the pre-Compton era (see the discussion in Bignami and Caraveo 1996 [52]). The identification of more sources with more distant interstellar gas clouds is difficult because of confusion as discussed already in Sect. 6.2.1.2. The biggest obstacle for an unambiguous identification is the limited angular resolution of the γ-ray detectors, which is orders of magnitude larger than the available resolution of radio, optical and X-ray instruments. As a consequence, one finds plenty of sources of lower frequency radiation within the error-box of a γ-ray source, and without additional information (e.g. the periodic time signal in the case of pulsars) an unambiguous identification cannot be made.

However, statistical clues are possible. Fig. 6.19 shows the location of the discovered COS-B point sources in the sky. The strong concentration of these

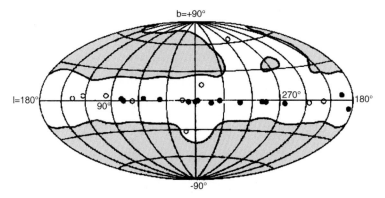

Fig. 6.19. γ-ray sources contained in the COS-B catalogue. **Filled circles**: sources with flux $> 1.3 \times 10^{-6}$ cm^{-2} s^{-1} above 100 MeV. **Open circles**: weaker sources. **Unshaded area**: region of sky searched. From Swanenburg et al. (1981 [533])

sources to the galactic plane is apparent. This concentration excludes that the sources are either close to or very far away since in both cases a more uniform sky distribution of the sources is to be expected, and suggests typical source distances between 0.5 and 4 kpc. The typical integral (> 100 MeV) γ-ray flux is 2×10^{-6} photons cm^{-2} s^{-1} resulting in a luminosity estimate in the range of $(0.1\text{--}6) \times 10^{35}$ erg s^{-1}, assuming isotropic emission.

7. Immediate Consequences of Galactic Cosmic Ray Observations

7.1 Cosmic Ray Energetics

Summarizing the observational material on galactic cosmic rays presented in the last three chapters we note that the solar system is bombarded by cosmic rays from all sides isotropically. The chemical and isotopic composition of the nucleonic component is similar to that of solar flare particles; all nucleons exhibit power law energy spectra over a wide range of kinetic energies. Cosmic ray electrons exhibit a break in their power law energy spectrum at about 20 GeV with spectral index $\simeq 2.2$ below the break energy and $\simeq 3.2$ above the break energy. Integrating these energy spectra over all energies and summing over all species i we can infer the energy density

$$w = 4\pi \sum_i \int_0^\infty dE_{\text{kin}} E_{\text{kin}}^2 N_i(E_{\text{kin}}) \quad \text{eV cm}^{-3} \tag{7.1.1}$$

and the pressure

$$p_c = (4\pi/3) \sum_i \int_0^\infty dE_{\text{kin}} E_{\text{kin}} v N_i(E_{\text{kin}}) \quad \text{dyne cm}^{-2} \tag{7.1.2}$$

of cosmic ray nucleons and electrons in the local galactic neighborhood. Using the measured spectra above 1 GeV/nucl., to avoid uncertainties due to the influence of solar modulation (Sect. 3.2) on the cosmic ray particle spectra below $\sim$1 GeV/nucl., one obtains (e.g. Wolfendale 1983 [566]) $w_N \simeq 0.5$ eV cm^{-3} and $p_N \simeq 3 \times 10^{-13}$ dyne cm^{-2} in cosmic ray nucleons and $w_e \simeq 0.006$ eV cm^{-3} and $p_e \simeq 3 \times 10^{-15}$ dyne cm^{-2} in cosmic ray electrons and positrons. Note that due to spectral index values larger than 2 these pressures and energy densities are determined by particles with energies of 1–30 GeV/nucl. in the case of nucleons and by electrons with energies of 400 MeV to 20 GeV, so that the effects of solar modulation cannot drastically change these estimates. Below 400 MeV the electron spectrum will flatten due to the onset of Coulomb and ionization losses. The very high energy ($E_{\text{kin}} > 10^{15}$ eV) cosmic rays do not influence the estimates of the energy density and pressure. It follows that in the local galactic neighborhood the cosmic ray pressure is comparable to the pressures of the interstellar gas and the galactic magnetic field (see Chap. 2).

The analysis of the diffuse radio background and galactic γ-ray emission has shown that these local values are fairly representative for the whole Galaxy of volume (Sect. 2.2) $V = 350$ kpc$^3 = 10^{67}$ cm^3. The measured abundances of radioactive cosmic ray clocks (Sect. 3.4.2.1) indicate a mean residence time of ~ 1 GeV/nucl. cosmic ray particles of about $\tau \simeq 10^7$ yr. We then obtain for the cosmic ray power in our Galaxy

$$Q_c = (w_N + w_e)V/\tau \simeq 3 \times 10^{40} \text{ erg s}^{-1}, \qquad (7.1.3)$$

which is distributed nearly uniformly over the Galaxy. The time history of cosmogenic nuclei (Sect. 3.4.4) established that within a factor of two the flux of extrasolar cosmic rays has been constant over the past 10^9 yr, indicating that the cosmic ray power (7.1.3) has had this value steadily at least over the last 10^9 yr. Since the mean residence time of cosmic rays in the Galaxy is two orders of magnitude shorter, possible cosmic ray sources have to inject steadily at least a source power equal to the estimate (7.1.3) in cosmic rays to account for this situation.

Besides the many observed elemental and isotopic subtilities, a successful model for the origin of galactic cosmic rays therefore has to explain the following key ingredients:

1. an over 10^9 yr constant cosmic ray power of $\sim 10^{40}$ erg s^{-1};
2. a nearly uniform and isotropic distribution of cosmic ray nucleons and electrons with energies below 10^{15} eV over the Galaxy;
3. elemental and isotopic composition similar to solar flare particles;
4. electron/nucleon ratio in relativistic cosmic rays at the same Lorentz factor of about 0.01;
5. the formation of power law energy spectra for all species of cosmic rays over large energy ranges accounting for the systematic differences in the spectral index values of primary and secondary cosmic ray nucleons and cosmic ray electrons.

The general problem of the origin of cosmic rays can be divided into two parts. The first part concerns the actual origin or injection of the cosmic rays into the Galaxy by sources which keep up the power (7.1.3) over a long time, while the second part concerns the subsequent behavior of the cosmic rays, their motion, transport and confinement in the Galaxy. We consider both in turn.

7.2 Global Cosmic Ray Source Energetics

Due to the large steady cosmic ray power (7.1.3) only four types of galactic objects can serve as potential cosmic ray source candidates, since their estimated total power output Q into the Galaxy is larger than the minimum requirement (7.1.3):

1. *Supernova explosions*: numerical simulations of the collapse of a star of 10 $M_\odot$ indicate that the resulting supernova explosion sets free an energy of $\sim 10^{51}$ erg. And with a supernova rate of 1 per 30 years in our Galaxy (Sect. 2.2) this yields a power input into the Galaxy of $Q \simeq 10^{42}$ erg s^{-1}.
2. *Neutron stars*: estimating the rate of rotational energy loss from known radio pulsars' periods P and their derivative $\dot{P}$ as

$$\dot{E}_{\rm rot} = -4\pi I \dot{P}/P^3 \,, \qquad (7.2.1)$$

   where $I \simeq 10^{45}$ g cm^2 denotes the neutron star's moment of inertia, summing over all observed pulsars and applying corrections for the fraction of neutron stars not seen due to their beamed radiation, one estimates a total power input into the Galaxy of $Q \simeq 10^{41}$ erg s^{-1}. This neutron star origin is supported by the detection of six pulsars as MeV–GeV γ-ray point sources (Sect. 6.3.4) and by the possible detection of Cyg X–3 as a PeV γ-ray source (Sect. 3.4.3.2).
3. *Stellar winds from young hot O/B stars*: hot stars of spectral type O, B and Wolf-Rayet stars drive powerful winds (Abbott 1988 [1]) by their strong radiation pressure with mass loss rates ($\dot{M} \simeq 10^{-7}-10^{-5} M_\odot/$ yr) that exceed the Sun's mass loss rate by some nine orders of magnitude. Although such stars only live $\sim 10^7$ yr on the main sequence, the large mass loss rate means that a substantial fraction ($\sim 50\%$) can be lost during its main sequence lifetime, implying an important contribution to the energy balance of the interstellar medium. Summing over their spatial distribution function a steady power input of $Q \simeq 10^{41}$ erg s^{-1} into the interstellar medium has been estimated (Casse and Paul 1982 [94]).
4. *Flare stars of spectral class K–M*: we know (Sect. 3.3.1) that our Sun produces cosmic rays with energies up to several tens of GeV during large flares. It has been observed that many late type stars of type M and K also flare in the optical and radio frequency band (Lyuten 1949 [310], Haro 1968 [211]). Lovell (1974 [307]) has estimated the total contribution of galactic flare stars to the interstellar energy flux in cosmic rays. He estimates a power input of $Q \simeq 3 \times 10^{40}$ erg s^{-1} from these objects which would go mainly into cosmic rays of energies below 300 MeV/nucl. In this model cosmic rays with higher energy result from further acceleration in the interstellar medium. This hypothesis is certainly supported by the similar isotopic and elemental composition of solar flare and galactic cosmic rays.

Note that in the case of the first three source candidates the power input estimates refer to the total power input into the Galaxy. They require a very efficient acceleration process for cosmic rays that ultimately converts a large percentage of the total power (1–10%) into cosmic rays. Also note that probably all four source candidates contribute to the cosmic ray power and that these object classes are not independent of each other: supernova explosions result from the collapse of young massive O/B stars and lead to the formation of neutron stars at the center of the explosion. A large star formation rate and the presence of many young hot stars will also lead to enhanced supernova explosions and an enhanced birth rate of neutron stars.

As a result we note that the global energetics requirements of cosmic rays can be met if an efficient acceleration process for cosmic rays can be found.

7.3 Cosmic Ray Scattering, Confinement and Isotropy

It is generally recognized that due to their small Larmor radii, for cosmic ray nucleons (of momentum p)

$$R_{\rm L} = \frac{(pc/{\rm eV})}{300(B/{\rm Gauss})} \text{ cm}, \tag{7.3.1}$$

as compared to galactic dimensions, the majority of cosmic rays with energies below $\sim 10^{15}$ eV propagate along the galactic magnetic field. Because of the observed isotropy and age of cosmic rays, it seems clear that the cosmic rays cannot propagate freely along the lines of force but must be continually scattered. If they were to propagate freely with the speed of light, they would leave the galaxy within 10^4 to 4×10^5 years, as the dimensions of our Galaxy suggest. But from the measured abundance of the cosmogenic cosmic ray clocks we know that their average lifetime in the Galaxy is $\sim 10^7$ yr. Moreover, if there was no scattering, we would expect a strong anisotropy towards the direction of the galactic center due to the peculiar location of the solar system, since there should be more sources of the type discussed in Sect. 7.2 towards the inner galaxy. Yet we do not see this anisotropy for cosmic rays with energies less than 10^{15} eV which apparently is washed out due to multiple scattering of the cosmic rays on their path from the sources to us.

The scattering cannot be by particles (Coulomb scattering) since the energies of cosmic rays are much higher than nuclear binding energies and such collisions would destroy all nuclear species heavier than protons in the interstellar medium, which is not the case (Kulsrud and Pearce 1969 [278]). Moreover, the mean free path for Coulomb collision of relativistic nucleons in the dilute interstellar medium of order $\sim 10^{23}\gamma/n_{\rm H}$ (cm^{-3}) is far too long. Thus, the most likely scattering mechanism is off *plasma waves*, i.e. fluctuating electromagnetic fields in the interstellar medium. How these plasma waves are produced and how they influence the dynamics of cosmic ray particles will be key topics of the theoretical chapters of this book and will be discussed in detail.

8. Statistical Mechanics of Charged Particles

As we have seen in Chap. 2 most of the non-stellar matter in the universe is ionized, and therefore the statistical mechanics of charged particles and their mutual interactions via electromagnetic fields, generated and maintained by their own streaming, is of particular importance not only for the cosmic rays but also the other components of the interstellar and intergalactic media. In this chapter we will consider the behavior of an electrically neutral (on the average) system of charged and neutral particles interacting with each other and with an external magnetic field. A gas containing so many free charged particles that their collective Lorentz forces influence the properties of the medium considerably is referred to as a *plasma*.

8.1 Basic Equations

There are many excellent monographs on the basic equations of plasma physics; here we follow closely the text by Hasegawa (1975 [216]).

8.1.1 Plasma Parameter

We start by demonstrating that plasmas in space, including the cosmic ray gas, can be considered effectively collision-free. Let us introduce the collision frequency. The time rate of collision, called the collision frequency ν, is given by

$$\nu = vn\sigma \quad \text{Hz} , \tag{8.1.1}$$

where v is the speed of the colliding particle, σ the cross-section of the target particle, and n the density of the target particles. In the case of collisions between charged particles, the cross-section σ can be obtained as follows. For electrons one might expect to use the classical electron radius $r_0 = e^2/(m_e c^2) = 2.82 \times 10^{-13}$ cm, but because of the Coulomb force, they cannot usually approach each other so closely. Let us see how close two particles with charge e and mass m_1 and m_2, respectively, can approach each other before being repelled by the Coulomb force. At a distance r the relative kinetic energy $\bar{m}\bar{v}^2/2$ (total kinetic energy minus that of the center of gravity) becomes comparable to the Coulomb potential energy e^2/r, where

$\bar{m} = m_1 m_2/(m_1+m_2)$ is the effective mass and $\bar{v}$ is the relative speed of the colliding particles. This distance r is the closest distance of approach and is given by

$$r = \frac{2e^2}{\bar{m}\bar{v}^2} \text{ cm}. \qquad (8.1.2)$$

The cross-section of the collision σ may then be obtained from r as $\sigma = \pi r^2$ implying for the collision frequency

$$\nu = \frac{4\pi e^4 n v}{\bar{m}^2 \bar{v}^4} \text{ Hz}. \qquad (8.1.3)$$

For the case of electron-electron collisions, (8.1.3) with $\bar{v} \simeq v$ can be reduced to

$$\nu_{ee} = \omega_{pe}/(4\pi n \lambda_{De}^3) = \omega_{pe}/3\lambda_s \text{ Hz}, \qquad (8.1.4)$$

where $\omega_{p,e}$ is the *electron plasma frequency* defined by

$$\omega_{pe} \equiv (4\pi e^2 n_e/m_e)^{1/2} = 5.64 \times 10^4 (n_e/1 \text{ cm}^{-3})^{1/2} \text{ Hz} \qquad (8.1.5)$$

and λ_{De} is called the *Debye length* given by

$$\lambda_{De} \equiv v_{th}/\omega_{pe} = \left(\frac{k_B T_e}{4\pi n_e e^2}\right)^{1/2} = 6.9\sqrt{\frac{T_e/1 \text{ K}}{n_e/1 \text{ cm}^{-3}}} \text{ cm}. \qquad (8.1.6)$$

The quantity

$$\lambda_s \equiv 4\pi n_e \lambda_{De}^3/3 \qquad (8.1.7)$$

in (8.1.4) represents the number of electrons in a sphere of radius λ_{De}, where λ_{De} represents the distance beyond which the Coulomb force of a single test ion is shielded by the surrounding plasma electrons. Any electrostatic fields originating outside this Debye sphere are effectively screened by the charged electrons and do not contribute significantly to the electric field existing at its center. As a result, each charge in a plasma interacts collectively only with the charges that lie inside its Debye sphere, its effect on more distant charges being negligibly small.

A necessary and obvious requirement for the existence of a plasma is that the physical dimensions of the system L are large compared to λ_{De},

$$L \gg \lambda_{De}, \qquad (8.1.8)$$

otherwise there would not be sufficient space for the collective shielding effect to take place, and a test charge would interact with other individual charges in 2-body Coulomb collisions as in vacuum. Note that the requirement (8.1.8) already implies overall charge neutrality since deviations from neutrality can occur only over distances of the order of λ_{De}. But overall charge neutrality obviously also requires that the density of positively charged particles equals the density of negatively charged particles, i.e. $\sum_i n_i = n_e$.

Because the shielding effect is a result of the collective particle behavior inside a Debye sphere, it is also necessary that the number of charges inside a Debye sphere be very large. A second criterion for the definition of a plasma is therefore that λ_s defined in (8.1.7) is very large, or that the *plasma parameter* defined as the inverse of λ_s is very small,

$$g \equiv 1/\lambda_s = 7.3 \times 10^{-4} \left(n_e/1 \text{ cm}^{-3}\right)^{1/2} (T_e/1 \text{ K})^{-3/2} \ll 1 \,. \tag{8.1.9}$$

The requirement (8.1.9) also guarantees that the potential energy of a typical particle due to its nearest neighbor is much smaller than its kinetic energy. If the opposite were true, then there would be a strong tendency for electrons and ions to bind together into atoms, thus destroying the plasma state. For a proof we calculate the average potential energy Φ of a particle due to its nearest neighbor in a plasma of density n_e. The distance between two particles is roughly $n_e^{-1/3}$, so that $|\Phi| \sim e^2 n_e^{1/3}$. Requiring that this potential energy be much less than the typical ion's kinetic energy $m_i v^2/2 = 3k_B T_i/2$ yields

$$n_e^{2/3} \left(\frac{k_B T_i}{n_e e^2}\right) \gg 1 \,. \tag{8.1.10}$$

Raising each side of (8.1.10) to the 3/2 power and introducing the Debye length (8.1.6) with $T_i = T_e$ leads to exactly condition (8.1.9).

From (8.1.4) we find that in a plasma (i.e. $g \ll 1$) in general the collision frequency is much less than the electron plasma frequency,

$$\nu_{ee} = \omega_{pe} g/3 \ll \omega_{pe} \,. \tag{8.1.11a}$$

Using (8.1.3) we can also derive the collision frequency between different particles. The electron-ion collision frequency $\nu_{ei} = \nu_{ee}/2\sqrt{2}$ is comparable to the electron-electron collision frequency ν_{ee}, while the ion-ion collision frequency is

$$\nu_{ii} = \nu_{ee} \left(\frac{m_e T_e^3}{m_i T_i^3}\right)^{1/2} , \tag{8.1.11b}$$

where $T_e = m_e v_{the}^2/k_B$ and $T_i = m_i v_{thi}^2/k_B$ are the electron and ion temperatures. For equal electron and ion temperatures the ion-ion collision frequency is smaller than the electron-electron collision frequency by the factor $(m_e/m_i)^{1/2}$. Finally, the ion-electron collision frequency ν_{ie} is smaller than the electron-ion collision frequency ν_{ei} by the mass ratio (m_e/m_i) because ions are difficult to deflect with electrons.

In Table 8.1 we have calculated the plasma parameters for various cosmic particle components including the interplanetary, interstellar and intergalactic medium and galactic cosmic rays. Note that these parameters refer to the interactions of the considered gas particles with other particles of the same gas, e.g. cosmic ray particles colliding elastically with other cosmic ray

Table 8.1. Estimate of cosmic plasma parameters

System	Interstellar gas	Molecular cloud	Solar corona	Active galactic nucleus	Cluster of galaxies	Cosmic rays
n_e (cm^{-3})	0.1	1	10^6	10^9	10^{-2}	10^{-9}
n_N (cm^{-3})	1	10^2	10^7	10^{10}	—	—
T_e (K)	10^4	10^2	10^6	10^7	10^7	10^{12}
B (G)	10^{-6}	10^{-5}	10^{-4}	1	10^{-7}	10^{-6}
L (cm)	10^{18}	10^{17}	10^{10}	10^{15}	10^{23}	10^{18}
ω_{pe} (s^{-1})	2×10^4	5×10^4	5×10^7	2×10^9	5×10^3	2
Ω_e (s^{-1})	20	200	2×10^3	2×10^7	2	20
λ_D (cm)	2×10^3	69	6.9	0.7	2×10^5	2×10^{11}
g	3×10^{-10}	7×10^{-7}	7×10^{-10}	7×10^{-10}	2×10^{-15}	2×10^{-26}
$\lambda_{ee} = \frac{3\lambda_D}{g}$ (cm)	2×10^{13}	3×10^8	3×10^{10}	3×10^9	3×10^{20}	3×10^{37}

particles, and *not* as in Chap. 6 to the interactions between different gases. Here we are concerned with the intrinsic interactions of the gas particles. In all cases the plasma parameter g is much less than unity as required by condition (8.1.9), and the Debye length is much smaller than the size of the system as required by condition (8.1.8). These estimates demonstrate that these cosmic gases are indeed in the plasma state. In Table 8.1 we also calculate the electron plasma frequency $\omega_{p,e}$ and the non-relativistic *electron and proton gyrofrequencies* $\Omega_{e,0}$ and $\Omega_{p,0}$, which are defined by

$$\Omega_{e,0} \equiv eB/(m_e c) = 1.78\times 10^7\ (B/1\ \mathrm{G})\ \mathrm{Hz}, \tag{8.1.12}$$
$$\Omega_{p,0} \equiv eB/(m_p c) = \Omega_e/1836 = 9.7\times 10^3\ (B/1\ \mathrm{G})\ \mathrm{Hz}, \tag{8.1.13}$$

respectively. These frequencies are characteristic for the interactions of the plasma particles with the electric ($\omega_{p,e}$) and magnetic ($\Omega_{e,0}$ and $\Omega_{p,0}$) field of the plasma. We also calculate the electron-electron collision length $\lambda_{ee} = v/\nu_{ee} = 3\lambda_{De}/g$ cm.

The estimates in Table 8.1 show that in all objects the collision frequency ν_{ee} is much smaller than both the electron plasma and gyrofrequency, stressing again the dominance of electromagnetic interactions over the elastic 2-body collisions. In this sense these cosmic plasmas can be regarded as effectively collision-free on the shortest time and length scale. However, we also note from Table 8.1 that for the solar corona, the interstellar medium, the intergalactic medium and the broad emission line region in quasars, the collision length λ_{ee} is orders of magnitude smaller than the typical size L of this system. Thus, although elastic collisions are not the fastest process one may well apply the ideal magnetohydrodynamic approach (see Sect. 8.1.3) to these systems if one is interested in the physics of these systems on scales large compared to λ_{ee} but small compared to L, because on these scales the elastic collisions will establish a Maxwellian velocity distribution of the gas particles. On the other hand, for the cosmic ray particles the elastic 2-body collisions are certainly negligible on all scales of interest since for cosmic rays $\lambda_{ee} = 3 \times 10^{37}$ cm, and the ideal magnetohydrodynamical approach is not appropriate.

8.1.2 Kinetic Description of Plasmas

We now derive the basic equations to represent the dynamics of a plasma with a small plasma parameter g.

8.1.2.1 The Non-relativistic Vlasov Equation.
Consider the motion of N non-relativistic electrons and ions. They obey equations of motion in which the acceleration is dominated by the electromagnetic Lorentz force:

$$m_j \frac{d\boldsymbol{v}_j}{dt} = q_j \left[\boldsymbol{E}(\boldsymbol{x}_j, t) + \frac{\boldsymbol{v}_j \times \boldsymbol{B}(\boldsymbol{x}_j, t)}{c} \right], \qquad (8.1.14a)$$

$$\frac{d\boldsymbol{x}_j}{dt} = \boldsymbol{v}_j, \qquad (8.1.14b)$$

where the subscript j denotes the jth particle with charge q_j and mass m_j located at position $\boldsymbol{x}_j$ at time t. The electric field $\boldsymbol{E}$ and magnetic field $\boldsymbol{B}$ are evaluated at the position of the particle j but are generated by all other particles and possibly by additional external sources.

Now assume for the moment that the electric and the magnetic fields are known as continuous functions of space and time. We can then integrate (8.1.14) for given initial positions $\boldsymbol{x}_j(0)$ and velocities $\boldsymbol{v}_j(0)$ to obtain the particle orbit $\boldsymbol{x}_j(t)$ at any later time. However, finding the initial positions and velocities of a large number of particles is impossible, therefore it is convenient to introduce a probability distribution function for the initial velocity such that the probability of finding the particle j with an initial velocity range $-\infty < \boldsymbol{v}_j(0) < \boldsymbol{v}$ at point $\boldsymbol{x}$ is defined by $F_j(\boldsymbol{x}, \boldsymbol{v})$ where $0 \leq F_j(\boldsymbol{x}, \boldsymbol{v}) \leq 1$. The probability density function $f_j(\boldsymbol{x}, \boldsymbol{v})$, also referred to as the *phase space density*, is obtained by

$$f_j(\boldsymbol{x}, \boldsymbol{v}) = \frac{\partial^3 F_j}{\partial v_x \partial v_y \partial v_z}. \qquad (8.1.15)$$

Here $\boldsymbol{x}$ is no longer the position of the particle $\boldsymbol{x}_j$, since we do not know exactly where the jth particle is located. Hence $\boldsymbol{x}$ and $\boldsymbol{v}$ in the above expression are independent variables.

Let us now consider how the phase space density f defined in (8.1.15) changes in time due to the action of the Lorentz force. Since no particles are lost or generated in phase space the time evolution of f follows from the equation of conservation of f_j in phase space,

$$\frac{\mathrm{d}f_j}{\mathrm{d}t} = \frac{\partial f_j}{\partial t} + \dot{\boldsymbol{x}} \cdot \frac{\partial f_j}{\partial \boldsymbol{x}} + \dot{\boldsymbol{v}} \cdot \frac{\partial f_j}{\partial \boldsymbol{v}} = 0, \qquad (8.1.16)$$

where $\dot{\boldsymbol{x}} = \boldsymbol{v}$ and $\dot{\boldsymbol{x}} \cdot \frac{\partial}{\partial \boldsymbol{x}} = \boldsymbol{v} \cdot \boldsymbol{\nabla}$.

The acceleration $\dot{\boldsymbol{v}}$ is given by the Lorentz force according to (8.1.14a). The Lorentz force consists of two kinds of fields. One is the electromagnetic field generated by the collective motion of all the other particles; the other is that generated by one or possibly more particles that are located within the colliding distance of the particle we are looking at. However, if the plasma parameter g is small, we can show that the effect due to colliding particles is small: the possibility that a neighboring particle is located within colliding distance r given by (8.1.2) is estimated by the ratio of r to the mean spacing of the particles $n^{-1/3}$ (with the density $n = N/L^3$) which becomes $g^{2/3}$, which is very small. Therefore it is rather unlikely that a neighboring particle is within the colliding distance. In this case, one can ignore the acceleration due to colliding particles and (8.1.16) reduces, with the help of the Lorentz force (8.1.14a), to

$$\frac{\mathrm{d}f_j}{\mathrm{d}t} = \frac{\partial f_j}{\partial t} + \dot{\boldsymbol{x}} \cdot \frac{\partial f_j}{\partial \boldsymbol{x}} + \frac{q}{m}\left[\boldsymbol{E}(\boldsymbol{x},t) + \frac{\boldsymbol{v} \times \boldsymbol{B}(\boldsymbol{x},t)}{c}\right] \cdot \frac{\partial f_j}{\partial \boldsymbol{v}} = S_j(\boldsymbol{x},\boldsymbol{v},t), \qquad (8.1.17)$$

where $\boldsymbol{E}$ and $\boldsymbol{B}$ are the average electric field intensity and magnetic flux density produced by the collective motion of plasma particles. In addition to (8.1.16) we have added on the right-hand side of (8.1.17) the source term $S_j(\boldsymbol{x},\boldsymbol{v},t)$ that represents additional sources and sinks of particles. The expression (8.1.17) is referred to as the *non-relativistic Vlasov equation* and is one of the basic equations we introduce in this section.

Note here that the coordinate variable $\boldsymbol{x}$ is no longer the position of the jth particle but is an *independent variable* fixed to the coordinate space. Although we have derived (8.1.17) for a specific particle j, one can now see that there is no need to distinguish the distribution function of any one of the particles in the plasma. Therefore we have deleted the subscript j here. Equation (8.1.17) describes the dynamic change of the phase space density f due to the collective field. The time scale of the change due to the collective electric field will be of the order $\omega_{\mathrm{p,e}}^{-1}$, while we ignore the change due to collisions whose time scale is ν^{-1}. Equation (8.1.4) has indicated that the ratio of these two time scales is again g^{-1} so that indeed the collision effect can be neglected as compared to the effect due to the collective electric field.

8.1.2.2 Maxwell's Equations. The electromagnetic field entering the non-relativistic Vlasov equation (8.1.17) can be obtained from *Maxwell's equations*. Let us consider a plasma that contains a different species of particles (electrons, protons, etc.). $\int_{-\infty}^{\infty} d^3v f_a(\boldsymbol{x}, \boldsymbol{v}, t)$ represents the total probability that a particle of species a can be found at $\boldsymbol{x}$ and t. If we multiply by the space-averaged number density n_a, the resulting quantity gives the number density of plasma particles of sort a, i.e.

$$n_a(\boldsymbol{x}, t) = n_a \int_{-\infty}^{\infty} d^3v f_a(\boldsymbol{x}, \boldsymbol{v}, t) \text{ cm}^{-3} \, . \tag{8.1.18}$$

Multiplying the number densities (8.1.18) by the respective charges q_a and summing over all species yields the total charge density

$$\rho(\boldsymbol{x}, t) = \sum_a n_a \, q_a \int_{-\infty}^{\infty} d^3v f_a(\boldsymbol{x}, \boldsymbol{v}, t) \, . \tag{8.1.19}$$

Similarly the total current density can be obtained by [1]

$$\boldsymbol{J}(\boldsymbol{x}, t) = \sum_a n_a \, q_a \int_{-\infty}^{\infty} d^3v \, \boldsymbol{v} f_a(\boldsymbol{x}, \boldsymbol{v}, t) \, . \tag{8.1.20}$$

The four Maxwell equations then read

$$\nabla \times \boldsymbol{B} = \frac{1}{c} \frac{\partial \boldsymbol{E}}{\partial t} + \frac{4\pi}{c} \sum_a n_a \, q_a \int_{-\infty}^{\infty} d^3v \, \boldsymbol{v} f_a(\boldsymbol{x}, \boldsymbol{v}, t) + \frac{4\pi}{c} \boldsymbol{J}_{\text{ext}} \, , \tag{8.1.21}$$

$$\nabla \cdot \boldsymbol{B} = 0 \, , \tag{8.1.22}$$

$$\nabla \times \boldsymbol{E} = -\frac{1}{c} \frac{\partial \boldsymbol{B}}{\partial t} \, , \tag{8.1.23}$$

$$\nabla \cdot \boldsymbol{E} = 4\pi \sum_a n_a \, q_a \int_{-\infty}^{\infty} d^3v f_a(\boldsymbol{x}, \boldsymbol{v}, t) + 4\pi \rho_{\text{ext}} \, . \tag{8.1.24}$$

8.1.2.3 The Nonlinear Field-Particle Problem. Besides possible external charges (ρ_{ext}) and currents ($\boldsymbol{J}_{\text{ext}}$) the electromagnetic fields $\boldsymbol{E}$ and $\boldsymbol{B}$ are generated by the charge density and current density produced by the moving plasma particles, which enter on the right-hand side of (8.1.24) and (8.1.21), respectively. Therefore the streaming plasma particles determine the resulting electromagnetic field according to the four Maxwell equations. The electromagnetic fields can be calculated by solving the four Maxwell equations using the solution of the kinetic equation (8.1.17) for each plasma sort a to evaluate the charge and current densities (8.1.19) and (8.1.20). However,

---

[1] Despite its obvious physical justification we want to emphasize that both (8.1.19) and (8.1.20) are an *ansatz* for the charge and current density in plasmas. As in nonlinear optics one may postulate more general equations for the charge and current densities. For example, near a strong ionizing radiation source charges may be produced by photoionization processes and/or electrons may be accelerated by Compton scattering by hot photons. In both cases the charge density is proportional to the intensity of the ionizing photons which leads to a nonlinear relation between the charge density and the phase space density.

the solution of the Vlasov equation (8.1.17) for the plasma particles requires knowledge of the electromagnetic fields that determine the Lorentz force. We immediately recognize the nonlinear coupling of these two problems: to determine the particle's phase space density from the Vlasov equation we have to know the collective electromagnetic field; but to determine the electromagnetic fields from Maxwell's equation we have to know the phase space densities of the plasma particles. We are facing the fundamental problem of plasma physics: the plasma particles determine the electromagnetic fields and vice versa in a highly nonlinear way. In order to proceed with a solution of this coupled problem two opposite points of view can be taken:

1. the *test wave approach*, in which the plasma particle distribution functions are assumed to be given in a prescribed initial state, so that the resulting electromagnetic field and their property can be discussed. We shall follow this approach in the next three chapters (Chaps. 9–11);
2. the *test particle approach*, in which the electromagnetic fields are assumed to be given, and the response of the particles can be discussed. This approach is studied in Chaps. 12 and 13.

A consistent theory should combine the two approaches at least to the level of avoiding dramatic contradictions.

Of course, using the combined system of Maxwell's equations and Vlasov equation is the most powerful for describing plasma dynamics. However, for the reasons just mentioned, this most powerful method is also most difficult to apply. Besides following the two extreme approaches (1) and (2) of the kinetic description, one therefore has developed more simplified dynamical equations than the Vlasov equation, namely the *magnetohydrodynamic equations* that we describe next.

8.1.3 Magnetohydrodynamics of Plasmas

The basic idea of the magnetohydrodynamic (MHD) approximation is to treat the plasma as if it were a continuous fluid. By taking velocity moments of the Vlasov equation (8.1.17) in seven-dimensional $(\boldsymbol{x}, \boldsymbol{v}, t)$ space, an infinite hierarchy of equations in four-dimensional $(\boldsymbol{x}, t)$ space can be derived. When an appropriate truncation of this infinite hierarchy is carried out, the closed set of MHD equations results that relates velocity-integrated quantities as the number density of plasma particles of sort a $n_a(\boldsymbol{x}, t)$ defined in (8.1.18) as the zeroth moment of the phase space distribution function, its mean velocity defined as the first velocity moment of the phase space distribution function

$$\boldsymbol{V}_a(\boldsymbol{x}, t) \equiv \frac{\int_{-\infty}^{\infty} \mathrm{d}^3 v \, \boldsymbol{v} f_a(\boldsymbol{x}, \boldsymbol{v}, t)}{n_a(\boldsymbol{x}, t)} , \qquad (8.1.25)$$

and its pressure tensor defined via the second moment of the phase space distribution function

$$\Pi_{a,ik}(\boldsymbol{x},t) \equiv m_a \int_{-\infty}^{\infty} d^3v f_a(\boldsymbol{x},\boldsymbol{v},t)(v_i - V_i)(v_k - V_k) \,. \tag{8.1.26}$$

The resulting moment equations, of course, are nothing more than the conservation equations for the flux of mass, momentum and energy which are correct statements regardless of the actual state of the plasma. The difficulty really lies in the fact that these moment equations are not closed, i.e. the conservation equation for the mass flux involves the mean velocity which can be obtained from the first-order moment equation. But that equation involves the pressure tensor that can be obtained only from the second-order moment equation that contains velocity-integrated quantities that can be obtained only from the third-order moment equation, and so on. Therefore some truncation of this infinite iteration scheme has to be performed, and that is precisely where the different approximations set in.

As an example, a truncation is possible in collision-dominated systems where the time scale for elastic 2-body interactions is much shorter than all other time scales of the problem. In such a case the 2-body interactions establish a Maxwellian distribution of particles in velocity space characterized by its temperature. Using this Maxwellian velocity distributions in the various velocity moments of the Vlasov equation yields the equations of *ideal MHD*. However, in (8.1.11) we have seen that in space plasmas the 2-body collision time scale is generally much longer than the time scale for interactions with the electromagnetic fields (see also Table 8.1), so that the equations of ideal MHD are appropriate for these systems only on scales large compared to the 2-body collision length or time. At shorter time or length scales a more careful justification for the MHD approach is required. Leaving this justification for the moment let us consider now in more detail the first velocity moments of the non-relativistic Vlasov equation.

8.1.3.1 Moments of the Vlasov Equation. Taking the zeroth velocity moment of the Vlasov equation (8.1.17) and using the definitions (8.1.18) and (8.1.25) immediately yields

$$\frac{\partial n_a}{\partial t} + \frac{\partial}{\partial \boldsymbol{x}} \cdot \left[n_a \boldsymbol{V}_a\right] + \frac{q_a}{m_a} \int_{-\infty}^{\infty} d^3v \left(\boldsymbol{E} + \frac{\boldsymbol{v} \times \boldsymbol{B}}{c}\right) \cdot \frac{\partial f_a}{\partial \boldsymbol{v}}$$
$$= \int_{-\infty}^{\infty} d^3v S_a \,. \tag{8.1.27}$$

Closer inspection of the third term on the left-hand side of (8.1.27) shows that this term vanishes. Consider for example the x component of this term:

$$\int dv_x dv_y dv_z \left[E_x + \frac{1}{c}(v_y B_z - v_z B_y)\right] \frac{\partial f_a}{\partial v_x}$$
$$= \int dv_y dv_z \left[E_x + \frac{1}{c}(v_y B_z - v_z B_y)\right] f_a \big|_{v_x = -\infty}^{v_x = \infty}$$
$$- \int dv_x f_a \frac{\partial \left[E_x + \frac{1}{c}(v_y B_z - v_z B_y)\right]}{\partial v_x} = 0 \,,$$

since f_a vanishes at $v_x = \pm\infty$ and the $\boldsymbol{E}$ and $\boldsymbol{B}$ as well as v_y and v_z do not depend on v_x. The y and z components vanish equally, and as a result (8.1.27) reduces to the mass flux continuity equation for the fluid species a

$$\frac{\partial n_a(\boldsymbol{x},t)}{\partial t} + \nabla \cdot (n_a(\boldsymbol{x},t)\boldsymbol{V}_a(\boldsymbol{x},t)) = Q_a(\boldsymbol{x},t), \tag{8.1.28}$$

where we introduce the mass source term $Q_a(\boldsymbol{x},t) \equiv \int_{-\infty}^{\infty} d^3v\, S_a(\boldsymbol{x},\boldsymbol{v},t)$.

Multiplying the Vlasov equation (8.1.17) with $\boldsymbol{v}$ and integrating over d^3v we obtain the first-order moment

$$\frac{\partial(n_a \boldsymbol{V}_a)}{\partial t} + \int d^3v\, \boldsymbol{v}\boldsymbol{v} \cdot \frac{\partial f_a}{\partial \boldsymbol{x}} + \frac{q_a}{m_a}\int_{-\infty}^{\infty} d^3v\, \boldsymbol{v}\left(\boldsymbol{E} + \frac{\boldsymbol{v}\times\boldsymbol{B}}{c}\right) \cdot \frac{\partial f_a}{\partial \boldsymbol{v}}$$
$$= \int_{-\infty}^{\infty} d^3v\, \boldsymbol{v} S_a. \tag{8.1.29}$$

By using the definitions (8.1.25) and (8.1.26) each component of the second term on the left-hand side of this equation reduces immediately to

$$m_a \sum_{i=1}^{3} \partial_{x_i} \int d^3v\, v_k v_i f_a = \sum_{i=1}^{3} \partial_{x_i} \Pi_{a,ki} + m_a \sum_{i=1}^{3} \partial_{x_i}(V_i V_k n_a), \tag{8.1.30}$$

where we use the abbreviation $\partial_x \equiv \partial/\partial x$.

With the Lorentz force $\boldsymbol{F} = (q_a/m_a)[\boldsymbol{E} + 1/c\ \boldsymbol{v}\times\boldsymbol{B}]$ we obtain, after partial integration for each component of the third term in (8.1.29),

$$\int d^3v\, v_k F_i \partial_{v_i} f_a = -\int d^3v\, f_a \partial_{v_i}(v_k F_i)$$
$$= -\int d^3v\, f_a\, F_i(\partial_{v_i} v_k)$$
$$= -\int d^3v\, f_a\, F_k$$
$$= -\frac{n_a q_a}{m_a}\left[\boldsymbol{E} + \frac{1}{c}\boldsymbol{V}_a\times\boldsymbol{B}\right]_k, \tag{8.1.31}$$

where we used $\partial_{v_i} F_i = 0$ and $\partial_{v_i} v_k = \delta_{ki}$. Collecting terms, we obtain for the left-hand side of (8.1.29)

$$m_a \partial_t(n_a\, V_k) + \sum_{i=1}^{3} \partial_{x_i} \Pi_{a,ki}$$
$$+ m_a \sum_{i=1}^{3} \partial_{x_i}(V_i V_k n_a)$$
$$- n_a\, q_a\left[\boldsymbol{E} + \frac{1}{c}\boldsymbol{V}_a\times\boldsymbol{B}\right]_k. \tag{8.1.32}$$

Using the mass continuity equation (8.1.28) in expression (8.1.32) yields the Euler force equation for species a

$$m_a n_a(\partial_t + \boldsymbol{V_a} \cdot \nabla)\boldsymbol{V_a} + \sum_{i=1}^{3} \partial_{x_i} \Pi_{a,ki}$$
$$= q_a n_a \left[\boldsymbol{E} + \frac{1}{c}\boldsymbol{V_a} \times \boldsymbol{B}\right] + Q\boldsymbol{P}_{a,k}(\boldsymbol{x}, t), \qquad (8.1.33)$$

where $Q\boldsymbol{P}_a(\boldsymbol{x}, t) \equiv m_a \int_{-\infty}^{\infty} d^3v (\boldsymbol{v} - \boldsymbol{V_a}) S_a(\boldsymbol{x}, \boldsymbol{v}, t)$ is the momentum source term. Equation (8.1.33) is also called the *momentum equation*, since it determines the time rate of change of momentum per unit volume.

Note that the continuity equation (8.1.28) for n_a involves the function $\boldsymbol{V_a}$, and the force equation (8.1.33) for $\boldsymbol{V_a}$ involves the pressure tensor Π_a. It is clear that every equation for n factors of $\boldsymbol{v}$ will involve a term with $n+1$ factors of $\boldsymbol{v}$ because the Vlasov equation (8.1.17) contains the term $\boldsymbol{v} \cdot \partial_{\boldsymbol{x}} f_a$. Thus, to obtain a complete description, we need an infinite number of moment equations as derived from the Vlasov equation. In practice, we seek to truncate this series of equations by using a physical argument to evaluate the term with $n+1$ factors of $\boldsymbol{v}$, rather than using the fluid equation for that term. For example, we can use physical arguments to evaluate the pressure tensor Π in (8.1.33), so that the force equation (8.1.33), the continuity equation (8.1.28) and Maxwell's equations become a complete description of the plasma. In a more advanced approach one may use the fluid equation for the pressure tensor and apply physical arguments to infer the terms with three components of velocity.

A popular simplification of the pressure tensor is the *cold plasma approximation* which assumes that all particles have the same macroscopic velocity $\boldsymbol{V_a}$. Then $f_a(\boldsymbol{x}, \boldsymbol{v}, t) = n_a(\boldsymbol{x}, t)\delta(\boldsymbol{v} - \boldsymbol{V_a})$ and according to (8.1.26)

$$\Pi_{a,\text{ cold}} = 0, \qquad (8.1.34)$$

the pressure tensor vanishes.

8.1.3.2 Anomalous MHD. In the last sections we established that the equations of ideal MHD fail to describe the dynamics of gases whose collision length is large compared to its physical size (as for galactic cosmic rays and the interplanetary medium) and the dynamics of plasmas at scales less than the collision length. For these systems the kinetic description based on the combination of the Vlasov equation and Maxwell's equations is most appropriate. After having solved for the phase space density f_a in such systems one may then use this solution in the moment equations (8.1.28) and (8.1.33) and work out the higher-order velocity terms such as the pressure tensor, etc. In such an approach the first three MHD moment equations correspond to the conservation equations for the mass, momentum and energy. However, the transport parameters in these MHD equations will differ from the ones of ideal MHD, which were based on the Maxwellian velocity distribution of particles, but are determined by the appropriate moments of the solution of the kinetic equation f_a which, depending on the dominating microscopic interaction processes, may differ considerably from the Maxwellian. These transport parameters are referred to as *anomalous* to differentiate them from their ideal values. Anomalous cosmic ray MHD has been used in connection with problems regarding the back-reaction of cosmic rays on the structure of shock waves and in work regarding the influence of cosmic rays on the large-scale galactic dynamics of interstellar matter. However, as emphasized this approach at first demands solution of the kinetic equations.

8.2 The Relativistic Vlasov Equation and Maxwell Equations

In Sect. 8.1 we have derived the basic plasma equations for non-relativistic particle velocities. For completeness we give here the relativistic versions of these equations for the electromagnetic fields $\boldsymbol{E}$ and $\boldsymbol{B}$ and the phase space density of plasma particles of sort a $f_a(\boldsymbol{x}, \boldsymbol{p}, t)$, where the particle momentum $\boldsymbol{p} = \gamma m_a \boldsymbol{v}$ and $\gamma = (1 - (v/c)^2)^{-1/2}$. The Vlasov equation is

$$\frac{\partial f_a}{\partial t} + \boldsymbol{v} \cdot \frac{\partial f_a}{\partial \boldsymbol{x}} + q_a \left[\boldsymbol{E}(\boldsymbol{x}, t) + \frac{\boldsymbol{v} \times \boldsymbol{B}(\boldsymbol{x}, t)}{c} \right] \cdot \frac{\partial f_a}{\partial \boldsymbol{p}} = S_a(\boldsymbol{x}, \boldsymbol{p}, t) , \quad (8.2.1)$$

and the Maxwell equations are

$$\nabla \times \boldsymbol{B} = \frac{1}{c} \frac{\partial \boldsymbol{E}}{\partial t} + \frac{4\pi}{c} \boldsymbol{J}$$

$$= \frac{1}{c} \frac{\partial \boldsymbol{E}}{\partial t} + \frac{4\pi}{c} \sum_a n_a q_a \int_{-\infty}^{\infty} d^3p \, \boldsymbol{v} f_a(\boldsymbol{x}, \boldsymbol{p}, t) + \frac{4\pi}{c} \boldsymbol{J}_{\text{ext}} , \quad (8.2.2)$$

$$\nabla \cdot \boldsymbol{B} = 0 , \quad (8.2.3)$$

$$\nabla \times \boldsymbol{E} = -\frac{1}{c} \frac{\partial \boldsymbol{B}}{\partial t} , \quad (8.2.4)$$

$$\nabla \cdot \boldsymbol{E} = 4\pi \rho = 4\pi \sum_a n_a q_a \int_{-\infty}^{\infty} d^3p \, f_a(\boldsymbol{x}, \boldsymbol{p}, t) + 4\pi \rho_{\text{ext}} , \quad (8.2.5)$$

Equations (8.2.3) and (8.2.5) are, in a sense, a consequence of (8.2.2) and (8.2.4); for, if we take the divergence of these latter two, there results

$$\frac{\partial}{\partial t} \nabla \cdot \boldsymbol{B} = 0 , \quad (8.2.6)$$

$$\frac{\partial}{\partial t} [\nabla \cdot \boldsymbol{E} - 4\pi \rho] = -4\pi \left[\frac{\partial \rho}{\partial t} + \nabla \cdot \boldsymbol{J} \right] = 0 , \quad (8.2.7)$$

because of charge conservation

$$\frac{\partial \rho}{\partial t} + \nabla \cdot \boldsymbol{J} = 0 \,. \tag{8.2.8}$$

Thus, if (8.2.3) and (8.2.5) are true at any particular time, (8.2.2) and (8.2.4) imply that they will be true for all time. Equations (8.2.3) and (8.2.5) may therefore be regarded as the initial condition for the electromagnetic field.

8.3 Kinetic Theory of Plasma Waves

It is difficult to solve the coupled system of the Vlasov equation and Maxwell equations analytically. In order to make some progress one investigates the behavior of small perturbations from a given known zeroth-order solution.

8.3.1 Linearization of Kinetic Plasma Equations

Let us assume we have found an equilibrium solution $(f_a^{(0)}, \boldsymbol{E}_0, \boldsymbol{B}_0)$ that exactly fulfills (8.2.1)–(8.2.5) with $\boldsymbol{J}_{\text{ext}} = 0$ and $\rho_{\text{ext}} = 0$. We then set

$$f_a(\boldsymbol{x}, \boldsymbol{p}, t) = f_a^{(0)}(\boldsymbol{x}, \boldsymbol{p}, t) + \epsilon \delta f_a(\boldsymbol{x}, \boldsymbol{p}, t) \,, \tag{8.3.1a}$$
$$\boldsymbol{E}(\boldsymbol{x}, t) = \boldsymbol{E}_0(\boldsymbol{x}, t) + \epsilon \delta \boldsymbol{E}(\boldsymbol{x}, t) \,, \tag{8.3.1b}$$
$$\boldsymbol{B}(\boldsymbol{x}, t) = \boldsymbol{B}_0(\boldsymbol{x}, t) + \epsilon \delta \boldsymbol{B}(\boldsymbol{x}, t) \,, \tag{8.3.1c}$$

with $\epsilon \ll 1$. Inserting (8.3.1) in (8.2.1)–(8.2.5), using the fact that the zeroth-order solution exactly fulfills the Vlasov-Maxwell equation systems, and neglecting all terms quadratic in the small number ϵ, we obtain the *linearized* Vlasov-Maxwell equations

$$\frac{\partial \delta f_a}{\partial t} + \boldsymbol{v} \cdot \frac{\partial \delta f_a}{\partial \boldsymbol{x}} + q_a \left[\boldsymbol{E}_0(\boldsymbol{x}, t) + \frac{\boldsymbol{v} \times \boldsymbol{B}_0(\boldsymbol{x}, t)}{c} \right] \cdot \frac{\partial \delta f_a}{\partial \boldsymbol{p}}$$
$$= - q_a \left[\delta \boldsymbol{E}(\boldsymbol{x}, t) + \frac{\boldsymbol{v} \times \delta \boldsymbol{B}(\boldsymbol{x}, t)}{c} \right] \cdot \frac{\partial f_a^{(0)}}{\partial \boldsymbol{p}} \,, \tag{8.3.2}$$

$$\nabla \times \delta \boldsymbol{B} = \frac{1}{c} \frac{\partial \delta \boldsymbol{E}}{\partial t} + \frac{4\pi}{c} \delta \boldsymbol{J}$$
$$= \frac{1}{c} \frac{\partial \delta \boldsymbol{E}}{\partial t} + \frac{4\pi}{c} \sum_a n_a q_a \int_{-\infty}^{\infty} \mathrm{d}^3 p \, \boldsymbol{v} \delta f_a(\boldsymbol{x}, \boldsymbol{p}, t) \,, \tag{8.3.3}$$

$$\nabla \cdot \delta \boldsymbol{B} = 0 \,, \tag{8.3.4}$$

$$\nabla \times \delta \boldsymbol{E} = -\frac{1}{c} \frac{\partial \delta \boldsymbol{B}}{\partial t} \,, \tag{8.3.5}$$

$$\nabla \cdot \delta \boldsymbol{E} = 4\pi \sum_a n_a q_a \int_{-\infty}^{\infty} \mathrm{d}^3 p \, \delta f_a(\boldsymbol{x}, \boldsymbol{p}, t) \,. \tag{8.3.6}$$

The linearized Vlasov-Maxwell equations relate the properties of the perturbations $(\delta f_a(\boldsymbol{x}, \boldsymbol{p}, t), \delta \boldsymbol{E}(\boldsymbol{x}, t), \delta \boldsymbol{B}(\boldsymbol{x}, t))$ to the prescribed zeroth-order solution $(f_a^{(0)}, \boldsymbol{E}_0, \boldsymbol{B}_0)$.

8.3.2 Solution of the Linearized Vlasov Equation

We follow the treatment by Baldwin et al. (1969 [26]). The structure of (8.3.2) permits one to construct a formal solution for δf_a in terms of $\delta \boldsymbol{E}$ and $\delta \boldsymbol{B}$. We note that the left-hand side of (8.3.2) can be integrated as a time derivative taken along the trajectory of a particle moving in the unperturbed electromagnetic field $\boldsymbol{E}_0$ and $\boldsymbol{B}_0$,

$$\frac{d\boldsymbol{p}}{dt} = q_a \left[\boldsymbol{E}_0(\boldsymbol{x},t) + \frac{\boldsymbol{v} \times \boldsymbol{B}_0(\boldsymbol{x},t)}{c} \right], \tag{8.3.7a}$$

$$\frac{d\boldsymbol{x}}{dt} = \boldsymbol{v} . \tag{8.3.7b}$$

This yields

$$\delta f_a(\boldsymbol{x},\boldsymbol{p},t) = \lim_{dt \to 0} \left[\delta f_a \right.$$
$$\left[\boldsymbol{x} + \boldsymbol{v} dt, \boldsymbol{p} + q_a \left(\boldsymbol{E}_0(\boldsymbol{x},t) + \frac{\boldsymbol{v} \times \boldsymbol{B}_0(\boldsymbol{x},t)}{c} \right) dt, t + dt \right]$$
$$\left. - \delta f_a(\boldsymbol{x},\boldsymbol{p},t) \right] . \tag{8.3.8}$$

which on Taylor expansion reduces in the limit $dt \to 0$ to the left-hand side of (8.3.2). Thus, if we introduce the vectors $\boldsymbol{x}'(\boldsymbol{x},\boldsymbol{p}; t'-t)$ and $\boldsymbol{p}'(\boldsymbol{x},\boldsymbol{p}; t'-t)$ such that

$$\frac{d\boldsymbol{p}'}{dt} = q_a \left[\boldsymbol{E}_0(\boldsymbol{x}') + \frac{\boldsymbol{v}' \times \boldsymbol{B}_0(\boldsymbol{x}')}{c} \right], \tag{8.3.9a}$$

$$\frac{d\boldsymbol{x}'}{dt} = \boldsymbol{v}' , \tag{8.3.9b}$$

and with $\boldsymbol{x}'(\boldsymbol{x},\boldsymbol{p};0) = \boldsymbol{x}$ and $\boldsymbol{p}'(\boldsymbol{x},\boldsymbol{p};0) = \boldsymbol{p}$ we can formally integrate (8.3.2) to obtain

$$\delta f_a(\boldsymbol{x},\boldsymbol{p},t)$$
$$= \delta f_a[\boldsymbol{x}', \boldsymbol{p}'(\boldsymbol{x},\boldsymbol{p},-t),0] - q_a \int_0^t dt'$$
$$\left[\delta \boldsymbol{E}[\boldsymbol{x}'(\boldsymbol{x},\boldsymbol{p},t'-t),t'] + \frac{\boldsymbol{v}'(\boldsymbol{x},\boldsymbol{p}; t'-t) \times \delta \boldsymbol{B}[\boldsymbol{x}'(\boldsymbol{x},\boldsymbol{p},t'-t),t']}{c} \right]$$
$$\cdot \frac{\partial f_a^{(0)}[\boldsymbol{x}'(\boldsymbol{x},\boldsymbol{p},t'-t),t']}{\partial \boldsymbol{p}'} . \tag{8.3.10}$$

Note that the boundary conditions on δf_a are automatically satisfied if they are satisfied by $f_a^{(0)}$.

Equation (8.3.10) expresses δf_a as a linear functional of $\delta \boldsymbol{E}$ and $\delta \boldsymbol{B}$ and one can construct therefrom a relation between the perturbation in the current density $\delta \boldsymbol{J}$ needed in (8.3.3) and the electric field perturbation $\delta \boldsymbol{E}$ as

$$\delta \boldsymbol{J} = \sum_a n_a\, q_a \int_{-\infty}^{\infty} \mathrm{d}^3 p\, \boldsymbol{v}\, \delta f_a(\boldsymbol{x},\boldsymbol{p},t) \stackrel{\mathrm{def}}{=} \boldsymbol{\sigma} \odot \delta \boldsymbol{E}\,. \tag{8.3.11}$$

where we introduced the *conductivity tensor* $\boldsymbol{\sigma}$. Equation (8.3.11) is referred to as the *generalized Ohm's law* relating the perturbations of the current density and the electric field components. Thus the problem has been reduced in principle to the joint solution of (8.3.3), (8.3.5), and (8.3.11).

8.3.3 Dispersion Relation

In general, (8.3.10) is not solvable because it is only under the simplest circumstances that the particle orbits are known explicitly. Since we have astrophysical applications in mind, we shall examine in great detail the case of an infinite homogeneous and stationary plasma, and treat large-scale spatial and temporal variations of the zeroth-order configuration ($f_a^{(0)}, \boldsymbol{E}_0, \boldsymbol{B}_0$) in the sense of a WKB approximation.

For the spatially homogeneous and stationary case the coefficients in the system of linear equations (8.3.3), (8.3.5), and (8.3.11) are independent of both $\boldsymbol{x}$ and t. By virtue of the independence of $\boldsymbol{x}$, we seek solutions whose space dependence is of the form of Fourier integrals

$$\delta \boldsymbol{E}(\boldsymbol{x},t) = \int_{-\infty}^{\infty} \mathrm{d}^3 k\, \boldsymbol{E}(\boldsymbol{k},t) \exp\,(\mathrm{i}\,\boldsymbol{k}\cdot\boldsymbol{x})\,, \tag{8.3.12a}$$

$$\delta \boldsymbol{B}(\boldsymbol{x},t) = \int_{-\infty}^{\infty} \mathrm{d}^3 k\, \boldsymbol{B}(\boldsymbol{k},t) \exp\,(\mathrm{i}\,\boldsymbol{k}\cdot\boldsymbol{x})\,, \tag{8.3.12b}$$

$$\delta f_a(\boldsymbol{x},\boldsymbol{p},t) = \int_{-\infty}^{\infty} \mathrm{d}^3 k\, f_a(\boldsymbol{k},\boldsymbol{p},t) \exp\,(\mathrm{i}\,\boldsymbol{k}\cdot\boldsymbol{x})\,, \tag{8.3.12c}$$

and, because of the independence of t, we introduce the Laplace transforms

$$\boldsymbol{E}_1(\boldsymbol{k},\omega) = \int_0^{\infty} \mathrm{d}t\, \boldsymbol{E}(\boldsymbol{k},t) \exp\,(\mathrm{i}\,\omega t)\,, \tag{8.3.13a}$$

$$\boldsymbol{B}_1(\boldsymbol{k},\omega) = \int_0^{\infty} \mathrm{d}t\, \boldsymbol{B}(\boldsymbol{k},t) \exp\,(\mathrm{i}\,\omega t)\,, \tag{8.3.13b}$$

$$f_{a,1}(\boldsymbol{k},\omega,\boldsymbol{p}) = \int_0^{\infty} \mathrm{d}t\, f_a(\boldsymbol{k},\boldsymbol{p},t) \exp\,(\mathrm{i}\,\omega t)\,, \tag{8.3.13c}$$

where the imaginary part of ω is chosen sufficiently positive so that the Laplace integral converges. The Maxwell equations (8.3.3) and (8.3.5) then yield

$$i\, c\mathbf{k}\times\mathbf{E}_1 = i\,\omega\mathbf{B}_1 + \mathbf{B}(\mathbf{k},0)\,, \tag{8.3.14}$$

$$i\, c\mathbf{k}\times\mathbf{B}_1 = 4\,\pi\mathbf{J}_1 - i\,\omega\mathbf{E}_1 - \mathbf{E}(\mathbf{k},0)\,, \tag{8.3.15}$$

where

$$\begin{aligned}\mathbf{J}_1(\mathbf{k},\omega) &= \frac{1}{2\pi^3}\int_0^\infty \mathrm{d}t \int_{-\infty}^\infty \mathrm{d}^3 k\,\delta\mathbf{J}(\mathbf{x},\mathbf{p},t)\exp\left[i\,(\omega t - \mathbf{k}\cdot\mathbf{x})\right] \\ &= \sum_a n_a\, q_a \int_{-\infty}^\infty \mathrm{d}^3 p\, \mathbf{v}\, f_{a,1}(\mathbf{k},\omega,\mathbf{p}) \end{aligned} \tag{8.3.16}$$

refers to the Fourier-Laplace transform of the current density perturbation $\delta\mathbf{J}$. In terms of the components of $\mathbf{J}_1(\mathbf{k},\omega)$ and $\mathbf{E}_1(\mathbf{k},\omega)$ Ohm's law (8.3.11) reads

$$J_{1,i}(\mathbf{k},\omega) = \sum_{j=1}^3 \sigma_{ij}(\mathbf{k},\omega) E_{1,j}(\mathbf{k},\omega)\,, \tag{8.3.17}$$

where $\sigma_{ij}(\mathbf{k},\omega)$ refers to the Fourier-Laplace transform of the conductivity tensor $\sigma_{ij}(\mathbf{x},t)$. Equation (8.3.14) readily yields

$$\mathbf{B}_1 = \frac{c}{\omega}\mathbf{k}\times\mathbf{E}_1 + \frac{i}{\omega}\mathbf{B}(\mathbf{k},0)\,. \tag{8.3.18}$$

Inserting (8.3.17) and (8.3.18) into (8.3.15) we obtain a single equation for $\delta\mathbf{E}(\mathbf{k},\omega)$:

$$(-\omega^2 + k^2 c^2 - 4\pi\, i\,\omega\sigma)\mathbf{E}_1 - c^2(\mathbf{k}\cdot\mathbf{E}_1)\mathbf{k} = i\, c\mathbf{k}\times\mathbf{B}(\mathbf{k},0) - i\,\omega\mathbf{E}(\mathbf{k},0)\,. \tag{8.3.19}$$

By introducing the normalized wave vector

$$\boldsymbol{\kappa} \equiv \mathbf{k}/|\mathbf{k}|\,, \tag{8.3.20}$$

(8.3.19) with Einstein's summation convention becomes

$$\Lambda_{nj} E_{1,j}(\mathbf{k},\omega) = A_n(\mathbf{k},\omega)\,, \tag{8.3.21}$$

with the *Maxwell operator*

$$\Lambda_{nj} \equiv \frac{k^2 c^2}{\omega^2}(\kappa_n \kappa_j - \delta_{nj}) + \psi_{nj} \tag{8.3.22}$$

defined in terms of the *dielectric tensor*

$$\psi_{nj} \equiv \delta_{nj} + \frac{4\pi i}{\omega}\sigma_{nj}\,, \tag{8.3.23}$$

that is determined by the (Fourier-Laplace transform of the) conductivity tensor $\sigma_{nj}(\mathbf{k},\omega)$. The initial electromagnetic field determines the components of the vector

$$A_n(\boldsymbol{k},\omega) \equiv \frac{\mathrm{i}}{\omega} E_n(\boldsymbol{k},0) - \frac{\mathrm{i}\, ck}{\omega^2} \epsilon_{nrs}\kappa_r B_s(\boldsymbol{k},0)\,. \tag{8.3.24}$$

δ_{nj} denotes the Kronecker symbol and ϵ_{nrs} the permutation symbol. Whereas the vector $\boldsymbol{A}$ contains the initial conditions of the perturbation, the Maxwell operator (8.3.22) is determined exclusively by the properties of the zeroth-order configuration $(f_a^{(0)}, \boldsymbol{E}_0, \boldsymbol{B}_0)$ via the conductivity tensor (8.3.17) entering the dielectric tensor (8.3.24).

Solving the wave equation (8.3.21) we obtain

$$\delta \boldsymbol{E}(\boldsymbol{k},\omega) = \boldsymbol{\Lambda}^{-1} \cdot \boldsymbol{A}\,, \tag{8.3.25}$$

where the inverse of the Maxwell operator with $\Lambda = \det[\Lambda_{ij}]$ is

$$\begin{aligned}\boldsymbol{\Lambda}^{-1} &= \frac{\boldsymbol{\Lambda}^T}{\Lambda}\\ &= \frac{1}{\Lambda} \begin{pmatrix} \Lambda_{22}\Lambda_{33} - \Lambda_{23}\Lambda_{32} & \Lambda_{31}\Lambda_{23} - \Lambda_{21}\Lambda_{33} & \Lambda_{21}\Lambda_{32} - \Lambda_{31}\Lambda_{22} \\ \Lambda_{32}\Lambda_{13} - \Lambda_{12}\Lambda_{33} & \Lambda_{11}\Lambda_{33} - \Lambda_{13}\Lambda_{31} & \Lambda_{12}\Lambda_{31} - \Lambda_{11}\Lambda_{32} \\ \Lambda_{12}\Lambda_{23} - \Lambda_{13}\Lambda_{22} & \Lambda_{21}\Lambda_{13} - \Lambda_{11}\Lambda_{23} & \Lambda_{11}\Lambda_{22} - \Lambda_{12}\Lambda_{21} \end{pmatrix}.\end{aligned} \tag{8.3.26}$$

There are two general properties of the dielectric tensor ψ_{nj} that we note (Melrose 1968 [338]). First, according to (8.3.12) and (8.3.13), $\psi_{nj}(\boldsymbol{k},\omega)$ is a Fourier-Laplace transform in space and time of a real tensor. The Fourier-Laplace transform $\mathrm{FL}(\boldsymbol{k},\omega)$ of any real quantity $F(\boldsymbol{x},t)$ satisfies

$$\mathrm{FL}\,(-\boldsymbol{k},-\omega) = \mathrm{FL}\,(\boldsymbol{k},\omega)^*\,, \tag{8.3.27}$$

where the asterisk denotes the complex conjugate. Hence

$$\psi_{nj}(-\boldsymbol{k},-\omega) = \psi_{nj}(\boldsymbol{k},\omega)^*\,. \tag{8.3.28}$$

The other property of ψ_{nj} is that in a magnetic field it satisfies the Onsager relations. These follow from arguments on the reality of ψ_{nj} on the transformation $\boldsymbol{B} \to -\boldsymbol{B}$, where $\boldsymbol{B}$ is the magnetic field. We will later [see (8.3.49)] choose our coordinate axis so that $\boldsymbol{B}$ is in the z direction and $\boldsymbol{k}$ in the x–z plane. Then the Onsager relations give

$$\psi_{xy} = -\psi_{yx}\,,\ \psi_{xz}^* = \psi_{zx}\,,\ \psi_{yz}^* = -\psi_{zy}\,. \tag{8.3.29}$$

Thus ψ_{xy}, ψ_{yz} are imaginary and all other components of ψ_{ij} are real.

Equation (8.3.25), together with (8.3.18), (8.3.10), (8.3.12) and (8.3.13), represent the formal solution to the given initial-value problem. Unfortunately, the integrals involved are much too difficult to perform analytically for any but the most trivial initial distribution functions $f_a^{(0)}$. For the special case of distributions which are analytic functions of momentum, several pieces of information can be extracted from (8.3.25) in the $t \to \infty$ limit, and we turn to this now.

8.3.4 The Landau Contour

The reason for demanding the initial distribution function be analytic is discussed in detail in Montgomery and Tidman (1964 [363], Sect.5.3), and we follow their discussion here. Consider a function $f(x)$ which is integrable along the real axis. Then consider the function $g(z)$, where z is a complex number, $\Im z \neq 0$,

$$g(z) = \int_{-\infty}^{\infty} \frac{f(x)\,\mathrm{d}x}{x - z} \ . \tag{8.3.30}$$

As we shall see below, the Maxwell operator $\Lambda(\mathbf{k}, \omega)$ depends on expressions of this type. Under very weak restrictions on $f(x)$, (8.3.30) for $g(z)$ defines a function which is analytic in *either* the upper or lower half plane, $\Im z > 0$ or $\Im z < 0$. Suppose, however, that $f(z)$ is an analytic function at all points $|z| < \infty$; such a function is said to be *entire*. If we define $h(z)$ for *all* z by the relations ($\imath = \sqrt{-1}$)

$$h(z) = \int_{-\infty}^{\infty} \frac{f(x)\,\mathrm{d}x}{x - z} \quad \text{for } \Im z > 0 \ , \tag{8.3.31a}$$

$$h(z) = \int_{-\infty}^{\infty} \frac{f(x)\,\mathrm{d}x}{x - z} + 2\pi\imath f(z) \quad \text{for } \Im z < 0 \tag{8.3.31b}$$

then we can make the considerably stronger statement that $h(z)$ itself is an entire function of z. Clearly, both expressions in (8.3.31) define functions analytic in their half planes of definition; *if* they approach the same limit as $\Im z \to 0$ from above (or below), then they are analytic continuations of each other. To demonstrate that they do approach the same limiting value, observe that the Plemelj formulae

$$\lim_{\epsilon \to 0} \frac{1}{x - (x' \pm \imath\epsilon)} = P\frac{1}{x - x'} \pm \imath\pi\delta(x - x') \tag{8.3.32}$$

show that, as $z \to y \pm \imath\epsilon, \epsilon \to 0$

$$h(y + \imath\epsilon) = P\int_{-\infty}^{\infty} \frac{f(x)\,\mathrm{d}x}{x - y} + \pi\imath f(y) \ , \tag{8.3.33a}$$

$$h(y - \imath\epsilon) = P\int_{-\infty}^{\infty} \frac{f(x)\,\mathrm{d}x}{x - y} - \pi\imath f(y) + 2\pi\imath f(y) \ . \tag{8.3.33b}$$

In (8.3.32) and (8.3.33), the standard notation P for the Cauchy principal-value integral has been adopted:

$$P\int_{-\infty}^{\infty} \frac{f(x)\,\mathrm{d}x}{x - y} \equiv \lim_{\eta \to 0} \left(\int_{-\infty}^{y-\eta} + \int_{y+\eta}^{\infty} \right) \frac{f(x)\,\mathrm{d}x}{x - y} \ . \tag{8.3.34}$$

Since the two analytic functions in (8.3.31) approach equal limits on the boundary of their regions of definition, each is the analytic continuation of the other.

8.3 Kinetic Theory of Plasma Waves

It is now readily seen what the implications are for analyticity of the initial distribution function $f_a^{(0)}$. Both the numerator and the denominator in the inverse of the Laplace transform (8.3.13a) with (8.3.25)

$$\begin{aligned} \boldsymbol{E}(\boldsymbol{k},t) &= \frac{1}{2\pi} \int_L d\omega \, \boldsymbol{E}_1(\boldsymbol{k},\omega) \exp(-\,\mathrm{i}\,\omega t) \\ &= \frac{1}{2\pi} \int_L d\omega \, \Lambda^{-1} \cdot \boldsymbol{A} \exp(-\,\mathrm{i}\,\omega t) \end{aligned} \qquad (8.3.35)$$

are entire functions of ω. The only possible singularities in the finite ω plane are poles, where the denominator Λ vanishes according to (8.3.26). This opens up the possibility of exploiting the freedom to deform the contour of the ω integration in (8.3.35) to any other path (Landau 1946 [280]), as long as poles are not crossed. We noted in the paragraph following (8.3.13) that the original path of integration L in the complex frequency plane is carried out along the Laplace contour as shown in Fig. 8.1. The Laplace contour must pass above all poles of $\boldsymbol{E}_1(\boldsymbol{k},\omega)$, which from (8.3.25) and (8.3.26) includes all zeros of $\Lambda(\boldsymbol{k},\omega)$,

$$\Lambda(\boldsymbol{k},\omega) \equiv \det[\Lambda_{nj}] = 0 , \qquad (8.3.36)$$

which relates values of the frequency ω to values of the wave vector $\boldsymbol{k}$, i.e.

$$\omega = \omega_m(\boldsymbol{k}), \quad m = 1, \ldots, N . \qquad (8.3.37)$$

Equations (8.3.36) and (8.3.37) are referred to as the *dispersion relation* that defines N possible wave modes.

Provided $\boldsymbol{E}_1(\boldsymbol{k},\omega)$ vanishes sufficiently rapidly as $|\omega| \to \infty$ along a line parallel to the real ω axis (as it always does in practice), we are free to deform the contour of integration from its original position in Fig. 8.1 to a different position as shown in Fig. 8.2, *so long as we pass around* the solutions of (8.3.36), i.e. $\omega_m(\boldsymbol{k})$ (see (8.3.37)). Equation (8.3.35) becomes, upon this change of contour,

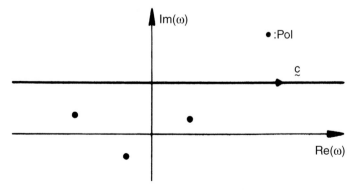

Fig. 8.1. The Laplace contour to invert the Laplace transform

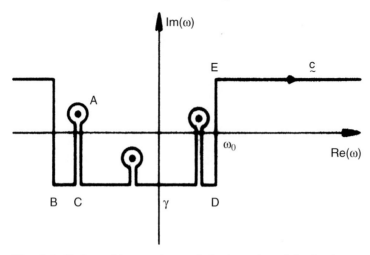

Fig. 8.2. Deformed integration path for inversion of the Laplace transform

$$\boldsymbol{E}(\boldsymbol{k},t) = \sum_m R_m \, e^{-i\,\omega_m(\boldsymbol{k})t} + \frac{1}{2\pi} \int_{-\infty-i\gamma}^{+\infty-i\gamma} d\omega \, \boldsymbol{E}_1(\boldsymbol{k},\omega) \exp(-i\omega t) \;, \tag{8.3.38}$$

where the contour has been pushed a distance γ *below* the real ω axis, as illustrated in Fig. 8.2, and where

$$R_m = i \lim_{\omega \to \omega_m(\boldsymbol{k})} (\omega - \omega_m(\boldsymbol{k})) \boldsymbol{E}_1(\boldsymbol{k},\omega) \tag{8.3.39}$$

is the residue at the pole of $1/\Lambda(\boldsymbol{k},\omega)$ at $\omega_m(\boldsymbol{k})$.

If $\boldsymbol{E}_1(\boldsymbol{k},\omega)$ satisfies some rather weak integrability conditions, the second term of $\boldsymbol{E}(\boldsymbol{k},t)$ in (8.3.38) is exponentially damped as $t \to \infty$, with respect to the pole terms. All poles with $\Im\omega_m < 0$, i.e. those below the real ω axis, lead to electric field terms $\boldsymbol{E}(\boldsymbol{k},t)$ that are eventually exponentially damped. The asymptotic $t \to \infty$ form of the electric field is therefore determined by the N possible plasma modes given by the dispersion relation (8.3.37),

$$\boldsymbol{E}(\boldsymbol{k}, t \to \infty) = \sum_{m=1}^{N} R_m \, e^{-i\,\omega_m(\boldsymbol{k})t} \;. \tag{8.3.40}$$

In general, the solutions $\omega_m(\boldsymbol{k})$ are complex; so, writing

$$\omega = \Re\omega + i\Im\omega \;, \tag{8.3.41}$$

we see that the time dependence $\boldsymbol{E}(\boldsymbol{k},t)$ becomes, for long times, a sinusoidal oscillation, which has also an exponential increase (if $\Im\omega_m > 0$) or decrease (if $\Im\omega_m < 0$) of the amplitude:

$$\boldsymbol{E}(\boldsymbol{k},t) \sim \sum_{m=1}^{N} R_m \, e^{-i\Re\omega_m(\boldsymbol{k})t} \, e^{\Im\omega_m(\boldsymbol{k})t} \;. \tag{8.3.42}$$

This damping of the field for $\Im\omega_m < 0$ is called *Landau damping* and is a dissipation process not associated with collisions. The *spatial* dependence of the mth Fourier component is a sinusoidal wave with phase velocity $\pm\Re\omega_m/|\boldsymbol{k}|$. Note that only in the limit $t \to \infty$ does there exist a dispersion relation which relates frequency to wavenumber (see (8.3.37)).

In the case of denominator zeros from (8.3.37) which lie *above* the real ω axis, i.e. with $\Im\omega_m > 0$, (8.3.42) would have predicted *growing* oscillations, or *instabilities*. Particularly interesting are those cases where the linear growth time of the instabilities, $\tau = (\Im\omega_m)^{-1}$, is short compared to the characteristic time scale of the initial configuration since then, after a few growth times, the growing oscillations will completely modify the initial configuration, so that our original ansatz of small fluctuations in the linearization (8.3.1) will become invalid.

However, in any case the investigation of the imaginary part of the frequency of the allowed plasma wave modes $\omega_m(\boldsymbol{k})$ will provide the important information whether the chosen initial state $(f_a^{(0)}, \boldsymbol{E}_0, \boldsymbol{B}_0)$ is *stable* (for $\Im\omega_m < 0$) or *unstable* (for $\Im\omega_m > 0$) to the growth of initially small perturbations. To what final configuration an unstable system develops cannot be answered by this investigation of linear stability.

8.3.5 Polarization Vector

The dispersion relation (8.3.36) is also useful to derive a general expression for the polarization state of the various wave modes. The polarization vector $\boldsymbol{e}$ for each mode is defined as

$$\boldsymbol{e} = \frac{\boldsymbol{E}_1}{|\boldsymbol{E}_1|}, \qquad (8.3.43a)$$

$$\boldsymbol{e}^* \cdot \boldsymbol{e} = 1, \qquad (8.3.43b)$$

where the asterisk * denotes the complex conjugate. Because of (8.3.14) $\boldsymbol{e} \perp \boldsymbol{k}$ (i.e. $\boldsymbol{e} \times \boldsymbol{k} = 0$). To calculate the polarization vector $\boldsymbol{e}$ we note, as in (8.3.26), that the determinant Λ can be expanded in terms of the matrix elements Λ_{ij} and the associated co-factors Λ_{ij}^T as

$$\Lambda_{ij}\Lambda_{jk}^T = \Lambda\delta_{ik}. \qquad (8.3.44)$$

Since for any real mode $\Lambda = 0$, (8.3.44) reduces to

$$\Lambda_{ij}\Lambda_{jk}^T = 0. \qquad (8.3.45)$$

Comparing (8.3.45) with (8.3.21) with its right hand side set to zero ($A_n = 0$) because of the dispersion relation (8.3.36), we see that for real modes it is possible to select

$$e_j = C\Lambda_{jk}^T a_k, \qquad (8.3.46)$$

as the polarization vector, where a_k is an arbitrary unit vector and C is a constant determined from the normalization condition (8.3.43b). The normalization constant in (8.3.46) may be written as (we use the Einstein sum convention throughout)

$$C = \left(\Lambda^T_{kl} a_k a_l \Lambda^T_{ss}\right)^{-1/2}, \qquad (8.3.47a)$$

where

$$\Lambda^T_{ss} = \Lambda^T_{11} + \Lambda^T_{22} + \Lambda^T_{33} \qquad (8.3.47b)$$

is the trace of Λ^T_{ij}. Therefore, the normalized polarization vector for waves with the dispersion relation (8.3.36) is

$$e_i = \frac{\Lambda^T_{ij} a_j}{(\Lambda^T_{kl} a_k a_l \Lambda^T_{ss})^{1/2}}. \qquad (8.3.48)$$

8.3.6 Conductivity Tensor in a Homogeneous Hot Magnetized Plasma

In order to calculate quantitatively the real and imaginary parts of the normal mode frequencies in terms of the zeroth-order distribution function $f_a^{(0)}$ we have to know the conductivity tensor $\boldsymbol{\sigma}$ that determines the Maxwell operator (8.3.22) through the dielectric tensor (8.3.23). This is accomplished, according to (8.3.11), by using the integration of the linearized Vlasov equation (8.3.2) along the unperturbed particle orbit, as indicated formally in (8.3.10). We restrict our discussion to the case of a uniform magnetic field and a vanishing uniform electric field $\boldsymbol{E}_0 = 0$. This case is of particular importance in astrophysics, because the high electrical conductivity of all space plasmas short out any large-scale electric fields.

Without loss of generality we choose a coordinate system (x, y, z) where the z axis is parallel to the magnetic field $\boldsymbol{B}_0$, and where the wave vector $\boldsymbol{k}$ lies in the $(x$–$z)$ plane only (see also Fig. 8.3), i.e.

$$\boldsymbol{k} = k_{\parallel} \boldsymbol{e}_z + k_{\perp} \boldsymbol{e}_x. \qquad (8.3.49a)$$

With this wave vector representation the components of the normalized wave vector (8.3.20) are

$$\kappa_x = \sin\theta, \quad \kappa_y = 0, \quad \kappa_z = \cos\theta, \qquad (8.3.49b)$$

where θ is the angle between $\boldsymbol{k}$ and $\boldsymbol{B}_0$. The Maxwell operator (8.3.22) then reads

$$\Lambda_{ij} = \psi_{ij} - N^2 \begin{pmatrix} \cos^2\theta & 0 & -\sin\theta\cos\theta \\ 0 & 1 & 0 \\ -\sin\theta\cos\theta & 0 & \sin^2\theta \end{pmatrix}, \qquad (8.3.50)$$

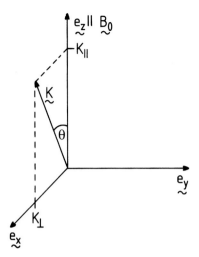

Fig. 8.3. Coordinate system chosen

where we have defined the *index of refraction*

$$N^2 \equiv \frac{k^2 c^2}{\omega^2} . \tag{8.3.51}$$

We note, that according to (8.3.50), (8.3.28) and (8.3.36), Λ is only a function of $k_\|^2$, $k_\perp^2$, and ω^2. Consequently, the solution $\omega = -\omega(\boldsymbol{k})$ does not define a new mode. Also we may chose to set $\omega(-\boldsymbol{k}) = \omega(\boldsymbol{k})$, or convert all negative frequencies into positive frequencies.

As shown in Appendix B.1 the Fourier-Laplace transform of the conductivity tensor in a hot plasma with a different sorts of particles in a uniform magnetic field is

$$\sigma_{ij}(k_\|, k_\perp, \omega) = \frac{2\pi}{i} \sum_a e_a^2 n_a \int_{-\infty}^\infty dp_\| \int_0^\infty dp_\perp p_\perp \sum_{n=-\infty}^\infty \frac{T_{ij}}{\omega - k_\| v_\| - n\Omega_a} , \tag{8.3.52}$$

with

$$T_{ij} \equiv \begin{pmatrix} \frac{n^2 J_n^2(z)}{z^2} v_\perp \boldsymbol{U} f_a^{(0)} & \frac{in J_n(z) J_n'(z)}{z} v_\perp \boldsymbol{U} f_a^{(0)} & \frac{n J_n^2(z)}{z} v_\perp \boldsymbol{W} f_a^{(0)} \\ \frac{-in J_n(z) J_n'(z)}{z} v_\perp \boldsymbol{U} f_a^{(0)} & (J_n'(z))^2 v_\perp \boldsymbol{U} f_a^{(0)} & -i J_n(z) J_n'(z) v_\perp \boldsymbol{W} f_a^{(0)} \\ \frac{n J_n^2(z)}{z} v_\| \boldsymbol{U} f_a^{(0)} & i J_n(z) J_n'(z) v_\| \boldsymbol{U} f_a^{(0)} & J_n^2(z) v_\| \boldsymbol{W} f_a^{(0)} \end{pmatrix} ,$$

$$\tag{8.3.53}$$

$$\mathbf{U} f_a^{(0)} = \frac{\partial f_a^{(0)}}{\partial p_\perp} + \frac{k_\parallel}{\gamma m_a \omega} \left(p_\perp \frac{\partial f_a^{(0)}}{\partial p_\parallel} - p_\parallel \frac{\partial f_a^{(0)}}{\partial p_\perp} \right), \qquad (8.3.54)$$

$$\mathbf{W} f_a^{(0)} = \frac{\partial f_a^{(0)}}{\partial p_\parallel} - \frac{n \Omega_a}{\omega p_\perp} \left(p_\perp \frac{\partial f_a^{(0)}}{\partial p_\parallel} - p_\parallel \frac{\partial f_a^{(0)}}{\partial p_\perp} \right), \qquad (8.3.55)$$

$$z \equiv \frac{k_\perp v_\perp}{\Omega_a}, \qquad (8.3.56)$$

where $\Omega_a = \epsilon_a \Omega_{0a}/\gamma$ is the relativistic gyrofrequency of particles of sort a, $\epsilon_a = e_a/|e_a|$ the charge sign, and

$$\Omega_{0a} = \frac{|e_a| B_0}{m_a c} \text{ Hz} \qquad (8.3.57)$$

the absolute value of the non-relativistic gyrofrequency of particles of sort a. $J_n(z)$ denotes the Bessel function of order n and $J'_n(z) \equiv \partial J_n(z)/\partial z$ its first derivative.

8.3.6.1 A Useful Identity for Sums of Bessel Functions.
Lerche (1966 [290], 1974 [292]) has given a useful method of summing the infinite series of Bessel functions occurring in expressions (8.3.52) and (8.3.53). [2] Let us consider the sum

$$S \equiv \sum_{n=-\infty}^{\infty} \frac{J_n^2(z)}{n - \alpha}. \qquad (8.3.58)$$

By using Neumann's integral representation (e.g. Watson 1966 [556], p.150) for the product of two Bessel functions,

$$J_\mu(z) J_\nu(z) = 2\pi^{-1} \int_0^{\pi/2} d\theta \, J_{\mu+\nu}(2z \cos\theta) \cos[(\mu - \nu)\theta], \qquad (8.3.59a)$$

which is valid if $\Re(\mu + \nu) > -1$ except when μ and ν are integers when

$$J_n^2(z) = 2\pi^{-1} (-1)^n \int_0^{\pi/2} d\theta \, J_0(2z \cos\theta) \cos(2n\theta), \qquad (8.3.59b)$$

the sum (8.3.58) reduces to

$$S = 2\pi^{-1} \int_0^{\pi/2} d\theta \, J_0(2z \cos\theta) \sum_{n=-\infty}^{\infty} \frac{(-1)^n \cos(2n\theta)}{n - \alpha}. \qquad (8.3.60)$$

Now

$$\sum_{n=-\infty}^{\infty} \frac{(-1)^n \cos(2n\theta)}{n - \alpha} = -\pi \csc(\pi \alpha) \cos(2\alpha\theta), \qquad (8.3.61)$$

---

[2] In his monograph, Swanson (1989 [534]) attributes this identity to a paper by Newberger (1982 [373]) who, independently from Lerche (1966 [290], 1974 [292]), has re-derived this sum rule.

so that

$$S(\alpha, z) = -\frac{\pi J_{-\alpha}(z) J_\alpha(z)}{\sin(\pi\alpha)} = -\Gamma[\alpha]\Gamma[1-\alpha]J_{-\alpha}(z)J_\alpha(z) , \quad (8.3.62)$$

where we have used Neumann's integral representation (8.3.59a) again.

From the series representation for the product of Bessel functions (Watson 1966 [556], p.147),

$$J_{-\alpha}(z)J_\alpha(z) = \sum_{k=0}^{\infty} \frac{(-1)^k \Gamma[2k+1](z/2)^{2k}}{k\Gamma[k+1]\Gamma[k+1+\alpha]\Gamma[k+1-\alpha]} , \quad (8.3.63)$$

one readily finds the asymptotic behavior of the function $S(\alpha, z)$ for small arguments $z \to 0$

$$S(\alpha, z \ll 1) \simeq -\frac{1}{\alpha} + \frac{z^2}{2\alpha(1-\alpha^2)} , \quad (8.3.64a)$$

$$\frac{\partial S(\alpha, z \ll 1)}{\partial z} \simeq \frac{z}{\alpha(1-\alpha^2)} , \quad (8.3.64b)$$

$$\frac{\partial^2 S(\alpha, z \ll 1)}{\partial z^2} \simeq \frac{1}{\alpha(1-\alpha^2)} . \quad (8.3.64c)$$

8.3.6.2 Alternative Expression for the Conductivity Tensor. All infinite sums occurring in the conductivity tensor expressions (8.3.52) and (8.3.53) can be expressed in terms of the closed analytical expression (8.3.62) as follows:

$$\sum_{n=-\infty}^{\infty} \frac{n^2 J_n^2(z)}{n-\alpha} = I_{11}(\alpha, z) = \alpha + \alpha^2 S(\alpha, z) , \quad (8.3.65a)$$

$$\sum_{n=-\infty}^{\infty} \frac{n J_n^2(z)}{n-\alpha} = I_{13}(\alpha, z) = 1 + \alpha S(\alpha, z) , \quad (8.3.65b)$$

$$\sum_{n=-\infty}^{\infty} \frac{J_n^2(z)}{n-\alpha} = I_{33}(\alpha, z) = S(\alpha, z) , \quad (8.3.65c)$$

$$\sum_{n=-\infty}^{\infty} \frac{J_n(z) J_n'(z)}{n-\alpha} = I_{23}(\alpha, z) = \frac{1}{2}\frac{\partial S(\alpha, z)}{\partial z} , \quad (8.3.65d)$$

$$\sum_{n=-\infty}^{\infty} \frac{n J_n(z) J_n'(z)}{n-\alpha} = I_{12}(\alpha, z) = \frac{\alpha}{2}\frac{\partial S(\alpha, z)}{\partial z} , \quad (8.3.65e)$$

$$\sum_{n=-\infty}^{\infty} \frac{(J_n'(z))^2}{n-\alpha} = I_{22}(\alpha, z) = \frac{1}{2}\frac{\partial^2 S(\alpha, z)}{\partial z^2}$$

$$+ \frac{1}{2z}\frac{\partial S(\alpha, z)}{\partial z} + \left(1 - \frac{\alpha^2}{z^2}\right)S(\alpha, z) - \frac{\alpha}{z^2} . \quad (8.3.65f)$$

Whereas the identities (8.3.65a)–(8.3.65e) follow readily from the respective differentiations on both sides of the identity (8.3.62), the last identity (8.3.65f) is obtained by considering $\partial I_{23}(\alpha, z)/\partial z$ and using Bessel's equation for $\partial^2 J_n/\partial z^2$.

Identifying
$$\alpha = \frac{\omega - k_\| v_\|}{\Omega_a} \tag{8.3.66}$$

and using the identities (8.3.65) reduces the conductivity tensor (8.3.52) to

$$\sigma_{ij}(k_\|, k_\perp, \omega) = 2\pi\imath \sum_a e_a^2 n_a \int_{-\infty}^{\infty} dp_\| \int_0^{\infty} dp_\perp \frac{p_\perp}{\Omega_a}$$

$$\begin{pmatrix} \frac{I_{11}(\alpha,z)}{z^2} v_\perp \boldsymbol{U} f_a^{(0)} & \frac{\imath I_{12}(\alpha,z)}{z} v_\perp \boldsymbol{U} f_a^{(0)} & \frac{v_\perp I_{13}(\alpha,z)}{z}\frac{\partial f_a^{(0)}}{\partial p_\|} - \frac{v_\perp I_{11}(\alpha,z)}{z} \boldsymbol{V} f_a^{(0)} \\ -\frac{\imath I_{12}(\alpha,z)}{z} v_\perp \boldsymbol{U} f_a^{(0)} & I_{22}(\alpha,z) v_\perp \boldsymbol{U} f_a^{(0)} - \imath v_\perp I_{23}(\alpha,z)\frac{\partial f_a^{(0)}}{\partial p_\|} & +\imath v_\perp I_{12}(\alpha,z) \boldsymbol{V} f_a^{(0)} \\ \frac{I_{13}(\alpha,z)}{z} v_\| \boldsymbol{U} f_a^{(0)} & \imath I_{23}(\alpha,z) v_\| \boldsymbol{U} f_a^{(0)} & v_\| I_{33}(\alpha,z)\frac{\partial f_a^{(0)}}{\partial p_\|} - v_\| I_{13}(\alpha,z) \boldsymbol{V} f_a^{(0)} \end{pmatrix}$$

$$\tag{8.3.67}$$

with
$$\boldsymbol{V} f_a^{(0)} = \frac{\Omega_a}{\omega p_\perp}\left(p_\perp \frac{\partial f_a^{(0)}}{\partial p_\|} - p_\| \frac{\partial f_a^{(0)}}{\partial p_\perp}\right). \tag{8.3.68}$$

In the next chapters we shall reduce both expressions (8.3.52) and (8.3.67) for the general conductivity tensor to simpler forms when we restrict our discussion to waves in cold plasmas, Maxwellian plasmas and wave propagation parallel or perpendicular to the uniform magnetic field.

8.3.7 Energy Transfer in Dispersive Media

The equation expressing the conservation of energy is obtained by multiplying the Maxwell equation (8.1.23) by $\boldsymbol{B}(\boldsymbol{x}, t)$ and the Maxwell equation (8.1.21) by $-\boldsymbol{E}(\boldsymbol{x}, t)$, adding and multiplying the result with $(c/(4\pi))$. This results in

$$\nabla \cdot \left(\frac{c}{4\pi}\boldsymbol{E}\times\boldsymbol{B}\right) + \frac{\partial}{\partial t}\left[\frac{B^2}{8\pi} + \frac{E^2}{8\pi}\right] = -\boldsymbol{J}\cdot\boldsymbol{E}. \tag{8.3.69}$$

The quantities on the left-hand side of (8.3.69) describe the total change of the electromagnetic field energy. Averaged over a long time T, the time-averaged rate of power absorption is

$$\langle P \rangle = \lim_{T\to\infty}\frac{1}{T}\int_{-T/2}^{T/2} dt \int d^3x \boldsymbol{J}(\boldsymbol{x}, t) \cdot \boldsymbol{E}(\boldsymbol{x}, t). \tag{8.3.70}$$

By expressing the perturbations of $\boldsymbol{J}$ and $\boldsymbol{E}$ in terms of their Fourier-Laplace transforms (8.3.12) and (8.3.13), (8.3.70) becomes

$$\langle P \rangle = \lim_{T \to \infty} \frac{1}{(2\pi)^4 T} \int d^3k \, d\omega \, \boldsymbol{J}_1(\boldsymbol{k},\omega)^* \cdot \boldsymbol{E}_1(\boldsymbol{k},\omega) \,. \tag{8.3.71}$$

As pointed out in Sect. 8.3.6, the negative and positive frequencies are equivalent physically, so separating $\int_{-\infty}^{0} + \int_{0}^{\infty}$ and using (8.3.27) yields

$$\langle P \rangle = \lim_{T \to \infty} \frac{1}{(2\pi)^4 T}$$
$$\times \int d^3k \int_0^\infty d\omega \left[\boldsymbol{J}_1(\boldsymbol{k},\omega)^* \cdot \boldsymbol{E}_1(\boldsymbol{k},\omega) + \boldsymbol{J}_1(\boldsymbol{k},\omega) \cdot \boldsymbol{E}_1(\boldsymbol{k},\omega)^* \right] \,. \tag{8.3.72}$$

In the absence of external currents we may use Ohm's law (8.3.17) and the definition of the dielectric tensor (8.3.23) to obtain

$$\langle P \rangle = \lim_{T \to \infty} \frac{1}{(2\pi)^4 T}$$
$$\times \int d^3k \int_0^\infty d\omega \, \frac{i\omega}{4\pi} \left[\psi_{ij}^* - \psi_{ji} \right] E_{1,i}(\boldsymbol{k},\omega) E_{1,j}(\boldsymbol{k},\omega)^* \,. \tag{8.3.73}$$

Equation (8.3.73) implies that energy transfer in a plasma is associated with a non-Hermitian dielectric tensor. If, on the other hand, the dielectric tensor is Hermitian, i.e.

$$\psi_{ij}^* = \psi_{ji} \,, \tag{8.3.74}$$

the plasma will be loss-free.

9. Test Wave Approach 1.
Waves in Cold Magnetized Plasmas

According to the kinetic theory of plasma waves (Sect. 8.3) the electromagnetic fluctuations in a given initial configuration ($f_a^{(0)}, \boldsymbol{E}_0, \boldsymbol{B}_0$), after an initial transit period, will consist of a spectrum of normal modes at different wavenumbers, see (8.3.42). In the following we will discuss the resulting plasma waves for different assumed initial configurations of magnetized plasma. After treating plasma waves in cold and hot plasmas in this and the next chapter we are ready to investigate the situation characteristics for cosmic rays in space plasmas, where relativistic cosmic rays co-exist with a background interstellar or intergalactic plasma.

9.1 Plasma Waves
in a Cold Moving Magnetized Plasma

We start from the conductivity tensor of a hot magnetized plasma written in a coordinate system where the ordered magnetic field lies parallel to the z axis and the wave vector is restricted to the (x–z) plane, which has been derived in Sect. 8.3.6. We consider a non-isotropic distribution of cold particles of density n_a that stream with velocity $U = P_0/(\Gamma m_a)$ parallel to the ordered magnetic field,

$$f_a^{(0)} = (2\pi p_\perp)^{-1}\delta(p_\perp)\delta(p_\| - P_0) \,. \tag{9.1.1}$$

The corresponding conductivity tensor follows by inserting into the general expression (8.3.67). Consider for example σ_{11}. Changing the momentum variables from $(p_\|, p_\perp)$ to $(w = p_\| - P_0, z = k_\perp v_\perp/\Omega_a)$ we obtain

$$\sigma_{11}^{\text{cold}} = \imath \sum_a \frac{e_a^2 n_a}{\Gamma m_a \Omega_a} \int_{-\infty}^{\infty} dw\, \delta(w) \int_0^\infty dz\, \delta(z)$$
$$\times \left(2\left[\frac{k_\|(P_0 + w)}{\Gamma m_a \omega} - 1\right] - \frac{k_\| \Gamma m_a \Omega_a^2 z^2}{\omega k_\perp^2 w} \right) \frac{I_{11}(\alpha, z)}{z^2}\,. \tag{9.1.2}$$

By using (8.3.65a), (8.3.64a) and (8.3.64c) we find for the two integrals

$$\int_0^\infty dz\, \delta(z) I_{11}(\alpha, z) = I_{11}(\alpha, z=0) = \alpha + \alpha^2 S(\alpha, 0) = 0 \,, \tag{9.1.3}$$

and

$$\int_0^\infty dz \delta(z) \frac{I_{11}(\alpha, z)}{z^2} = \lim_{z \to 0} \frac{I_{11}(\alpha, z)}{z^2}$$
$$= \frac{1}{2} \frac{\partial^2 I_{11}(\alpha, z)}{\partial z^2}\Big|_{z=0} = \frac{\alpha}{2(1-\alpha^2)}, \quad (9.1.4)$$

where in the latter we have employed L'Hopital's rule. Inserting the definition of α according to (8.3.66) and doing the remaining w-integration we arrive at

$$\sigma_{11}^{\text{cold}} = \frac{i}{4\pi} \sum_a \frac{\omega_{p,a}^2}{\omega \Gamma} \frac{(\omega - k_\| U)^2}{(\omega - k_\| U)^2 - \Omega_a^2}, \quad (9.1.5)$$

where we have introduced the *plasma frequency* of particles of species a with mass m_a, charge $e_a = Z_a e$ and density n_a,

$$\omega_{p,a} \equiv \left(\frac{4\pi Z_a^2 e^2 n_a}{m_a}\right)^{1/2} \text{ Hz}, \quad (9.1.6)$$

which, in the case of electrons, reduces to (8.1.5).

The other eight components of the conductivity tensor are evaluated likewise. Inserting the result into (8.3.23) we obtain for the dielectric tensor

$$\psi_{ij}^{\text{cold}} = \begin{pmatrix} \psi_{11} & \psi_{12} & \psi_{13} \\ -\psi_{12} & \psi_{11} & \psi_{23} \\ \psi_{13} & -\psi_{23} & \psi_{33} \end{pmatrix}, \quad (9.1.7)$$

where

$$\psi_{11} = 1 - \frac{1}{\omega^2} \sum_a \frac{\omega_{p,a}^2}{\Gamma} \frac{(\omega - k_\| U)^2}{(\omega - k_\| U)^2 - \Omega_a^2}, \quad (9.1.8)$$

$$\psi_{12} = -\frac{i}{\omega^2} \sum_a \frac{\omega_{p,a}^2}{\Gamma} \frac{\Omega_a(\omega - k_\| U)}{(\omega - k_\| U)^2 - \Omega_a^2}, \quad (9.1.9)$$

$$\psi_{13} = \frac{1}{\omega^2} \sum_a \frac{\omega_{p,a}^2}{\Gamma} \frac{k_\perp U(\omega - k_\| U)}{(\omega - k_\| U)^2 - \Omega_a^2}, \quad (9.1.10)$$

$$\psi_{23} = \frac{i}{\omega^2} \sum_a \frac{\omega_{p,a}^2}{\Gamma} \frac{k_\perp U \Omega_a}{(\omega - k_\| U)^2 - \Omega_a^2}, \quad (9.1.11)$$

$$\psi_{33} = 1 - \frac{1}{\omega^2} \sum_a \frac{\omega_{p,a}^2}{\Gamma} \left[\frac{\omega^2}{\Gamma^2(\omega - k_\| U)^2} + \frac{k_\perp^2 U^2}{(\omega - k_\| U)^2 - \Omega_a^2}\right]. \quad (9.1.12)$$

Calculating the Maxwell operator (8.3.50) with this dielectric tensor and evaluating its determinant yields the dispersion relation in this system according to (8.3.36). The dielectric tensor (9.1.7) will be useful to analyze the allowed plasma modes for the case of two particle distributions streaming with respect to each other (see Sect. 11.1). As the first non-trivial case we consider now the resulting plasma waves in a cold plasma at rest.

9.2 Plasma Waves in a Cold Magnetized Plasma at Rest

If the cold plasma is at rest,[1] i.e. $U = P_0 = 0$ in the distribution function (9.1.1), the dielectric tensor (9.1.7) reduces to the form given by Stix (1962 [525])

$$\psi_{ij}^{\text{cold, rest}} = \begin{pmatrix} S & -iD & 0 \\ iD & S & 0 \\ 0 & 0 & P \end{pmatrix}, \qquad (9.2.1)$$

where we have introduced

$$S \equiv (R+L)/2, \qquad (9.2.2a)$$
$$D \equiv (R-L)/2, \qquad (9.2.2b)$$

with

$$R \equiv 1 - \sum_a \frac{\omega_{p,a}^2}{\omega^2} \frac{\omega}{\omega + \Omega_a}, \qquad (9.2.2c)$$

$$L \equiv 1 - \sum_a \frac{\omega_{p,a}^2}{\omega^2} \frac{\omega}{\omega - \Omega_a}, \qquad (9.2.2d)$$

$$P \equiv 1 - \sum_a \frac{\omega_{p,a}^2}{\omega^2}. \qquad (9.2.2e)$$

Evidently $\psi_{ij}^{\text{cold, rest}}$ is Hermitian, so that according to (8.3.74) a cold plasma at rest is loss-free.

With (9.2.1) the Maxwell operator (8.3.50) reads

$$\Lambda_{ij}^{\text{cold, rest}} = \begin{pmatrix} S - N^2 \cos^2\theta & -iD & N^2 \sin\theta \cos\theta \\ iD & S - N^2 & 0 \\ N^2 \sin\theta \cos\theta & 0 & P - N^2 \sin^2\theta \end{pmatrix}$$

$$= \begin{pmatrix} S - N_{\parallel}^2 & -iD & N_{\perp} N_{\parallel} \\ iD & S - N^2 & 0 \\ N_{\perp} N_{\parallel} & 0 & P - N_{\perp}^2 \end{pmatrix}, \qquad (9.2.3)$$

where we have defined

$$N_{\parallel} \equiv N \cos\theta \quad N_{\perp} \equiv N \sin\theta. \qquad (9.2.4)$$

The dispersion relation (8.3.36) for plasma waves in a cold magnetized plasma at rest is then

---

[1] This special case also applies to the physical situation of a plasma steaming with one bulk velocity U, if we transform to the comoving coordinate system.

$$\Lambda^{\text{cold, rest}} = \det\left[\Lambda_{ij}^{\text{cold, rest}}\right] = aN^4 - bN^2 + c = 0, \tag{9.2.5}$$

with

$$a \equiv S\sin^2\theta + P\cos^2\theta, \tag{9.2.6a}$$
$$b \equiv (S^2 - D^2)\sin^2\theta + PS(1 + \cos^2\theta)$$
$$= RL\sin^2\theta + PS(1 + \cos^2\theta), \tag{9.2.6b}$$
$$c \equiv P(S^2 - D^2) = PRL. \tag{9.2.6c}$$

Equation (9.2.5) was first derived by Aström (1950 [19]). Solving this biquadratic equation yields two cold plasma modes (provided $a \neq 0$)

$$N^2 = N_\pm^2 = \frac{b \pm f}{2a}, \tag{9.2.7a}$$

where

$$f \equiv \sqrt{b^2 - 4ac} = \sqrt{(RL - PS)^2 \sin^4\theta + 4P^2D^2\cos^2\theta}. \tag{9.2.7b}$$

Note that f^2 is positive definite for all values of P, S, R, L, D, and θ.

Equation (9.2.5) can be put in yet another form, which is particularly suited for studying the various wave modes propagating along ($\theta = 0$) or across ($\theta = \pi/2$) the external magnetic field. Expressing $\sin\theta$ and $\cos\theta$ in terms of $\tan\theta$ in (9.2.5) we find, after some simple manipulations

$$\tan^2\theta = -\frac{P(N^2 - R)(N^2 - L)}{S(N^2 - P)[N^2 - (RL/S)]}, \tag{9.2.8}$$

from which one immediately obtains as solutions for

$$\theta = 0: P = 0, \ N^2 = R, \ N^2 = L, \tag{9.2.9}$$
$$\theta = \pi/2: N^2 = P, \ N^2 = (RL/S). \tag{9.2.10}$$

The physical significance of these waves will be discussed later.

From (9.2.5) we note that for certain values of the plasma parameters, N^2 goes to zero or infinity. In the literature, the former case is referred to as a *cutoff*, whereas the latter as a *resonance*. Cutoffs occur, according to (9.2.5) and (9.2.6c), when $c = RPL = 0$, i.e. $P = 0$, or $R = 0$, or $L = 0$. On the other hand, a resonance occurs for propagation at the angle θ_r which satisfies the criterion

$$\tan\theta_r = -P/S, \tag{9.2.11}$$

where we have used (9.2.8) in the limit of infinite $N \to \infty$.

The co-factors Λ^T can be calculated according to (8.3.26) using the Maxwell operator (9.2.3),

$$\Lambda^{T,\text{cold, rest}} = \begin{pmatrix} (S-N^2)(P-N_\perp^2) & -\imath D\,(P-N_\perp^2) & -N_\perp N_\|(S-N^2) \\ \imath D\,(P-N_\perp^2) & SP - PN_\|^2 - SN_\perp^2 & -\imath D N_\perp N_\| \\ -N_\perp N_\|(S-N^2) & \imath D N_\perp N_\| & D^2 - (S-N^2)(S-N_\|^2) \end{pmatrix}.$$

(9.2.12)

Choosing without any loss of generality $\boldsymbol{a} = (0,1,0)$ as arbitrary unit vector the polarization vectors $\boldsymbol{e}$ can be constructed according to (8.3.46) as

$$\boldsymbol{e} = \left(C\Lambda_{12}^T, C\Lambda_{22}^T, C\Lambda_{23}^T\right) = C \begin{pmatrix} -\imath D\,(P-N_\perp^2) \\ S\,(P-N_\perp^2) - PN_\|^2 \\ \imath D\, N_\perp N_\| \end{pmatrix}, \quad (9.2.13)$$

with the normalization

$$C = \left(\Lambda_{22}^T \Lambda_{ss}^T\right)^{-1/2} = \left[\left(SP - PN_\|^2 - SN_\perp^2\right) \right.$$
$$\left. \times \left(SP - PN_\|^2 - SN_\perp^2 + (S-N^2)\left(P - S + N_\|^2 - N_\perp^2\right) + D^2\right)\right]^{-1/2}.$$

(9.2.14)

In cases where $\boldsymbol{e} \cdot \boldsymbol{k} = 0$, the plasma waves are referred to as *transverse* waves, whereas in the opposite case ($\boldsymbol{e} \times \boldsymbol{k} = 0$) the plasma waves are referred to as *longitudinal* waves. [2]

After these general formulas for cold plasmas at rest, let us now discuss several important special cases.

9.2.1 Cold Unmagnetized Plasma at Rest

We start by discussing the limiting case of a vanishing background magnetic field $\boldsymbol{B}_0 = 0$ so that all gyrofrequencies are zero, $\Omega_a = 0$. From (9.2.2) we obtain

$$S = R = L = P = 1 - \sum_a \frac{\omega_{p,a}^2}{\omega^2}, \quad D = 0, \quad (9.2.15)$$

so that according to (9.2.6) and (9.2.7b)

$$a = P, \; b = 2P^2, \; c = P^3, \; f = 0. \quad (9.2.16)$$

From (9.2.5) we obtain the two solutions

$$P = 0 \quad (9.2.17a)$$

and, provided $P \neq 0$,

---

[2] We follow the nomenclature of Stix (1962 [525]) using "longitudinal" and "transverse" to refer to propagation with $\boldsymbol{k} \parallel \boldsymbol{\delta E}$ and $\boldsymbol{k} \perp \boldsymbol{\delta E}$, while "parallel" and "perpendicular" are used to describe propagation with $\boldsymbol{k} \parallel \boldsymbol{B}_0$ and $\boldsymbol{k} \perp \boldsymbol{B}_0$.

$$N^2 = (kc/\omega)^2 = P = 1 - \sum_a \frac{\omega_{p,a}^2}{\omega^2}, \qquad (9.2.17b)$$

where we used the definition (8.3.51).

The solution (9.2.17b) leads to two linearly independent transverse [3] modes $e = (0, 1, 0)$ with the dispersion relation

$$\omega_{1,2} = \pm \sqrt{\sum_a \omega_{p,a}^2 + k^2 c^2}. \qquad (9.2.18)$$

In vacuum ($\sum_a \omega_{p,a}^2 = 0$) the two transverse waves become ordinary electromagnetic radiation that travel with phase velocity $V_{\rm ph} \equiv \omega/k = \pm c$ which equals the group velocity $V_{\rm G} \equiv \partial\omega/\partial k = \pm c$. However, in the presence of the plasma it will only propagate, for a given density, for frequencies such that $\omega > \left(\sum_a \omega_{p,a}^2\right)^{1/2}$. Below this frequency the phase velocity becomes imaginary. The group velocity as a function of frequency is

$$V_{\rm G} = \pm c \sqrt{1 - \sum_a \frac{\omega_{p,a}^2}{\omega^2}} \qquad (9.2.19)$$

is drastically reduced, and becomes zero at $\omega = \left(\sum_a \omega_{p,a}^2\right)^{1/2}$. Note that $V_{\rm ph} = c^2/V_{\rm G}$ is superluminal. At frequencies below $\left(\sum_a \omega_{p,a}^2\right)^{1/2}$ the plasma prevents the propagation of the waves. The frequency dependence of the wave's group velocity is referred to as dispersion and can be used as a diagnostic tool for space plasmas (see Sect. 9.2.1.1).

The second solution (9.2.17a), on the other hand, does not correspond in any limit to a vacuum phenomenon and is a true plasma effect. The phase velocity of these waves is $V_{\rm ph} = \left(\sum_a \omega_{p,a}^2\right)^{1/2}/k$. They have the peculiar property that at all wavenumbers they oscillate at the same frequency.

9.2.1.1 Dispersion of Pulsar Signals. Pulsars are periodic radio sources that exhibit sharp signals of electromagnetic radiation (see Sect. 2.4) that have to propagate through the partially ionized interstellar and interplanetary medium to reach terrestrial telescopes. Both space plasmas predominantly contain electron and protons, so that with order of magnitude densities of these plasmas

$$\sum_a \omega_{p,a}^2 \simeq \left(1 + \frac{m_{\rm e}}{m_{\rm p}}\right) \omega_{p,e}^2 \sim 3 \times 10^9 \left(n_{\rm e}/1~{\rm cm}^{-3}\right)~{\rm Hz}^2. \qquad (9.2.20)$$

While at optical frequencies ($\omega \simeq 10^{15}$ Hz) the reduction in group velocity from (9.2.19) is too small to be observable, there is a measurable effect at

---

[3] Note that in the unmagnetized plasma the z axis should be identified with the wave vector direction so that $N_\parallel^2 = N^2 = P$ for this mode.

radio frequencies ($\omega \simeq 10^9$ Hz). Consider the difference in the signal propagation time from the pulsar position (A) to the terrestrial observer (B) at two neighboring radio frequencies ($\omega, \omega + \delta\omega$) with $\delta\omega \ll \omega$:

$$\delta t = t_\omega - t_{\omega+\delta\omega} = \int_A^B \frac{ds}{V_{G,\omega}} - \int_A^B \frac{ds}{V_{G,\omega+\delta\omega}} . \tag{9.2.21}$$

Using the frequency-dependent group velocity (9.2.19) one finds to lowest order in the ratio $\delta\omega/\omega \ll 1$ that

$$\delta t = -\frac{1}{c}\frac{\delta\omega}{\omega^3} \int_A^B ds \sum_a \omega_{p,a}^2 , \tag{9.2.22}$$

yielding with (9.2.20)

$$\frac{\delta\omega}{\delta t} = c_1 \frac{\omega^3}{DM} \tag{9.2.23}$$

with the constant $c_1 = m_e c/(4\pi e^2)$ and the *dispersion measure*

$$DM \equiv \int_A^B ds\ n_e \ \text{cm}^{-2} , \tag{9.2.24}$$

which is the column density of thermal electrons along the line of sight between the pulsar and the observer. The cubic frequency dependence of the frequency drift of pulsar signals (9.2.23) has indeed been verified: Fig. 9.1 shows the results for the pulsar CP1919 (Maran and Cameron 1968 [318]). And, from the absolute level of the frequency drift, one can infer the value of the dispersion measure. If the distance of the pulsar is known or assumed one can estimate the mean thermal electron density along the line of sight. This effect has been observed for the few hundred pulsars in our Galaxy in different directions and at different distances, and has become a powerful diagnostic to infer the density of ionized thermal electrons in the interstellar medium (compare Sect. 2.2).

9.2.2 Cold Electron-Proton Plasma

We now want to infer the properties of a cold magnetized plasma that consists of electrons and protons of equal densities to ensure overall charge neutrality. Generalization to a multi-component plasma is straightforward. For the calculation it is convenient to introduce mass and frequency ratios:

$$\xi \equiv m_e/m_p , \tag{9.2.25a}$$

$$X \equiv \frac{\omega_{pe}^2}{\omega^2}(1+\xi) , \tag{9.2.25b}$$

$$Y \equiv \Omega_e/\omega , \tag{9.2.25c}$$

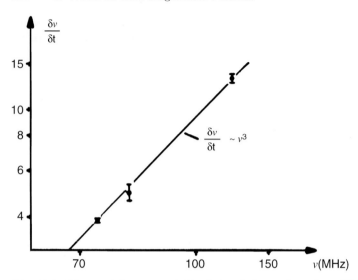

Fig. 9.1. Observed frequency drift of signals from the pulsar CP1919. From Kippenhahn and Möllenhoff (1975 [260]) after Maran and Cameron (1968 [318])

where we abbreviate $\Omega_e = \Omega_{oe}$ for the absolute value of the electron gyrofrequency (8.3.57). From (9.2.2) we obtain

$$R = 1 - \frac{\omega_{p,i}^2}{\omega(\omega + \Omega_p)} - \frac{\omega_{p,e}^2}{\omega(\omega - \Omega_e)}, \qquad (9.2.26a)$$

$$L = 1 - \frac{\omega_{p,i}^2}{\omega(\omega - \Omega_p)} - \frac{\omega_{p,e}^2}{\omega(\omega + \Omega_e)}, \qquad (9.2.26b)$$

$$P = 1 - \frac{\omega_{p,i}^2 + \omega_{p,e}^2}{\omega^2} = 1 - X, \qquad (9.2.26c)$$

$$S = 1 - \frac{X(1 - \xi Y^2)}{(1 - Y^2)(1 - \xi^2 Y^2)}, \qquad (9.2.26d)$$

$$D = -\frac{XY(1-\xi)}{(1-Y^2)(1-\xi^2 Y^2)}. \qquad (9.2.26e)$$

These plasma parameters are determined by four characteristic frequencies: the electron gyro- and plasma frequency $(\Omega_e, \omega_{p,e})$ and the proton gyro- and plasma frequency $(\Omega_p, \omega_{p,i})$. Their respective ratios

$$\omega_{p,i}/\omega_{p,e} = \xi^{1/2}, \quad \Omega_p/\Omega_e = \xi, \qquad (9.2.26f)$$

are constants which can be expressed in terms of the electron-proton mass ratio (9.2.25a). Alternatively the plasma/gyrofrequency ratio

$$\frac{\omega_{p,e}}{\Omega_e} = \frac{\xi^{1/2} \omega_{p,i}}{\Omega_p} = \frac{\sqrt{4\pi m_e n_e c^2}}{B_0} = \sqrt{\frac{\xi}{1+\xi}} \frac{c}{V_A} \qquad (9.2.27)$$

can, in principle, attain any value. According to (9.2.27) the ratio can be expressed in terms of the mass ratio ξ and a new characteristic velocity, referred to as the *Alfvén velocity* V_A which is defined as

$$V_A \equiv \frac{B_0}{\sqrt{4\pi(m_p + m_e)n_e}}$$
$$= 2.18 \times 10^{11}(B_0/\ 1\ \mathrm{G})(n_e/\ 1\ \mathrm{cm}^{-3})^{-1/2}\ \mathrm{cm/s}\ . \quad (9.2.28)$$

Large and small values of the ratio $\omega_{p,e}/\Omega_e$ as compared to unity are referred to as the high and low density cases, respectively. Using the parameter estimates of Table 8.1 one finds that almost all space plasmas are high-density plasmas since $\beta_A = V_A/c \ll \xi^{1/2} \simeq 0.023$.

The dispersion relation for this uniform two-component system of protons and electrons at zero temperature implies the existence of a variety of plasma modes. We are going to discuss the full dispersion relation for propagation parallel and perpendicular to the ordered magnetic field, whereas at intermediate propagation angles we shall consider the low and high frequency regimes separately.

9.2.3 Propagation Along the Static Magnetic Field

For this case $\theta = N_\perp = 0$, $N_\parallel = N$ and the three plasma modes follow from (9.2.9). The first mode $P = 0$ is the same as the plasma oscillations (9.2.17a) discussed in Sect. 9.2.1. The other two modes $N^2 = R$ and $N^2 = L$ correspond, according to (9.2.13) to polarization vectors $\boldsymbol{e}_{[N^2=R]} = (\imath/\sqrt{2}, 1/\sqrt{2}, 0)$ and $\boldsymbol{e}_{[N^2=L]} = (-\imath/\sqrt{2}, 1/\sqrt{2}, 0)$ which are both purely transverse waves ($\boldsymbol{k} \cdot \boldsymbol{e}_{R,L} = 0$).

9.2.3.1 Plasma Physics Definition of Polarization.
We follow the *plasma physics definition of polarization*, which defines right- or left-handed to be the sense of rotation in time at a fixed point in space of the polarization vector (9.2.13) when viewed in the *direction parallel to the ordered magnetic field*, i.e. for an arbitrary sign of the real frequency

$$\rho = \frac{e_2}{e_1}\frac{\omega}{|\omega|} = \imath\frac{S(P - N_\perp^2) - PN_\parallel^2}{DP}\frac{\omega}{|\omega|}, \quad (9.2.29)$$

where the second equality follows from (9.2.13).

It is important to realize that the *astronomical definition of polarization* refers to the sense of rotation in time of the wave's fluctuating electric field when viewed in the *direction parallel to the wave propagation direction* $\boldsymbol{\kappa}$. Only for parallel propagating waves do these two definitions coincide.

The wave polarization (9.2.29) in general, is a complex number which shows how the transverse electric field component varies with time if the wave propagation direction is parallel to the z axis. If $\rho = -\imath$, e_1 and e_2 are in quadrature and have equal amplitudes then the polarization is circular. To an observer looking in the direction of the wave normal, that is in the direction of positive z, the electric field vector observed at a fixed point would appear to rotate clockwise, *if the wave frequency is positive, $\omega > 0$*. This is called right-handed circular polarization. If $\rho = -i$ and the wave's frequency $\omega < 0$ is negative, the electric field vector would appear to rotate counterclockwise, which is called left-handed circular polarization. The latter definition of left-handed circular polarization is equivalent to that traditionally used, i.e. $\rho = i$ *and* positive wave frequency.

With this definition we have two choices to discuss the two dispersion relations $N^2 = R$ and $N^2 = L$:

1. We either restrict our frequency range to positive values $\omega > 0$ only, and discuss both relations individually. This is the approach used in most treatments of plasma waves.
2. We take either one of the two dispersion relations, for definiteness say $N^2 = L$, and allow also for negative wave frequencies. The physically left-handed waves will be those with positive frequency, the right-handed are those with negative frequency. The mathematical proof for this redundancy of information follows from the symmetry property

$$N_L^2(-\omega, k) = N_R^2(\omega, k) \,, \qquad (9.2.30)$$

exhibited by (9.2.26a) and (9.2.26b). Here we take the second approach.

9.2.3.2 The Dispersion Relation of Parallel Waves.
The dispersion relation (9.2.26b) reads

$$N^2 = L = 1 - \frac{\omega_{p,i}^2}{\omega(\omega - \Omega_p)} - \frac{\omega_{p,e}^2}{\omega(\omega + \Omega_e)} \,, \qquad (9.2.31)$$

so the resonances are clearly at $\omega = \Omega_p$ for the left-handed and at $\omega = -\Omega_e$ for the right-handed waves, respectively. The cutoff frequencies, where $N^2 = 0$, are given by

$$\omega_c = \pm \left[\sqrt{(1+\xi)\omega_{p,e}^2 + \left(\frac{\Omega_e + \Omega_p}{2}\right)^2} \mp \frac{\Omega_e - \Omega_p}{2} \right]$$

$$\simeq_{\text{(hd)}} \pm \left[\omega_{p,e} \mp \frac{1}{2}\Omega_e \right] \,, \qquad (9.2.32)$$

where the latter approximation holds for high-density plasmas.

If we consider only low-frequency waves with $|\omega| \ll \Omega_e$ and subluminal phase speeds,

$$V_{\text{ph}} = \omega/k \ll c \,, \qquad (9.2.33)$$

where the latter condition has to be checked a posteriori, the dispersion relation (9.2.31) can be well approximated by

$$\frac{k^2 c^2}{\omega^2} \simeq \frac{\omega_{p,i}^2}{\Omega_p(\Omega_p - \omega)} = \left(\frac{c}{V_A}\right)^2 \frac{\Omega_p}{(\Omega_p - \omega)}, \qquad (9.2.34)$$

where we used $\xi \ll 1$. The solution of (9.2.34) is

$$\omega_\pm = \pm V_A k \sqrt{1 + \frac{k^2}{k_c^2}} - dk^2 = V_A k \left[\pm\sqrt{1 + \frac{k^2}{k_c^2}} - \frac{k}{k_c}\right], \qquad (9.2.35)$$

where the main contribution to the wave dispersion is given by

$$d \equiv V_A^2/(2\Omega_p) = V_A/k_c, \qquad (9.2.36)$$

and where we have introduced the characteristic wavenumber

$$k_c \equiv 2\Omega_p/V_A = 2\,(\omega_{p,e}/c)\,\xi^{1/2}(1+\xi)^{1/2}$$
$$= 8.78 \times 10^{-8} (n_e/1\ \mathrm{cm}^{-3})^{1/2}\ \mathrm{cm}^{-1}. \qquad (9.2.37)$$

For small wavenumbers $|k| \ll k_c$ (9.2.35) reduces to

$$\omega_\pm \simeq \pm V_A k, \qquad (9.2.38)$$

whereas in the opposite case of large wavenumbers $|k| \gg k_c$ the frequencies are approximately

$$\omega_\pm \simeq \begin{cases} \Omega_p & \text{for } \pm V_A k > 0 \\ -(2dk^2 + \Omega_p) & \text{for } \pm V_A k < 0 \end{cases}. \qquad (9.2.39)$$

We can identify three plasma modes with different dispersion relation:

1. Equation (9.2.38) at small wavenumbers $|k| \ll k_c$ and small frequencies $|\omega| \ll \Omega_p$ describes non-dispersive (i.e. $V_{ph} = $ const.) forward ($V_{ph} > 0$) and backward ($V_{ph} < 0$) moving *Alfvén waves* which are either right-handed ($\omega > 0$) or left-handed ($\omega < 0$) circularly polarized. Equation (8.3.18) shows that for these four Alfvén waves the magnetic field component is much larger than the electric field component,

$$|\boldsymbol{B}_{1,A}(\boldsymbol{k},\omega)| = \frac{c}{V_A}|\boldsymbol{E}_{1,A}(\boldsymbol{k},\omega)|. \qquad (9.2.40)$$

2. At large wavenumbers $|k| \gg k_c$ the left-handed Alfvén branch develops into the left-handed *ion cyclotron wave* branch, which can propagate both forward and backward and which, according to (9.2.39b) at all values of k have the same frequency $\omega = \Omega_p$.

3. At large wavenumbers $|k| \gg k_c$ and frequencies between $\Omega_p < |\omega| < \Omega_e$ the right-handed Alfvén branch develops into the right-handed *whistler wave* branch, which is dispersive ($V_{ph} \neq$const.), because of the quadratic wavenumber dependence (9.2.40b), and which propagates both forward and backward.

In the intermediate frequency range $|\omega| \gg \Omega_p$ but still $|\omega| \leq \Omega_e$ and superluminal phase speeds (9.2.33) we can approximate the dispersion relation (9.2.31) by

$$\frac{k^2 c^2}{\omega^2} \simeq -\frac{\omega_{p,e}^2}{\omega(\omega + \Omega_e)}, \qquad (9.2.41)$$

yielding

$$\omega = \frac{-\Omega_e k^2 c^2}{\omega_{p,e}^2 + k^2 c^2} \simeq \begin{cases} -2dk^2 & \text{for } |k| \ll \omega_{p,e}/c = 21.4 k_c \\ -\Omega_e & \text{for } |k| \gg \omega_{p,e}/c = 21.4 k_c \end{cases}. \qquad (9.2.42)$$

Equation (9.2.42a) agrees with (9.2.39a) and describes the right-handed whistler waves, while (9.2.42b) indicates that at very large $k \geq k_c/[2\xi^{1/2}(1+\xi)^{1/2}] = 21.4 k_c$ the right-handed whistler wave branch develops into the right-handed *electron cyclotron wave* branch, which again can propagate forward and backward.

These right-handed oscillations, for which ω is near the electron cyclotron frequency, exhibit some curious properties (Montgomery and Tidman 1964 [363]). Rewriting their dispersion relation (9.2.31) in terms of the phase speed $V_{ph} = \omega/k$, but allowing now also superluminal phase speeds, so that the 1 in (9.2.31) can no longer be neglected, gives

$$\frac{V_{ph}^2}{c^2} = \frac{1 - (\Omega_e/\omega_0)}{1 - (\Omega_e/\omega_0) - (\omega_{p,e}^2/\omega_0^2)}, \qquad (9.2.43)$$

which for positive frequencies ($\omega_0 \equiv -\omega$) has been plotted as a function of ω_0 in Fig. 9.2. In the frequency range

$$-\Omega_e > \omega > -\frac{1}{2}\left[\Omega_e + \sqrt{\Omega_e^2 + 4\omega_{p,e}^2}\right] \simeq_{(hd)} \omega_{c,L} \qquad (9.2.44)$$

V_{ph} is imaginary, which corresponds to waves which will not propagate. The plasma acts as a filter for these frequencies. In the ordered magnetic field directed along the positive z axis, the electrons of the background plasma rotate in a counterclockwise (right-hand) sense. The right-hand polarized electron cyclotron wave (9.2.42b) has the property that its electric field vector also rotates in a counterclockwise sense. Thus, near $\omega \simeq -\Omega_e$, the electrons see a nearly constant electric field in their own coordinate system. As a consequence, considerable accelerations and attendant power transfer occur in the neighborhood of $\omega \sim -\Omega_e$, and this coincides with the erratic behavior on the graph in Fig. 9.2. This phenomenon is called "electron cyclotron resonance".

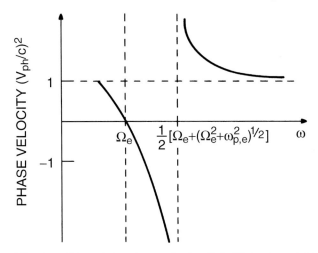

Fig. 9.2. Plot of the phase velocity (9.2.43) in the neighborhood of the cutoff

Repeating the same analysis for positive frequencies indicates that a similar "proton cyclotron resonance" phenomenon occurs for the left-handed ion cyclotron waves (9.2.39a) near $\omega = \Omega_p$ in the frequency range

$$\left[\sqrt{\omega_{p,e}^2 + \frac{\Omega_e^2}{4}} - \frac{\Omega_e}{2}\right] > \omega > \Omega_p , \qquad (9.2.45)$$

which, of course, is caused by the acceleration of the background protons by the co-rotating left-handed ion cyclotron waves and the associated power transfer from the waves to the protons.

At very high frequencies $|\omega| \gg \Omega_e$ the dispersion relation (9.2.31) reduces to

$$N^2 \simeq 1 - \frac{\omega_{p,e}^2 + \omega_{p,i}^2}{\omega^2} + \frac{\omega_{p,e}^2 \Omega_e}{\omega^3} , \qquad (9.2.46)$$

yielding

$$\omega_\pm \simeq \pm\sqrt{k^2 c^2 + \omega_{p,e}^2 + \omega_{p,i}^2} , \qquad (9.2.47)$$

which agrees exactly with (9.2.18), i.e. ordinary left- and right-handed electromagnetic waves are recovered.

The full solutions (9.2.35), and in particular the asymptotic modes (9.2.38)–(9.2.42), are illustrated in Fig. 9.3. Each curve with positive frequencies (first and fourth quadrants if counted clockwise) represents a left-handed plasma mode; each with negative frequencies (second and third quadrants) a right-handed mode. Curves in the first and third quadrants have positive phase speeds and are referred to as forward moving; those in the second and fourth quadrants are backward moving modes since their phase speeds are negative. Curves with positive (negative) slope have positive (negative) group speeds.

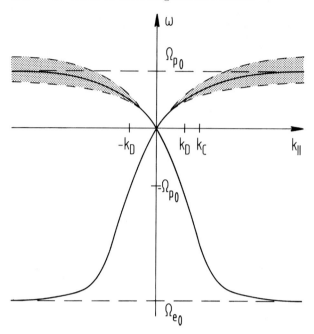

Fig. 9.3. Qualitative picture of the dispersion relation of right- and left-handed parallel propagating waves in a cold electron-proton plasma

9.2.4 Propagation Across the Static Magnetic Field

In this case $\theta = N_\parallel = 0$, $N_\perp = N$ and the two plasma modes follow from (9.2.10) as $N^2 = P$ and $N^2 = RL/S$ which are referred to as "ordinary" and "extraordinary" waves, respectively.

9.2.4.1 The Ordinary Wave. For this mode use of the dispersion relation $N^2 = N_\perp^2 = P$ directly in (9.2.13) would lead to the indeterminate polarization vector $e = (0,0,0/0)$; however, inserting the dispersion relation into the co-factor matrix (9.2.12) shows that $\Lambda_{33}^{T,\text{cold, rest}} = D^2 - S(S-P)$ is the only non-vanishing co-factor element, implying the polarization vector $e_{[N^2=P]} = (0,0,1)$ which lies perpendicular to the wave vector (8.3.49b) for $\theta = \pi/2$, $\boldsymbol{\kappa} = (1,0,0)$. The ordinary waves are transverse plasma waves propagating across the static magnetic field; their electric field is parallel to the static magnetic field. The dispersion relation is the same as in an unmagnetized plasma (see (9.2.18)), and is simply

$$\omega_{1,2} = \pm\sqrt{\sum_a \omega_{\text{p},a}^2 + k^2c^2} \ . \tag{9.2.48}$$

According to our convention we may refer to the two solutions at positive and negative frequencies as left-handed and right-handed, respectively. From the dispersion relation it is clear that there is no resonance since $P \to \infty$ is a trivial solution (either $\omega \to 0$, so no wave at all, or $\omega_{p,e} \to \infty$ which is impossible). A cutoff occurs at $\omega = \sqrt{\sum_a \omega_{p,a}^2}$.

9.2.4.2 The Extraordinary Wave. The polarization vector for this mode follows from (9.2.13),

$$e_{[N^2=RL/S]} = (-\imath\,\mathrm{sgn}(H)\sin\Psi,\,\mathrm{sgn}(H)\cos\Psi,\,0)\,, \qquad (9.2.49\mathrm{a})$$

where $H \equiv PS - RL$, $\mathrm{sgn}(x) = +1$ for $x > 0$ and $\mathrm{sgn}(x) = -1$ for $x < 0$, and where we have introduced the angle Ψ as

$$\cos\Psi = \frac{1}{\sqrt{1+(D^2/S^2)}}, \quad \sin\Psi = \frac{(D/S)}{\sqrt{1+(D^2/S^2)}}\,. \qquad (9.2.49\mathrm{b})$$

It follows that

$$\mathbf{e}\cdot\boldsymbol{\kappa} = -\imath\,\mathrm{sgn}(H)\sin\Psi\,, \qquad (9.2.50\mathrm{a})$$

and

$$\mathbf{e}\times\boldsymbol{\kappa} = (0,\,0,\,-\mathrm{sgn}(H)\cos\Psi)\,. \qquad (9.2.50\mathrm{b})$$

These waves, propagating across the static magnetic field with their electric fields perpendicular to it, are a mixed type of wave, i.e. they are neither transverse nor purely longitudinal.

According to (9.2.11) resonances occur at $S \to 0$, yielding with (9.2.26)

$$\omega_{r1,2}^2 = \frac{w_e^2 + w_p^2}{2} \pm \sqrt{\left(\frac{w_e^2 - w_p^2}{2}\right)^2 + \omega_{p,e}^2\omega_{p,i}^2}\,, \qquad (9.2.51\mathrm{a})$$

where we have introduced the abbreviations

$$w_e^2 \equiv \Omega_e^2 + \omega_{p,e}^2\,, \qquad (9.2.51\mathrm{b})$$
$$w_p^2 \equiv \Omega_p^2 + \omega_{p,i}^2\,. \qquad (9.2.51\mathrm{c})$$

Because of their complicated dependence on plasma and gyrofrequencies, these resonances are referred to as *hybrid resonances*. To lowest order in the small parameter $\xi = m_e/m_p$ we obtain for the two values of (9.2.51a)

upper hybrid resonance $\quad \omega_{UH}^2 = \omega_{p,e}^2 + \Omega_e^2\,, \qquad (9.2.52)$

and

lower hybrid resonance $\quad \omega_{LH}^2 = \Omega_e\Omega_p\left(\dfrac{\omega_{p,e}^2 + \Omega_e\Omega_p}{\omega_{p,e}^2 + \omega_{p,i}^2}\right) \simeq \Omega_e\Omega_p\,, \qquad (9.2.53)$

where the latter approximation holds for high-density plasmas. Since both (9.2.52) and (9.2.53) imply resonances at $\pm\omega_{\text{UH}}$ and $\pm\omega_{\text{LH}}$, we find that the hybrid resonances occur for both handedness (left and right) waves.

The cutoffs for the extraordinary wave are given by the roots of $RL = 0$, leading to

$$\omega_{\text{c1,2}}^2 = (1+\xi)\omega_{\text{p,e}}^2 + \frac{(1+\xi^2)}{2}\Omega_e^2$$
$$\pm (1-\xi)\Omega_e \sqrt{(1+\xi)\omega_{\text{p,e}}^2 + \frac{(1+\xi)^2}{4}\Omega_e^2} \,, \qquad (9.2.54)$$

which in the high-density limit to lowest order in ξ reduces to

$$\omega_{\text{c1,2,hd}}^2 \simeq \omega_{\text{p,e}}(\omega_{\text{p,e}} \pm \Omega_e) \,, \qquad (9.2.55)$$

while in the low-density limit we obtain

$$\omega_{\text{c1,ld}}^2 \simeq \Omega_e^2 + 2\omega_{\text{p,e}}^2 \,, \qquad (9.2.56a)$$
$$\omega_{\text{c2,ld}}^2 \simeq \Omega_p^2 + 2\omega_{\text{p,i}}^2 \,. \qquad (9.2.56b)$$

Again, (9.2.54) implies cutoffs at $\pm\omega_{\text{c1,2}}$, so that the cutoffs occur for both handedness waves.

With the four characteristic resonance and cutoff frequencies ω_{UH}, ω_{LH}, ω_{c1} and ω_{c2} we may write the dispersion relation for the extraordinary waves as

$$\frac{V_{\text{ph}}^2}{c^2} = \frac{\omega^2}{k^2 c^2} = \frac{(\omega^2 - \omega_{\text{UH}}^2)(\omega^2 - \omega_{\text{LH}}^2)}{(\omega^2 - \omega_{\text{c1}}^2)(\omega^2 - \omega_{\text{c2}}^2)} \,. \qquad (9.2.57)$$

Because of the apparent symmetries in $\pm k$ and $\pm\omega$, each solution of the dispersion relation (9.2.57) gives rise to both right- and left-handed plasma waves each propagating forward and backward.

The variation of the phase velocity (9.2.57) as a function of $|\omega|$ is illustrated in Fig. 9.4 in the high-density limit $\omega_{\text{p,e}} \gg |\Omega_e|$. In the two frequency ranges $\omega_{\text{LH}} < |\omega| < \omega_{\text{c2}}$ and $\omega_{\text{UH}} < |\omega| < \omega_{\text{c1}}$, V_{ph} is imaginary, so that at these frequencies the waves will not propagate. Again the plasma acts as a filter for these perpendicular waves.

The full solutions of (9.2.57) are illustrated in Fig. 9.5 together with the ordinary wave solutions (9.2.48) at high frequencies. From the asymptotic behavior of the extraordinary dispersion relation we may identify the following individual modes:

1. For small frequencies $|\omega| \to 0$ (9.2.57) with $V_{\text{ph}}^2 = \omega^2/k^2$ approaches

$$\omega^2 = V_A^2 k^2 \,, \qquad (9.2.58)$$

which is formally the same dispersion relation as for parallel Alfvén waves (9.2.38). For reasons that will become obvious later, these waves are referred to as *compressional Alfvén waves*. However, besides the noted differences in their polarization properties (see (9.2.50)) there is an important distinction between the parallel and the compressional Alfvén waves

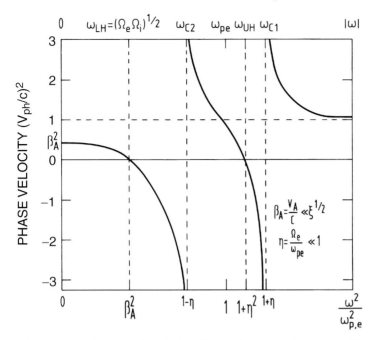

Fig. 9.4. Qualitative plot of the phase velocity of the extraordinary wave in the high-density limit

as $|\omega| \to \Omega_p$. The character of the parallel Alfvén mode alters drastically with the left-handed branch degenerating into an ion cyclotron wave (see (9.2.39a)) and the right-handed branch developing into a whistler wave (see (9.2.39b). The compressional Alfvén wave, on the other hand, in general undergoes no radical change in character at this resonance. As can be seen from Fig. 9.5, in the high-density limit compressional Alfvén waves propagate not only below $|\omega| = \Omega_p$ but also through the ion cyclotron resonance. For this mode there is no resonance until the lower hybrid frequency $\omega_{LH} \simeq 43\Omega_p$ is reached.

2. As the lower hybrid frequency is approached the compressional Alfvén waves develop into the *lower hybrid waves* which at all values of k have nearly the same frequency $\omega = \pm\omega_{LH}$.

3. In the stop band $\omega_{LH} < |\omega| < \omega_{c2}$ no waves propagate in the high-density limit, but in the small frequency band $\omega_{c2} < |\omega| < \omega_{UH}$ the *upper hybrid waves* exist which at all values of k have nearly the same frequency $\omega = \pm\omega_{UH}$.

4. At high frequencies the extraordinary ($|\omega| > \omega_{c1}$) and ordinary ($|\omega| > \omega_{p,e}$) electromagnetic waves exist.

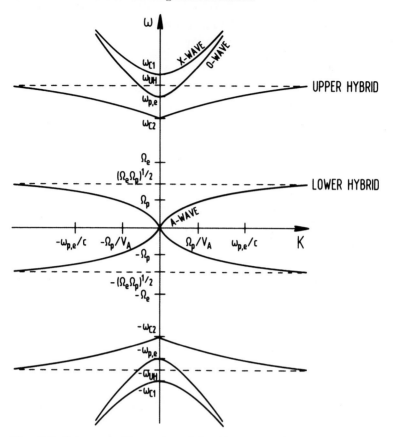

Fig. 9.5. Schematic illustration of the dispersion relation of perpendicular propagating waves in a cold electron-proton plasma in the high-density limit

9.2.4.3 Interlude. One could continue indefinitely with a discussion of dispersion relations at different propagation angles and in many limits. From the above examples we hope the reader has already understood the wealth of wave motions possible in a magneto-active plasma. In the following two subsections we shall restrict our analysis to the oblique propagation at low ($|\omega| \ll \Omega_p$) and high ($|\omega| \gg \Omega_p$) frequencies following Stix (1962 [525], Chap. 2).

9.2.5 Oblique Propagation of Low-Frequency Waves

We consider first frequencies well below the lower hybrid frequency, $\omega \ll \omega_{\text{LH}} = 43\Omega_p$, and we restrict our analysis throughout to the high-density limit $\omega_{\text{pe}} \gg \Omega_e$. From (9.2.25) and (9.2.26) we obtain with $X \gg 1$, $Y \gg 1$, $\xi Y^2 = (\omega_{\text{LH}}/\omega)^2 \gg 1$, $|P| \simeq X \gg 1$, that

9.2 Plasma Waves in a Cold Magnetized Plasmas at Rest

$$S/P \simeq \begin{cases} (\omega/\omega_{\mathrm{LH}})^2 \ll 1 & \text{for } |\omega| \ll \Omega_{\mathrm{p}} \\ \xi \ll 1 & \text{for } \Omega_{\mathrm{p}} \ll |\omega| \ll \omega_{\mathrm{LH}} \end{cases}, \quad (9.2.59)$$

so that at all frequencies different from Ω_{p} we may neglect S as compared to P. The quantity D from (9.2.26e) behaves asymptotically as

$$D \simeq \begin{cases} -(c/V_{\mathrm{A}})^2 (\omega/\Omega_{\mathrm{p}}) & \text{for } |\omega| \ll \Omega_{\mathrm{p}} \\ (c/V_{\mathrm{A}})^2 (\Omega_{\mathrm{p}}/\omega) & \text{for } \Omega_{\mathrm{p}} \ll |\omega| \ll \omega_{\mathrm{LH}} \end{cases}, \quad (9.2.60)$$

9.2.5.1 Magnetosonic and Alfvén Waves.

At very low frequencies, $\omega \ll (V_{\mathrm{A}}/c)^2 \Omega_{\mathrm{p}}$, $D \to 0$ tends to zero. In this limit

$$R \simeq L \simeq S \simeq (c/V_{\mathrm{A}})^2 ,$$

implying according to (9.2.6) and (9.2.7b)

$$a \simeq P\cos^2\theta, \ b \simeq PS(1+\cos^2\theta), \ c \simeq PS^2, \ f \simeq PS\sin^2\theta . \quad (9.2.61)$$

From (9.2.7a) one obtains the two plasma modes

$$N_{1,2}^2 = \begin{cases} S \\ S/\cos^2\theta \end{cases} . \quad (9.2.62)$$

With the definition of the index of refraction (8.3.51) and (9.2.60) the two plasma modes read

$$\omega_{1,2}^2 = \begin{cases} V_{\mathrm{A}}^2 k^2 & \text{Magnetosonic waves} \\ V_{\mathrm{A}}^2 k_{\parallel}^2 & \text{Alfvén waves} \end{cases} . \quad (9.2.63)$$

In the same limit $D \to 0$ and $S/P \to 0$ the only non-vanishing elements of the co-factor matrix (9.2.12) are $\Lambda_{11}^T = P(S-N^2)$ and $\Lambda_{22}^T = P(S-N_{\parallel}^2)$. For the magnetosonic waves with the dispersion relation $N^2 = S$ we obtain $\Lambda_{11,\mathrm{M}}^T = 0$, so that the polarization vector is $\boldsymbol{e}_{\mathrm{M}} = (0,1,0)$, indicating that the magnetosonic waves are linearly polarized and purely transversal $\boldsymbol{e}_{\mathrm{M}} \cdot \boldsymbol{k} = 0$.

For the Alfvén waves with the dispersion relation $N_{\parallel}^2 = S$ we obtain $\Lambda_{22,\mathrm{A}}^T = 0$ implying for the polarization vector $\boldsymbol{e}_{\mathrm{A}} = (1,0,0)$, so that these waves are linearly polarized. Since $\boldsymbol{e}_{\mathrm{A}} \cdot \boldsymbol{k} = k\sin\theta$ they are purely transversal waves for $\theta = 0$ but become increasingly more longitudinal with respect to the direction of $\boldsymbol{B}_0$.

9.2.5.2 Ion Cyclotron Waves.

Let us now consider frequencies just below the ion cyclotron frequency. We use $|\omega| \ll \Omega_{\mathrm{e}}$ and $V_{\mathrm{A}} \ll c$ to write

$$R \simeq 1 + \left(\frac{c}{V_{\mathrm{A}}}\right)^2 \frac{\Omega_{\mathrm{p}}}{\Omega_{\mathrm{p}}+\omega} \simeq \left(\frac{c}{V_{\mathrm{A}}}\right)^2 \frac{\Omega_{\mathrm{p}}}{\Omega_{\mathrm{p}}+\omega} , \quad (9.2.64\mathrm{a})$$

$$L \simeq 1 + \left(\frac{c}{V_{\mathrm{A}}}\right)^2 \frac{\Omega_{\mathrm{p}}}{\Omega_{\mathrm{p}}-\omega} \simeq \left(\frac{c}{V_{\mathrm{A}}}\right)^2 \frac{\Omega_{\mathrm{p}}}{\Omega_{\mathrm{p}}-\omega} . \quad (9.2.64\mathrm{b})$$

Again we take advantage of the high ion-to-electron-mass ratio and approximate $|P| \simeq X \gg 1$, $X \gg R$, and $X \gg L$. The coefficients a, b, c in Aström's (9.2.5) may then be approximated as

$$a \simeq -X\cos^2\theta, \quad b \simeq -\frac{X}{2}(R+L)(1+\cos^2\theta), \quad c \simeq -XRL, \qquad (9.2.65)$$

where we avoid values of θ in the immediate vicinity of $\pi/2$. Aström's equation then reduces to

$$N^4\cos^2\theta - N^2\left(\frac{c}{V_A}\right)^2\frac{\Omega_p^2}{\Omega_p^2-\omega^2}(1+\cos^2\theta) + \left(\frac{c}{V_A}\right)^4\frac{\Omega_p^2}{\Omega_p^2-\omega^2} = 0. \qquad (9.2.66)$$

For $\omega \ll \Omega_p$ (9.2.66) reduces to the factorable Alfvén case

$$\left[N^2 - \left(\frac{c}{V_A}\right)^2\right]\left[N^2\cos^2\theta - \left(\frac{c}{V_A}\right)^2\right] \simeq 0 \qquad (9.2.67)$$

discussed in Sect. 9.2.5.1 with the two modes (9.2.63). For frequencies close to the ion cyclotron frequency we may factor (9.2.66) as

$$\left[\left(\frac{c}{V_A}\right)^2\frac{\Omega_p^2}{\Omega_p^2-\omega^2} - \frac{N^4\cos^2\theta}{N^2(1+\cos^2\theta)-(c/V_A)^2}\right]\left[\left(\frac{c}{V_A}\right)^2 - N^2(1+\cos^2\theta)\right] = 0. \qquad (9.2.68)$$

For frequencies close to the ion cyclotron frequency $(\omega/k) \ll V_A$, so that in the last term of (9.2.68) we may neglect $(c/V_A)^2$ with respect to $N^2(1+\cos^2\theta)$, we find the two asymptotic dispersion relations

$$N^2 \simeq \frac{(c/V_A)^2}{1+\cos^2\theta} \qquad (9.2.69a)$$

and

$$N^2\cos^2\theta \simeq \left(\frac{c}{V_A}\right)^2\frac{\Omega_p^2}{\Omega_p^2-\omega^2}(1+\cos^2\theta). \qquad (9.2.69b)$$

Equation (9.2.69a) is the dispersion relation (9.2.63b) for the compressional Alfvén wave for oblique propagation angle. As stated already in Sect. 9.2.4.2, this mode is not affected by ion cyclotron resonance, and we note that (9.2.69a) is not greatly different from the low-frequency relation (9.2.58) for this mode.

Equation (9.2.69b) is the ion cyclotron wave, and this equation may be rewritten

$$\omega^2 \simeq \Omega_p^2 \bigg/ \left[1 + \frac{\omega_{p,i}^2}{k_\|^2 c^2} + \frac{\omega_{p,i}^2}{k_\|^2 c^2 + k_\perp^2 c^2} \right]^{-1}. \qquad (9.2.70)$$

In this form it may be seen that $k_\|^2 \to \infty$ as $|\omega| \to \Omega_p$, so that resonance at the ion cyclotron frequency takes places at $\theta = 0$. This is the resonance predicted by (9.2.11) for the case $S \to \infty$.

9.2.6 Oblique Propagation of High-Frequency Waves

At frequencies well above the proton gyrofrequency $\omega \gg \Omega_p$, the heavy plasma protons do not respond to the perturbation and the dispersion relation is entirely determined by the motion of the plasma electrons. The dispersion relation for the electromagnetic electron modes was first derived by Hartree (1931 [214]) and Appleton (1932 [16]). Adding the term aN^2 on both sides of Aström's (9.2.5) we readily find

$$N^2 = \frac{aN^2 - c}{aN^2 + a - b} = 1 - \frac{2(a-b+c)}{2a - b \pm f}, \qquad (9.2.71)$$

where we have inserted (9.2.7a) on the right-hand side. Defining $u \equiv \Omega_e^2/\omega^2$ we may approximate (9.2.26a)–(9.2.26c)

$$R \simeq 1 - \frac{X}{1 - u^{1/2}}, \quad L \simeq 1 - \frac{X}{1 + u^{1/2}}, \quad P = 1 - X, \qquad (9.2.72)$$

and obtain, after straightforward arithmetic, the Appleton-Hartree dispersion relation

$$N^2 = 1 - \frac{2X(1-X)}{2(1-X) - u\sin^2\theta \pm u^{1/2}\Delta} = 1 - hX, \qquad (9.2.73a)$$

where

$$\Delta \equiv \left[u\sin^4\theta + 4(1-X)^2 \cos^2\theta \right]^{1/2}, \qquad (9.2.73b)$$

A convenient factoring of (9.2.73a) was obtained by Booker (1936 [73]) with approximations based on the relative size of the two terms in Δ. The two possibilities are

$$u\sin^4\theta \gg 4(1-X)^2 \cos^2\theta \qquad \text{QT}, \qquad (9.2.74)$$
$$u\sin^4\theta \ll 4(1-X)^2 \cos^2\theta \qquad \text{QL}, \qquad (9.2.75)$$

for the quasi-transverse (QT) and quasi-longitudinal (QL) cases.

With the first of these approximations, we have the ordinary (O) and extraordinary (X) modes according to the $+, -$ choice of sign in (9.2.73a),

$$N^2 \simeq \frac{1-X}{1-X\cos^2\theta} \qquad \text{QT}-\text{O}\,, \tag{9.2.76}$$

$$N^2 \simeq \frac{(1-X)^2 - u\sin^2\theta}{(1-X) - u\sin^2\theta} \qquad \text{QT}-\text{X}\,. \tag{9.2.77}$$

We note that the QT-O mode (9.2.76) reduces to the electromagnetic plasma mode

$$N^2 = P = 1 - X \tag{9.2.78}$$

for $\theta = \pi/2$. The dispersion relation (9.2.76) is seen to be independent of the magnetic field strength not only at $\theta = \pi/2$ but also for all values of θ obeying (9.2.74).

Use of condition (9.2.75) yields the quasi-longitudinal dispersion relation

$$N^2 \simeq 1 - \frac{X}{1 \pm u^{1/2}\cos\theta} = 1 - \frac{X\omega}{\omega \pm \Omega_\text{e}\cos\theta} \qquad \text{QL} - \substack{\text{L}\\\text{R}}\,. \tag{9.2.79}$$

Comparison of (9.2.79) with the second equation (9.2.72) shows immediately that the upper sign in the denominator corresponds to the $N^2 = L$ left-hand solution at $\theta = 0$, see also (9.2.9c). We have seen in Sect. 9.2.3 that this wave does not propagate between the proton gyrofrequency and the electron plasma frequency, as follows from its dispersion relation

$$N^2_{\text{QL}-\text{L}} = \frac{(1-X)\omega + \Omega_\text{e}\cos\theta}{\omega + \Omega_\text{e}\cos\theta}\,, \tag{9.2.80}$$

which yields positive $N^2 > 0$ only for frequencies

$$\omega > \sqrt{(1+\xi)\,\omega^2_{\text{p,e}} + \frac{\Omega^2_\text{e}\cos^2\theta}{4}} - \Omega_\text{e}\cos\theta/2\,, \tag{9.2.81}$$

which, in the usual high-density plasma $\omega_{\text{p,e}} \gg \Omega_\text{e}$, for all values of θ is larger than the electron plasma frequency $\omega_{\text{p,e}}$. At frequencies larger than the electron plasma frequencies we may approximate (9.2.80) as

$$N^2_{\text{QL}-\text{L}}(\omega \gg \omega_{\text{p,e}}) \simeq 1 - \frac{(1+\xi)\,\omega^2_{\text{p,e}}}{\omega^2}\left(1 - \frac{\Omega_\text{e}\cos\theta}{\omega}\right)\,. \tag{9.2.82}$$

The dispersion relation (9.2.82) describes the propagation of left-hand polarized electromagnetic waves at oblique quasi-longitudinal (i.e. consistent with restriction (9.2.75)) angles to the background magnetic field.

The second solution of (9.2.79), obtained with the $-$ sign in the denominator, is

$$N^2_{\text{QL}-\text{R}} \simeq \frac{(1-X)\omega - \Omega_\text{e}\cos\theta}{\omega - \Omega_\text{e}\cos\theta} = \frac{(1+\xi)\,\omega^2_{\text{p,e}} + \Omega_\text{e}\omega\cos\theta - \omega^2}{\omega(\Omega_\text{e}\cos\theta - \omega)}\,. \tag{9.2.83}$$

9.2 Plasma Waves in a Cold Magnetized Plasmas at Rest

This dispersion relation yields positive $N^2 > 0$ in two frequency ranges:

(i) for $\omega < \Omega_e \cos\theta \ll \omega_{p,e}^2$, (9.2.81) reduces to

$$\omega \simeq \frac{-\Omega_e k^2 c^2 \cos\theta}{\omega_{p,e}^2 + k^2 c^2}, \qquad (9.2.84)$$

which for $\theta = 0$ equals (9.2.42). The dispersion relation (9.2.84) describes the whistler-electron cyclotron wave plasma mode for oblique propagation;

(ii) for frequencies larger than the electron plasma frequency (9.2.83) can be approximated as

$$N_{\text{QL-R}}^2(\omega \gg \omega_{p,e}) \simeq 1 - \frac{(1+\xi)\,\omega_{p,e}^2}{\omega^2}\left(1 + \frac{\Omega_e \cos\theta}{\omega}\right). \qquad (9.2.85)$$

The dispersion relation (9.2.85) describes the propagation of right-hand polarized electromagnetic waves at oblique quasi-longitudinal (i.e. consistent with restriction (9.2.75)) angles to the background magnetic field.

To calculate the polarization vector for these oblique waves it is convenient to express the components of vector (9.2.13) in terms of the parameters X and u. One obtains

$$e_1 = \frac{\imath X u^{1/2} C}{1-u}\left[1 - X - N^2 \sin^2\theta\right], \qquad (9.2.86a)$$

$$e_2 = \frac{C}{1-u}\left[(1-X-u)(1-X-N^2) - uXN^2 \cos^2\theta\right], \qquad (9.2.86b)$$

$$e_3 = \frac{-\imath X u^{1/2} C}{1-u} N^2 \sin\theta \cos\theta. \qquad (9.2.86c)$$

Inserting the Appleton–Hartree dispersion relation (9.2.73a) yields

$$e_1 = \frac{\imath X u^{1/2} C}{1-u}\left[\cos^2\theta + X(h\sin^2\theta - 1)\right], \qquad (9.2.87a)$$

$$e_2 = \frac{CX}{1-u}\left[(1-X-u)(h-1) - u\cos^2\theta(1-hX)\right], \qquad (9.2.87b)$$

$$e_3 = \frac{-\imath X u^{1/2} C}{1-u}(1-hX)\sin\theta\cos\theta. \qquad (9.2.87c)$$

For QL-propagation

$$h_{\text{QL}} \simeq \left[1 \pm u^{1/2}\cos\theta\right]^{-1}, \qquad (9.2.88)$$

so that

$$
\begin{aligned}
e_{\mathrm{QL}} \simeq {}& \frac{Cxu^{1/2}\cos\theta}{(1-u)(1\pm u^{1/2}\cos\theta)} \\
& \times \left(\imath\left[\cos\theta(1-X)\pm u^{1/2}\left(\cos^2\theta - X\right)\right],\right. \\
& \times (1-X)\left(\mp 1 - u^{1/2}\cos\theta\right)\pm u\sin^2\theta, \\
& \left. -\imath\sin\theta\left[1-X\pm u^{1/2}\cos\theta\right]\right).
\end{aligned}
\tag{9.2.89}
$$

At high frequencies $\omega \gg \omega_{\mathrm{p,e}}$, where the parameters X and u are negligibly small compared to unity, (9.2.89) becomes with the normalization (8.3.43b)

$$
e_{\mathrm{QL}}(\omega \gg \omega_{\mathrm{p,e}}) \simeq \frac{\imath}{\sqrt{2}}(\cos\theta, \pm\imath, -\sin\theta).
\tag{9.2.90}
$$

It can be readily seen that the upper sign in (9.2.90) describes left-handed polarized waves while the lower sign describes right-handed polarized waves. With (8.3.49) we immediately prove $\boldsymbol{e}\cdot\boldsymbol{k} = 0$, i.e. these waves are purely transverse electromagnetic waves.

For whistler waves we choose the lower sign in (9.2.89) and use the approximation $X \gg 1$ to obtain

$$
\begin{aligned}
e_{\mathrm{W}} &= \frac{\left(\imath\left(u^{1/2}-\cos\theta\right),\ u^{1/2}\cos\theta - 1,\ \imath\sin\theta\right)}{\sqrt{u(1+\cos^2\theta)+2-4u^{1/2}\cos\theta}} \\
&\simeq \frac{1}{\sqrt{1+\cos^2\theta}}\left(\imath,\ \cos\theta,\ \frac{\imath\sin\theta}{u^{1/2}}\right).
\end{aligned}
\tag{9.2.91}
$$

since $u \gg 1$ in the whistler wave range. The electric field component parallel to the background magnetic field is a factor of order $u^{-1/2} = \omega/\Omega_{\mathrm{e}} \ll 1$ smaller than the perpendicular electric field components at oblique longitudinal propagation angles, and vanishes for parallel propagation. From (9.2.91) and (8.3.49) we find

$$
e_{\mathrm{W}}\cdot\boldsymbol{k} = \frac{\imath u^{1/2}\sin\theta}{\sqrt{u(1+\cos^2\theta)+2-4u^{1/2}\cos\theta}} \simeq \frac{\imath\sin\theta}{\sqrt{1+\cos^2\theta}},
\tag{9.2.92}
$$

indicating that parallel propagating whistler waves are purely transverse waves, and become increasingly more longitudinal with increasing propagation angle.

9.2.7 More Rigorous Treatment of Dispersion Relation and Polarization of Low-Frequency Waves

While the approximations in Sect. 9.2.5 are valid at very low frequencies and close to the ion cyclotron frequency, we consider now a more rigorous solution of Aström's (9.2.5) at frequencies smaller than the lower hybrid frequency, $|\omega| < \omega_{\text{LH}} = 43\Omega_{\text{p}}$ and for subluminal Alfvén speeds, $\alpha \equiv (c/V_A)^2 \gg 1$. Introducing the dimensionless frequency

$$v \equiv \Omega_{\text{p}}^2/\omega^2, \qquad (9.2.93)$$

the solution of Aström's equation (9.2.7a) can be written as

$$N_\pm^2 = \frac{S\left(1 + \cos^2\theta + g\sin^2\theta\right) \pm |S(1-g)|\sqrt{\sin^4\theta + (2D/S(1-g))^2 \cos^2\theta}}{2\cos^2\theta\left(1 + S/P\tan^2\theta\right)}, \qquad (9.2.94)$$

where the quantity

$$g \equiv \frac{RL}{PS} = \frac{(1+\alpha)^2 v - 1}{(1-X)[(1+\alpha)v - 1]} \simeq -\frac{\alpha}{X} = -\left(\frac{\omega}{\omega_{\text{LH}}}\right)^2 = -\frac{\xi}{v}. \qquad (9.2.95)$$

Obviously, $|g| \ll 1$ is very small compared to unity at these frequencies since $\alpha \gg 1$. Equation (9.2.94) then becomes

$$N_\pm^2 \simeq \frac{S\left(1 + \cos^2\theta\right) \pm |S|\sqrt{\sin^4\theta + (2D/S)^2 \cos^2\theta}}{2\cos^2\theta\left(1 + (S/P)\tan^2\theta\right)}. \qquad (9.2.96)$$

The behavior of (9.2.94) and (9.2.96) as a function of propagation angle θ and frequency v is controlled by two ratios. The absolute value of the ratio

$$\frac{2D}{S} = -\frac{2\alpha v^{1/2}}{(1+\alpha)v - 1} \simeq -2v^{-1/2} \qquad (9.2.97)$$

is small at frequencies below the proton gyrofrequency Ω_{p} and large compared to unity at frequencies above Ω_{p}. The second ratio in (9.2.96)

$$\frac{S}{P} = \frac{(1+\alpha)v - 1}{(v-1)(1-X)} \simeq -\frac{\alpha v}{X(v-1)} \simeq \frac{\xi}{v-1} \qquad (9.2.98)$$

can attain any value between $-\infty$ and $+\infty$. Obviously, as indicated in Fig. 9.6, we can divide the frequency-propagation angle-plane into four regions 1, 2, 3, 4, in which the dispersion relation (9.2.94) approaches different limits. Let us define two frequency-dependent propagation angles θ_{c} and θ_{D} as

$$\theta_{\text{c}} = \arccos\left[\sqrt{1 + v^{-1}} - v^{-1/2}\right] \simeq \begin{cases} \sqrt{2\omega/\Omega_{\text{p}}} & \text{for } \omega \ll \Omega_{\text{p}} \\ (\pi/2) - (\Omega_{\text{p}}/2\omega) & \text{for } \omega \gg \Omega_{\text{p}} \end{cases}, \qquad (9.2.99)$$

and

$$\theta_{\text{D}} = \arctan\left[43\sqrt{|v-1|}\right] \simeq \begin{cases} (\pi/2) - (\omega/\Omega_{\text{p}})/43 & \text{for } \omega \ll \Omega_{\text{p}} \\ \arctan(43) = 88.67° & \text{for } \omega \gg \Omega_{\text{p}} \end{cases}. \qquad (9.2.100)$$

Let us consider each region in turn.

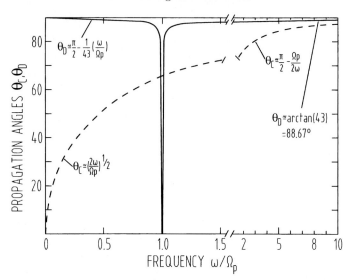

Fig. 9.6. Division of the frequency-propagation angle-plane into four asymptotic regions where the propagation angle is much smaller or larger than the characteristic angles θ_c and θ_D defined in (9.2.99) and (9.2.100)

9.2.7.1 Region 1: Small Propagation Angles $\theta \ll \theta_c$ and $\theta \ll \theta_D$.

For small angles $\theta \ll \theta_c$ we have $\sin^4\theta \ll (2D/S)\cos^2\theta$ in the argument of the square root in (9.2.96), while for angles $\theta \ll \theta_D$ the denominator in (9.2.96) is $2\cos^2\theta$. We then find

$$N_\pm^2(|\theta| \ll \theta_c, \theta_D) \simeq \frac{S\left(1 + \cos^2\theta\right) \pm 2D\cos\theta}{2\cos^2\theta}, \qquad (9.2.101)$$

which for $\theta = 0$ reduces to solutions (9.2.9b, 9.2.9c). For the polarization we readily find by expanding (9.2.29) to lowest order in the small parameter $(1/|P|)$

$$\begin{aligned}\rho(\theta \ll \theta_c, \theta_D) &\simeq \mp i\left[\cos\theta \mp \frac{S}{2D}\sin^2\theta\right] \\ &\simeq \mp i\left[\cos\theta \mp \frac{v^{1/2}}{2}\sin^2\theta\right] \simeq \mp i\cos\theta\,. \end{aligned} \qquad (9.2.102)$$

For small propagation angles the polarization of the two wave modes is left- and right-handed elliptical becoming circular at $\theta = 0$, again in agreement with the result in Sect. 9.2.3.

9.2.7.2 Region 2: Intermediate Propagation Angles $\theta_c \ll |\theta| \ll \theta_D$.

In region 2 (see Fig. 9.6) of intermediate propagation angles the denominator in (9.2.96) can again be approximated by $2\cos^2\theta$ while the argument of the square root is approximately $\sin^2\theta$, so that (9.2.96) becomes

$$N_\pm^2(\theta_c \ll |\theta| \ll \theta_D) \simeq \frac{S(1 + \cos^2\theta \pm \sin^2\theta)}{2\cos^2\theta} \simeq \begin{cases} S \\ S/\cos^2\theta \end{cases}, \tag{9.2.103}$$

which agrees with the dispersion relation for Alfvén waves and magnetosonic waves (9.2.62) derived in Sect. 9.2.5.1. For the polarization of these waves we obtain with (9.2.103) used in (9.2.29)

$$\rho(\theta_c \ll |\theta| \ll \theta_D) \simeq \imath \begin{cases} 0 \\ -v^{1/2}\sin^2\theta \end{cases}. \tag{9.2.104}$$

At intermediate propagation angles the Alfvén mode is linearly polarized and the magnetosonic mode approximately linearly polarized, which agrees with the earlier results in Sect. 9.2.5.1.

9.2.7.3 Region 3: Large Propagation Angles $|\theta| \gg \theta_c, \theta_D$.
In region 3 (see Fig. 9.6) of large propagation angles the denominator in (9.2.94) is approximately $2S\sin^2\theta/P$ and the argument of the square root is $\sin^2\theta$ yielding

$$\begin{aligned}N_\pm^2(|\theta| \gg \theta_c \gg \theta_D) &\simeq \frac{P\left[1 + \cos^2\theta + g\sin^2\theta \pm |1 - g|\sin^2\theta\right]}{2\sin^2\theta} \\ &= \frac{P}{\sin^2\theta}\begin{cases} 1 \\ \cos^2\theta + g\sin^2\theta \end{cases}.\end{aligned} \tag{9.2.105}$$

For $\theta = \pi/2$ the two solutions (9.2.105) approach the ordinary and extraordinary wave solutions (9.2.10) whose properties have been discussed in detail in Sect. 9.2.4. The modes (9.2.105) obviously are the generalizations to nearly perpendicular propagation of the ordinary and extraordinary waves.

9.2.7.4 Region 4: Propagation Angles $|\theta| \ll 65°$ Near the Proton Gyrofrequency.
For frequencies very close to the proton gyrofrequency $\omega \simeq \Omega_p(1 \pm \epsilon)$, with $\epsilon \ll 1/3672$ we find a very narrow region of finite propagation angles smaller than 65° (see Fig. 9.6) where (9.2.94) can be approximated as

$$N_\pm^2(\theta \ll \theta_c \leq 65°) \simeq \frac{P}{2}\frac{\left[1 + \cos^2\theta \mp 2(1+\epsilon)\cos\theta\right]}{\sin^2\theta - (2\epsilon P/\alpha)\cos^2\theta}, \tag{9.2.106}$$

which for parallel propagation ($\theta = 0$) readily reduces to the two solutions $N_+^2 = R$ and $N_-^2 = L$. For small but finite and fixed propagation angles we may approximate the two solutions of (9.2.106) as

$$N_+^2(\theta < 2.93°) \simeq \frac{P\left[\frac{5}{48}\theta^4 - \epsilon\right]}{\theta^2 - \frac{2\epsilon P}{\alpha}} \simeq \begin{cases} R & \text{for } \epsilon \geq \epsilon_2 \\ -\frac{5R\theta^4}{48\epsilon} & \text{for } \epsilon_2 \geq \epsilon \geq \epsilon_1 \\ \frac{5P\theta^2}{48} & \text{for } \epsilon < \epsilon_1 \end{cases}, \tag{9.2.107a}$$

$$N_+^2(2.93° \leq \theta \ll 57°) \simeq \begin{cases} R & \text{for } \epsilon \geq \epsilon_1 \\ -\frac{P\epsilon}{\theta^2} & \text{for } \epsilon_1 \geq \epsilon \geq \epsilon_2 \\ \frac{5P\theta^2}{48} & \text{for } \epsilon < \epsilon_2 \end{cases}, \quad (9.2.107b)$$

where we have introduced (with θ in radians)

$$\epsilon_1 \equiv \frac{\theta^2}{3672}, \quad \epsilon_2 \equiv 0.104\theta^4. \quad (9.2.108)$$

As one can see, the dispersion relation $N_+^2 = R$ ultimately becomes modified at small but finite propagation angle as one approaches the proton gyrofrequency $\epsilon \to 0$.

Likewise we obtain for the second solution of (9.2.106) at small propagation angles

$$N_-^2(\theta \ll 57°) \simeq \frac{2P}{\theta^2 - (2\epsilon P/\alpha)} \simeq \begin{cases} L & \text{for } \epsilon \geq \epsilon_2 \\ (2P/\theta^2) & \text{for } \epsilon < \epsilon_2 \end{cases}. \quad (9.2.109)$$

Again the dispersion relation $N_-^2 = L$ is modified at small but finite propagation angles as one approaches the proton gyrofrequency.

For the corresponding polarization of these waves we obtain

$$\rho_+(\theta < 2.93°) \simeq -\imath \begin{cases} 1 & \text{for } \epsilon \geq \epsilon_2 \\ 1 - \frac{5\theta^4}{48} & \text{for } \epsilon_2 \geq \epsilon \geq \epsilon_1 \\ 1 + 382.5\epsilon\theta^2 & \text{for } \epsilon < \epsilon_1 \end{cases}, \quad (9.2.110a)$$

$$\rho_+(2.93° \leq \theta \ll 57°) \simeq -\imath \begin{cases} 1 & \text{for } \epsilon \geq \epsilon_1 \\ 1 - \frac{3672\epsilon^2}{\theta^2} & \text{for } \epsilon_1 \geq \epsilon \geq \epsilon_2 \\ 1 + 382.5\epsilon\theta^2 & \text{for } \epsilon < \epsilon_2 \end{cases}, \quad (9.2.110b)$$

and

$$\rho_-(\theta \ll 57°) \simeq \imath \begin{cases} 1 & \text{for } \epsilon \geq \epsilon_2 \\ 1 - \frac{7344\epsilon}{\theta^2} & \text{for } \epsilon < \epsilon_2 \end{cases}. \quad (9.2.110c)$$

9.2.8 Faraday Rotation

Let us now consider the propagation of one stable monochromatic high-frequency ($\omega \gg \omega_{p,e}, \Omega_e$) wave mode through a magneto-ionic plasma. According to (8.3.42) its electric field changes with position and time as

$$\boldsymbol{E}_1(\boldsymbol{x},t) = \boldsymbol{R}_1(\boldsymbol{k},\omega)e^{\imath(\boldsymbol{k}\cdot\boldsymbol{x}-\omega t)} = \boldsymbol{R}_1(\boldsymbol{k},\omega)e^{\imath\omega[Nr/c\,-\,t]}, \quad (9.2.111)$$

where r is the distance measured in the direction of wave propagation (see Fig. 8.3). At high frequencies ($X \ll 1$) we may approximate the Appleton–Hartree dispersion relation (9.2.73a) as

$$N_{\mathrm{L,R}}^2 \simeq 1 - X\left[1 + \frac{u\sin^2\theta}{2} \mp \frac{u^{1/2}}{2}\sqrt{u\sin^4\theta + 4\cos^2\theta}\right], \quad (9.2.112)$$

which reduces to the limits (9.2.76), (9.2.77) and (9.2.79) in the quasi-transverse and quasi-longitudinal cases.

Faraday rotation refers to the propagation of a high-frequency linearly polarized wave. Such a polarized wave may be regarded as the superposition of two circularly polarized waves of equal amplitude and opposite handiness. Since according to (9.2.112) left- and right-handed circularly polarized waves have a different index of refraction, the plane of the resultant linearly polarized wave rotates as the wave propagates.

For the quasi-longitudinal (small θ) case let us consider two circularly polarized waves traveling out of the page (see Fig. 9.7). The elemental angular rotation of the left-handed polarized wave is given by

$$\mathrm{d}\phi_{\mathrm{L}} = \omega N_{\mathrm{L}}\,\mathrm{d}r/c \qquad (9.2.113\mathrm{a})$$

and of the right-handed wave by

$$\mathrm{d}\phi_{\mathrm{R}} = \omega N_{\mathrm{R}}\,\mathrm{d}r/c\,, \qquad (9.2.113\mathrm{b})$$

where $N_{\mathrm{L,R}}$ refer to the two values of (9.2.112). The net rotation angle then is

$$\mathrm{d}\phi = (N_{\mathrm{L}} - N_{\mathrm{R}})\,\omega\,\mathrm{d}r/2c = \sqrt{u\sin^4\theta + 4\cos^2\theta}\,Xu^{1/2}\frac{\omega\mathrm{d}r}{4c}\,, \quad (9.2.114)$$

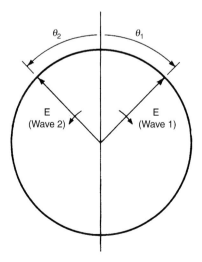

Fig. 9.7. Representation of a linearly polarized wave as two circularly polarized waves (L and R) of opposite handiness. From Kraus (1967 [273])

where we have introduced (9.2.112). In the quasi-longitudinal limit we obtain

$$d\phi_{QL} = \frac{\omega_{p,e}^2 \Omega_e \cos\theta \, dr}{2c\omega^2} = \frac{e^3 \lambda^2}{2\pi m_e^2 c^4} n_e B_0 \cos\theta \, dr . \tag{9.2.115}$$

In the opposite case of quasi-transverse propagation we obtain

$$d\phi_{QT} = \frac{\omega_{p,e}^2 \Omega_e^2 \sin^2\theta \, dr}{4c\omega^3} = \frac{e^4 \lambda^3}{8\pi^3 m_e^3 c^6} n_e B_0^2 \sin^2\theta \, dr . \tag{9.2.116}$$

The ratio of the transverse and longitudinal rotations is

$$\frac{d\phi_{QT}}{d\phi_{QL}} \simeq \frac{\Omega_e}{2\omega} \ll 1 \tag{9.2.117}$$

always very small in the high-frequency limit, so that the Faraday rotation effects may be described entirely in terms of the quasi-longitudinal case (9.2.115).

If the linearly polarized wave travels from its point of origin (A) to the observer (B) through a magneto-ionic medium the total Faraday rotation at wavelength λ is

$$\phi = \text{RM } \lambda^2 \text{ rad} , \tag{9.2.118}$$

where the *rotation measure*

$$\text{RM} = 8.1 \times 10^5 \int_A^B dr n_e(r) B_{0,\parallel}(r) \tag{9.2.119}$$

involves the line-of-sight integral over the product of the thermal electron density in number per cm^3 and the parallel component of the ordered magnetic field in units of Gauss. The numerical factor in (9.2.119) holds if the wavelength λ is measured in m and the path length in units of pc. Combined with the dispersion measure (see Sect. 9.2.1.1) the measurements of the rotation measure (9.2.119) towards pulsars has become a powerful tool to infer the strength of galactic magnetic fields (see our discussion in Sect. 2.4).

10. Test Wave Approach 2. Waves in Hot Magnetized Isotropic Plasmas

According to the kinetic theory of plasma waves (Sect. 8.3) the electromagnetic fluctuations in a given initial configuration $(f_a^{(0)}, \boldsymbol{E}_0, \boldsymbol{B}_0)$, after an initial transit period, will consist of a spectrum of normal modes at different wavenumbers, see (8.3.42). After discussing the cold plasma limit in the last chapter we now turn our attention to hot plasmas.

10.1 Conductivity Tensor for Isotropic Distribution Functions

We start from the conductivity tensor (8.3.52) of a hot magnetized plasma in a coordinate system where the ordered magnetic field lies parallel to the z axis and the wave vector is restricted to the $(x\text{--}z)$ plane, and we restrict our discussion to isotropic background distribution functions $f_a^{(0)}(\boldsymbol{p}) = f_a^{(0)}(p)$ only. For isotropic distribution functions

$$p_\perp \frac{\partial f_a^{(0)}}{\partial p_\|} - p_\| \frac{\partial f_a^{(0)}}{\partial p_\perp} = 0 \;,$$

implying

$$U f_a^{(0)} = \frac{\partial f_a^{(0)}}{\partial p_\perp}, \quad W f_a^{(0)} = \frac{\partial f_a^{(0)}}{\partial p_\|} \;,$$

so that the conductivity tensor for isotropic distribution functions becomes

$$\sigma_{ij}^{\text{iso}} = \frac{2\pi}{i} \sum_a e_a^2 n_a \int_{-\infty}^{\infty} dp_\| \int_0^{\infty} dp_\perp p_\perp \sum_{n=-\infty}^{\infty} \frac{T_{ij}}{\omega - k_\| v_\| - n\Omega_a} \;, \quad (10.1.1)$$

with

$$T_{ij} = \begin{pmatrix} \frac{n^2 J_n^2(z)}{z^2} v_\perp \frac{\partial f_a^{(0)}}{\partial p_\perp} & \frac{i n J_n(z) J_n'(z)}{z} v_\perp \frac{\partial f_a^{(0)}}{\partial p_\perp} & \frac{n J_n^2(z)}{z} v_\perp \frac{\partial f_a^{(0)}}{\partial p_\|} \\ \frac{-i n J_n(z) J_n'(z)}{z} v_\perp \frac{\partial f_a^{(0)}}{\partial p_\perp} & (J_n'(z))^2 v_\perp \frac{\partial f_a^{(0)}}{\partial p_\perp} & -i J_n(z) J_n'(z) v_\perp \frac{\partial f_a^{(0)}}{\partial p_\|} \\ \frac{n J_n^2(z)}{z} v_\| \frac{\partial f_a^{(0)}}{\partial p_\perp} & i J_n(z) J_n'(z) v_\| \frac{\partial f_a^{(0)}}{\partial p_\perp} & J_n^2(z) v_\| \frac{\partial f_a^{(0)}}{\partial p_\|} \end{pmatrix} \;.$$

$$(10.1.2)$$

We recall that $z = k_\perp v_\perp/\Omega_a$, where $\Omega_a = \epsilon_a \Omega_{0a}/\gamma$ is the relativistic gyrofrequency of particles of sort a, and $\epsilon_a = e_a/|e_a|$ the charge sign. $J_n(z)$ denotes the Bessel function of order n and $J'_n(z) \equiv \partial J_n(z)/\partial z$ its first derivative.

We also limit our discussion here to wave propagation along the ordered magnetic field ($k_\perp = 0$). The conductivity tensor (10.1.1) then simplifies considerably, and we obtain for the dielectric tensor (8.3.23)

$$\psi_{ij}^{\text{iso}}(k_\parallel, k_\perp = 0, \omega) = \delta_{ij} + \frac{2\pi}{\omega} \sum_a \omega_{p,a}^2 \int_{-\infty}^{\infty} dp_\parallel \int_0^{\infty} dp_\perp \frac{p_\perp}{\gamma}$$

$$\times \begin{pmatrix} \dfrac{(\omega - k_\parallel v_\parallel)p_\perp}{2\left[(\omega - k_\parallel v_\parallel)^2 - \Omega_a^2\right]} \dfrac{\partial F_a}{\partial p_\perp} & \imath \dfrac{\Omega_a p_\perp}{2\left[(\omega - k_\parallel v_\parallel)^2 - \Omega_a^2\right]} \dfrac{\partial F_a}{\partial p_\perp} & 0 \\ -\imath \dfrac{\Omega_a p_\perp}{2\left[(\omega - k_\parallel v_\parallel)^2 - \Omega_a^2\right]} \dfrac{\partial F_a}{\partial p_\perp} & \dfrac{(\omega - k_\parallel v_\parallel)p_\perp}{2\left[(\omega - k_\parallel v_\parallel)^2 - \Omega_a^2\right]} \dfrac{\partial F_a}{\partial p_\perp} & 0 \\ 0 & 0 & \dfrac{p_\parallel}{\omega - k_\parallel v_\parallel} \dfrac{\partial F_a}{\partial p_\parallel} \end{pmatrix}, \quad (10.1.3)$$

where we abbreviate

$$F_a(p) = f_a^{(0)}(p). \quad (10.1.4)$$

For the study of the possible fluctuations we determine the Maxwell operator according to (8.3.22)

$$\Lambda_{ij}^+ = N^2 \left(\frac{k_i k_j}{k^2} - \delta_{ij}\right) + \psi_{ij}, \quad (10.1.5)$$

since the dispersion relation is provided by (8.3.36)

$$\Lambda^+ = \det\left(\Lambda_{ij}^+\right) = 0. \quad (10.1.6)$$

$N = kc/\omega$ denotes the index of refraction. As shown in Chap. 8 the dispersion relation (10.1.6) has been derived from the coupled system of the linearized Vlasov and Maxwell equations, applying the Fourier transformation with respect to the spatial coordinates $\boldsymbol{x}$, and the Laplace transform in time, i.e. the fluctuating electromagnetic fields and phase space densities and their Fourier–Laplace transforms are related as

$$\begin{Bmatrix} \boldsymbol{E}_1(\boldsymbol{k},\omega) \\ \boldsymbol{B}_1(\boldsymbol{k},\omega) \\ f_1(\boldsymbol{p},\boldsymbol{k},\omega) \end{Bmatrix} = (2\pi)^{-4} \int_{-\infty}^{\infty} d^3x \int_0^{\infty} dt \begin{Bmatrix} \delta\boldsymbol{E}(\boldsymbol{x},t) \\ \delta\boldsymbol{B}(\boldsymbol{x},t) \\ \delta f(\boldsymbol{x},\boldsymbol{p},t) \end{Bmatrix} e^{\imath(\omega t - \boldsymbol{k}\cdot\boldsymbol{x})},$$

$$(10.1.7\text{a})$$

$$\begin{Bmatrix} \delta\boldsymbol{E}(\boldsymbol{x},t) \\ \delta\boldsymbol{B}(\boldsymbol{x},t) \\ \delta f(\boldsymbol{x},\boldsymbol{p},t) \end{Bmatrix} = \int_{-\infty}^{\infty} d^3k \int_L d\omega \begin{Bmatrix} \boldsymbol{E}_1(\boldsymbol{k},\omega) \\ \boldsymbol{B}_1(\boldsymbol{k},\omega) \\ f_1(\boldsymbol{p},\boldsymbol{k},\omega) \end{Bmatrix} e^{\imath(\boldsymbol{k}\cdot\boldsymbol{x} - \omega t)},$$

$$(10.1.7\text{b})$$

10.1 Conductivity Tensor for Isotropic Distribution Functions

Since we consider an infinitely extended physical system we chose the wavenumber k to be real but pay the price that, in general, the time Fourier transform gives rise to complex frequencies $\omega = \omega_R + i\Gamma$, implying also a complex index of refraction. In the derivation it is assumed that the imaginary part of the complex frequency is sufficiently positive,

$$\Gamma > 0, \tag{10.1.8}$$

so that the Laplace integrals in (10.1.7a) converge. The index $+$ in (10.1.5) and (10.1.6) indicates that the condition (10.1.8) holds.

For parallel wave propagation $\boldsymbol{k} = (0, 0, k)$, where $k = k_{\parallel}$, the Maxwell operator (10.1.5) reduces to

$$\Lambda_{ij}^+ = N^2 \left(\delta_{3i}\delta_{3j} - \delta_{ij} \right) + \psi_{ij}, \tag{10.1.9}$$

and we obtain for the dispersion relation (10.1.6)

$$\Lambda^+ = \Lambda_L^+ \Lambda_1^+ \Lambda_2^+ = 0, \tag{10.1.10}$$

yielding as solutions:

- the longitudinal mode

$$\Lambda_L^+(k,\omega) = 1 + \frac{2\pi}{\omega} \sum_a \omega_{p,a}^2 \int_{-\infty}^{\infty} dp_{\parallel} \int_0^{\infty} dp_{\perp} \frac{p_{\perp} p_{\parallel}}{\gamma(\omega - kv_{\parallel})} \frac{\partial F_a}{\partial p_{\parallel}} = 0, \tag{10.1.11}$$

and

- the two transverse modes

$$\Lambda_{1,2}^+(k,\omega) = \eta_{1,2} + 1 - N^2 = 0, \tag{10.1.12}$$

with

$$\eta_{1,2}(k,\omega) = \frac{\pi}{\omega} \sum_a \omega_{p,a}^2 \int_{-\infty}^{\infty} dp_{\parallel} \int_0^{\infty} dp_{\perp} \frac{p_{\perp}^2}{\gamma(\omega - kv_{\parallel} \pm \Omega_a)} \frac{\partial F_a}{\partial p_{\perp}}. \tag{10.1.13}$$

Note that in an unmagnetized ($B_0 = \Omega_a = 0$) plasma

$$\eta_1 = \eta_2 = \eta_T = \frac{\pi}{\omega} \sum_a \omega_{p,a}^2 \int_{-\infty}^{\infty} dp_{\parallel} \int_0^{\infty} dp_{\perp} \frac{p_{\perp}^2}{\gamma(\omega - kv_{\parallel})} \frac{\partial F_a}{\partial p_{\perp}}, \tag{10.1.14}$$

so that the two transverse modes (10.1.12) become identical.

Moreover, in an unmagnetized plasma the longitudinal and transverse dispersion relations hold for all propagation directions. We discuss both modes in turn.

10.2 The Longitudinal Mode

It is convenient (Lerche 1967 [291]) to transform to new variables of integration in (10.1.11),

$$y = p_\|/(m_a c), \quad E = \sqrt{1 + \frac{p_\|^2 + p_\perp^2}{m_a^2 c^2}}, \qquad (10.2.1)$$

implying

$$\frac{\partial F}{\partial p_\|} = (m_a c)^{-1} \left[\frac{\partial F}{\partial y} + \frac{y}{E} \frac{\partial F}{\partial E} \right],$$

$$\frac{\partial F}{\partial p_\perp} = (m_a c)^{-1} \frac{\sqrt{E^2 - 1 - y^2}}{E} \frac{\partial F}{\partial E}. \qquad (10.2.2)$$

The longitudinal dispersion relation (10.1.11) then becomes

$$\Lambda_L^+(k, \omega) = 1 + \frac{2\pi}{\omega} \sum_a \omega_{p,a}^2 (m_a c)^3 \int_1^\infty dE$$

$$\times \int_{-\sqrt{E^2-1}}^{+\sqrt{E^2-1}} dy \frac{y^2}{\omega E - kcy} \frac{\partial F_a}{\partial E}$$

$$= 1 - \frac{2\pi N}{k^2 c^2} \sum_a \omega_{p,a}^2 (m_a c)^3 \int_1^\infty dE \frac{\partial F_a}{\partial E}$$

$$\times \int_{-\sqrt{E^2-1}}^{+\sqrt{E^2-1}} dy \frac{y^2}{y - (E/N)} = 0. \qquad (10.2.3)$$

Defining a function $u(E)$ such that

$$\frac{\partial u}{\partial E} = E^2 \frac{\partial F_a}{\partial E}, \qquad (10.2.4)$$

the dispersion relation (10.2.3) can be cast in the form

$$\Lambda_L^+(k, \omega) = 1 - \frac{4\pi \sum_a \omega_{p,a}^2 (m_a c)^3}{k^2 c^2} \int_1^\infty dE \sqrt{1 - E^{-2}} \frac{\partial u}{\partial E}$$

$$+ \frac{2\pi \sum_a \omega_{p,a}^2 (m_a c)^3}{k^2 c^2 N} \int_1^\infty dE \frac{\partial u}{\partial E} H_+(E, N), \qquad (10.2.5)$$

where

$$H_+(E, N) \equiv - \int_{-\sqrt{E^2-1}}^{+\sqrt{E^2-1}} \frac{dy}{y - (E/N)} = \int_1^S \frac{dy}{y}, \qquad (10.2.6)$$

with

$$S \equiv \frac{1 + N\sqrt{1 - E^{-2}}}{1 - N\sqrt{1 - E^{-2}}} = \Re S + \imath \Im S , \tag{10.2.7}$$

$$\Re S = \frac{\omega_{\rm R}^2 + \Gamma^2 - k^2 c^2 (1 - E^{-2})}{(\omega_{\rm R} - kc\sqrt{1 - E^{-2}})^2 + \Gamma^2} \tag{10.2.8a}$$

$$\Im S = -\frac{2kc\Gamma\sqrt{1 - E^{-2}}}{(\omega_{\rm R} - kc\sqrt{1 - E^{-2}})^2 + \Gamma^2} , \tag{10.2.8b}$$

where we have introduced the complex frequency $\omega = \omega_{\rm R} + \imath \Gamma$. Equation (10.2.8b) indicates that (for positive wavenumbers k) the upper ($\Im(\omega) = \Gamma > 0$) complex frequency half-plane corresponds to S-values with $\Im S < 0$, while the lower ($\Im(\omega) = \Gamma < 0$) complex frequency half-plane corresponds to S-values with $\Im S > 0$. The integral H_+ and thus the dispersion relation (10.2.7) have to be calculated for $\Im(\omega) = \Gamma > 0$ as demanded by (10.1.8). However, as discussed already in Sect. 8.3.4 the Laplace inversion (10.1.7b) is most easily performed along the Landau contour lying in the negative ($\Im(\omega) = \Gamma < 0$) complex frequency half-plane. We therefore have to extend the range of definition of the dispersion relation and determine its analytic continuation in the lower frequency half-plane.

To find the analytic continuation of $\Lambda_{\rm L}^+$ in the lower half-plane with $\Gamma \leq 0$, the limit

$$\lim_{\Gamma \to 0^+} \Lambda_{\rm L}^+ = \lim_{\Gamma \to 0^-} \Lambda_{\rm L}^- \tag{10.2.9a}$$

has to be fulfilled, which is identical to

$$\lim_{\Gamma \to 0^+} H_+ = \lim_{\Gamma \to 0^-} H_- , \tag{10.2.9b}$$

where H_- denotes the integral (10.2.6) for negative values of $\Gamma \leq 0$.

The integral (10.2.6) is the complex logarithmic function which is infinitely many-valued,

$$H_+(E, N) = \ln|S| + \imath \Theta = \ln|S| + \imath \arg S . \tag{10.2.10}$$

We select the particular argument of S, which lies in the interval $-\pi < \Theta_{\rm p} \leq \pi$, as the principal argument and denote this as $\Theta_{\rm p} = {\rm Arg} S$. The resulting principal branch

$$H_{\rm p}(E, N) = \ln|S| + \imath \, {\rm Arg} S \tag{10.2.11}$$

is continuous and analytic at all points in the complex S-plane except the points on the non-positive real axis. This means that the analytic continuations (10.2.9) are fulfilled for all values of S not lying on the non-positive real axis. However, we have to be careful with the evaluation of the integral (10.2.6) and (10.2.11) in cases where

$$\lim_{\Gamma \to 0^\pm} (\Re(S)) \leq 0 . \tag{10.2.12}$$

According to (10.2.8a) we derive

$$\lim_{\Gamma \to 0^\pm} (\Re(S)) = \frac{\omega_R^2 - k^2 c^2 (1 - E^{-2})}{\left(\omega_R - kc\sqrt{1 - E^{-2}}\right)^2} . \quad (10.2.13)$$

From (10.2.13) it is evident that case (10.2.12) never occurs for superluminal waves $\omega_R^2 \geq k^2 c^2$, so that here the dispersion relation (10.2.5) holds for all values of Γ with

$$H_+ = H_- = H_p \quad (10.2.14)$$

and H_p given by (10.2.11).

For subluminal waves $\omega_R^2 < k^2 c^2$ the condition (10.2.12) is fulfilled provided $E > E_c(\omega_R)$ where

$$E_c(\omega_R) = 1/\sqrt{1 - (\omega_R/kc)^2} . \quad (10.2.15)$$

To guarantee that H_+ and H_- according to (10.2.9b) indeed do approach the same limiting value as $\Gamma \to 0^\pm$, respectively, and that they are analytic continuations of each other, we have to choose H_- and H_+ for general ω (or N) as

$$H_+ = H_p , \quad H_- = H_p - 2\pi \imath \theta \left[E - E_c(N)\right] , \quad (10.2.16)$$

where $\theta[x] = 1(0)$ for $x \geq (<) 0$ denotes the step function, and

$$E_c(N) = N/\sqrt{N^2 - 1} . \quad (10.2.17)$$

Introducing the real part of the index of refraction $N_R = \Re(N) = kc/\omega_R$ we can combine (10.2.14) and (10.2.16) as

$$H_+ = H_p , \quad H_- = H_p - 2\pi \imath \theta \left[N_R - 1\right] \theta \left[E - E_c(N)\right] . \quad (10.2.18)$$

As a consequence of (10.2.18) we obtain for the values of the imaginary parts of the functions $H_\pm$ and $\Lambda_L^\pm$ at the real frequency axis for subluminal waves,

$$\lim_{\Gamma \to 0^+} \Im(H_+) = \lim_{\Gamma \to 0^-} \Im(H_-) + 2\pi \theta \left[N_R - 1\right] \theta \left[E - E_c(N)\right] , \quad (10.2.19)$$

implying

$$\lim_{\Gamma \to 0^+} \Im\left(\Lambda_L^+\right) = \lim_{\Gamma \to 0^-} \Im\left(\Lambda_L^-\right)$$
$$- \frac{4\pi^2 \sum_a \omega_{p,a}^2 (m_a c)^3}{k^2 c^2 N_R} \theta \left[N_R - 1\right] u \left(E_c(N)\right) . \quad (10.2.20)$$

The two analytic functions in (10.2.18) approach equal limits on the common boundary of their regions of definition, and therefore each is an analytical continuation of the other. Having thus extended the region of domain for the dispersion relation, we can, in particular, study solutions of the dispersion relation with a negative imaginary part.

10.2.1 Superluminal Waves

We start our analysis with superluminal waves whose phase speed $V_{\rm ph} = \omega_{\rm R}/k = c/N_{\rm R} \geq c$ implying values of $N_{\rm R} < 1$. In this case (10.2.14) holds and we obtain for all values of Γ the dispersion relation of superluminal longitudinal waves as

$$\Lambda_{\rm L}(k, N_{\rm R} < 1, \Gamma) = 1 - \frac{4\pi \sum_a \omega_{{\rm p},a}^2 (m_a c)^3}{k^2 c^2} \int_1^\infty dE \sqrt{1 - E^{-2}} \, \frac{\partial u}{\partial E}$$
$$- \frac{2\pi \sum_a \omega_{{\rm p},a}^2 (m_a c)^3}{k^2 c^2 N_{\rm R}} \int_1^\infty dE \frac{\partial u}{\partial E}$$
$$\times \int_{-N_{\rm R}\sqrt{1-E^{-2}}}^{N_{\rm R}\sqrt{1-E^{-2}}} ds \, \frac{1 + \imath \beta}{s - (1 + \imath \beta)} \, .$$
(10.2.21)

We partially integrate both integrals in (10.2.21) and obtain with (10.2.11) the alternative expression

$$\Lambda_{\rm L}(k, N_{\rm R} < 1, \Gamma) = 1 + \frac{4\pi \sum_a \omega_{{\rm p},a}^2 (m_a c)^3}{c^2 k^2} \int_1^\infty dE \, \frac{u(E)}{E^2 \sqrt{E^2 - 1}}$$
$$+ \frac{4\pi \sum_a \omega_{{\rm p},a}^2 (m_a c)^3}{c^2 k^2} \int_1^\infty dE$$
$$\times \frac{u(E)}{E^2 \sqrt{E^2 - 1} \left[N^2 (1 - E^{-2}) - 1 \right]} = 0 \, .$$
(10.2.22)

In both equations β denotes the ratio

$$\beta = \frac{\Gamma}{\omega_{\rm R}} \, . \tag{10.2.23}$$

It is straightforward to separate (10.2.21) (or (10.2.22)) into real and imaginary parts and to derive

$$\Re \Lambda_{\rm L}(k, N_{\rm R} < 1, \beta) = 1 + \frac{4\pi \sum_a \omega_{{\rm p},a}^2 (m_a c)^3}{c^2 k^2} \int_1^\infty dE \frac{u(E)}{E^2 \sqrt{E^2 - 1}}$$
$$- \frac{2\pi \sum_a \omega_{{\rm p},a}^2 (m_a c)^3}{N_{\rm R} c^2 k^2} \int_1^\infty dE \frac{\partial u}{\partial E}$$
$$\times \int_{-N_{\rm R}\sqrt{1-E^{-2}}}^{N_{\rm R}\sqrt{1-E^{-2}}} ds \, \frac{s - 1 - \beta^2}{(s-1)^2 + \beta^2} = 0 \, ,$$
(10.2.24)

and

$$\Im \Lambda_{\mathrm{L}}(k, N_{\mathrm{R}} < 1, \beta)$$
$$= -\frac{2\pi \sum_a \omega_{\mathrm{p},a}^2 (m_a c)^3 \beta}{N_{\mathrm{R}} c^2 k^2} \int_1^\infty dE \frac{\partial u}{\partial E}$$
$$\times \int_{-N_{\mathrm{R}}\sqrt{1-E^{-2}}}^{N_{\mathrm{R}}\sqrt{1-E^{-2}}} ds \frac{s}{(s-1)^2 + \beta^2}$$
$$= -\frac{8\pi \sum_a \omega_{\mathrm{p},a}^2 (m_a c)^3 \beta}{N_{\mathrm{R}} c^2 k^2} \int_1^\infty dE E^2 \frac{\partial F_a}{\partial E}$$
$$\times \int_0^{N_{\mathrm{R}}\sqrt{1-E^{-2}}} ds \frac{s^2}{[(s-1)^2 + \beta^2][(s+1)^2 + \beta^2]}$$
$$= \frac{8\pi \sum_a \omega_{\mathrm{p},a}^2 (m_a c)^3 \beta}{N_{\mathrm{R}} c^2 k^2} \int_1^\infty dE F_a(E)$$
$$\times \left(2E \int_0^{N_{\mathrm{R}}\sqrt{1-E^{-2}}} ds \frac{s^2}{[(s-1)^2 + \beta^2][(s+1)^2 + \beta^2]} \right.$$
$$\left. + \frac{N_{\mathrm{R}}^2 (E^2 - 1)}{\left[\left(N_{\mathrm{R}}\sqrt{1-E^{-2}} - 1\right)^2 + \beta^2 \right] \left[\left(N_{\mathrm{R}}\sqrt{1-E^{-2}} + 1\right)^2 + \beta^2 \right]} \right) = 0 .$$
(10.2.25)

A first important result follows from (10.2.25): since the integrand for all values of E is a positive function, the only possible solution to (10.2.25) for the imaginary part of the dispersion relation of superluminal longitudinal waves is $\beta = 0$ implying
$$\Gamma = 0 \qquad (10.2.26)$$
for any isotropic particle distribution function. We conclude:

(i) *parallel propagating superluminal longitudinal waves undergo no Landau damping in isotropic magnetized plasmas;*

(ii) *superluminal longitudinal waves undergo no Landau damping in isotropic unmagnetized plasmas.*

$\beta = 0$ used in (10.2.24) yields for the real part
$$\Re \Lambda_{\mathrm{L}}(k, N_{\mathrm{R}} < 1)$$
$$= 1 + \frac{4\pi \sum_a \omega_{\mathrm{p},a}^2 (m_a c)^3}{c^2 k^2} \int_1^\infty dE \frac{u(E)}{E^2 \sqrt{E^2 - 1}}$$
$$+ \frac{4\pi \sum_a \omega_{\mathrm{p},a}^2 (m_a c)^3}{c^2 k^2} \int_1^\infty dE \frac{u(E)}{E^2 \sqrt{E^2 - 1} \left[N_{\mathrm{R}}^2 (1 - E^{-2}) - 1 \right]}$$
$$= 1 - \frac{4\pi \sum_a \omega_{\mathrm{p},a}^2 (m_a c)^3}{\omega_{\mathrm{R}}^2} \int_1^\infty dE \frac{u(E)\sqrt{E^2 - 1}}{E^2 \left[N_{\mathrm{R}}^2 + (1 - N_{\mathrm{R}}^2) E^2 \right]} = 0 .$$
(10.2.27)

It is convenient to introduce the quantity

$$\sigma \equiv (1 - N_\mathrm{R}^2)^{-1/2}, \qquad (10.2.28)$$

which has values between $1 \leq \sigma \leq \infty$. Equation (10.2.27) can then be cast in the compact form

$$\begin{aligned}&\Re\Lambda_\mathrm{L}(k, N_\mathrm{R} < 1)\\&= 1 - \frac{4\pi \sum_a \omega_{\mathrm{p},a}^2 (m_a c)^3 (\sigma^2 - 1)}{c^2 k^2} \int_1^\infty dE \, \frac{u(E)\sqrt{E^2 - 1}}{E^2 \left[E^2 + \sigma^2 - 1\right]} = 0.\end{aligned}$$
$$(10.2.29)$$

After prescribing the composition of the plasma and the phase space distribution function F_a, so that according to (10.2.4) the function $u(E)$ can be inferred, the dispersion relation (10.2.29) will yield the possible superluminal longitudinal modes $N_\mathrm{R} = N_\mathrm{R}(k)$.

10.2.2 Subluminal Waves

For subluminal phase speeds $N_\mathrm{R} \geq 1$ (10.2.16) holds.

10.2.2.1 Growing Subluminal Waves $\beta > 0$. Analyzing first the case of $\beta > 0$, i.e. using $H_+ = H_\mathrm{p}$ in (10.2.3) and (10.2.5), we obtain

$$\begin{aligned}&\Lambda_\mathrm{L}(k, N_\mathrm{R} \geq 1, \Gamma > 0)\\&= 1 + \frac{4\pi \sum_a \omega_{\mathrm{p},a}^2 (m_a c)^3}{c^2 k^2} \int_1^\infty dE \, \frac{u(E)}{E^2 \sqrt{E^2 - 1}}\\&\quad + \frac{4\pi \sum_a \omega_{\mathrm{p},a}^2 (m_a c)^3}{c^2 k^2} \int_1^\infty dE \, \frac{u(E)}{E^2 \sqrt{E^2 - 1}\left[N^2 (1 - E^{-2}) - 1\right]} = 0,\end{aligned}$$
$$(10.2.30)$$

which agrees exactly with (10.2.22). For the imaginary part of the dispersion relation (10.2.30) we find exactly the same result (10.2.25) as before, yielding the solution (10.2.26) $\Gamma = 0$. As a second important general result we thus state that no longitudinal solutions with positive $\Gamma > 0$ exist in isotropic unmagnetized plasmas, and that no parallel propagating longitudinal solutions with positive $\Gamma > 0$ exist in isotropic magnetized plasmas. This is commonly referred to as the Newcomb–Gardner theorem (Bernstein 1958 [47], Gardner 1963 [182], Krall and Trivelpiece 1973 [272]):

(i) *all isotropic particle distribution functions in an unmagnetized plasma are stable against the excitation of longitudinal waves;*
(ii) *all isotropic particle distribution functions in a magnetized plasma are stable against the excitation of parallel propagating longitudinal waves.*

10.2.2.2 Damped Subluminal Waves.

We now consider the case of negative $\beta < 0$ and use (10.2.18b) in (10.2.3) and (10.2.5). We derive for the dispersion relation

$$\Lambda_{\mathrm{L}}(k, N_{\mathrm{R}} \geq 1, \beta < 0)$$
$$= \Re\Lambda_{\mathrm{L}}(k, N_{\mathrm{R}} \geq 1, \beta < 0) + i\Im\Lambda_{\mathrm{L}}(k, N_{\mathrm{R}} \geq 1, \beta < 0)$$
$$= 1 - \frac{4\pi \sum_a \omega_{\mathrm{p},a}^2 (m_a c)^3}{c^2 k^2} \int_1^\infty dE \sqrt{1 - E^{-2}} \frac{\partial u}{\partial E}$$
$$+ \frac{4\pi^2 i \sum_a \omega_{\mathrm{p},a}^2 (m_a c)^3}{Nc^2 k^2} u(E_c) + \frac{2\pi \sum_a \omega_{\mathrm{p},a}^2 (m_a c)^3}{Nc^2 k^2} \int_1^\infty dE \frac{\partial u}{\partial E} H_{\mathrm{p}}$$
$$= 1 + \frac{4\pi^2 i \sum_a \omega_{\mathrm{p},a}^2 (m_a c)^3 u(E_c)}{Nc^2 k^2} + \frac{4\pi \sum_a \omega_{\mathrm{p},a}^2 (m_a c)^3}{c^2 k^2} \int_1^\infty dE \frac{u(E)}{E^2 \sqrt{E^2 - 1}}$$
$$+ \frac{4\pi \sum_a \omega_{\mathrm{p},a}^2 (m_a c)^3}{c^2 k^2} \int_1^\infty dE \frac{u(E)}{E^2 \sqrt{E^2 - 1} \left[N^2 (1 - E^{-2}) - 1\right]} = 0 \, . \tag{10.2.31}$$

For given wavenumber k the solution of (10.2.31) will yield the variations $\omega = \omega(k)$ of damped subluminal longitudinal waves, once the plasma composition and the form of the isotropic distribution function F_a or $u(E)$ are prescribed.

10.3 Illustrative Example: Longitudinal Waves in an Equilibrium Electron Plasma

For illustration of our general results we consider the example of a pure electron plasma with an equilibrium Maxwell–Boltzmann–Jüttner distribution function

$$F(E) = C \exp(-\mu E) \, , \tag{10.3.1}$$

where the dimensionless parameter μ characterizes the electron plasma temperature

$$\mu = \frac{m_e c^2}{k_{\mathrm{B}} T_e} \, . \tag{10.3.2}$$

With the transformations (10.2.1) the normalization

$$\int d^3 p F(E) = 2\pi \int_{-\infty}^\infty dp_\| \int_0^\infty dp_\perp p_\perp F(E)$$
$$= 4\pi (m_e c)^3 \int_1^\infty dE E \sqrt{E^2 - 1} F(E) = 1$$

fixes the constant C to be

10.3 Longitudinal Waves in an Equilibrium Electron Plasma

$$C = \frac{\mu}{4\pi(m_ec)^3 K_2(\mu)}, \quad (10.3.3)$$

where $K_\nu(z)$ denotes the modified Bessel function of index ν. From (10.3.3) we obtain for the combination

$$4\pi C \sum_a \omega_{p,a}^2 (m_ac)^3 = \frac{\omega_{p,e}^2 \mu}{K_2(\mu)}. \quad (10.3.4)$$

It is convenient to scale the wavenumber k in units of the inverse of the electron skin length $\omega_{p,e}/c$, i.e. we introduce the dimensionless wavenumber

$$\kappa = \frac{ck}{\omega_{p,e}}. \quad (10.3.5)$$

This reduces the combination (10.3.4) to

$$\frac{4\pi C \sum_a \omega_{p,a}^2 (m_ac)^3}{c^2 k^2} = \frac{\mu}{\kappa^2 K_2(\mu)}. \quad (10.3.6)$$

The chosen distribution function (10.3.1) yields, according to (10.2.4)

$$u(E) = Ce^{-\mu E}\left[E^2 + \frac{2E}{\mu} + \frac{2}{\mu^2}\right]. \quad (10.3.7)$$

Collecting terms we then obtain for the real part of the dispersion relation of superluminal longitudinal waves (10.2.27)

$$\Re \Lambda_L(\kappa, N_R < 1) = 1 + \frac{\mu}{\kappa^2} - \frac{\mu}{\kappa^2 K_2(\mu)} J(\mu, N = N_R) = 0, \quad (10.3.8)$$

and for the dispersion relation of subluminal longitudinal waves (10.2.31)

$$\Lambda_L^-(k, N_R \geq 1, \beta < 0) = 1 +$$
$$\frac{\mu}{\kappa^2}\left[1 + \frac{\pi \imath N e^{-\mu E_c}}{(N^2-1) K_2(\mu)}\left(1 + \frac{2}{\mu E_c(N)} + \frac{2}{\mu^2 E_c^2(N)}\right) - \frac{J(\mu, N)}{K_2(\mu)}\right] = 0, \quad (10.3.9)$$

with the integral

$$J(\mu, N) \equiv \int_1^\infty dE \, \frac{e^{-\mu E}\left[1 + (2/\mu E) + (2/\mu^2 E^2)\right]}{\sqrt{E^2-1}\left[1 - N^2(1-E^{-2})\right]}. \quad (10.3.10)$$

The properties of the integral (10.3.10) are discussed in detail in Appendix C.

10.3.1 Superluminal Waves

Here we find according to (C.22) that

$$J(\mu, N_{\rm R}) = \sigma \int_0^\infty dt\, e^{-\sigma t}$$
$$\times \left[\frac{2K_0\left(\sqrt{\mu^2+t^2}\right)}{\mu^2} + \frac{K_1\left(\sqrt{\mu^2+t^2}\right)}{\sqrt{\mu^2+t^2}} + \frac{\mu^2 K_2\left(\sqrt{\mu^2+t^2}\right)}{\mu^2+t^2} \right], \tag{10.3.11}$$

with σ given in (10.2.28). For non-relativistic plasma temperatures $\mu \gg 1$ we may use the series expansion given in (C.23) which converges rapidly in this case since $\mu\sigma^2 \gg 1$. With the first three terms we obtain

$$J(\mu \gg 1)$$
$$= \sum_{i=0}^\infty \mu^{-i} \left[\left(1 + \frac{2}{\mu^2}\right) K_i(\mu) + \frac{2i+3}{\mu} K_{i+1}(\mu) \right] \left(\frac{\partial}{\partial\sigma}\sigma^{-1}\right)^{(i)}$$
$$\simeq \left(1 + \frac{2}{\mu^2}\right) K_0(\mu) + \frac{3}{\mu} K_1(\mu) - \frac{1}{\mu\sigma^2}\left[\left(1 + \frac{2}{\mu^2}\right) K_1(\mu) + \frac{5}{\mu} K_2(\mu)\right]$$
$$+ \frac{3}{\mu^2\sigma^4}\left[\left(1 + \frac{2}{\mu^2}\right) K_2(\mu) + \frac{7}{\mu} K_3(\mu)\right], \tag{10.3.12}$$

so that the dispersion relation (10.3.8) reduces to

$$\kappa^2 \simeq \kappa_c^2 - \frac{A(\mu)}{\sigma^2} + \frac{B(\mu)}{\sigma^4} \simeq 0, \tag{10.3.13}$$

with the characteristic wavenumber

$$\kappa_c^2 = \frac{K_1(\mu)}{K_2(\mu)} + \frac{2}{\mu}\frac{K_0(\mu)}{K_2(\mu)} \simeq 1 + \frac{1}{2\mu}, \tag{10.3.14}$$

$$A(\mu) = \frac{5}{\mu} + \left(1 + \frac{2}{\mu^2}\right)\frac{K_1(\mu)}{K_2(\mu)} \simeq 1 + \frac{7}{2\mu}, \tag{10.3.15}$$

and

$$B(\mu) = \frac{3}{\mu}\left[1 + \frac{30}{\mu^2} + \left(\frac{7}{\mu}\frac{K_1(\mu)}{K_2(\mu)}\right)\right] \simeq \frac{3}{\mu}, \tag{10.3.16}$$

where the respective approximations hold for non-relativistic plasma temperatures $\mu \gg 1$. Since according to (10.2.28) $\sigma^{-2} = 1 - (\kappa/f)^2$ the superluminal dispersion relation (10.3.13) with the non-relativistic approximations for κ_c^2, $A(\mu)$, and $B(\mu)$ reduces to the biquadratic equation

$$f^4 - \left(1 - \frac{5}{2\mu}\right) f^2 = \frac{3\kappa^2}{\mu}. \tag{10.3.17}$$

10.3 Longitudinal Waves in an Equilibrium Electron Plasma

Equation (10.3.17) has the solution

$$2f^2 = \left(1 - \frac{5}{2\mu}\right) + \sqrt{\left(1 - \frac{5}{2\mu}\right)^2 + \frac{12\kappa^2}{\mu}}, \qquad (10.3.18)$$

which is superluminal for all wavenumbers less than the characteristic wavenumber κ_c.

10.3.2 Subluminal Waves

As shown in Appendix C (see (C.17) and (C.29)) for subluminal waves we obtain with $y = 1/\sqrt{1-N^2}$ for the integral (10.3.10)

$$J = W(\mu, \imath y)$$
$$+ \frac{\pi}{2}\text{sgn}\,(\Re y)y\sqrt{1-y^2}\,e^{-\mu\sqrt{1-y^2}}\left[1 + \frac{2}{\mu\sqrt{1-y^2}} + \frac{2}{\mu^2(1-y^2)}\right], \qquad (10.3.19)$$

where

$$W(\mu, \imath y) = -\frac{2y^2}{\mu^2}\int_0^\infty dt\left[1 + \frac{\mu^2}{4t} + \frac{\mu^4}{8t^2}\right]M\left(1, \frac{3}{2}, y^2 t\right)\exp\left(-t - \frac{\mu^2}{4t}\right). \qquad (10.3.20)$$

The subluminal dispersion relation (10.3.9) then becomes

$$\Lambda_L^-(k, N_R \geq 1, \beta < 0)$$
$$= 1 - \frac{\mu}{\kappa^2}\left[\frac{W(\mu, \imath y)}{K_2(\mu)} - 1 - \imath\frac{\pi E_c e^{-\mu E_c}}{2\sqrt{N^2 - 1}K_2(\mu)}[2 - \text{sgn}\,(\Re y)]\right.$$
$$\left.\times \left(1 + \frac{2}{\mu E_c(N)} + \frac{2}{\mu^2 E_c^2(N)}\right)\right] = 0, \qquad (10.3.21)$$

since $E_c = \sqrt{1-y^2}$. It has been shown by Schlickeiser and Mause (1995 [468]) that (10.3.20) can be expressed as

$$W(\mu, \imath y) = -y^2\int_0^1 du\,[K_2(z(u)) + z(u)K_1(z(u))], \qquad (10.3.22)$$

where

$$z(u) \equiv \mu\sqrt{1 - y^2 + y^2 u^2}. \qquad (10.3.23)$$

The integral (10.3.10) can also be expressed as the difference of two series (see (C.37))

$$J = \sqrt{\pi \mu y^2/2} \left[\sum_{n=0}^{\infty} \frac{(\mu y^2/2)^n}{\Gamma(1+n)} \left[\left(1 + \frac{2}{\mu^2}\right) K_{n+\frac{1}{2}}(\mu) - \frac{2(n-1)}{\mu} K_{n-\frac{1}{2}}(\mu) \right] \right.$$
$$\left. - \sqrt{\mu y^2/2} \sum_{n=0}^{\infty} \frac{(\mu y^2/2)^n}{\Gamma((3/2)+n)} \left[\left(1 + \frac{2}{\mu^2}\right) K_{n+1}(\mu) - \frac{2n-1}{\mu} K_n(\mu) \right] \right].$$
(10.3.24)

The representation (10.3.24) is particularly useful for non-relativistic ($\mu \gg 1$) and ultrarelativistic ($\mu \ll 1$) plasma temperatures, since one may derive simpler approximations by using the appropriate asymptotic expansions of the modified Bessel functions.

10.3.3 Non-relativistic Plasma Temperatures

For non-relativistic temperatures $\mu \gg 1$ we may use the asymptotic expansion of the modified Bessel functions for large argument to approximate (10.3.22) to lowest order in the small parameter $1/\mu \ll 1$ as

$$\frac{W(\mu, \imath y)}{K_2(\mu)}$$
$$\simeq -\mu y^2 \int_0^1 du \, [1 - y^2(1-u^2)]^{1/4} \exp\left(-\mu \left[\sqrt{1 - y^2(1-u^2)} - 1\right]\right).$$
(10.3.25)

10.3.3.1 Large Index of Refraction $|N^2| > 2$. For absolute values of the index of refraction $|N^2| > 2$ we have $|-y^2|(1-u^2) < 1$ for all values of u, so that we may approximate (10.3.25) as

$$\frac{W(|N^2| > 2)}{K_2(\mu)} \simeq -\mu y^2 \int_0^1 du \, \exp\left(\frac{\mu y^2(1-u^2)}{2}\right)$$
$$= -\sqrt{2\mu} y e^{\mu y^2/2} \int_0^{\sqrt{\mu/2}y} dt \, e^{-t^2}.$$
(10.3.26)

In terms of the variable

$$x = \imath \sqrt{\mu/2} y = \sqrt{\frac{\mu}{2(N^2-1)}},$$
(10.3.27)

and the plasma dispersion function (Fried and Conte 1961 [177])

$$Z(x) = 2\imath e^{-x^2} \int_{-\infty}^{\imath x} dt \, e^{-t^2} = \imath \pi^{1/2} e^{-x^2} - 2e^{-x^2} \int_0^x dt \, e^{t^2},$$
(10.3.28)

(10.3.26) can be expressed in the form

10.3 Longitudinal Waves in an Equilibrium Electron Plasma

$$\frac{W\left(|N^2|>2\right)}{K_2(\mu)} \simeq i\pi^{1/2}xe^{-x^2} - xZ(x) \,. \tag{10.3.29}$$

With this approximation the dispersion relation (10.3.21) of subluminal waves in this range, to lowest order in the small value $\mu^{-1} \ll 1$, becomes

$$\begin{aligned}\Lambda_{\mathrm{L}}^{-}\left(\kappa, |N^2| > 2\right) &= 1 + \frac{\mu}{\kappa^2}\left[1 + xZ(x) + i\pi^{1/2}xe^{-x^2}\left[1 + \mathrm{sgn}\left(\Im x\right)\right]\right]\\ &= 1 + \frac{\mu}{\kappa^2}\left[1 + xZ(x) + 2i\pi^{1/2}xe^{-x^2}\theta\left(\Im x\right)\right] = 0 \,,\end{aligned}$$
(10.3.30)

where θ, as before, denotes the step function. Equation (10.3.30) can also be derived according to (C.48) which results from the approximation (C.38) of the series representation (10.3.24).

Using the properties (C.42) and (C.44) of the plasma dispersion function (see Appendix C) we find from (10.3.30) that if $x = q + is$ is a root of the dispersion relation (10.3.31), then $-x^* = -q + is$ is also a root of this relation. Both roots have positive imaginary parts, corresponding correctly to non-positive values of β and negative values of Γ.

Introducing the normalized Debye length $d \equiv \mu^{-1/2}$, (10.3.30) for $\Im(x) < 0$ yields

$$0 = \kappa^2 d^2 + 1 + xZ(x) \,. \tag{10.3.31}$$

In functional form the dispersion relation (10.3.30) or (10.3.31) agrees with the earlier results of Jackson (1960 [235], (Equation A2.10)) and Fried and Gould (1961 [178], (Equation 19)). There are, however, two significant differences:

(i) First, we have established that (10.3.30) is strictly true only for values of the index of refraction $|N^2| > 2$ while in earlier work no restriction on the index of refraction has been noted.
(ii) Second, the respective arguments x of the plasma dispersion function are different. While we find according to (10.3.27) that $x = \sqrt{\mu/2(N^2-1)}$, in the former work of Jackson (1960 [235]) and Fried and Gould (1961 [178]) x is simply $(\omega_{\mathrm{R}} + i\Gamma)/kv_{\mathrm{th}}$, where

$$v_{\mathrm{th}} = \sqrt{2k_{\mathrm{B}}T_{\mathrm{e}}/m_{\mathrm{e}}} \,. \tag{10.3.32}$$

As we shall demonstrate below (Sect. 10.3.4) we can reproduce the earlier result only if we require, additionally, the limit of an infinitely large speed of light, $c \to \infty$.

Writing the dispersion relation (10.3.31) in the form

$$xZ(x) = -(1 + \kappa^2 d^2) \,, \tag{10.3.33}$$

we note that the general results obtained by Backus (1960 [23]), Hayes (1961 [222], 1963 [223]) and Saenz (1963 [442]) on the number, order, and location of the roots of the dispersion relation (10.3.33) hold. In particular, Picard's theorem implies for a given value of $\kappa \neq 0$ that the dispersion relation (10.3.33) has an infinite number of roots, and (see Roos 1969 [439], p. 463) that these roots are simple.

10.3.3.2 Index of Refraction $1 < |N^2| < 2$. For absolute values of the complex index of refraction close to unity ($1 < |N^2| < 2$) the evaluation of (10.3.25) is more complicated, and it is more advantageous to use the series approximation (C.23)–(C.25) of Appendix C, see also (10.3.12), since for nonrelativistic temperatures $|\mu - y^2| \gg 1$ is always fulfilled in this case. With the first three terms in the series we obtain

$$\begin{aligned}
&J\left(\mu, 1 < |N^2| < 2\right) \\
&\simeq \left(1 + \frac{2}{\mu^2}\right) K_0(\mu) \\
&\quad + \frac{3}{\mu} K_1(\mu) - \frac{1}{\mu y^2}\left[\left(1 + \frac{2}{\mu^2}\right) K_1(\mu) + \frac{5}{\mu} K_2(\mu)\right] \\
&\quad + \frac{3}{\mu^2 y^4}\left[\left(1 + \frac{2}{\mu^2}\right) K_2(\mu) + \frac{7}{\mu} K_3(\mu)\right].
\end{aligned} \qquad (10.3.34)$$

By using (10.3.14)–(10.3.16) we derive

$$\frac{J\left(\mu, 1 < |N^2| < 2\right)}{K_2(\mu)} \simeq 1 + \frac{1}{\mu}\left[\kappa_c^2 - \frac{A(\mu)}{y^2} + \frac{B(\mu)}{y^4}\right], \qquad (10.3.35)$$

and the subluminal dispersion relation (10.3.21) becomes

$$\begin{aligned}
&\Lambda_L^-\left(k, 1 < |N^2| < 2\right) \\
&= 1 - \frac{1}{\kappa^2}\left[\kappa_c^2 - \frac{A(\mu)}{y^2} + \frac{B(\mu)}{y^4}\right. \\
&\quad \left. + \imath\frac{\pi\mu E_c e^{-\mu E_c}}{\sqrt{N^2 - 1} K_2(\mu)}\left(1 + \frac{2}{\mu E_c(N)} + \frac{2}{\mu^2 E_c^2(N)}\right)\right] = 0.
\end{aligned} \qquad (10.3.36)$$

In terms of the variable (10.3.27) the dispersion relation (10.3.36) becomes to lowest order in $1/\mu \ll 1$

$$\begin{aligned}
&\Lambda_L^-\left(k, 1 < |N^2| < 2\right) \\
&= 1 - \frac{1}{\kappa^2}\left[\kappa_c^2 + \frac{\mu A(\mu)}{2x^2} + \frac{\mu^2 B(\mu)}{4x^4} - 2\imath\sqrt{2\pi\mu} x^2 e^{-\sqrt{2\mu} x}\right] \\
&\simeq 1 - \frac{1}{\kappa^2}\left[1 + \frac{1}{2\mu} + \frac{\mu + (7/2)}{2x^2} + \frac{3(\mu + 7)}{4x^4} - 2\imath\sqrt{2\pi\mu} x^2 e^{-\sqrt{2\mu} x}\right] = 0.
\end{aligned} \qquad (10.3.37)$$

10.3.4 Infinitely Large Speed of Light, $c \to \infty$

Previous non-relativistic work (Landau 1946 [280], Jackson 1960 [235], Fried and Gould 1961 [178]) corresponds to the formal limit of an infinitely large value of the speed of light $c \to \infty$. In this limit all wave phase speeds are subluminal, and (10.3.8) does not apply. In the limit $c \to \infty$ we find

$$\lim_{c \to \infty} \left(\kappa^2 d^2 \right) = k^2 / k_D^2 , \qquad (10.3.38)$$

where

$$k_D = \left[\frac{m_e \omega_{p,e}^2}{k_B T_e} \right]^{1/2} = \sqrt{2} \, \frac{\omega_{p,e}}{v_{th}} \qquad (10.3.39)$$

is the Debye wavenumber. From (10.2.17) we establish

$$\lim_{c \to \infty} E_c = 1 , \qquad (10.3.40)$$

and (10.3.27) yields

$$\lim_{c \to \infty} (x) \equiv x_0 = \frac{\omega_R + i\Gamma}{v_{th} k} . \qquad (10.3.41)$$

Collecting terms the subluminal dispersion relation (10.3.31) in this limit reduces to

$$\frac{k^2}{k_D^2} + 1 + x_0 Z(x_0) = \frac{k^2}{k_D^2} - \frac{1}{2} Z'(x_0) = 0 , \qquad (10.3.42)$$

and is valid for all waves. Now the dispersion relation (10.3.42) agrees exactly with the earlier results of Jackson (1960 [235], (A.2.10)) and Fried and Gould (1961 [178], (Equation 19)), proving our proposition above (see discussion following (10.3.31)). Apparently, the earlier expressions can only be reproduced if besides the non-relativistic approximation $\mu \gg 1$ the limit $c \to \infty$ is demanded. Since the latter assumption is certainly incorrect, (10.3.30) and (10.3.37) with x defined according to (10.3.27) holding in their respective limits–and not (10.3.42)–are the correct non-relativistic dispersion relations of subluminal longitudinal waves.

10.3.5 Solution of the Non-relativistic Dispersion Relation

In the range $|N^2| > 2$ the functional form of the non-relativistic dispersion relation (10.3.30) is identical to the relation (10.3.42) discussed previously, and we can adopt the convenient solution method developed by Fried and Gould (1961 [178]). Separating (10.3.30) into real and imaginary parts we obtain with

$$x = q + is \qquad (10.3.43)$$

the two equations

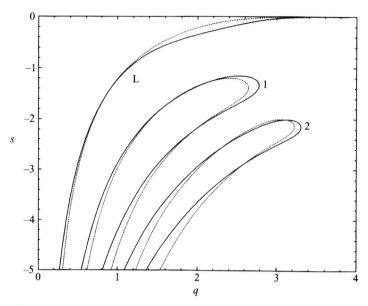

Fig. 10.1. Loci in the complex $x(=q+\imath s)$-plane of solutions of the subluminal longitudinal dispersion relation in an electron plasma. The **full curves** are calculated numerically from the imaginary part of the relativistically correct dispersion relation (10.3.44) for a non-relativistic electron temperature value $\mu = 200$. The **dashed curves** are calculated from the non-relativistic approximation (10.3.42) and are valid for any non-relativistic electron temperature $\mu \gg 1$. The labels "L", "1", and "2" refer to the principal Landau mode and the first two electron sound wave modes, respectively

$$\Im[xZ(x)] = q\Im Z(q+\imath s) + s\Re Z(q+\imath s) = 0 , \qquad (10.3.44)$$

and

$$\kappa^2 d^2 + 1 + \Re[xZ(x)] = \kappa^2 d^2 + 1 + q\Re Z(q+\imath s) - s\Im Z(q+\imath s) = 0 . \quad (10.3.45)$$

Regarded as a function of the variables q and s, (10.3.44) does not involve the wavenumber κ and the plasma temperature μ (or d) explicitly. Therefore from (10.3.44) we obtain the loci of points in the x plane, shown as dashed curves in Fig. 10.1, fixing the relation between q_l and s_l, $l = L, 1, 2, \cdots$), for any value of κ and μ. For comparison, the full curves in Fig. 10.1 show the loci of points obtained by calculating numerically the imaginary part of the relativistically exact dispersion relation (10.3.9)–(10.3.10) for a value of $\mu = 200$ without making the additional assumption of $|N^2| > 2$. Although quantitative deviations occur, one finds the same qualitative result that different loci of points, labeled as "L", "1" and "2", can be derived that fix the relation between q_l and s_l.

10.3 Longitudinal Waves in an Equilibrium Electron Plasma

At any one of these loci, ($l = L, 1, 2, ...$), we can then compute the wavenumber values from (10.3.45) in the limit $|N^2| > 2$ as

$$\kappa^2 d^2 = s_l \Im Z(q_l + \imath s_l) - q_l \Re Z(q_l + \imath s_l) - 1 \qquad (10.3.46)$$

in units of d^{-1}. Only if $\kappa^2 \geq 0$ we have a bona fide solution of the dispersion relation (10.3.30). As emphasized, the resulting values of $(q, s, \kappa^2 d^2)$ are the same for any non-relativistic plasma temperature $\mu \gg 1$. Of course, in the relativistic case (10.3.46) corresponds to

$$\Re\left[\Lambda\left(q_l + is_l\right)\right] = 0 \ . \qquad (10.3.47)$$

Then, using (10.3.27) written in the form

$$\omega^2 = \frac{2x^2}{\mu + 2x^2} k^2 c^2 \ , \qquad (10.3.48)$$

we derive the variation of the real, $\omega_{R,l} = \omega_{R,l}(k)$, and imaginary, $\Gamma_l = \Gamma_l(k)$, parts for the different possible plasma modes, after the plasma temperature value μ has been specified. The solutions for the principal Landau ("L") mode and the first ("1") mode are shown in Figs. 10.2 and 10.4, respectively.

In the case of the principal Landau mode (Fig. 10.2) we show the curves obtained with the exact relativistic dispersion relation and with the non-relativistic approximation. We notice from Fig. 10.2 that the principal mode

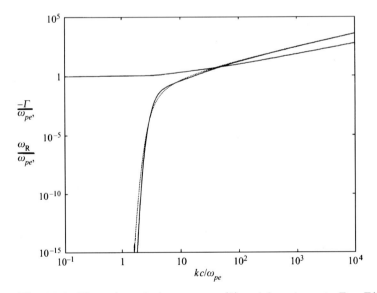

Fig. 10.2. Dispersion relation $\omega_R = \omega_R(k)$ and damping rate $\Gamma = \Gamma(k)$ of the principal Landau mode calculated from the relativistically correct dispersion relation (**full curve**) and its non-relativistic approximation (**dotted curve**) for an electron temperature $\mu = 200$

is weakly ($\Gamma \ll \omega_R$) damped at small wavenumbers and strongly ($\Gamma \gg \omega_R$) damped at large wavenumbers. This is in qualitative agreement with the non-relativistic analysis, but as Fig. 10.2 also shows there are quantitative differences from the exact curves. Particularly noteworthy is the overestimation of the damping rates at small wavenumbers by the non-relativistic approximation. This is seen more clearly in Fig. 10.3 where we compare the exact damping rate with its non-relativistic approximation for a value of $\mu = 20$ in the vicinity of $\kappa = 1$. While the exact theory correctly describes that no damping occurs for superluminal waves below the critical wavenumber κ_c, the non-relativistic theory yields small but finite damping rates in this wavenumber region. Also, for wavenumber values slightly larger than κ_c the two damping rates exhibit orders of magnitude differences.

Compared to the principal Landau mode the first mode shown in Fig. 10.4 is subluminal for all wavenumbers. We find that this mode is weakly ($\Gamma \ll \omega_R$) damped at small wavenumbers and strongly damped ($\Gamma \gg \omega_R$) at large wavenumbers.

Similarly to (10.3.44) and (10.3.45) we can treat the dispersion relation (10.3.37) for indices of refraction close to unity $1 < |N^2| < 2$. Inserting (10.3.43) and separating real and imaginary parts of the dispersion relation we derive the two equations

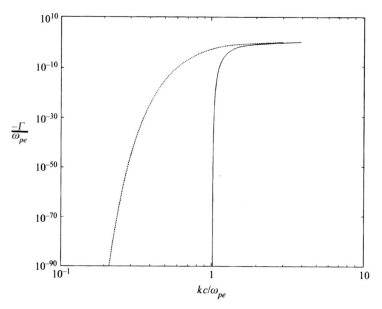

Fig. 10.3. Damping rate of the principal Landau mode for an electron temperature $\mu = 200$ calculated from the relativistically correct dispersion relation (**full curve**) and its non-relativistic approximation (**dotted curve**). Note the dramatic differences at superluminal wavenumbers

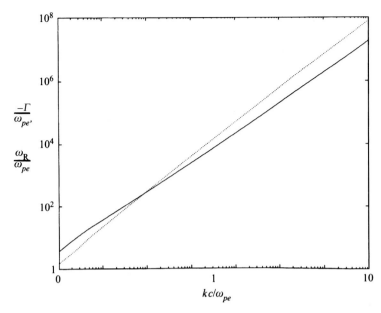

Fig. 10.4. Dispersion relation $\omega_R = \omega_R(k)$ (**full curve**) and damping rate $\Gamma = \Gamma(k)$ (**dotted curve**) of the fist electron sound wave mode "1" calculated from the relativistically correct dispersion relation for an electron temperature $\mu = 200$

$$\Im\left(\kappa^2 \Lambda_L^-(k, 1 < |N^2| < 2)\right)$$
$$= \frac{[\mu + (7/2)]qs}{(q^2 + s^2)^2} + \frac{3qs\left(q^2 - s^2\right)(\mu + 7)}{(q^2 + s^2)^4}$$
$$- 2\sqrt{2\pi}\mu e^{-\sqrt{2\mu}q}\left[\left(q^2 - s^2\right)\cos\left(\sqrt{2\mu}s\right) + 2qs\sin\left(\sqrt{2\mu}s\right)\right] = 0,$$
(10.3.49)

and

$$\Re\left(\kappa^2 \Lambda_L^-(k, 1 < |N^2| < 2)\right)$$
$$= \kappa^2 - 1 - \frac{[\mu + (7/2)]\left(q^2 - s^2\right)}{2(q^2 + s^2)^2} + \frac{3\left(q^4 - 6q^2s^2 + s^4\right)(\mu + 7)}{4(q^2 + s^2)^4}$$
$$- 2\sqrt{2\pi}\mu e^{-\sqrt{2\mu}q}\left[\left(q^2 - s^2\right)\sin\left(\sqrt{2\mu}s\right) - 2qs\cos\left(\sqrt{2\mu}s\right)\right] = 0,$$
(10.3.50)

whose solution yields the combination $(q, s, \kappa^2 d^2)$ in this range of index of refraction. Employing again (10.3.48) we obtain the corresponding various plasma modes in this range.

10.3.6 Landau Damping of Subluminal Solutions

In the case of subluminal longitudinal waves ($N^2 > 1$) in isotropic plasmas we have found negative values of the damping rate $\Gamma_m = I_m(\omega_m) < 0$ in the case of an unmagnetized plasma, or in case of parallel propagation. Recalling, see (8.3.42) and (10.1.7b), that the inverse Laplace transform yields the time dependence of the fluctuating fields as a sum over all j roots of the dispersion relation

$$\delta \boldsymbol{E}(\boldsymbol{k},t) \sim \sum_{m=1}^{j} R_m e^{\Im\omega_m(\boldsymbol{k})t - i\Re\omega_m(\boldsymbol{k})t}, \qquad (10.3.51)$$

we find that all subluminal modes are collisionless damped in the plasma (*Landau damping*).

10.3.7 Weak Damping Approximation

In general, dispersion relations are complex integral relations and it is convenient to calculate approximative frequency-wavenumber relations in the limit of weak damping $\Gamma \to 0^{\pm}$. Separating real and imaginary parts

$$\Lambda = \Re\Lambda + i\Im\Lambda = 0 \qquad (10.3.52)$$

approximate solutions to the dispersion relation can be found provided $|\Im\Lambda| \ll |\Re\Lambda|$. Equating the real and imaginary parts of (10.3.52) to zero and making a Taylor-expansion around $\Gamma = 0$ we obtain

$$\Re\Lambda(\boldsymbol{k},\omega_R,\Gamma) \simeq \Re\Lambda(\boldsymbol{k},\omega_R,\Gamma=0) + \Gamma\left[\frac{\partial\Re\Lambda(\boldsymbol{k},\omega_R,\Gamma)}{\partial\Gamma}\right]_{\Gamma=0} = 0, \qquad (10.3.53)$$

$$\Im\Lambda(\boldsymbol{k},\omega_R,\Gamma) \simeq \Im\Lambda(\boldsymbol{k},\omega_R,\Gamma=0) + \Gamma\left[\frac{\partial\Im\Lambda(\boldsymbol{k},\omega_R,\Gamma)}{\partial\Gamma}\right]_{\Gamma=0} = 0. \qquad (10.3.54)$$

Since $\Lambda(\boldsymbol{k},\omega)$ is a meromorphic function of the complex variable ω, we may use the Cauchy-Riemann relations

$$\frac{\partial\Re\Lambda(\boldsymbol{k},\omega_R,\Gamma)}{\partial\omega_R} = \frac{\partial\Im\Lambda(\boldsymbol{k},\omega_R,\Gamma)}{\partial\Gamma}, \qquad (10.3.55)$$

$$\frac{\partial\Re\Lambda(\boldsymbol{k},\omega_R,\Gamma)}{\partial\Gamma} = -\frac{\partial\Im\Lambda(\boldsymbol{k},\omega_R,\Gamma)}{\partial\omega_R} \qquad (10.3.56)$$

to obtain from (10.3.54) that $\Im\Lambda$ is, to first order in Γ,

$$\Im\Lambda(\boldsymbol{k},\omega_R,\Gamma=0) + \Gamma\frac{\partial\Re\Lambda(\boldsymbol{k},\omega_R,0)}{\partial\omega_R} = 0. \qquad (10.3.57)$$

Using (10.3.56) in (10.3.53) yields

$$\Re \Lambda(\mathbf{k}, \omega_R, \Gamma = 0) - \Gamma \frac{\partial \Im \Lambda(\mathbf{k}, \omega_R, 0)}{\partial \omega_R} = 0 , \qquad (10.3.58)$$

which in combination with (10.3.57) yields to lowest order in the small quantity $|\Gamma|/\omega_R \ll 1$ that the real part of the dispersion relation satisfies

$$\Re \Lambda(\mathbf{k}, \omega_R, \Gamma = 0) = 0 , \qquad (10.3.59)$$

while (10.3.57) provides the corresponding imaginary part of the plasma mode frequency

$$\Gamma = - \frac{\Im \Lambda(\mathbf{k}, \omega_R, 0)}{[\partial \Re \Lambda(\mathbf{k}, \omega_R, 0)/\partial \omega_R]} . \qquad (10.3.60)$$

Equations (10.3.59) and (10.3.60) can be used to calculate approximately the real and imaginary part of the frequencies that obey the dispersion relation. The results from such an approximation, however, only apply as long as the damping rate is much smaller than the real part of the frequency, i.e. $|\Gamma| \ll \omega_R$. In the case of longitudinal waves this has been demonstrated by the work of Schlickeiser and Mause (1995 [468]) as well as Schlickeiser and Kneller (1997 [465]) who studied the longitudinal dispersion relation with and without the weak-damping approximation, respectively.

10.4 The Transverse Modes

Applying the integration variable transformation (10.2.1) and using the notation (10.2.4) and (10.2.6) we can cast the transverse dispersion relation (10.1.12) in the form

$$\Lambda_{1,2}^+ = 1 - N^2 + \frac{2\pi \sum_a \omega_{p,a}^2 (m_a c)^3}{k^2 c^2} \int_1^\infty dE\, E^2 \sqrt{1 - E^{-2}} \left[1 \pm \frac{\epsilon_a \Omega_{0a}}{\omega E}\right] \frac{\partial F_a}{\partial E}$$

$$+ \frac{\pi \sum_a \omega_{p,a}^2 (m_a c)^3}{k^2 c^2 N} \int_1^\infty dE\, E^2 \left[N^2 \left(1 - E^{-2}\right)\right.$$

$$\left. - \left(1 \pm \frac{\epsilon_a \Omega_{0a}}{\omega E}\right)^2\right] \frac{\partial F_a}{\partial E} H_+(E, N)$$

$$= 1 - N^2 + \frac{2\pi \sum_a \omega_{p,a}^2 (m_a c)^3}{k^2 c^2} \int_1^\infty dE \sqrt{1 - E^{-2}} \left[1 \pm \frac{\epsilon_a \Omega_{0a}}{\omega E}\right] \frac{\partial u}{\partial E}$$

$$+ \frac{\pi \sum_a \omega_{p,a}^2 (m_a c)^3}{k^2 c^2 N} \int_1^\infty dE \left[N^2 \left(1 - E^{-2}\right)\right.$$

$$\left. - \left(1 \pm \frac{\epsilon_a \Omega_{0a}}{\omega E}\right)^2\right] \frac{\partial u}{\partial E} H_+(E, N) = 0 , \qquad (10.4.1)$$

with
$$H_+(E,N) = \int_1^{T_{1,2}} \frac{dy}{y}, \qquad (10.4.2)$$
where
$$T_{1,2} \equiv \frac{1 \pm (\epsilon_a \Omega_{0a}/\omega E) + N\sqrt{1-E^{-2}}}{1 \pm (\epsilon_a \Omega_{0a}/\omega E) - N\sqrt{1-E^{-2}}} = \Re T_{1,2} + i\Im T_{1,2}. \qquad (10.4.3)$$

From (10.4.3) we derive
$$\Re T_{1,2} = \frac{(\omega_R \pm (\epsilon_a \Omega_{0a}/E))^2 + \Gamma^2 - k^2 c^2 (1 - E^{-2})}{\left(\omega_R \pm (\epsilon_a \Omega_{0a}/E) - kc\sqrt{1-E^{-2}}\right)^2 + \Gamma^2}, \qquad (10.4.4a)$$

$$\Im T_{1,2} = -\frac{2kc\Gamma\sqrt{1-E^{-2}}}{\left(\omega_R \pm (\epsilon_a \Omega_{0a}/E) - kc\sqrt{1-E^{-2}}\right)^2 + \Gamma^2}. \qquad (10.4.4b)$$

Again (for positive wavenumbers k) the upper ($\Im(\omega) = \Gamma > 0$) complex frequency half-plane corresponds to $T_{1,2}$-values with $\Im T_{1,2} < 0$, while the lower ($\Im(\omega) = \Gamma < 0$) complex frequency half-plane corresponds to $T_{1,2}$-values with $\Im T_{1,2} > 0$. In complete analogy to the discussion in Sect. 10.2 we have to be careful with the analytic continuation of the integral H_+ and the dispersion relation (10.4.1) into the lower frequency half-plane in cases where

$$\lim_{\Gamma \to 0^\pm} (\Re(T_{1,2})) = \frac{(\omega_R \pm (\epsilon_a \Omega_{0a}/E))^2 - k^2 c^2 (1 - E^{-2})}{\left(\omega_R \pm (\epsilon_a \Omega_{0a}/E) - kc\sqrt{1-E^{-2}}\right)^2} \leq 0. \qquad (10.4.5)$$

Condition (10.4.5) is equivalent to the inequality

$$\left[E \mp \frac{\epsilon_a \Omega_{0a}}{\omega_R (N_R^2 - 1)}\right]^2 \geq \frac{N_R^2}{(N_R^2 - 1)^2} \frac{\Omega_{0a}^2}{\omega_R^2}\left[1 + \frac{\omega_R^2 (N_R^2 - 1)}{\Omega_{0a}^2}\right]. \qquad (10.4.6)$$

While the left hand side of the inequality (10.4.6) is always positive, the right hand side is negative for index of refractions smaller than $N_R^2 < N_c^2$ where

$$N_c^2 = 1 - \frac{\Omega_{0a}^2}{\omega_R^2}. \qquad (10.4.7)$$

Hence for superluminal waves with $N_R^2 \leq N_c^2 < 1$ the case (10.4.5) never occurs, so that here the dispersion relation (10.4.1) holds for all values of Γ with
$$H_+ = H_- = H_p, \qquad (10.4.8)$$
where
$$H_p = \ln|T_{1,2}| + i\,\text{Arg}\,T_{1,2}. \qquad (10.4.9)$$
However, for subluminal ($N_R^2 > 1$) waves and superluminal waves with $N_c^2 < N_R^2 \leq 1$ the condition (10.4.5) is fulfilled provided $E \geq E_{c,1,2}(\omega_R)$ where

$$E_{c,1,2}(\omega_R) = \frac{\Omega_{0a}}{\omega_R \left(N_R^2 - 1\right)} \left[\pm \epsilon_a + N_R \sqrt{1 + \frac{\omega_R^2 \left(N_R^2 - 1\right)}{\Omega_{0a}^2}} \right]. \quad (10.4.10)$$

To guarantee that H_+ and H_- according to (10.2.9b) indeed do approach the same limiting value as $\Gamma \to 0^{\pm}$, respectively, and that they are analytic continuations of each other, we have to choose H_- and H_+ for general ω (or N) as

$$H_+ = H_p, \quad H_- = H_p - 2\pi\imath\theta\left[E - E_{c,1,2}(N)\right], \quad (10.4.11)$$

where $\theta[x] = 1(0)$ for $x \geq (<)0$ denotes the step function, and

$$E_{c,1,2}(N) = \frac{\Omega_{0a}}{\omega \left(N^2 - 1\right)} \left[\pm \epsilon_a + N \sqrt{1 + \frac{\omega^2 \left(N^2 - 1\right)}{\Omega_{0a}^2}} \right]. \quad (10.4.12)$$

Again we can combine (10.4.8) and (10.4.11) as

$$H_+ = H_p, \quad H_- = H_p - 2\pi\imath\theta\left[N_R - 1\right]\theta\left[E - E_{c,1,2}(N)\right]. \quad (10.4.13)$$

Work on the general transverse dispersion relation (10.4.1) is still pending. We therefore restrict our discussion here to the weak-damping, but relativistically correct, analysis of parallel propagating waves in thermal equilibrium plasmas.

10.4.1 Collisionless Damping of Transverse Oscillations

From the relativistically correct weak-damping analysis for transverse waves, propagating parallel to the ordered magnetic field, one obtains for the weak-damping rate (Schlickeiser et al. 1997b [474])

$$\Gamma_{l,r}(k) = -\operatorname{sgn}(\omega_R) \left[\left(\frac{\partial \operatorname{Re}\Lambda_L(k,\omega)}{\partial \omega_R} \right)_{\omega = \omega_R} \right]^{-1} \frac{\pi}{2} \sum_a \frac{\omega_{p,a}^2 \sigma \sqrt{1 + \frac{\sigma^2 \Omega_{0,a}^2}{\omega_R^2}}}{\omega_R^2 \left(1 + \sigma^2\right) K_2(\mu_a)}$$

$$\times \left(1 + \frac{1}{\mu_a} \left[(1+\sigma^2) \left(1 + \frac{\sigma^2 \Omega_{0,a}^2}{\omega_R^2} \right) \right]^{-1/2} \right)$$

$$\times \exp\left(-\mu_a \left[\mp \frac{\Omega_{0,a}\sigma^2}{\omega_R} + \sqrt{(1+\sigma^2)\left(1 + \frac{\sigma^2 \Omega_{0,a}^2}{\omega_R^2}\right)} \right] \right). \quad (10.4.14)$$

This formula gives the damping rate of transverse left- and right-handed oscillations in a hot, magnetized equilibrium plasma. The actual damping of a specific wave mode can be obtained according to (10.3.60) by inserting the corresponding dispersion curve $\omega_R(k)$ from the solution of the real part equation (10.3.59).

10.4.2 An Illustrative Example

In order to investigate the differences between the non-relativistic and the relativistic theory of transverse oscillations we analyze the damping of subluminal waves at the example of the interplanetary plasma. From in-situ satellite measurements at the orbit of the Earth we know the circumsolar plasma to have an electron temperature of about $T_\mathrm{e} = 10^5$ K and a proton temperature of about $T_\mathrm{p} = 5 \times 10^4$ K. In terms of the dimensionless reciprocal temperatures μ_a these values translate into $\mu_\mathrm{e} = 6 \times 10^4$ and $\mu_\mathrm{p} = 2.2 \times 10^8$. For the electron frequencies one finds $\omega_{\mathrm{p,e}} \simeq 2 \times 10^5$ Hz and $|\Omega_\mathrm{e}| \simeq 2 \times 10^3$ Hz resulting in $\omega_{\mathrm{p,i}} = \sqrt{m_\mathrm{e}/m_\mathrm{p}}\,\omega_{\mathrm{p,e}} \simeq 5 \times 10^3$ Hz and $\Omega_{\mathrm{p},0} = m_\mathrm{e}/m_\mathrm{p}|\Omega_{\mathrm{e},0}| \simeq 1.1$ Hz for the protons.

Distinguishing the two cases of left-handed low-frequency waves ($|\omega_\mathrm{R}(k)| \ll |\Omega_{\mathrm{e},0}|$) and right-handed high-frequency waves ($|\omega_\mathrm{R}(k)| \gg |\Omega_{\mathrm{p},0}|$) the dispersion equations read (see (9.2.35) and (9.2.42)):

$$\omega_\mathrm{R}(k) = -\frac{V_\mathrm{A}^2 k^2}{2\Omega_{\mathrm{p},0}} + V_\mathrm{A} k \sqrt{\frac{v_\mathrm{A}^2 k^2}{2\Omega_{\mathrm{p},0}^2} + 1} \qquad |\omega_\mathrm{R}(k)| \ll |\Omega_{\mathrm{e},0}| \qquad (10.4.15)$$

and

$$\omega_\mathrm{R}(k) = \Omega_{\mathrm{e},0} \frac{\omega_{\mathrm{p,p}}^2 + k^2 c^2}{\omega_{\mathrm{p,e}}^2 + k^2 c^2} \qquad |\omega_\mathrm{R}(k)| \gg \Omega_{\mathrm{p},0} \qquad |k| \gg \frac{\Omega_{\mathrm{p},0}}{V_\mathrm{A}} \qquad (10.4.16)$$

and are valid for waves propagating forwardly, i.e. in the direction of the magnetic field. The first relation contains the Alfvén as well as ion-cyclotron waves and the second the whistler as well as electron-cyclotron waves. Both cases fulfill the inequality $|\omega_\mathrm{R}|/k < c \Leftrightarrow N_\mathrm{R} > 1$, i.e. the waves are subluminal and damped.

10.4.2.1 Left-Handed Low-Frequency Waves.
Using the above mentioned parameter values being characteristic for the interplanetary medium one obtains the results shown in Fig. 10.5.

The line indicating $N = 1$ divides the diagram into the region of sub- and superluminal waves. Both the non-relativistic damping rate and the damping rate from the relativistically correct expression (10.4.14) are shown. We observe two significant differences between the non-relativistic and the relativistic treatment. First, for the latter the damping sets in at higher wavenumbers, and, second, in the physically meaningful range limited by the condition $|\Gamma/\omega_\mathrm{R}| \ll 1$, the damping at a given wavenumber is lower. The comparison of the relativistically correct treatment with the non-relativistic theory yields decreased damping rates for transverse, left-handed low-frequency waves.

10.4.2.2 Right-Handed High-Frequency Waves.
Figure 10.5 also illustrates the dispersion curves and damping rates following from equations (10.4.14) and (10.4.16). Since the validity of (10.4.16) requires $|\omega_\mathrm{R}(k)| \gg |\Omega_{\mathrm{p},0}|$ and $|k| \gg \Omega_\mathrm{p}/V_\mathrm{A}$ the physically relevant k-interval is at large k values.

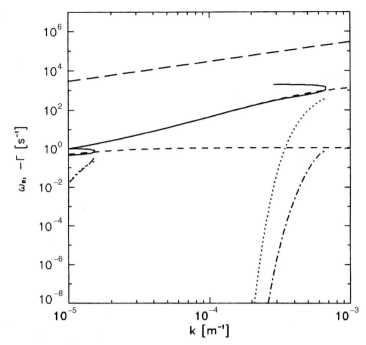

Fig. 10.5. The dispersion curves $\omega_R(k)$ and damping rates $\Gamma(k)$ for subluminal transverse oscillations in an electron-proton plasma with $\mu_e = 6 \times 10^4$ and $\mu_p = 2 \times 10^8$. Shown are the dispersion curves $\omega_R(k)$ (the two **solid lines**) for right- and left-handed waves together with the damping rates derived from the relativistic (**dash-dotted lines**) and from the non-relativistic theory (**dotted lines**). The two left damping curves at low k values correspond to the left-handed mode and the two right damping curves at higher k values correspond to the right-handed mode. The long-dashed line indicates $N = 1$, and the two short-dashed lines represent the dispersion curves of the cold plasma. From Schlickeiser et al. (1997b [474])

We again come to the conclusion that the damping derived from a relativistic theory (dash-dotted line) is significantly lower than the non-relativistic result (dotted line). Contrary to the latter the relativistic damping curve always fulfills $|\Gamma/\omega_R| \ll 10^{-2}$.

11. Test Wave Approach 3. Generation of Plasma Waves

In this chapter we will study particle distribution functions $f_a^0(p)$ that lead to solutions of the dispersion relation with *positive* values of the imaginary part of the wave frequency, $\Gamma > 0$, and thus give rise to growing fluctuations. We will start out from the classical two-stream instability and then study in detail cosmic-ray-induced instabilities.

11.1 Two-Stream Instability

We investigate the stability of a plasma system consisting of a stream of electrons of density n moving with momentum P_e parallel to an ordered magnetic field $\boldsymbol{B}_0$ with respect to a cold background gas of ions of density N_i at rest. The unperturbed particle distribution function then is

$$f_0(p_\perp, p_\parallel) = \frac{1}{2\pi p_\perp (N_i + n)} [N_i \delta(p_\perp)\delta(p_\parallel) + n\delta(p_\perp)\delta(p_\parallel - P_e)]. \qquad (11.1.1)$$

According to (9.1.7) the dielectric tensor for this distribution is

$$\psi_{ij} = \begin{Bmatrix} S & -\imath D & 0 \\ \imath D & S & 0 \\ 0 & 0 & P \end{Bmatrix} \qquad (11.1.2a)$$

with

$$S = 1 - \frac{\omega_{p,i}^2}{\omega^2 - \Omega_i^2} - \frac{\omega_{p,e}^2 (\omega - k_\parallel U)^2}{\Gamma \omega^2 \left[(\omega - k_\parallel U)^2 - \Omega_e^2\right]}, \qquad (11.1.2b)$$

$$D = \frac{\omega_{p,i}^2 \Omega_i}{\omega (\omega^2 - \Omega_i^2)} - \frac{\omega_{p,e}^2 (\omega - k_\parallel U) \Omega_e}{\Gamma \omega^2 \left[(\omega - k_\parallel U)^2 - \Omega_e^2\right]}, \qquad (11.1.2c)$$

$$P = 1 - \frac{\omega_{p,i}^2}{\omega^2} - \frac{\omega_{p,e}^2}{\Gamma^3 (\omega - k_\parallel U)^2}, \qquad (11.1.2d)$$

where

$$U = \frac{P_e}{\Gamma m_e} \;;\; \Gamma = \left[1 - \frac{U^2}{c^2}\right]^{-1/2} \;;\; \omega_{p,e}^2 = \frac{4\pi e^2 n}{m_e} \;;\; \omega_{p,i}^2 = \frac{4\pi Z_i^2 e^2 N_i}{m_i} \;.$$
(11.1.2e)

We also limit our discussion to plasma waves moving parallel to the ordered magnetic field $\mathbf{B}_0$. Using (11.1.2a) in the Maxwell operator (8.3.50) with $\theta = 0$ we obtain from (8.3.36) the dispersion relation

$$P\left[(S - N^2)^2 - D^2\right] = 0$$
(11.1.3)

yielding the longitudinal mode

$$P = 0$$
(11.1.4)

and two transverse modes

$$N^2 = S \pm D \;.$$
(11.1.5)

For a discussion of the transverse modes (11.1.5) we refer the reader to the work of Achatz et al. (1990 [5]) and Achatz and Schlickeiser (1993 [4]). Here we only consider the longitudinal mode (11.1.4), analyzed first by Bunemann (1959 [81]) in the non-relativistic limit $\Gamma \to 1$.

Combining (11.1.4) and (11.1.2d) we obtain

$$L(k_\parallel, \omega) = 1 \;,$$
(11.1.6)

where we have introduced the function

$$L(k_\parallel, \omega) \equiv \omega_{p,e}^2 \left[\frac{A}{\omega^2} + \frac{1}{\Gamma^3(\omega - k_\parallel U)^2}\right] \;,$$
(11.1.7)

with

$$A \equiv \left(\frac{\omega_{p,i}}{\omega_{p,e}}\right)^2 = \frac{m_e Z_i^2 N}{m_i n} \;.$$
(11.1.8)

The function $L(\omega)$ is sketched in Fig. 11.1.

The solutions of (11.1.6) are the intersections of the curve $L(\omega)$ with the line $L = 1$. In general there are four solutions, and they are real if the minimum of the function $L_{\min} = L(\omega_m) \leq 1$ between $\omega = 0$ and $\omega = k_\parallel U$ is smaller than unity. In the opposite case

$$L_{\min} > 1 \;,$$
(11.1.9)

there are two real solutions and two conjugate complex solutions of the form

$$\omega_0 = \omega_R \pm \imath \alpha \;,$$
(11.1.10)

so that according to (8.3.42) one solution gives rise to fluctuations growing with time, an instability. The position of the minimum can be derived from

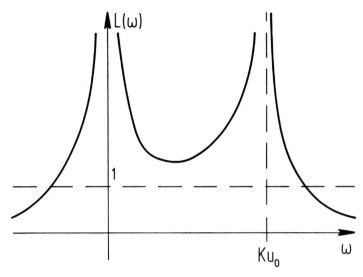

Fig. 11.1. Dispersion function for longitudinal waves with a cold electron stream in a cold ion background

$$0 = \frac{\partial L}{\partial \omega}\bigg|_{\omega=\omega_m}$$

yielding

$$\omega_m = k_\parallel U \left[1 + \left(\Gamma A^{1/3}\right)^{-1}\right]^{-1} \quad (11.1.11)$$

and

$$L_{\min} = \frac{\omega_{p,e}^2}{\Gamma^3 k_\parallel^2 U^2}\left[1 + A^{1/3}\Gamma\right]^3. \quad (11.1.12)$$

The condition for instability (11.1.9) is therefore equivalent to

$$|k_\parallel U| < \omega_{p,e}\left[\frac{1 + A^{1/3}\Gamma}{\Gamma}\right]^{3/2}, \quad (11.1.13)$$

implying that waves are always excited for any non-zero streaming velocity U over a broad range of wavenumbers.

We now derive the growth rate α for frequencies

$$\omega^2 \gg \omega_{p,i}^2 \quad (11.1.14)$$

and introduce the normalized frequency

$$x \equiv \frac{\omega}{\left(\omega_{p,e}\omega_{p,i}^2\right)^{1/3}\Gamma^{-1/2}} \quad (11.1.15)$$

and the normalized wavenumber

$$y \equiv \frac{\left|k_\| U - \omega_{p,e}\Gamma^{-3/2}\right|}{\left(\omega_{p,e}\omega_{p,i}^2\right)^{1/3}\Gamma^{-1/2}}. \tag{11.1.16}$$

The biquadratic (11.1.6) then reduces to the cubic equation

$$x^3 - yx^2 + \frac{1}{2} = 0. \tag{11.1.17}$$

Equation (11.1.17) has one real and two conjugate complex solutions provided $y < 3/2$ which according to (11.1.16) is equivalent to

$$\left|k_\| U\right| < \frac{\omega_{p,e}}{\Gamma^{3/2}}\left[1 + \frac{3}{2}\Gamma A^{1/3}\right], \tag{11.1.18}$$

which, under our approximation (11.1.14), corresponds to condition (11.1.13). The complex conjugate solutions are

$$x_{1,2} = \frac{y}{3} - \frac{1}{2}\left[\left(\Delta^{1/2} - 2\Delta - \frac{1}{8}\right)^{1/3} + \left(-\Delta^{1/2} - 2\Delta - \frac{1}{8}\right)^{1/3}\right]$$

$$\pm i\frac{\sqrt{3}}{2}\left[\left(\Delta^{1/2} - 2\Delta - \frac{1}{8}\right)^{1/3} - \left(-\Delta^{1/2} - 2\Delta - \frac{1}{8}\right)^{1/3}\right], \tag{11.1.19}$$

where

$$\Delta = \frac{1}{2}\left[\frac{1}{8} - \frac{y^3}{27}\right]. \tag{11.1.20}$$

The complex solutions $\omega = \omega_R \pm i\alpha$ are plotted against $k_\| U$ in Fig. 11.2.

The maximum growth rate occurs for $y = 0$ and is given by

$$\Im\omega_{max} = \alpha_{max} = \alpha(y = 0) = \frac{\sqrt{3}}{2}\frac{\omega_{p,e}}{\Gamma^{1/2}}\left[\frac{A}{2}\right]^{1/3} \propto \frac{\omega_{p,e}}{\Gamma^{1/2}}\left[\frac{N_i}{n}\right]^{1/3}. \tag{11.1.21}$$

We have thus established that a two-stream initial configuration, as described by (11.1.1) will lead to a longitudinal plasma mode whose electric field exponentially increases with time $\propto \exp(\alpha_{max}t)$. After a few e-folding times, being of the order of a few $\Gamma^{1/2}/\omega_{p,e}$, this growing electric field will completely modify the initial electromagnetic field and thus the initial configuration. In particular, our starting assumptions on the smallness of electric fields as compared to the ordered magnetic field $\boldsymbol{B}_0$ will be invalidated and, as a consequence, our linear stability analysis does not allow any conclusions to be drawn concerning the state into which the initial configuration will develop.

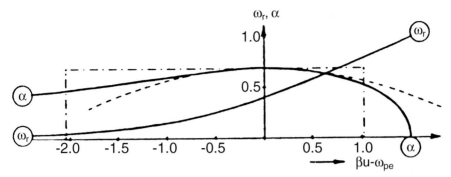

Fig. 11.2. Dispersion relation. Real part ω_R and imaginary part α of ω versus $\beta u - \omega_{p,e} \equiv k_\| U - \omega_{p,e} \Gamma^{-3/2}$. The horizontal and vertical scales are in units of $(\omega_{p,e}\omega_{p,i}^2)^{1/3}\Gamma^{-1/2}$. From Bunemann (1959 [81])

11.2 Cosmic-Ray-Induced Instabilities

We now discuss the possible fluctuations in a plasma that resembles closely the co-existence of cosmic ray particles with ions and electrons in interplanetary and interstellar gases. We consider mobile cosmic rays which are immersed in an infinite homogeneous and uniform magnetic field $\boldsymbol{B}_0$. We assume the co-existence of a background plasma, either cold or in thermal state, whose sole purpose is to preserve space charge neutrality in equilibrium and which takes no part in any motion. Thus we may split the distribution function of the whole plasma into two parts:

$$F(\boldsymbol{r},\boldsymbol{p},t) = n_a(\boldsymbol{r})f_a(\boldsymbol{p},t) + N_r(\boldsymbol{r})f_r(\boldsymbol{p},t) , \qquad (11.2.1)$$

where f_a and f_r denote the distribution functions of the background plasma and the cosmic rays, respectively, and n_a and N_r the respective number densities. As our discussion of cosmic ray observations (Chaps. 3–5) has shown, in most astrophysical systems the number density of cosmic ray particles N_r is orders of magnitude smaller than the number density of the background gas particles n_a.

Using the distribution function (11.2.1) we may split the dielectric tensor (8.3.23) for this system as

$$\psi_{ij}(\boldsymbol{k},\omega) = \psi_{ij}^a(\boldsymbol{k},\omega) + \psi_{ij}^r(\boldsymbol{k},\omega) , \qquad (11.2.2)$$

where ψ_{ij}^a refers to the dielectric tensor of the background plasma distribution: e.g. in the case of a cold background at rest ψ_{ij}^a is given by (9.2.1). The contribution to the dielectric tensor from the relativistic cosmic ray particles is given by (see (8.3.52) ff.)

$$\psi_{ij}^r(\boldsymbol{k},\omega) = \frac{4\pi i}{\omega}\sigma_{ij}^r = \frac{2\pi}{\omega}\sum_r \frac{\omega_{p,a}^2 N_r}{n_a}\frac{m_a\,e_r^2}{e_a^2}$$

$$\times \int_{-\infty}^{\infty} dp_\parallel \int_0^{\infty} dp_\perp p_\perp \sum_{n=-\infty}^{\infty}\frac{\gamma_r^{-1} T_{ij}^r}{\omega - k_\parallel v_\parallel - n\Omega_a}, \qquad (11.2.3)$$

with

$$T_{ij}^r \equiv \begin{pmatrix} \frac{n^2 J_n^2(z)}{z^2} v_\perp U f_{\mathrm r} & \frac{i n J_n(z) J_n'(z)}{z} v_\perp U f_{\mathrm r} & \frac{n J_n^2(z)}{z} v_\perp W f_{\mathrm r} \\ \frac{-in J_n(z) J_n'(z)}{z} v_\perp U f_{\mathrm r} & (J_n'(z))^2 v_\perp U f_{\mathrm r} & -i J_n(z) J_n'(z) v_\perp W f_{\mathrm r} \\ \frac{n J_n^2(z)}{z} v_\parallel U f_{\mathrm r} & i J_n(z) J_n'(z) v_\parallel U f_{\mathrm r} & J_n^2(z) v_\parallel W f_{\mathrm r} \end{pmatrix},$$
(11.2.4)

$$U f_{\mathrm r} = \frac{\partial f_r}{\partial p_\perp} + \frac{k_\parallel}{\gamma m_a \omega}\left(p_\perp \frac{\partial f_r}{\partial p_\parallel} - p_\parallel \frac{\partial f_r}{\partial p_\perp}\right), \qquad (11.2.5)$$

$$W f_{\mathrm r} = \frac{\partial f_r}{\partial p_\parallel} - \frac{n\Omega_a}{\omega p_\perp}\left(p_\perp \frac{\partial f_r}{\partial p_\parallel} - p_\parallel \frac{\partial f_r}{\partial p_\perp}\right), \qquad (11.2.6)$$

$$z \equiv \frac{k_\perp v_\perp}{\Omega_r}, \qquad (11.2.7)$$

where $\Omega_r = \epsilon_r \Omega_{0r}/\gamma$ is the relativistic gyrofrequency of cosmic rays of sort r. $J_n(z)$ denotes the Bessel function of order n and $J_n'(z) \equiv \partial J_n(z)/\partial z$ its first derivative.

In the following sections we will treat in detail the case of parallel propagating waves but only quote the results of studies for oblique propagating waves.

11.2.1 Parallel Propagating Waves

In the special case of waves propagating parallel to the ordered magnetic field $\boldsymbol{B}_0$, i.e. $z = 0$, the dielectric tensor for the cosmic ray particles (11.2.3) reduces to

$$\psi_{ij}^r(k_\parallel, k_\perp = 0, \omega)$$

$$= \frac{2\pi}{\omega}\sum_r \frac{\omega_{p,a}^2 N_r}{n_a}\frac{m_a\,e_r^2}{e_a^2}\int_{-\infty}^{\infty} dp_\parallel \int_0^{\infty} dp_\perp p_\perp \begin{pmatrix} T_1^r & i T_2^r & 0 \\ -i T_2^r & T_1^r & 0 \\ 0 & 0 & T_3^r \end{pmatrix},$$

$$\equiv \begin{pmatrix} \psi_{11}^r & i\psi_{12}^r & 0 \\ -i\psi_{12}^r & \psi_{11}^r & 0 \\ 0 & 0 & \psi_{33}^r \end{pmatrix}, \qquad (11.2.8)$$

where

$$T_1 = \frac{v_\perp(\omega - k_\| v_\|)\boldsymbol{U} f_{\rm r}}{2\left[(\omega - k_\| v_\|)^2 - \Omega_r^2\right]} , \qquad (11.2.9a)$$

$$T_2 = \frac{v_\perp \Omega_r \boldsymbol{U} f_{\rm r}}{2\left[(\omega - k_\| v_\|)^2 - \Omega_r^2\right]} , \qquad (11.2.9b)$$

$$T_3 = \frac{v_\|}{\omega - k_\| v_\|} \frac{\partial f_{\rm r}}{\partial p_\|} . \qquad (11.2.9c)$$

The dielectric tensor for the background particles becomes

$$\psi_{ij}^a(k_\|, k_\perp = 0, \omega) = \begin{pmatrix} \psi_{11}^a & \imath\psi_{12}^a & 0 \\ -\imath\psi_{12}^a & \psi_{11}^a & 0 \\ 0 & 0 & \psi_{33}^a \end{pmatrix} , \qquad (11.2.10)$$

where the individual tensor components follow readily from (9.2.1) in the case of a cold background and from the corresponding form in the case of a thermal equilibrium background distribution.

For the dispersion relation (8.3.36) for parallel propagating waves we obtain

$$\Lambda = \det \Lambda_{ij} = [\psi_{33}^a + \psi_{33}^r]\left[(\psi_{11}^a + \psi_{11}^r - N^2)^2 - (\psi_{12}^a + \psi_{12}^r)^2\right] = 0 \qquad (11.2.11)$$

with three possible solutions:

1. The longitudinal mode : $\Lambda_{\rm L} = \psi_{33}^a + \psi_{33}^r = 0$. $\qquad (11.2.12)$
2. The two transverse modes : $\Lambda_{\rm T} = (\psi_{11}^a \pm \psi_{12}^a) + (\psi_{11}^r \pm \psi_{12}^r) - N^2 = 0$.
$\qquad (11.2.13)$

According to (11.2.11), the propagation of parallel waves with electric field vectors polarized along $\boldsymbol{B}_0$ is decoupled from the propagation of waves with electric field vectors polarized perpendicular to $\boldsymbol{B}_0$. The physical reason for this decoupling is that the particle motion associated with longitudinal waves is unaffected by the ordered magnetic field $\boldsymbol{B}_0$, whereas for the transverse wave the influence of the magnetic field becomes decisive.

We calculate the dispersion relation and the linear growth rate for each mode in the weak damping approximation (see Sect. 10.3.7), i.e.

$$\Re\Lambda(\boldsymbol{k}, \omega_{\rm R}, \Gamma = 0) = 0 , \qquad (11.2.14)$$

and

$$\Gamma = -\frac{\Im\Lambda(\boldsymbol{k}, \omega_{\rm R}, 0)}{[\partial\Re\Lambda(\boldsymbol{k}, \omega_{\rm R}, 0)/\partial\omega_{\rm R}]} . \qquad (11.2.15)$$

Applying again the integration variable transformations (10.2.1) implying (10.2.2) we obtain for the elements of the tensor (11.2.8) appearing in (11.2.12) and (11.2.13)

$$\psi^r_{33}(k_\parallel, k_\perp = 0, \omega = \omega_R, \Gamma = 0) = \frac{2\pi}{\omega_R} \sum_r \frac{\omega^2_{p,a} N_r (m_r c)^3}{n_a}$$

$$\times \int_1^\infty dE\, E \int_{-\sqrt{E^2-1}}^{+\sqrt{E^2-1}} dy\, \frac{y\left[(\partial f_r/\partial y) + (y/E)(\partial f_r/\partial E)\right]}{\omega_R E - k_\parallel c y}, \quad (11.2.16a)$$

$$\psi^r_{R,L} \equiv \psi^r_{11}(k_\parallel, k_\perp = 0, \omega = \omega_R, \Gamma = 0)$$
$$\pm \psi^r_{12}(k_\parallel, k_\perp = 0, \omega = \omega_R, \Gamma = 0)$$
$$= \frac{\pi}{\omega_R} \sum_r \frac{\omega^2_{p,a} N_r (m_r c)^3}{n_a} \int_1^\infty dE \int_{-\sqrt{E^2-1}}^{+\sqrt{E^2-1}} dy$$

$$(E^2 - 1 - y^2) \frac{\left[(\partial f_r/\partial E) + (k_\parallel c/\omega_R)(\partial f_r/\partial y)\right]}{\omega_R E - k_\parallel c y \mp \Omega_{r,0}}, \quad (11.2.16b)$$

where $\Omega_r = \Omega_{r,0}/E$ denotes the gyrofrequency of the respective cosmic ray particle.

11.2.1.1 Longitudinal Mode. For $k_\parallel$ real and positive we see that as $\Gamma \to 0^+$ from above (compare with (8.3.32)) in expression (11.2.16a) we must understand

$$w^{-1} = P_C\left(w^{-1}\right) + \imath \pi \delta(w), \quad (11.2.17a)$$

where

$$w \equiv y - \frac{\omega_R E}{k_\parallel c}, \quad (11.2.17b)$$

and we regard w as a function of y. Here $P_C(w^{-1})$ denotes Cauchy's principal value of w^{-1}. We then obtain

$$\psi^r_{33} = \Re \psi^r_{33} + \imath \Im \psi^r_{33}, \quad (11.2.18)$$

with

$$\Re \psi^r_{33}(k_\parallel, k_\perp = 0, \omega = \omega_R, \Gamma = 0) = -\frac{2\pi}{\omega_R} \sum_r \frac{\omega^2_{p,a} N_r (m_r c)^3}{n_a k_\parallel c}$$

$$\times P_C \left[\int_1^\infty dE\, E \int_{-\sqrt{E^2-1}}^{+\sqrt{E^2-1}} dy\, \frac{y\left[(\partial f_r/\partial y) + (y/E)(\partial f_r/\partial E)\right]}{y - (\omega_R E/k_\parallel c)} \right]$$

$$(11.2.19)$$

and

$$\Im \psi^r_{33}(k_\parallel, k_\perp = 0, \omega = \omega_R, \Gamma = 0) = -\frac{2\pi^2}{\omega_R} \sum_r \frac{\omega^2_{p,a} N_r (m_r c)^3}{n_a k_\parallel c}$$

$$\times \int_1^\infty dE\, E \int_{-\sqrt{E^2-1}}^{+\sqrt{E^2-1}} dy\, \delta\left(y - \frac{\omega_R E}{k_\parallel c}\right) y \left[(\partial f_r/\partial y) + (y/E)(\partial f_r/\partial E)\right].$$

$$(11.2.20)$$

We see from (11.2.20) that a non-zero value of $\Im\psi_{33}^r$ can be obtained provided that

$$-\sqrt{E^2-1} \leq \frac{\omega_R E}{k_\| c} \leq \sqrt{E^2-1}, \qquad (11.2.21)$$

which is equivalent to

$$E \geq E_0 = \frac{N}{\sqrt{N^2-1}}, \qquad (11.2.22)$$

where $N = k_\| c/\omega_R$ is the refractive index of the waves. Of course, E_0 agrees with E_c defined in (10.2.17). It is a simple matter to show that $E_0 > 1$, provided $N > 1$, i.e. the waves are subluminal $\omega_R/k_\| < c$. We then find

$$\Im\psi_{33}^r(k_\|, k_\perp = 0, \omega = \omega_R, \Gamma = 0)$$
$$= -\frac{2\pi^2}{(k_\| c)^2} \sum_r \frac{\omega_{p,a}^2 N_r (m_r c)^3}{n_a} \int_{E_0}^\infty dE E^2 \left[\frac{\partial f_r}{\partial y}\bigg|_{y=E/N} + \frac{1}{N}\frac{\partial f_r}{\partial E}\right].$$
(11.2.23)

Using (10.3.53) and (11.2.23) in (11.2.12) we find

$$\Lambda_L = \Re\psi_{33}^a + \Re\psi_{33}^r + i[\Im\psi_{33}^a + \Im\psi_{33}^r]$$
$$\simeq \Re\psi_{33}^a + i[\Im\psi_{33}^a + \Im\psi_{33}^r] = 0, \qquad (11.2.24)$$

as the longitudinal dispersion relation. Here we have made use of the fact that the number density of cosmic ray particles N_r is much smaller than the number density of the background gas particles n_a, so that because of the small ratio $\sum_r N_r/n_a \ll 1$ in (11.2.24) we may neglect the principal value contribution in $\Re\psi_{33}^r$ to the real part of the dispersion relation (11.2.12). For the special case of the cold background distribution we obtain for the latter with (9.2.2e)

$$\omega_R = \pm\sqrt{\left(\sum_a \omega_{p,a}^2\right)^{1/2}} \qquad (11.2.25)$$

and for the associated imaginary frequency

$$\Gamma = -\frac{\omega_R}{2}\Im\psi_{33}^r = \frac{\pi^2 \omega_R^2}{(k_\| c)^3} \sum_r \frac{\omega_{p,a}^2 N_r (m_r c)^3}{n_a}$$
$$\times \int_{E_0}^\infty dE E^2 \left[N\frac{\partial f_r}{\partial y}\bigg|_{y=E/N} + \frac{\partial f_r}{\partial E}\right]. \qquad (11.2.26)$$

In the case of a hot background distribution (11.2.26) generalizes to

$$\Gamma_L = -\frac{[\Im\psi_{33}^a + \Im\psi_{33}^r]}{[\partial\Re\Lambda_L/\partial\omega_R]}, \qquad (11.2.27)$$

where the appropriate real part $\Re\Lambda_L(\omega_R, \Gamma = 0)$ has to be inserted. Of course, this adds the imaginary Landau damping term (10.3.60) to the growth rate

(11.2.27) but does not affect our results on ψ_{33}^r. As an aside we note that the real part of the dispersion relation is modified for astrophysical systems where more relativistic than non-relativistic particles ($N_r/n_a \gg 1$) exist; an extensive study of this case has been performed by Mikhailovski (1980 [352]).

It is evident from (11.2.26) that in the case of *isotropic* cosmic ray particle distribution functions, i.e. $f_r(p_\parallel, p_\perp) = f(p)$ where $p = \sqrt{p_\perp{}^2 + p_\parallel{}^2}$ implying $\partial f_r/\partial y = 0$, we obtain

$$\Gamma = \frac{\pi^2 \omega_R^2}{(k_\parallel c)^3} \sum_r \frac{\omega_{p,a}^2 N_r (m_r c)^3}{n_a} \int_{E_0}^\infty dE E^2 \frac{\partial f_r}{\partial E}$$

$$= -\frac{\pi^2 \omega_R^2}{(k_\parallel c)^3} \sum_r \frac{\omega_{p,a}^2 N_r (m_r c)^3}{n_a} \left[E_0^2 f_r(E_0) + 2 \int_{E_0}^\infty dE E f_r(E) \right] < 0 ,$$

(11.2.28)

which is negative for all possible distribution function $f_r(E)$. This is in full agreement with the relativistic Newcomb–Gardner theorem that *all isotropic particle distribution functions are stable against the excitation of longitudinal waves propagating parallel to the ordered magnetic field* $\boldsymbol{B}_0$, derived in Sect. 10.2.2.1. However, this result does not hold for *oblique*, with respect to $\boldsymbol{B}_0$, propagating longitudinal waves, as demonstrated by Lesch et al. (1989 [301]).

11.2.1.2 Transverse Mode. Applying the same limiting process (11.2.17) to (11.2.16b) where now

$$w = y - \frac{\omega_R E}{k_\parallel c} \pm \frac{\Omega_{0,r}}{k_\parallel c} .$$

(11.2.29)

we obtain

$$\psi_{R,L}^r = \Re \psi_{R,L}^r + i \Im \psi_{R,L}^r ,$$

(11.2.30)

with

$$\Re \psi_{R,L}^r(k_\parallel, k_\perp = 0, \omega = \omega_R, \Gamma = 0) = -\frac{\pi}{\omega_R k_\parallel c} \sum_r \frac{\omega_{p,a}^2 N_r(m_r c)^3}{n_a} P_C$$

$$\times \left[\int_1^\infty dE \int_{-\sqrt{E^2-1}}^{+\sqrt{E^2-1}} dy\, (E^2 - 1 - y^2) \frac{[(\partial f_r/\partial E) + (k_\parallel c/\omega_R)(\partial f_r/\partial y)]}{y - (\omega_R E/k_\parallel c) \pm (\Omega_{r,0}/k_\parallel c)} \right] ,$$

and

$$\Im \psi_{R,L}^r(k_\parallel, k_\perp = 0, \omega = \omega_R, \Gamma = 0) = -\frac{\pi^2}{\omega_R k_\parallel c} \sum_r \frac{\omega_{p,a}^2 N_r(m_r c)^3}{n_a}$$

$$\times \int_1^\infty dE \int_{-\sqrt{E^2-1}}^{+\sqrt{E^2-1}} dy (E^2 - 1 - y^2) \delta(y - (\omega_R E/k_\parallel c) \pm (\Omega_{r,0}/k_\parallel c))$$

$$\left[(\partial f_r/\partial E) + (k_\parallel c/\omega_R)(\partial f_r/\partial y) \right] .$$

(11.2.31)

A non-zero value of $\Im\psi^r_{R,L}$ can be obtained provided that

$$-ck_\parallel\sqrt{E^2-1} \leq \mp\Omega_{r,0} + \omega_R E \leq ck_\parallel\sqrt{E^2-1}. \tag{11.2.32}$$

Introducing

$$x_r = \frac{\Omega_{r,0}}{ck_\parallel}, \tag{11.2.33}$$

and the index of refraction $N = k_\parallel c/\omega_R$, the inequality (11.2.32) is equivalent to

$$-\sqrt{E^2-1} \leq \frac{E}{N} \mp x_r \leq \sqrt{E^2-1}. \tag{11.2.34}$$

To reduce condition (11.2.34) further we have to specify the polarization state of the plasma wave.

For LH circularly polarized waves the minus sign applies in the inequality (11.2.34), and in order to have a non-vanishing contribution for the integral (11.2.31) in this case, we have to demand that

$$\left(EN^{-1} - x_r\right)^2 \leq E^2 - 1, \tag{11.2.35}$$

implying

$$E \geq E_L = \frac{-x_r N^{-1} + \sqrt{1 - N^{-2} + x_r^2}}{1 - N^{-2}}, \tag{11.2.36}$$

The condition $E_L > 1$ requires

$$x_r N^{-1} + 1 - N^{-2} < \sqrt{x_r^2 + 1 - N^{-2}}. \tag{11.2.37}$$

Equation (11.2.37) is valid provided $x_r N^{-1} < -1$ since then the left-hand side of this inequality is negative. If $x_r N^{-1} > -1$ the inequality (11.2.37) reduces to

$$x_r^2 - 2x_r N^{-1} + N^{-2} = (x_r - N^{-1})^2 > 0, \tag{11.2.38}$$

which also is always valid. The integral (11.2.31) reduces to

$$\Im\psi^r_L = -\frac{\pi^2}{\omega_R^2 N} \sum_r \frac{\omega_{p,a}^2 N_r (m_r c)^3}{n_a} \int_{E_L}^\infty dE$$
$$\left[E^2 - 1 - \left(EN^{-1} - x_r\right)^2\right] \left(\frac{\partial f_r}{\partial E} + N\frac{\partial f_r}{\partial y}\right) \delta(y - (EN^{-1} - x_r)). \tag{11.2.39}$$

Again using the fact that the number density of cosmic ray particles N_r is much smaller than the number density of the background gas particles n_a, we may neglect the principal value contribution of $\Re\psi^r_{R,L}$ to the real part of the dispersion relation (11.2.13). We derive

$$\Lambda_{R,L} \simeq (\Re\psi_{11}^a \pm \Re\psi_{12}^a) - N^2 + i\left[\Im\psi_{11}^a \pm \Im\psi_{12}^a + \Im\psi_{R,L}^r\right] = 0 \quad (11.2.40)$$

for the two transverse dispersion relations.

In the weak-damping approximation (11.2.14/15) we obtain for the real part of the dispersion relation

$$\Re\Lambda_{R,L} = (\Re\psi_{11}^a \pm \Re\psi_{12}^a) - N^2 = 0, \quad (11.2.41)$$

which is equivalent to (9.2.9) and (9.2.31) in the case of a cold background distribution, and for the associated growth rates

$$\Gamma_{R,L} = -\frac{[(\Im\psi_{11}^a \pm \Im\psi_{12}^a) + \Im\psi_{R,L}^r]}{[\partial\Re\Lambda_{R,L}/\partial\omega_R]}. \quad (11.2.42)$$

For example, with (11.2.39) we deduce in the case of a cold background plasma

$$\Gamma_L = \frac{\pi^2}{\omega_R^2 N}\left[\frac{\partial\Re\Lambda_L}{\partial\omega_R}\right]^{-1}\sum_r \frac{\omega_{p,a}^2 N_r(m_r c)^3}{n_a}\int_{E_L}^{\infty} dE$$
$$\left[E^2 - 1 - (EN^{-1} - x_r)^2\right]\delta(y - (EN^{-1} - x_r))\left(\frac{\partial f_r}{\partial E} + N\frac{\partial f_r}{\partial y}\right). \quad (11.2.43)$$

If the background plasma consists of electrons and protons, supporting left-handed and right-handed transverse plasma waves with dispersion relations (9.2.30) and (9.2.31), respectively, we find according to (11.2.41)

$$\Re\Lambda_L = L - N^2 = 0, \quad \Re\Lambda_R = R - N^2 = 0, \quad (11.2.44)$$

yielding

$$\frac{\partial\Re\Lambda_L}{\partial\omega_R} = \frac{\partial(L - N^2)}{\partial\omega_R} = \frac{2k_\parallel^2 c^2}{\omega_R^3} + \frac{\omega_{p,i}^2(2\omega_R - \Omega_p)}{\omega^2(\omega_R - \Omega_p)^2} + \frac{\omega_{p,e}^2(2\omega_R + \Omega_e)}{\omega^2(\omega_R + \Omega_e)^2} \quad (11.2.45a)$$

and

$$\frac{\partial\Re\Lambda_R}{\partial\omega_R} = \frac{\partial(R - N^2)}{\partial\omega_R} = \frac{2k_\parallel^2 c^2}{\omega_R^3} + \frac{\omega_{p,i}^2(2\omega_R + \Omega_p)}{\omega^2(\omega_R + \Omega_p)^2} + \frac{\omega_{p,e}^2(2\omega_R - \Omega_e)}{\omega^2(\omega_R - \Omega_e)^2}. \quad (11.2.45b)$$

We restrict our analysis to wave frequencies $\omega_R \ll \Omega_p$ much less than the proton gyrofrequency, so that for both polarization states the solution of (11.2.44) reduces to the Alfvénic relation (9.2.38)

$$\omega_R \simeq \pm V_A k, \quad N \simeq \pm\frac{c}{V_A}, \quad (11.2.46)$$

for $V_A \ll c$. Equations (11.2.45) become in this limit

$$\frac{\partial \Re \Lambda_{\mathrm{R,L}}}{\partial \omega_{\mathrm{R}}} \simeq \frac{2}{\omega_{\mathrm{R}}} \frac{c^2}{V_A^2} \left[1 \pm \frac{\omega_{\mathrm{R}}}{2\Omega_{\mathrm{p}}}\right] . \tag{11.2.47}$$

In the frequency range of Alfvén waves, $\omega_{\mathrm{R}} \ll \Omega_{\mathrm{p}}$, we obtain from (11.2.47) that

$$\frac{\partial \Re \Lambda_{\mathrm{L}}}{\partial \omega_{\mathrm{R}}} \simeq \frac{2N^2}{\omega_{\mathrm{R}}} . \tag{11.2.48}$$

It then follows from (11.2.43) that

$$\Gamma_{\mathrm{L}} = \frac{\pi^2}{2\omega_{\mathrm{R}} N^3} \sum_r \frac{\omega_{\mathrm{p},a}^2 N_r (m_r c)^3}{n_a} \int_{E_{\mathrm{L}}}^{\infty} dE \left[E^2 - 1 - (EN^{-1} - x_r)^2\right] \frac{\partial f_r}{\partial E} < 0 , \tag{11.2.49}$$

if f_r is a function of E only, indicating that the system is stable with respect to parallel propagating Alfvén wave generation, no matter whether forward or backward moving.

Moreover, we notice from the form of the bracket

$$\left[N \frac{\partial f_r}{\partial y} + \frac{\partial f_r}{\partial E}\right] \tag{11.2.50}$$

appearing in (11.2.43), that a slight *anisotropic* cosmic ray distribution can lead to instability since the first term $\propto (\partial F/\partial y)$ is enhanced by the factor $|N| \gg 1$ (for subluminal waves) as compared to the second term. For Alfvén waves this yields for forward (f) and backward (b) moving waves

$$\Gamma_{\mathrm{L}}^{\mathrm{f,b}} \simeq \pm \frac{\pi^2 V_A}{2c^2 k_{\|}} \sum_r \frac{\omega_{\mathrm{p},a}^2 N_r (m_r c)^3}{n_a}$$
$$\times \int_{E_{\mathrm{L}}}^{\infty} dE \left[E^2 - 1 - (EN^{-1} - x_r)^2\right] \frac{\partial f_r}{\partial y} \delta(y - (EN^{-1} - x_r)) , \tag{11.2.51}$$

so that depending on the sign of the anisotropy one of the two waves will be amplified while the other is damped.

An analysis analogous to (11.2.35)–(11.2.51) can be carried out for right-hand circularly polarized waves, for which the plus sign in the inequality (11.2.34) applies. Again (see also Lerche 1967 [291]) one finds that for isotropic distribution functions the system is stable with respect to parallel propagating waves. For anisotropic distribution functions the growth rate of RH parallel propagating Alfvén waves becomes

$$\Gamma_{\mathrm{R}}^{\mathrm{f,b}} \simeq \pm \frac{\pi^2 V_A}{2c^2 k_{\|}} \sum_r \frac{\omega_{\mathrm{p},a}^2 N_r (m_r c)^3}{n_a}$$
$$\times \int_{E_{\mathrm{R}}}^{\infty} dE \left[E^2 - 1 - (EN^{-1} + x_r)^2\right] \frac{\partial f_r}{\partial y} \delta(y - (EN^{-1} + x_r)) , \tag{11.2.52}$$

where
$$E_R = \frac{x_r N^{-1} + \sqrt{1 - N^{-2} + x_r^2}}{1 - N^{-2}}. \quad (11.2.53)$$

Again (11.2.52) indicates that either the forward or backward waves are amplified depending on the sign of the anisotropy.

Hence, a slightly anisotropic cosmic ray distribution function effectively generates plasma waves. Such a slight anisotropy is easily produced, for example, if the cosmic rays have a small streaming velocity with respect to the rest frame of the background medium. By choosing particular anisotropic cosmic ray trial distribution functions this has been quantitatively demonstrated in many investigations following the pioneering work of Lerche (1967 [291]).

11.2.1.3 Streaming Instability for Transverse Magnetosonic Waves.

Let us consider first the particularly illustrating cosmic ray nucleon distribution function of Kulsrud and Pearce (1969 [278])

$$f_r(E, y) = [f_0(E) + \mu f_1(E)] H[E - E_c]$$
$$\times \left[f_0(E) + \frac{y}{\sqrt{E^2 - 1}} f_1(E) \right] H[E - E_c], \quad (11.2.54)$$

with a first harmonic anisotropy ($\mu = p_\parallel / p$), with

$$f_0 = K_1 E^{-s}, \quad f_1 = K_2 E^{-r}, \quad (11.2.55)$$

above some minimum Lorentz factor E_c and $s > 3$, $r > 3$. The constant K_1 is fixed by the normalization $\int d^3 p f_r(p) = 1$ yielding

$$K_1 = \frac{s-3}{4\pi} \left[m_r c \sqrt{E_c^2 - 1} \right]^{(s-3)/2}. \quad (11.2.56)$$

We represent the interstellar medium as a dense cold electron-proton plasma so that we can use (11.2.45b). At wave frequencies $\omega_R \ll \Omega_p$ much less than the proton gyrofrequency, (11.2.46) and (11.2.47) hold, so that for right-handed polarized magnetosonic waves, evaluating the integrals in (11.2.39) to leading order only, we obtain

$$\Gamma_R \simeq \pm \frac{\pi s(s-3)}{8(s-2)\omega_R E_R} \left(\frac{V_A}{c} \right)^4 \frac{\omega_{p,a}^2 N_r(> E_R)}{n_a}$$
$$\times \left[-1 + N \frac{s-2}{s(r-2)} \frac{f_1(E_R)}{f_0(E_R)} \right], \quad (11.2.57)$$

where E_R is defined in (11.2.53) and $N_r(> E_R)$ denotes the number of cosmic ray protons above Lorentz factor E_R. We see that the cosmic rays lead to growth of forward moving magnetosonic waves ($N = +c/V_A$) for anisotropies

$$\frac{1}{f_r} \frac{\partial f_r}{\partial \mu} \propto f_1/f_0 \gtrsim \frac{1}{N} = V_A/c, \quad (11.2.58)$$

whereas backward moving (for which $N < 0$) magnetosonic waves are damped. Thus if f_1 corresponds to a net streaming of cosmic rays with velocity U with respect to the background medium, $U \gtrsim V_A$ leads to growth of these forward moving waves. Since within diffusion theory (see (12.3.18) ff.) the differential flux, or streaming, parallel to the ordered magnetic field is $-\kappa_\| \partial f_r/\partial z$, we can relate the streaming anisotropy of cosmic rays f_1/f_0 in (11.2.58) to the observed cosmic ray spatial diffusion coefficient ($\kappa_\|$) and the measured cosmic ray spatial gradient in the Galaxy ($\partial f_r/\partial z$). Doing so, Kulsrud and Pearce (1969 [278]) have calculated from (11.2.57) a value for the growth rate of left-handed polarized forward moving waves in the interstellar medium of about $\Gamma_L \simeq 2.7 \times 10^{-11}$ s^{-1}. If this exceeds the damping rate of the waves, then these waves will grow with an e-folding time of $\tau_L = 1/\Gamma_L = 1170$ yr, which is one order of magnitude shorter than the typical time it takes a cosmic ray particle to escape freely from the Galaxy (see Sect. 17.2.2). Similar and even shorter values for the cosmic-ray-induced instability growth time have been calculated by Lerche (1967 [291]), Tademaru (1969 [537]), and Melrose and Wentzel (1970 [339]).

11.2.2 Growth Rates at Arbitrary Angles of Propagation

By chosing the particular normalized cosmic ray proton trial distribution function in the rest frame of the background plasma

$$f_r(E,y) = \frac{4(1-\beta_0^2)^{1/2} b^3 \pi^{-2}}{\left[p_\perp^2 + b^2 + (p_\| - \beta_0 \epsilon)^2/(1-\beta_0^2)\right]^3}, \quad (11.2.59)$$

where $\epsilon = \sqrt{1+p^2}$ and $V_{CR} = \beta_0 c$ is the mean streaming velocity, Tademaru (1969 [537]) has calculated the growth rates of low-frequency transverse waves propagating at arbitrary angles with respect to the ordered magnetic field in a cold background plasma. The particular distribution function (11.2.59) has the advantage that it greatly facilitates the mathematics, and that it also behaves in a similar manner to the observed high-energy cosmic ray proton spectrum in our Galaxy. At the same time it gives an idea of the response due to the inclusion of a fixed amount of anisotropy through the parameter β_0. The distribution function (11.2.59) possesses the essential features of the observed cosmic ray spectrum.

The dispersion relations for these magnetosonic and Alfvén waves are given in (9.2.62). For parallel propagation the Alfvén mode corresponds to the left-hand polarized mode, while the magnetosonic mode corresponds to the right-hand polarized mode at $\theta = 0$. According to Tademaru (1969 [537]) the condition for instability of the Alfvén mode is $V_{CR}^2 > V_A^2$, while for the magnetosonic mode the instability condition is $V_{CR}^2 > V_A^2/\cos^2(\theta)$. The growth rates have maxima at frequencies of about $\omega_R = V_A \Omega_{p,0}/bc$, much less than the cyclotron frequency of the cold protons, for all allowed angles of propagation. The maximum growth rate of the Alfvén mode is given to a good approximation by

$$\Gamma_A(\theta) \simeq \Gamma_0 \left[T_A^{(1)}(\theta) + T_A^{(2)}(\theta) \right], \qquad (11.2.60)$$

where Γ_0 is the maximum growth rate at $\theta = 0$ (calculated in Sect. 11.2.1.3), and

$$T_A^{(1)}(\theta) = 8 \left[I_1(z_1)K_1(z_1) + (z_1/2)^2 I_0(z_1)K_0(z_1) \right.$$
$$\left. - \left(\frac{3}{2} + \frac{z_1^2}{2} \right) I_2(z_1)K_2(z_1) \right], \qquad (11.2.61)$$

$$T_A^{(2)}(\theta) = \frac{4}{3} \left[z_2^2 I_1(z_2)K_1(z_2) + 8 I_2(z_2)K_2(z_2) \right.$$
$$\left. - (12 + z_2^2) I_3(z_2)K_3(z_2) \right], \qquad (11.2.62)$$

where $z_1 = 2\tan\theta$, $z_2 = \sqrt{7}\tan\theta$. I and K denote modified Bessel functions of the first and second kind, respectively. The variations of $T_A^{(1)}$ and $T_A^{(2)}$ with angle are shown in Fig. 11.3.

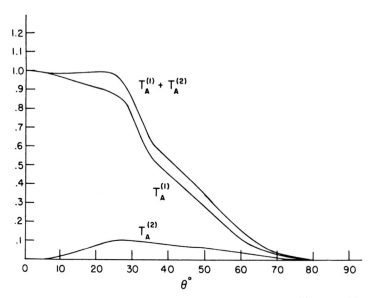

Fig. 11.3. Variation of the dimensionless functions $T_A^{(1)}$ and $T_A^{(2)}$ and their sum with the wave propagation angle θ. From Tademaru (1969 [537])

The total growth rate of Alfvén waves is a slowly varying function of $\tan\theta$ and decreases to zero as the angle approaches $90°$. The growth rate is not less than one-tenth the value at $\theta = 0$ until $\theta \gtrsim 60°$.

Likewise the growth rate of the magnetosonic mode has a maximum at

$$\omega_R^2 \simeq 3\frac{V_A^2 \Omega_{p,0}^2}{b^2 c^2} \left[\cos 2(\theta) + 3\sin^2(\theta)\right]^{-1}, \qquad (11.2.63)$$

and is approximately given by

$$\Gamma_M(\theta) \simeq \Gamma_0 \left[T_M^{(0)}(\theta) + T_M^{(1)}(\theta) + T_M^{(2)}(\theta)\right] \frac{V_{CR} - [V_A/\cos(\theta)]}{V_{CR} - V_A}, \qquad (11.2.64)$$

with

$$T_M^{(0)}(\theta) = 6\left[z_0 I_1(z_0) K_1(z_0) + \frac{z_0^3}{2}[I_2(z_0)K_2(z_1) - I_0(z_0)K_0(z_0)]\right], \qquad (11.2.65)$$

$$T_M^{(1)}(\theta) = 128(1 + 2\sin^2\theta)^{1/2}(4 + 3\tan^2\theta)^{-2}\left[I_1(z_3)K_1(z_3) \right.$$
$$\left. + \frac{z_3^2}{4}(1+z_3^2)I_0(z_3)K_0(z_3) - \left(\frac{3}{2} + \frac{3}{4}z_3^2 + \frac{1}{4}z_3^4\right)I_2(z_3)K_2(z_3)\right], \qquad (11.2.66)$$

and

$$T_M^{(2)}(\theta) = 512(1 + 2\sin^2\theta)^{1/2}(7 + 12\tan^2\theta)^{-2}$$
$$\times \left[\left(4 + \frac{7}{8}z_4^2 + \frac{1}{16}z_4^4\right)I_2(z_4)K_2(z_4) - \left(\frac{3}{8}z_4^2 + \frac{1}{16}z_4^4\right)I_0(z_4)K_0(z_4)\right.$$
$$\left. - \left(\frac{3}{2} + \frac{1}{4}z_4^2\right)I_1(z_4)K_1(z_4) - \left(\frac{3}{2} + \frac{1}{8}z_4^2\right)I_3(z_4)K_3(z_4)\right], \qquad (11.2.67)$$

with $z_0 = \sqrt{3/[3+\cot^2\theta]}$, $z_3 = \tan\theta\,[1 + 3/(1+3\tan^2\theta]^{1/2}$, and $z_4 = \tan\theta\,[4 + 3/(1+3\tan^2\theta]^{1/2}$. The variations of $T_M^{(0)}$, $T_M^{(1)}$, $T_M^{(2)}$ and their sum with angle are shown in Fig. 11.4.

The sum $T_M^{(0)} + T_M^{(1)} + T_M^{(2)}$ is always of order unity for all angles. However, as the last term in (11.2.64) indicates, the instability is cut off at angles $\theta_c = \cos^{-1}(V_A/V_{CR})$; for example, if $V_{CR} = 2V_A$, then $\theta_c = 60°$.

Tademaru (1969 [537]) has shown that the fast cosmic ray instabilities found for parallel propagating waves persist over a very large range of angles. In his approximations of a rigorously cold plasma background he completely neglected Landau damping that would occur in an isotropic background plasma at a finite temperature (cf. Sect. 10.4.1). For small propagation angles the damping by a thermal plasma is indeed quite small, and it

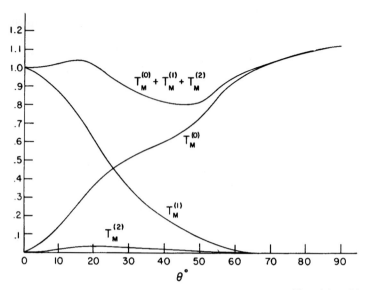

Fig. 11.4. Variation of the dimensionless functions $T_M^{(0)}$, $T_M^{(1)}$, $T_M^{(2)}$ and their sum with the wave propagation angle θ. From Tademaru (1969 [537])

is a reasonable approximation to neglect Landau damping. However, Landau damping increases as the angle increases. Especially for the magnetosonic mode the damping is strong at finite angles if the streaming velocity is less than the thermal electron velocity (Barnes 1967 [31], Achterberg 1981 [8]). In this case it can hardly be said that these waves propagate as a mode at all.

11.2.3 Cosmic Ray Self-Confinement in Galaxies

From the calculations presented in the preceeding sections the picture of "cosmic ray self-confinement" in galaxies emerges. We have considered here the stability of an initial state consisting of a uniform and infinitely extended magnetic field parallel to the galactic plane which couples the relativistic cosmic ray gas to a cold background plasma of electrons and ions. By a normal mode analysis of the linearized Vlasov–Maxwell equations we established that this initial state is unstable with growth time ~ 1000 yr if the cosmic ray gas streams with a net velocity $V_{\rm CR} > V_{\rm A}$ in excess of the Alfvén velocity with respect to the thermal plasma. The growing low-frequency Alfvén waves modify the original electromagnetic field within a short time compared to the mean cosmic ray residence time in our Galaxy, and due to the nonlinear relaxation processes then operating a final state results in magnetic and electric field irregularities superposed on the regular field component. As we shall see in the next section the magnetic irregularities scatter the cosmic rays and reduce their streaming values to values smaller than $V_{\rm A}$. This means that cosmic rays as a gas cannot travel with a velocity faster than V_A with respect to the thermal gas away from their sources, although the individual cosmic ray particle speeds are much higher. For the same reasons also anisotropic cosmic ray processes are forbidden.

11.3 Competition of Wave Growth and Damping

Whether low-frequency transverse waves will grow or damp in time depends sensitively, in the absence of wave cascading processes, on the total decrement (11.2.42)

$$\Gamma_{\text{tot}} = \Gamma_{\text{growth}} + \Gamma_{\text{damp}} \tag{11.3.1}$$

being positive or negative, respectively. The total decrement is the sum of all positive growth rates (including the cosmic ray streaming instability) and all negative damping rates (including the Landau damping rate in the hot interstellar medium).

11.3.1 Damping by Ion-Neutral Collisions

Additional damping mechanisms for the waves in the interstellar medium have been discussed by Kulsrud and Pearce (1969 [278]). These are ion-neutral collisions, resistivity, electron viscosity, electron thermal conductivity and ion resistivity. Kulsrud and Pearce (1969 [278]) demonstrated that for interstellar gas parameters typical for the HI phase ion-neutral collisions are most important, and they derived the overlapping damping rates

$$\Gamma_n \simeq -\frac{1}{2} \begin{cases} \nu_{\text{in}} & \text{for } |\omega_R| \gg \nu_{\text{in}}\sqrt{n_{\text{CR}}/n_{\text{HI}}} \\ \frac{\omega_R^2}{\nu_{\text{in}}} & \text{for } \omega_R \ll \nu_{\text{in}} \end{cases}, \tag{11.3.2}$$

where

$$\nu_{\text{in}} \simeq 8 \times 10^{-9} n_{\text{HI}} \text{ s}^{-1}. \tag{11.3.3}$$

11.3.2 Nonlinear Landau Damping

At appreciable wave intensities, nonlinear damping processes set in. While a single Alfvén wave traveling parallel to the ordered magnetic field can be damped only through gyroresonance (cf. Sect. 10.4), the situation is different when many waves are present. Specifically, two circularly polarized Alfvén waves can form a beat wave which drives a second-order (in wave amplitude) longitudinal electric field, and which consequently suffers Landau damping through the modified Landau resonance $(\omega_{R,1} - \omega_{R,2}) - (k_1 - k_2)v_\parallel = 0$, where the subscripts refer to waves 1 and 2. (Here ω_R and k are signed quantities: $\omega_R > 0$ and $\omega_R < 0$ denote waves with left-hand and right-hand polarization, respectively, while $\omega_R/k > 0$ and $\omega_R/k < 0$ denote propagation parallel or antiparallel to the ordered magnetic field, respectively.) Resonance occurs when the phase speed of the beat wave is in phase with the parallel particle velocity, i.e.

$$v_\phi = \frac{\omega_{R,1} - \omega_{R,2}}{k_1 - k_2} = v_\| \ . \tag{11.3.4}$$

This damping is referred to as nonlinear Landau damping and has been treated in detail by Lee and Völk (1973 [286]) and Miller (1991 [353]).

If the waves propagate in the same direction, then (11.3.4) reduces to $v_\| = \pm V_A$ which is independent of k_1 and k_2. In a low-beta ($\beta = 2c_s^2/V_A^2 \ll 1$) isothermal plasma, where the Alfvén speed $V_A \gg c_s$ is much larger than the sound speed, only thermal electrons can satisfy this resonance condition. As the plasma beta increases, however, thermal protons will attain speeds greater than V_A, and participate significantly in this damping process. This is because the damping depends sensitively upon the number of particles having speeds slightly less than V_A relative to the number of particles having speeds slightly greater than V_A, that is the slope of the particle distribution function: the greater the magnitude of the slope, the stronger the damping. In a high-beta plasma ($\beta \gg 1$) the electron distribution function around V_A is quite flat, while the proton distribution function is relatively steep. Thus even if there are more resonant electrons than protons, the protons dominate the damping. It is for this physical reason that in a low-beta plasma the parallel propagating waves will damp on the electrons, whereas in a high-beta plasma damping will occur on the protons.

If the waves propagate in opposite directions, the phase speed in (11.3.4) depends on the wavenumbers k_1 and k_2. If they have the same polarization then $|v_\phi|$ is always less than V_A, and can be much less than V_A if $k_1 \simeq k_2$. In this case, nonlinear Landau damping by thermal protons will dominate in either a low- or high-beta plasma. If the waves have opposite polarizations, then $|v_\phi|$ is always greater than V_A, and the damping will be on the electrons in a low-beta plasma.

These very elementary considerations imply that in a low-beta plasma:

1. interactions between waves propagating in the same direction will lead to electron heating;
2. interactions between waves of opposite directions propagating in opposite directions will also lead to electron heating;
3. however, interactions between waves of the same polarization propagating in opposite directions will lead to proton heating.

In the first interaction (1) the electron distribution will be quite flat near V_A, and as a consequence the damping rate will be small. For the second type of interaction (2) the largest damping rate for a wave will occur when it interacts with another wave such that $|v_\phi| = c_{s,e}$, where the slope of the electron distribution is greatest. For the last type of interaction (3) the largest damping rate for a wave occurs when it interacts with a second wave of comparable wavenumber, in which case $|v_\phi| \simeq c_{s,p}$. Since the maximum slope of a Maxwellian distribution scales in proportional to particle mass over temperature ($\propto m/T$), the maximum slope of the proton distribution is 1836 times greater than that of the electron distribution for a given temperature. The maximum damping rate on the protons is thus much greater than that on the electrons if there are many waves present in a spectrum, i.e. if the wave's spectrum is broad. The dissipation of broad wave spectra by nonlinear Landau damping results primarily from interactions of antiparallel waves of the same polarization and leads to efficient proton heating. These qualitative considerations are in accord with the more quantitative calculations of Miller (1991 [353]).

11.3.3 Wave Cascading

Another relevant process that sets in at appreciable wave intensities is the cascading of waves. As a consequence of wave steepening spectral wave energy cascades to higher frequencies and wavenumbers (e.g. Marsch 1991 [321]). The idea to describe the evolution of turbulence by a diffusion of energy in wavenumber space was pioneered by Leith (1967 [289]) in hydrodynamics, and subsequently introduced to magnetohydrodynamics by Zhou and Matthaeus (1990 [574]). This approach provides a simple framework to take into account, at least approximately, turbulence evolution in space physics applications. Zhou and Matthaeus (1990 [574]) present a general transport equation for the wave spectral density in the case of isotropic turbulence, which includes terms for spatial convection and propagation, nonlinear transfer of energy across the wavenumber spectrum, and a source and sink of wave energy. The kinetic equation for a 3-dimensional spectral density $\bar{W}_i(\mathbf{k})$ of mode i, which denotes the wave energy density per unit volume of wavenumber space of the plasma mode i, is given by the usual conservation equation

$$\frac{\partial}{\partial t} \bar{W}_i(\mathbf{k}) = -\frac{\partial}{\partial \mathbf{k}} \cdot \mathbf{F}(\mathbf{k}), \qquad (11.3.5a)$$

where the flux

$$\mathbf{F}(\mathbf{k}) = -D \frac{\partial \bar{W}_i(\mathbf{k})}{\partial \mathbf{k}} \qquad (11.3.5b)$$

is expressed as a diffusive term with the diffusion coefficient in wavenumber space

$$D = k^2/\tau_s(k) \qquad (11.3.5c)$$

and the spectral energy transfer time scale $\tau_s(k)$. For isotropic turbulence one obtains for the associated one-dimensional spectral density $W_i(k) = 4\pi k^2 \bar{W}_i(\mathbf{k})$ the simplified diffusion equation

$$\frac{\partial W_{\rm i}}{\partial t} = \frac{\partial}{\partial k}\left[\frac{k^4}{\tau_{\rm s}(k)}\frac{\partial}{\partial k}\left(k^{-2}W_{\rm i}\right)\right] + \Gamma_{\rm i}W_{\rm i} + S_{\rm i}(k) , \qquad (11.3.6)$$

where we included a term for the damping or growth of waves (see (11.3.1)) and a wave energy injection and/or sink term $S_{\rm i}(k)$.

The spectral energy transfer time scale (or the wavenumber diffusion coefficient (11.3.5c)) depends upon the cascade phenomenology. In the Kolmogorov treatment the spectral energy transfer time at a particular wavelength λ is the eddy turnover time $\lambda/\delta v$, where δv is the velocity fluctuation of the wave. In the Kraichnan treatment the transfer time is longer by a factor $V_{\rm A}/\delta v$. Both phenomenologies are further discussed in Zhou and Matthaeus (1990 [574]) and yield

$$\tau_{\rm s}(k) \simeq \frac{1}{V_{\rm A}k^{3/2}}\begin{cases} \sqrt{2U_{\rm B}/W_{\rm i}} & \text{(Kolmogorov)} \\ 2U_{\rm B}/k^{1/2}W_{\rm i} & \text{(Kraichnan)} \end{cases}, \qquad (11.3.7)$$

where $U_{\rm B} = B_0^2/8\pi$ denotes the energy density of the ordered magnetic field. Substituting these transfer time scales into (11.3.6), and assuming a steady state with no damping, we obtain $W_{\rm i} = W_0 k^{-s}$, where $s = 5/3$ for the Kolmogorov case and $s = 3/2$ for the Kraichnan phenomenology. The diffusion equation (11.3.6) in either case is nonlinear.

Besides the noted generation of turbulence by kinetic instabilities (Sect. 11.2), very often wave cascading from low to high wavenumbers is an important way of producing broadband wave spectra. One possibility, discussed, for example, in the context of solar flares (Miller and Roberts 1995 [354]), is that long-wavelength turbulence results from the rearrangement of large-scale magnetic fields and/or a shear flow instability, so that it is reasonable to assume the deposition of wave energy peaked at long wavelength, probably comparable to the physical size of the system, as the primary energy release. Cascading as described by the nonlinear diffusion term in (11.3.6) will then transfer this spectral energy to higher wavenumbers, where the waves will be able to resonate with progressively lower energy cosmic ray particles, until they eventually interact with the charged particles in the tail of the background thermal distribution, which is described in (11.3.6) by the various wave damping rates.

11.3.4 Super-Alfvénic Cosmic Ray Propagation?

The picture of cosmic ray self-confinement in galaxies discussed in Sect. 11.2.3 has been contested by Holman et al. (1979 [230]) and Davila and Scott (1984 [117]) on the basis of (11.3.1). They repeated the analysis of the cosmic ray streaming instability using, however, instead of a cold background plasma as Tademaru (1969 [537]) did, the hot thermal coronal phase of the interstellar medium. They found that in this medium at very high wavenumbers the (negative) ion-cyclotron damping of Alfvén waves by the hot magnetized

11.3 Competition of Wave Growth and Damping

interstellar medium is much stronger than the (positive) growth rate by the streaming cosmic rays, so that consequently the total damping rate (11.3.1) is negative. They concluded therefore that the very short-wavelength waves, which as we shall see in Chap. 13 are needed to scatter cosmic rays of pitch angle very close to 90° and isotropize them, are absent in the hot interstellar medium. As a consequence these missing waves for isotropization in the wave frame lead to net streaming speeds in excess of the Alfvén speed: the streaming speed may be as large as the ion sound speed of the coronal medium, $c_s = 130(T/10^6 \text{ K})^{1/2}$, before the instability develops.

A related argument also based on (11.3.1) has been made by Kulsrud and Cesarsky (1971 [277]) who noted that at some finite small wavenumber in the partially neutral hydrogen interstellar medium the streaming growth rate is smaller than the absolute value of the ion-neutral damping rate (11.3.2), so that there would be no Alfvén waves present to confine cosmic ray particles above a certain energy which they estimated to be of order 10^{14} eV.

It is evident that these difficulties vanish immediately if finite wave cascading (yielding wave diffusion according to (11.3.6)) exists in the interstellar medium. It is one property of the diffusion equation (11.3.6) that it yields finite positive wave spectral energy densities at all wavenumbers k even if in some wavenumber intervals the net damping rate (11.3.1) is negative. Due to the large amplitude of interstellar turbulence we have no doubt that wave cascading exists, accordingly we see no difficulty with the cosmic ray self-confinement picture.

12. Test Particle Approach 1. Hierarchy of Transport Equations

After having established the nature and sources of cosmic plasma waves we now turn to the effect of these plasma waves on the cosmic rays.

12.1 Quasilinear Theory

We start from the relativistic Vlasov equations (8.2.1)

$$\frac{\partial f_a}{\partial t} + \boldsymbol{v} \cdot \frac{\partial f_a}{\partial \boldsymbol{x}} + \dot{\boldsymbol{p}} \cdot \frac{\partial f_a}{\partial \boldsymbol{p}} = S_a(\boldsymbol{x}, \boldsymbol{p}, t) , \qquad (12.1.1)$$

with the equations of motion

$$\dot{\boldsymbol{p}} = q_a \left[\boldsymbol{E}_T(\boldsymbol{x}, t) + \frac{\boldsymbol{v} \times \boldsymbol{B}_T(\boldsymbol{x}, t)}{c} \right] , \qquad (12.1.2a)$$

$$\dot{\boldsymbol{x}} = \boldsymbol{v} = \frac{\boldsymbol{p}}{\gamma m_a} . \qquad (12.1.2b)$$

S_a in (12.1.1) denotes sources and sinks of particles. Because of the high conductivity of cosmic plasmas we neglect any large-scale electric field in the system, so that the total electromagnetic field entering (12.1.2a) is a superposition of the uniform magnetic field $\boldsymbol{B}_0 = B_0 \boldsymbol{e}_z$ and the plasma turbulence $(\boldsymbol{\delta E}, \boldsymbol{\delta B})$, i.e.

$$\boldsymbol{B}_T = \boldsymbol{B}_0 + \boldsymbol{\delta B}(\boldsymbol{x}, t), \quad \boldsymbol{E}_T = \boldsymbol{\delta E}(\boldsymbol{x}, t) . \qquad (12.1.3)$$

Because of the gyrorotation of the particles in the uniform magnetic field we are not so much interested in their actual position as in the coordinates of the guiding center

$$\boldsymbol{R} = (X, Y, Z) = \boldsymbol{x} + \frac{\boldsymbol{v} \times \boldsymbol{e}_z}{\epsilon \Omega} , \qquad (12.1.4)$$

where Ω as before denotes the absolute value of the particlet's gyrofrequency in the uniform field and $\epsilon = q_a/|q_a|$ the charge sign. It is also convenient to use again spherical coordinates (p, μ, ϕ) in momentum space defined by

$$p_x = p \cos\phi \sqrt{1-\mu^2}, \quad p_y = p \sin\phi \sqrt{1-\mu^2}, \quad p_z = p\mu . \qquad (12.1.5)$$

Consequently

$$X = x + \frac{v\sqrt{1-\mu^2}}{\epsilon\Omega}\sin\phi, \quad Y = y - \frac{v\sqrt{1-\mu^2}}{\epsilon\Omega}\cos\phi, \quad Z = z. \quad (12.1.6)$$

Transforming (12.1.1) to the coordinate set

$$x_\sigma = (p, \mu, \phi, X, Y, Z) \quad (12.1.7)$$

by using (12.1.2)–(12.1.6) one obtains the appropriate form of the Vlasov equation (using the Einstein summation convention) (e.g. Hall and Sturrock 1968 [208], Urch 1977 [545])

$$\frac{\partial f_a}{\partial t} + v\mu\frac{\partial f_a}{\partial Z} - \epsilon\Omega\frac{\partial f_a}{\partial \phi} + \frac{1}{p^2}\frac{\partial}{\partial x_\sigma}(p^2 g_{x_\sigma} f_a) = S_a(\boldsymbol{x}, \boldsymbol{p}, t), \quad (12.1.8)$$

where the generalized force term g_σ includes the effects of the randomly fluctuating electromagnetic fields. Solutions of (12.1.8) with $g_\sigma = 0$ will be called "unperturbed orbits". The components of the fluctuating force term are

$$g_p = \dot{p} = \frac{m_a c \gamma \epsilon \Omega}{p B_0} \boldsymbol{p} \cdot \boldsymbol{\delta E} = \frac{\epsilon\Omega pc}{vB_0}\left[\mu\delta E_\| + \sqrt{\frac{1-\mu^2}{2}}(\delta E_L e^{-\imath\phi} + \delta E_R e^{\imath\phi})\right],$$
$$(12.1.9a)$$

$$g_\mu = \dot{\mu} = \frac{\epsilon\Omega\sqrt{1-\mu^2}}{B_0}\left[\frac{c}{v}\sqrt{1-\mu^2}\delta E_\| + \frac{\imath}{\sqrt{2}}\left[e^{\imath\phi}\left(\delta B_R + \imath\mu\frac{c}{v}\delta E_R\right)\right.\right.$$
$$\left.\left. - e^{-\imath\phi}\left(\delta B_L - \imath\mu\frac{c}{v}\delta E_L\right)\right]\right], \quad (12.1.9b)$$

$$g_\phi = -\epsilon\Omega\frac{\delta B_\|}{B_0} + \frac{\epsilon\Omega}{\sqrt{2(1-\mu^2)}B_0}\left[e^{\imath\phi}\left(\mu\delta B_R + \frac{\imath c}{v}\delta E_R\right)\right.$$
$$\left. + e^{-\imath\phi}\left(\mu\delta B_L - \frac{\imath c}{v}\delta E_L\right)\right], \quad (12.1.9c)$$

$$g_X = -v\sqrt{1-\mu^2}\cos\phi\frac{\delta B_\|}{B_0} + \frac{\imath c}{\sqrt{2}B_0}\left[\delta E_R - \delta E_L - \frac{\imath\mu v}{c}(\delta B_L + \delta B_R)\right],$$
$$(12.1.9d)$$

$$g_Y = -v\sqrt{1-\mu^2}\sin\phi\frac{\delta B_\|}{B_0} - \frac{c}{\sqrt{2}B_0}\left[\delta E_L + \delta E_R + \frac{\imath\mu v}{c}(\delta B_L - \delta B_R)\right],$$
$$(12.1.9e)$$

$$g_Z = 0. \quad (12.1.9f)$$

In the derivation of (12.1.9) we have introduced

$$\delta B_{L,R} \equiv \frac{1}{\sqrt{2}}(\delta B_x \pm \imath \delta B_y), \quad \delta B_\| = \delta B_z, \quad (12.1.10a)$$

and
$$\delta E_{\text{L,R}} \equiv \frac{1}{\sqrt{2}}(\delta E_x \pm \imath \delta E_y), \quad \delta E_\| = \delta E_z, \tag{12.1.10b}$$

which are related to the left-handed and right-handed polarized field components (see also Chap. 9).

The function f_a develops in an irregular way under the influence of g_σ, but the detailed fluctuations are not of interest. An expectation value of f_a must be found in terms of the statistical properties of g_σ, so we consider an *ensemble* of distribution functions all beginning with identical values at some time $t = t_0$. Let each of these functions be subject to a different member of an ensemble of realizations of g_σ, i.e. fluctuating field histories which are independent of one another in detail, but identical as to statistical averages. At any time $t > t_0$, the various functions differ from each other, and we require an equation for

$$\langle f_a(\boldsymbol{x}, \boldsymbol{p}, t) \rangle = F_a(\boldsymbol{x}, \boldsymbol{p}, t), \tag{12.1.11}$$

the average of f_a over all members of the ensemble. This is obtained by taking the average of (12.1.8) using

$$\langle \boldsymbol{\delta B}(\boldsymbol{x}, t) \rangle = \langle \boldsymbol{\delta E}(\boldsymbol{x}, t) \rangle = 0, \tag{12.1.12a}$$

implying

$$\langle \boldsymbol{B}(\boldsymbol{x}, t) \rangle = \boldsymbol{B}_0, \quad \langle \boldsymbol{E}(\boldsymbol{x}, t) \rangle = 0. \tag{12.1.12b}$$

We find

$$\frac{\partial F_a}{\partial t} + v\mu \frac{\partial F_a}{\partial Z} - \epsilon\Omega \frac{\partial F_a}{\partial \phi} = S_a(\boldsymbol{x}, \boldsymbol{p}, t) - \frac{1}{p^2}\frac{\partial}{\partial x_\sigma}\left(\langle p^2 g_{x_\sigma} \delta f_a \rangle\right), \tag{12.1.13}$$

where

$$\delta f_a(\boldsymbol{x}, \boldsymbol{p}, t) = f_a(\boldsymbol{x}, \boldsymbol{p}, t) - F_a(\boldsymbol{x}, \boldsymbol{p}, t). \tag{12.1.14}$$

Subtracting (12.1.13) from (12.1.8) gives an equation for the fluctuation

$$\frac{\partial \delta f_a}{\partial t} + v\mu \frac{\partial \delta f_a}{\partial Z} - \epsilon\Omega \frac{\partial \delta f_a}{\partial \phi} = -g_{x_\sigma}\frac{\partial F_a}{\partial x_\sigma} - g_{x_\sigma}\frac{\partial \delta f_a}{\partial x_\sigma} + \left\langle g_{x_\sigma}\frac{\partial \delta f_a}{\partial x_\sigma}\right\rangle, \tag{12.1.15}$$

where we used

$$\frac{1}{p^2}\frac{\partial}{\partial x_\sigma}\left(p^2 g_{x_\sigma}\right) = 0. \tag{12.1.16}$$

We use a perturbation method (referred to as the *quasilinear approximation*) to solve these equations, and the remainder of the analysis is based on the assumption that the fluctuations are of small amplitude: g_σ must be significantly small so that there exists a time scale T satisfying

12. Hierarchy of Transport Equations

$$t_c \ll T \ll t_F , \tag{12.1.17}$$

where t_F represents the time scale on which g_σ affects the evolution of the distribution function. From the equation of motion t_F can be estimated as

$$t_F = F_a / \left| g_{x_\sigma} \frac{\partial F_a}{\partial x_\sigma} \right| . \tag{12.1.18}$$

Then according to (12.1.15) the variation δf_a generated by g_σ within a time T must remain much smaller than F_a, and the right-hand side of (12.1.15) may be approximated by its first term, leading to

$$\frac{\partial \delta f_a}{\partial t} + v\mu \frac{\partial \delta f_a}{\partial Z} - \epsilon\Omega \frac{\partial \delta f_a}{\partial \phi} \simeq - g_{x_\sigma} \frac{\partial F_a}{\partial x_\sigma} . \tag{12.1.19}$$

Equation (12.1.19) can readily be solved by the method of characteristics and we obtain

$$\delta f_a(t) = \delta f_a(t_0) - \int_{t_0}^{t} ds \left[g_{x_\sigma}(x_\nu^*, s) \frac{\partial F_a(x_\nu, s)}{\partial x_\sigma} \right]' , \tag{12.1.20}$$

where the prime indicates that the bracketed quantities are to be evaluated along the characteristics, i.e. an unperturbed particle orbit in the uniform magnetic field, given by

$$\bar{X} = x_0, \qquad \bar{Y} = y_0 , \qquad \bar{Z} = z_0 + v\mu (s - t) ,$$
$$\bar{x} = x_0 - \frac{v\sqrt{1 - \mu^2}}{\epsilon\Omega} \sin(\bar{\phi}) ,$$
$$\bar{y} = y_0 + \frac{v\sqrt{1 - \mu^2}}{\epsilon\Omega} \cos(\bar{\phi}) ,$$
$$\bar{z} = z_0 + v\mu (s - t)$$
$$\bar{p} = p, \qquad \bar{\mu} = \mu, \qquad \bar{\phi} = \phi_0 - \epsilon\Omega (s - t) , \tag{12.1.21}$$

where all $(x_0, y_0, z_0, \phi_0, p, \mu)$ denote the initial phase space coordinate values at time t_0. Moreover, we demand that at the initial time t_0 the particle's phase space density is completely uncorrelated to the turbulent field, so that the ensemble average

$$\langle \delta f_a g_{x_\sigma} \rangle = 0 \tag{12.1.22}$$

vanishes. Then, inserting (12.1.20) into the averaged (12.1.13) leads to

$$\frac{\partial F_a}{\partial t} + v\mu \frac{\partial F_a}{\partial Z} - \epsilon\Omega \frac{\partial F_a}{\partial \phi} = S_a(\bm{x}, \bm{p}, t)$$
$$+ \frac{1}{p^2} \frac{\partial}{\partial x_\sigma} \left(\left\langle p^2 g_{x_\sigma} \int_{t_0}^{t} ds \left[g_{x_\nu}(x_\nu, s) \frac{\partial F_a(x_\nu, s)}{\partial x_\nu} \right]' \right\rangle \right) .$$
$$\tag{12.1.23}$$

(As an aside, note that the third term on the right-hand side of (12.1.15), if kept, would vanish in (12.1.23) because $\langle g_{x_\sigma} \rangle = 0$.) Here $(t - t_0) \sim T$, where T satisfies the restriction (12.1.17). Under certain physical conditions the integrodifferential equation (12.1.23) reduces to a differential equation for F_a. The second term on the right-hand side of (12.1.23) can be rearranged as

$$M_1 \equiv \frac{1}{p^2} \frac{\partial}{\partial x_\sigma} \left(\left\langle p^2 g_{x_\sigma} \int_{t_0}^{t} ds \left[g_{x_\nu}(x_\nu, s) \frac{\partial F_a(x_\nu, s)}{\partial x_\nu} \right]' \right\rangle \right)$$

$$= \frac{1}{p^2} \frac{\partial}{\partial x_\sigma} \left(p^2 \int_{t_0}^{t} ds \left\langle g_{x_\sigma} g'_{x_\nu}(x_\nu, s) \right\rangle \left[\frac{\partial F_a(x_\nu, s)}{\partial x_\nu} \right]' \right). \quad (12.1.24)$$

Suppose that there exists a correlation time t_c so that the correlation function of magnetic and electric irregularities that determine the correlation function $\left\langle g_{x_\sigma} g'_{x_\nu} \right\rangle$, falls to a negligible magnitude for times $t-s > t_c$. This means that the important contribution to the right-hand side of (12.1.24) comes from the integral from $t-t_c$ to t which can be assumed to be finite. Moreover, suppose that the variation of $[\partial F_a/\partial x_\nu]'$ during this time interval is so small that the value is nearly equal to that at $s = t$. Then the term (12.1.24) reduces to

$$M_1 \simeq \frac{1}{p^2} \frac{\partial}{\partial x_\sigma} \left(p^2 \left[\int_{t-t_c}^{t} ds \left\langle g_{x_\sigma} g_{x_\nu}(x_\nu, s) \right\rangle \right]' \frac{\partial F_a(x_\nu, t)}{\partial x_\nu} \right)$$

$$= \frac{1}{p^2} \frac{\partial}{\partial x_\sigma} \left(p^2 \left[\int_{0}^{t} ds \left\langle g_{x_\sigma} g_{x_\nu}(x_\nu, s) \right\rangle \right]' \frac{\partial F_a(x_\nu, t)}{\partial x_\nu} \right). \quad (12.1.25)$$

The requirement (12.1.17), i.e. $t-t_0 \gg t_c$, makes the term (12.1.25) a function of t alone, eliminating any dependence upon conditions at t_0 which justifies the replacement 0 for $t - t_c$ in the lower boundary of the integration.

With these rearrangements (12.1.23) reduces to a diffusion equation, involving only second-order correlation functions of the fluctuating field g_{x_σ} integrated along the unperturbed orbit. This equation, hereafter referred to as the *Fokker-Planck equation*, can be written

$$\frac{\partial F_a}{\partial t} + v\mu \frac{\partial F_a}{\partial Z} - \epsilon \Omega \frac{\partial F_a}{\partial \phi} = S_a(\boldsymbol{x}, \boldsymbol{p}, t) + \frac{1}{p^2} \frac{\partial}{\partial x_\sigma} \left(p^2 D_{x_\sigma \, x_\nu} \frac{\partial F_a}{\partial x_\nu} \right), \quad (12.1.26)$$

with the *Fokker-Planck coefficients*

$$D_{\sigma \, \nu}(x_\eta, t) = \int_0^t ds \left\langle \bar{g}_{x_\sigma}(t) \bar{g}_{x_\nu}(s) \right\rangle, \quad (12.1.27)$$

being homogeneous integrals along unperturbed particle orbits of the fluctuating force field's correlation functions.

The force fields g_{x_σ} are real quantities, so for their complex conjugates we have $g^*_{x_\sigma} = g_{x_\sigma}$. With $\xi = t - s$ we can write (12.1.27) also as

$$\begin{aligned}
D_{x_\sigma\, x_\nu}(x_\eta, t) &= \Re \int_0^t ds\, \langle \bar{g}_{x_\sigma}(t) \bar{g}^*_{x_\nu}(s) \rangle \\
&= -\Re \int_t^0 d\xi\, \langle \bar{g}_{x_\sigma}(t) \bar{g}^*_{x_\nu}(t-\xi) \rangle \\
&= \Re \int_0^t d\xi\, \langle \bar{g}_{x_\sigma}(t) \bar{g}^*_{x_\nu}(t-\xi) \rangle\,. \quad (12.1.28)
\end{aligned}$$

For stationary fluctuations the correlation function $\langle \bar{g}_{x_\sigma}(t) \bar{g}^*_{x_\nu}(t-\xi) \rangle$ depends only on the value of the difference $|\xi|$. Recalling condition (12.1.17), that we consider times long compared to the correlation time t_c, beyond which the correlation function becomes negligibly small, we may finally replace the upper integration boundary in (12.1.28) by infinity to obtain for the Fokker-Planck coefficients

$$D_{x_\sigma\, x_\nu} = \Re \int_0^\infty d\xi\, \langle \bar{g}_{x_\sigma}(t) \bar{g}^*_{x_\nu}(t+\xi) \rangle\,. \quad (12.1.29)$$

The bar notation in (12.1.29) indicates that the force fields have to be calculated along the unperturbed orbit of the particles.

This quasilinear derivation of the Fokker-Planck equation (12.1.26) and its coefficients (12.1.29) for magnetized plasma follows closely the work of Kennel and Engelmann (1966 [258]), Roelof (1968 [436]) and Hall and Sturrock (1968 [208]); modifications suggested in the work of Newman (1973 [374]) and Urch (1977 [545]) were examined by Achatz et al. (1991 [6]) who found mathematical and physical inconsistencies when using these modifications.

12.2 Quasilinear Fokker-Planck Coefficients

12.2.1 Unperturbed Particle Orbits

The first step in calculating the quasilinear Fokker-Planck coefficients (12.1.29) is to determine the fluctuating force fields (12.1.9) along the unperturbed particle orbit (12.1.21). In order to do so we use the Fourier representation of the fluctuating electric and magnetic fields (8.3.12) and approximate the true orbit $\boldsymbol{x}(t)$ by the unperturbed orbit $\bar{\boldsymbol{x}}(t)$ according to (12.1.21):

$$\delta \boldsymbol{E}(\boldsymbol{x},t) = \int_{-\infty}^{\infty} d^3k\, \boldsymbol{E}(\boldsymbol{k},t) e^{i\boldsymbol{k}\cdot\boldsymbol{x}(t)} \simeq \int_{-\infty}^{\infty} d^3k\, \boldsymbol{E}(\boldsymbol{k},t) e^{i\boldsymbol{k}\cdot\bar{\boldsymbol{x}}(t)}\,, \quad (12.2.1a)$$

$$\delta \boldsymbol{B}(\boldsymbol{x},t) = \int_{-\infty}^{\infty} d^3k\, \boldsymbol{B}(\boldsymbol{k},t) e^{i\boldsymbol{k}\cdot\boldsymbol{x}(t)} \simeq \int_{-\infty}^{\infty} d^3k\, \boldsymbol{B}(\boldsymbol{k},t) e^{i\boldsymbol{k}\cdot\bar{\boldsymbol{x}}(t)}\,. \quad (12.2.1b)$$

12.2 Quasilinear Fokker-Planck Coefficients

With the wavenumber representation $k_\| = k_z$, $\psi = \cot^{-1}(k_x/k_y)$, $k_\perp = \sqrt{k_x^2 + k_y^2}$, and the unperturbed orbit (12.1.21) starting from the initial position $x_0 = y_0 = z_0 = 0$ at time $t = 0$, we derive

$$\exp(\imath \boldsymbol{k} \cdot \bar{\boldsymbol{x}}(t)) = \sum_{n=-\infty}^{\infty} J_n(W) \exp\left[\imath k_\| v_\| t + \imath n(\psi - \phi_0 + \epsilon \Omega t)\right] , \tag{12.2.2a}$$

where we used the abbreviation

$$W \equiv \frac{k_\perp v \sqrt{1-\mu^2}}{\Omega} , \tag{12.2.2b}$$

and where we have made use of the Bessel-function identity (B.13a), see Appendix B.1. For the electromagnetic field components (12.1.10) we obtain with (12.2.1) and (12.2.2)

$$\delta B_{\mathrm{L,R},\|} \simeq \sum_{n=-\infty}^{\infty} \int_{-\infty}^{\infty} d^3k \; B_{\mathrm{L,R},\|}(\boldsymbol{k},t) J_n(W) e^{\imath n(\psi-\phi_0) + \imath(k_\| v_\| + n\epsilon\Omega)t} , \tag{12.2.3a}$$

and

$$\delta E_{\mathrm{L,R},\|} \simeq \sum_{n=-\infty}^{\infty} \int_{-\infty}^{\infty} d^3k \; E_{\mathrm{L,R},\|}(\boldsymbol{k},t) J_n(W) e^{\imath n(\psi-\phi_0) + \imath(k_\| v_\| + n\epsilon\Omega)t} . \tag{12.2.3b}$$

With these representations we derive for the fluctuating force fields (12.1.9), calculated for the unperturbed particle orbit, that

$$\bar{g}_{\mathrm{p}} = \frac{\epsilon \Omega pc}{vB_0} \sum_{n=-\infty}^{\infty}$$
$$\times \int_{-\infty}^{\infty} d^3k \; \exp\left[\imath n(\psi - \phi_0) + \imath(k_\| v_\| + n\epsilon\Omega)t\right]$$
$$\times \left(\mu J_n(W) E_\|(\boldsymbol{k},t) + \sqrt{\frac{1-\mu^2}{2}}\right.$$
$$\left. \times \left(E_{\mathrm{L}}(\boldsymbol{k},t) e^{-\imath\psi} J_{n-1}(W) + E_{\mathrm{R}}(\boldsymbol{k},t) e^{\imath\psi} J_{n+1}(W)\right)\right) , \tag{12.2.4a}$$

$$\bar{g}_\mu = \frac{\epsilon\Omega\sqrt{1-\mu^2}}{B_0} \sum_{n=-\infty}^{\infty} \int_{-\infty}^{\infty} d^3k \, \exp\left[in(\psi - \phi_0) + i(k_\| v_\| + n\epsilon\Omega)t\right]$$

$$\times \left(\frac{c}{v}\sqrt{1-\mu^2} J_n(W) E_\|(\mathbf{k},t) + \frac{i}{\sqrt{2}}\left[\left[B_R(\mathbf{k},t) + i\mu \frac{c}{v} E_R(\mathbf{k},t)\right]\right.\right.$$

$$\left.\left.\times e^{i\psi} J_{n+1}(W) - \left[B_L(\mathbf{k},t) - i\mu\frac{c}{v}E_L(\mathbf{k},t)\right] e^{-i\psi} J_{n-1}(W)\right]\right),$$

(12.2.4b)

$$\bar{g}_\phi = \frac{\epsilon\Omega}{B_0} \sum_{n=-\infty}^{\infty} \int_{-\infty}^{\infty} d^3k \, \exp\left[in(\psi - \phi_0) + i(k_\| v_\| + n\epsilon\Omega)t\right]$$

$$\times \left(-J_n(W)B_\|(\mathbf{k},t) + \frac{1}{\sqrt{2(1-\mu^2)}}\left[\left[\mu B_R(\mathbf{k},t) + i\frac{c}{v} E_R(\mathbf{k},t)\right]\right.\right.$$

$$\left.\left.\times e^{i\psi} J_{n+1}(W) + \left[\mu B_L(\mathbf{k},t) - i\frac{c}{v}E_L(\mathbf{k},t)\right] e^{-i\psi} J_{n-1}(W)\right]\right),$$

(12.2.4c)

$$\bar{g}_X = -\frac{v\sqrt{1-\mu^2}}{2B_0} \sum_{n=-\infty}^{\infty} \int_{-\infty}^{\infty} d^3k \, \exp\left[in(\psi - \phi_0) + i(k_\| v_\| + n\epsilon\Omega)t\right]$$

$$\times B_\|(\mathbf{k},t)\left(J_{n+1}(W)e^{i\psi} + J_{n-1}(W)e^{-i\psi}\right)$$

$$+ \frac{ic}{\sqrt{2}B_0}\sum_{n=-\infty}^{\infty}\int_{-\infty}^{\infty} d^3k \, \exp\left[in(\psi-\phi_0)+i(k_\| v_\|+n\epsilon\Omega)t\right] J_n(W)$$

$$\times \left[E_R - \frac{i\mu v}{c}B_R - \left(E_L + \frac{i\mu v}{c}B_L\right)\right],$$

(12.2.4d)

$$\bar{g}_Y = \frac{iv\sqrt{1-\mu^2}}{2B_0}\sum_{n=-\infty}^{\infty}\int_{-\infty}^{\infty} d^3k \, \exp\left[in(\psi-\phi_0)+i(k_\| v_\|+n\epsilon\Omega)t\right]$$

$$\times B_\|(\mathbf{k},t)\left(J_{n+1}(W)e^{i\psi} - J_{n-1}(W)e^{-i\psi}\right)$$

$$- \frac{c}{\sqrt{2}B_0}\sum_{n=-\infty}^{\infty}\int_{-\infty}^{\infty} d^3k \, \exp\left[in(\psi-\phi_0)+i(k_\| v_\|+n\epsilon\Omega)t\right] J_n(W)$$

$$\times \left[E_R - \frac{i\mu v}{c}B_R + \left(E_L + \frac{i\mu v}{c}B_L\right)\right],$$

(12.2.4e)

$$\bar{g}_Z = 0.$$ (12.2.4f)

Using these unperturbed equations of motion we now proceed to calculate the Fokker-Planck coefficients according to (12.1.29). Because of lack of space we cannot write down here all 25 coefficients in detail; we rather illustrate the necessary computations for one of them, $D_{\mu\mu}$. By using (12.2.4b) we obtain

12.2 Quasilinear Fokker-Planck Coefficients

$$D_{\mu\mu} = \frac{\Omega^2}{B_0^2}(1-\mu^2)\,\Re \sum_{n=-\infty}^{\infty} \int_{-\infty}^{\infty} d^3k \int_0^{\infty} d\xi \exp\left[-\imath(k_\| v_\| + n\epsilon\Omega)\xi\right]$$

$$\times \left\{ \frac{c^2}{v^2}(1-\mu^2)J_n^2(W)R_{\|\|}(\boldsymbol{k},\xi) + \frac{1}{2}J_{n+1}^2(W) \right.$$

$$\times \left(P_{\mathrm{RR}}(\boldsymbol{k},\xi) + \mu^2\frac{c^2}{v^2}R_{\mathrm{RR}}(\boldsymbol{k},\xi) + \imath\mu\frac{c}{v}\left[Q_{\mathrm{RR}}(\boldsymbol{k},\xi) - T_{\mathrm{RR}}(\boldsymbol{k},\xi)\right]\right)$$

$$+ \frac{1}{2}J_{n-1}^2(W)\left(P_{\mathrm{LL}}(\boldsymbol{k},\xi) + \mu^2\frac{c^2}{v^2}R_{\mathrm{LL}}(\boldsymbol{k},\xi)\right.$$

$$\left. + \imath\mu\frac{c}{v}\left[T_{\mathrm{LL}}(\boldsymbol{k},\xi) - Q_{\mathrm{LL}}(\boldsymbol{k},\xi)\right]\right) - \frac{1}{2}J_{n-1}(W)J_{n+1}(W)\left[e^{2\imath\psi}\right.$$

$$\times \left(P_{\mathrm{RL}}(\boldsymbol{k},\xi) - \mu^2\frac{c^2}{v^2}R_{\mathrm{RL}}(\boldsymbol{k},\xi) + \imath\mu\frac{c}{v}\left[T_{\mathrm{RL}}(\boldsymbol{k},\xi) + Q_{\mathrm{RL}}(\boldsymbol{k},\xi)\right]\right)$$

$$+ e^{-2\imath\psi}\left(P_{\mathrm{LR}}(\boldsymbol{k},\xi) - \mu^2\frac{c^2}{v^2}R_{\mathrm{LR}}(\boldsymbol{k},\xi)\right.$$

$$\left.\left. - \imath\mu\frac{c}{v}\left[T_{\mathrm{LR}}(\boldsymbol{k},\xi) + Q_{\mathrm{LR}}(\boldsymbol{k},\xi)\right]\right)\right] + \frac{\imath c\sqrt{1-\mu^2}}{\sqrt{2}v}J_n(W)$$

$$\times \left[J_{n+1}(W)\left(e^{\imath\psi}T_{\mathrm{R}\|}(\boldsymbol{k},\xi) - e^{-\imath\psi}Q_{\|\mathrm{R}}(\boldsymbol{k},\xi)\right.\right.$$

$$\left. + \imath\mu\frac{c}{v}\left[R_{\mathrm{R}\|}(\boldsymbol{k},\xi)e^{\imath\psi} + R_{\|\mathrm{R}}(\boldsymbol{k},\xi)e^{-\imath\psi}\right]\right)$$

$$+ J_{n-1}(W)\left(e^{\imath\psi}Q_{\|\mathrm{L}}(\boldsymbol{k},\xi) - e^{-\imath\psi}T_{\mathrm{L}\|}(\boldsymbol{k},\xi)\right.$$

$$\left.\left.\left. + \imath\mu\frac{c}{v}\left[R_{\|\mathrm{L}}(\boldsymbol{k},\xi)e^{\imath\psi} + R_{\mathrm{L}\|}(\boldsymbol{k},\xi)e^{-\imath\psi}\right]\right)\right]\right\}, \qquad (12.2.5)$$

where we have made the assumption that Fourier components at different wave vectors are uncorrelated, so that we introduced the correlation tensors of the electric and magnetic field fluctuations

$$\left\langle B_\alpha(\boldsymbol{k},t)B_\beta^*(\boldsymbol{k}',t+\xi)\right\rangle = \delta(\boldsymbol{k}-\boldsymbol{k}')P_{\alpha,\beta}(\boldsymbol{k},\xi),$$

$$\left\langle E_\alpha(\boldsymbol{k},t)B_\beta^*(\boldsymbol{k}',t+\xi)\right\rangle = \delta(\boldsymbol{k}-\boldsymbol{k}')Q_{\alpha,\beta}(\boldsymbol{k},\xi),$$

$$\left\langle B_\alpha(\boldsymbol{k},t)E_\beta^*(\boldsymbol{k}',t+\xi)\right\rangle = \delta(\boldsymbol{k}-\boldsymbol{k}')T_{\alpha,\beta}(\boldsymbol{k},\xi),$$

$$\left\langle E_\alpha(\boldsymbol{k},t)E_\beta^*(\boldsymbol{k}',t+\xi)\right\rangle = \delta(\boldsymbol{k}-\boldsymbol{k}')R_{\alpha,\beta}(\boldsymbol{k},\xi). \qquad (12.2.6)$$

We also have averaged in (12.2.5) over the initial phase ϕ_0.

12.2.2 Fokker-Planck Coefficients for Plasma Wave Turbulence

Further reduction of the Fokker-Planck coefficient (12.2.5) can be achieved only if additional assumptions are made about the properties of the plasma turbulence, especially the time dependence of the correlation functions, in order to perform the ξ-integration. Guided by our test wave analysis (Chaps. 8–11) we follow here the "wave viewpoint" approach for the electromagnetic turbulence that represents the Fourier transforms of the magnetic and electric field fluctuations as superpositions of N individual plasma modes of frequencies (compare with e.g. (10.3.51))

$$\omega = \omega_j(\boldsymbol{k}) = \omega_{R,j}(\boldsymbol{k}) - \imath \Gamma_j(\boldsymbol{k}) , \qquad (12.2.7)$$

$j = 1, ..., N$, which can have both real and imaginary parts, so that

$$[\boldsymbol{B}(\boldsymbol{k}, t), \boldsymbol{E}(\boldsymbol{k}, t)] = \sum_{j=1}^{N} \left[\boldsymbol{B}^j(\boldsymbol{k}), \boldsymbol{E}^j(\boldsymbol{k})\right] \exp[-\imath \omega_j t] . \qquad (12.2.8)$$

Note that throughout this chapter with (12.2.7) we count damping of the waves with a positive $\Gamma^j > 0$. Maxwell's induction law relates

$$\boldsymbol{B}^j(\boldsymbol{k}) = \frac{c}{\omega_j} \boldsymbol{k} \times \boldsymbol{E}^j(\boldsymbol{k}) . \qquad (12.2.9)$$

As a consequence of the representation (12.2.8), the magnetic correlation tensor (12.2.6a) becomes

$$P_{\alpha\beta}(\boldsymbol{k}, \xi) = \sum_{j=1}^{N} P^j_{\alpha\beta}(\boldsymbol{k}) \exp(+\imath \omega^j_{r,\beta}(\boldsymbol{k})\xi - \Gamma^j_\beta(\boldsymbol{k})\xi) , \qquad (12.2.10a)$$

where

$$P^j_{\alpha\beta}(\boldsymbol{k}) = \left\langle B^j_\alpha(\boldsymbol{k}) B^{j*}_\beta(\boldsymbol{k}_s) \right\rangle \delta(\boldsymbol{k} - \boldsymbol{k}_s) . \qquad (12.2.10b)$$

Similar relations hold for the other three correlation tensors (12.2.6b)–(12.2.6d). Through (12.2.7)–(12.2.10) we mathematically describe the coupling of the cosmic ray particles to the background plasma: the coupling proceeds via the properties of the possible plasma modes of the electromagnetic turbulence that affect the cosmic rays via the Fokker-Planck coefficients, but whose properties are determined by the wave-carrying background medium, as we have discussed in detail in Chaps. 9 and 10. [As an aside we note that different representations of the plasma turbulence, as, for example, the dynamical magnetic turbulence approach of Bieber and Matthaeus (1992 [50]), would modify the correlation tensor description (12.2.9), see Schlickeiser and Achatz 1993a [462].]

12.2 Quasilinear Fokker-Planck Coefficients

Using the plasma wave turbulence approach (12.2.10) and making the assumption that the Fourier components of different plasma modes are uncorrelated, the Fokker-Planck coefficient (12.2.5) reduces to

$$D_{\mu\mu} = \frac{\Omega^2}{B_0^2}(1-\mu^2) \sum_{j=1}^{N} \sum_{n=-\infty}^{\infty} \int_{-\infty}^{\infty} d^3k \, \mathcal{R}_j(\boldsymbol{k}, \omega_j)$$

$$\times \left\{ \frac{c^2}{v^2}(1-\mu^2) J_n^2(W) R_{\|\|}^j(\boldsymbol{k}) + \frac{1}{2} J_{n+1}^2(W) \right.$$

$$\times \left(P_{RR}^j(\boldsymbol{k}) + \mu^2 \frac{c^2}{v^2} R_{RR}^j(\boldsymbol{k}) + i\mu \frac{c}{v} \left[Q_{RR}^j(\boldsymbol{k}) - T_{RR}^j(\boldsymbol{k}) \right] \right)$$

$$+ \frac{1}{2} J_{n-1}^2(W) \left(P_{LL}^j(\boldsymbol{k}) + \mu^2 \frac{c^2}{v^2} R_{LL}^j(\boldsymbol{k}) + i\mu \frac{c}{v} \left[T_{LL}^j(\boldsymbol{k}) - Q_{LL}^j(\boldsymbol{k}) \right] \right)$$

$$- \frac{1}{2} J_{n-1}(W) J_{n+1}(W) \left[e^{2i\psi} \right.$$

$$\times \left(P_{RL}^j(\boldsymbol{k}) - \mu^2 \frac{c^2}{v^2} R_{RL}^j(\boldsymbol{k}) + i\mu \frac{c}{v} \left[T_{RL}^j(\boldsymbol{k}) + Q_{RL}^j(\boldsymbol{k}) \right] \right)$$

$$+ e^{-2i\psi} \left(P_{LR}^j(\boldsymbol{k}) - \mu^2 \frac{c^2}{v^2} R_{LR}^j(\boldsymbol{k}) - i\mu \frac{c}{v} \left[T_{LR}^j(\boldsymbol{k}) + Q_{LR}^j(\boldsymbol{k}) \right] \right) \right]$$

$$+ \frac{ic\sqrt{1-\mu^2}}{\sqrt{2}v} J_n(W) \left[J_{n+1}(W) \left(e^{i\psi} T_{R\|}^j(\boldsymbol{k}) - e^{-i\psi} Q_{\|R}^j(\boldsymbol{k}) \right) \right.$$

$$+ i\mu \frac{c}{v} \left[R_{R\|}^j(\boldsymbol{k}) e^{i\psi} + R_{\|R}^j(\boldsymbol{k}) e^{-i\psi} \right] \right) + J_{n-1}(W) \left(e^{i\psi} Q_{\|L}^j(\boldsymbol{k}) \right.$$

$$\left. \left. - e^{-i\psi} T_{L\|}^j(\boldsymbol{k}) + i\mu \frac{c}{v} \left[R_{\|L}^j(\boldsymbol{k}) e^{i\psi} + R_{L\|}^j(\boldsymbol{k}) e^{-i\psi} \right] \right) \right] \right\}, \quad (12.2.11)$$

with the Breit-Wigner type resonance function

$$\mathcal{R}_j(\boldsymbol{k}, \omega_j) \equiv \frac{\Gamma_j(\boldsymbol{k})}{\Gamma_j^2(\boldsymbol{k}) + (v\mu k_\| - \omega_{R,j} + n\Omega)^2} . \quad (12.2.12)$$

In the case of negligible damping $\Gamma_j \to \infty$, use of the δ-function representation

$$\lim_{\Gamma \to 0} \frac{\Gamma}{\Gamma^2 + \chi^2} = \pi \delta(\chi)$$

reduces the resonance function (12.2.12) to

$$\mathcal{R}_j(\Gamma_j = 0) = \pi \delta(v\mu k_\| - \omega_{R,j} + n\Omega) . \quad (12.2.13)$$

Performing the same manipulations on the other Fokker-Planck coefficients we derive for a selected few

$$D_{\mu p} = D_{p\mu} = \frac{\Omega^2 (1-\mu^2)^{1/2}}{v B_0^2} pc \, \Re \sum_j \sum_{n=-\infty}^{\infty} \int d^3k \, \mathcal{R}_j(\boldsymbol{k}, \omega_j)$$

$$\times \frac{v\mu k_\| - \omega_{R,j} + n\Omega}{\Gamma_j} \left\{ -\imath \frac{c}{v} \mu \left(1-\mu^2\right)^{1/2} J_n^2(W) R_{\|\|}^j(\boldsymbol{k}) \right.$$

$$+ \frac{\left(1-\mu^2\right)^{1/2}}{2} \left[J_{n+1}^2(W) \left(T_{RR}^j(\boldsymbol{k}) + \imath\mu \frac{c}{v} R_{RR}^j(\boldsymbol{k})\right) \right.$$

$$- J_{n-1}^2 \left(\frac{k_\perp v_\perp}{|\Omega|}\right) \left(T_{LL}^j(\boldsymbol{k}) - \imath\mu \frac{c}{v} R_{LL}^j(\boldsymbol{k})\right)$$

$$+ J_{n+1}(W) J_{n-1}(W) \left[e^{2\imath\psi} \left(T_{RL}^j(\boldsymbol{k}) + \imath\mu \frac{c}{v} R_{RL}^j(\boldsymbol{k})\right) \right.$$

$$\left.\left. - e^{-2\imath\psi} \left(T_{LR}^j(\boldsymbol{k}) - \imath\mu \frac{c}{v} R_{LR}^j(\boldsymbol{k})\right) \right] \right]$$

$$+ \frac{1}{\sqrt{2}} J_{n-1}(W) J_n(W) \left[\mu e^{-\imath\psi} \left(-T_{L\|}^j(\boldsymbol{k}) + \imath\mu \frac{c}{v} R_{L\|}^j(\boldsymbol{k})\right) \right.$$

$$\left. - \imath\left(1-\mu^2\right) \frac{c}{v} e^{\imath\psi} R_{\|L}^j(\boldsymbol{k}) \right]$$

$$+ \frac{1}{\sqrt{2}} J_{n+1}(W) J_n(W) \left[\mu e^{\imath\psi} \left(T_{R\|}^j(\boldsymbol{k}) + \imath\mu \frac{c}{v} R_{R\|}^j(\boldsymbol{k})\right) \right.$$

$$\left.\left. - \imath\left(1-\mu^2\right) \frac{c}{v} e^{-\imath\psi} R_{\|R}^j(\boldsymbol{k}) \right] \right\} , \qquad (12.2.14)$$

$$D_{pp} = \frac{\Omega^2 p^2 c^2}{B_0^2 v^2} \Re \sum_j \sum_{n=-\infty}^{\infty} \int d^3 k \, \mathcal{R}_j(\boldsymbol{k}, \omega_j) \left\{ \mu^2 J_n^2(W) R_{\|\|}^j(\boldsymbol{k}) \right.$$

$$+ \frac{1-\mu^2}{2} \left[J_{n+1}^2(W) R_{RR}^j(\boldsymbol{k}) + J_{n-1}^2(W) R_{LL}^j(\boldsymbol{k}) \right.$$

$$\left. + J_{n-1}(W) J_{n+1}(W) \left[R_{LR}^j(\boldsymbol{k}) e^{-2\imath\psi} + R_{RL}^j(\boldsymbol{k}) e^{2\imath\psi} \right] \right]$$

$$+ \frac{\mu\left(1-\mu^2\right)^{1/2}}{\sqrt{2}} J_n(W) \left[J_{n-1}(W) \left(e^{\imath\psi} R_{\|L}^j(\boldsymbol{k}) + e^{-\imath\psi} R_{L\|}^j(\boldsymbol{k})\right) \right.$$

$$\left.\left. + J_{n+1}(W) \left(e^{-\imath\psi} R_{\|R}^j(\boldsymbol{k}) + e^{\imath\psi} R_{R\|}^j(\boldsymbol{k})\right) \right] \right\} , \qquad (12.2.15)$$

$$D_{YY} = \frac{1}{2B_0^2} \sum_{j=1}^{N} \sum_{n=-\infty}^{\infty} \int_{-\infty}^{\infty} d^3 k \, \mathcal{R}_j(\boldsymbol{k}, \omega_j) \left(\frac{v^2}{2}\left(1-\mu^2\right) P_{\|\|}^j(\boldsymbol{k}) \right.$$

$$\times \left[J_{n+1}^2(W) + J_{n-1}^2(W) - J_{n+1}(W) J_{n-1}(W) \left(e^{2\imath\psi} + e^{-2\imath\psi}\right) \right]$$

$$+ \imath c v \sqrt{(1-\mu^2)/2} J_n(W) \left\{ \left(J_{n+1}(W) e^{\imath\psi} - J_{n-1}(W) e^{-\imath\psi}\right) \right.$$

12.2 Quasilinear Fokker-Planck Coefficients

$$\times \left[T^j_{\|\mathrm{R}}(\boldsymbol{k}) + T^j_{\|\mathrm{L}}(\boldsymbol{k}) + \imath\mu \frac{v}{c} \left[P^j_{\|\mathrm{R}}(\boldsymbol{k}) - P^j_{\|\mathrm{L}}(\boldsymbol{k}) \right] \right]$$

$$+ \left(J_{n-1}(W)\mathrm{e}^{\imath\psi} - J_{n+1}(W)\mathrm{e}^{-\imath\psi} \right)$$

$$\times \left[Q^j_{\mathrm{R}\|}(\boldsymbol{k}) + Q^j_{\mathrm{L}\|}(\boldsymbol{k}) + \imath\mu \frac{v}{c} \left[P^j_{\mathrm{L}\|}(\boldsymbol{k}) - P^j_{\mathrm{R}\|}(\boldsymbol{k}) \right] \right] \Bigg\} + c^2 J_n^2(W)$$

$$\times \left(R^j_{\mathrm{RR}}(\boldsymbol{k}) + \imath\mu \frac{v}{c} \left[Q^j_{\mathrm{RR}}(\boldsymbol{k}) - T^j_{\mathrm{RR}}(\boldsymbol{k}) \right] + \mu^2 \frac{v^2}{c^2} P^j_{\mathrm{RR}}(\boldsymbol{k}) \right.$$

$$+ R^j_{\mathrm{LL}}(\boldsymbol{k}) + \imath\mu \frac{v}{c} \left[T^j_{\mathrm{LL}}(\boldsymbol{k}) - Q^j_{\mathrm{LL}}(\boldsymbol{k}) \right] + \mu^2 \frac{v^2}{c^2} P^j_{\mathrm{LL}}(\boldsymbol{k})$$

$$+ R^j_{\mathrm{RL}}(\boldsymbol{k}) - \imath\mu \frac{v}{c} \left[T^j_{\mathrm{RL}}(\boldsymbol{k}) + Q^j_{\mathrm{RL}}(\boldsymbol{k}) \right] - \mu^2 \frac{v^2}{c^2} P^j_{\mathrm{RL}}(\boldsymbol{k})$$

$$\left. + R^j_{\mathrm{LR}}(\boldsymbol{k}) + \imath\mu \frac{v}{c} \left[T^j_{\mathrm{LR}}(\boldsymbol{k}) + Q^j_{\mathrm{LR}}(\boldsymbol{k}) \right] - \mu^2 \frac{v^2}{c^2} P^j_{\mathrm{LR}}(\boldsymbol{k}) \right) \Bigg),$$
(12.2.16)

$$D_{XX} = \frac{1}{2B_0^2} \sum_{j=1}^{N} \sum_{n=-\infty}^{\infty} \int_{-\infty}^{\infty} \mathrm{d}^3 k \, \mathcal{R}_j(\boldsymbol{k}, \omega_j) \left(\frac{v^2}{2} (1-\mu^2) P^j_{\|\|}(\boldsymbol{k}) \right.$$

$$\times \left[J_{n+1}^2(W) + J_{n-1}^2(W) + J_{n+1}(W) J_{n-1}(W) \left(\mathrm{e}^{2\imath\psi} + \mathrm{e}^{-2\imath\psi} \right) \right]$$

$$+ \imath c v \sqrt{(1-\mu^2)/2}\, J_n(W) \Bigg\{ \left(J_{n+1}(W)\mathrm{e}^{\imath\psi} + J_{n-1}(W)\mathrm{e}^{-\imath\psi} \right)$$

$$\times \left[T^j_{\|\mathrm{R}}(\boldsymbol{k}) - T^j_{\|\mathrm{L}}(\boldsymbol{k}) + \imath\mu \frac{v}{c} \left[P^j_{\|\mathrm{R}}(\boldsymbol{k}) + P^j_{\|\mathrm{L}}(\boldsymbol{k}) \right] \right]$$

$$= - \left(J_{n-1}(W)\mathrm{e}^{\imath\psi} + J_{n+1}(W)\mathrm{e}^{-\imath\psi} \right)$$

$$\times \left[Q^j_{\mathrm{R}\|}(\boldsymbol{k}) - Q^j_{\mathrm{L}\|}(\boldsymbol{k}) - \imath\mu \frac{v}{c} \left[P^j_{\mathrm{L}\|}(\boldsymbol{k}) + P^j_{\mathrm{R}\|}(\boldsymbol{k}) \right] \right] \Bigg\} + c^2 J_n^2(W)$$

$$\times \left(R^j_{\mathrm{RR}}(\boldsymbol{k}) + \imath\mu \frac{v}{c} \left[Q^j_{\mathrm{RR}}(\boldsymbol{k}) - T^j_{\mathrm{RR}}(\boldsymbol{k}) \right] + \mu^2 \frac{v^2}{c^2} P^j_{\mathrm{RR}}(\boldsymbol{k}) \right.$$

$$+ R^j_{\mathrm{LL}}(\boldsymbol{k}) + \imath\mu \frac{v}{c} \left[T^j_{\mathrm{LL}}(\boldsymbol{k}) - Q^j_{\mathrm{LL}}(\boldsymbol{k}) \right] + \mu^2 \frac{v^2}{c^2} P^j_{\mathrm{LL}}(\boldsymbol{k})$$

$$- R^j_{\mathrm{RL}}(\boldsymbol{k}) + \imath\mu \frac{v}{c} \left[T^j_{\mathrm{RL}}(\boldsymbol{k}) + Q^j_{\mathrm{RL}}(\boldsymbol{k}) \right] + \mu^2 \frac{v^2}{c^2} P^j_{\mathrm{RL}}(\boldsymbol{k})$$

$$\left. - R^j_{\mathrm{LR}}(\boldsymbol{k}) - \imath\mu \frac{v}{c} \left[T^j_{\mathrm{LR}}(\boldsymbol{k}) + Q^j_{\mathrm{LR}}(\boldsymbol{k}) \right] + \mu^2 \frac{v^2}{c^2} P^j_{\mathrm{LR}}(\boldsymbol{k}) \right) \Bigg),$$
(12.2.17)

$$D_{\phi\phi} = \frac{\Omega^2}{B_0^2} \sum_{j=1}^{N} \sum_{n=-\infty}^{\infty} \int_{-\infty}^{\infty} \mathrm{d}^3 k \, \mathcal{R}_j(\boldsymbol{k}, \omega_j) \left\{ J_n^2(W) P^j_{\|\|}(\boldsymbol{k}) + \frac{1}{2(1-\mu^2)} \right.$$

$$\times \left[J_{n+1}^2(W) \left(\mu^2 P_{\mathrm{RR}}^j(\boldsymbol{k}) + \frac{c^2}{v^2} R_{\mathrm{RR}}^j(\boldsymbol{k}) + \imath\mu \frac{c}{v} \left[Q_{\mathrm{RR}}^j(\boldsymbol{k}) - T_{\mathrm{RR}}^j(\boldsymbol{k}) \right] \right) \right.$$

$$+ J_{n-1}^2(W) \left(\mu^2 P_{\mathrm{LL}}^j(\boldsymbol{k}) + \frac{c^2}{v^2} R_{\mathrm{LL}}^j(\boldsymbol{k}) - \imath\mu \frac{c}{v} \left[Q_{\mathrm{LL}}^j(\boldsymbol{k}) - T_{\mathrm{LL}}^j(\boldsymbol{k}) \right] \right)$$

$$+ J_{n-1}(W) J_{n+1}(W) \left[e^{2\imath\psi} \right.$$

$$\times \left(\mu^2 P_{\mathrm{RL}}^j(\boldsymbol{k}) - \frac{c^2}{v^2} R_{\mathrm{RL}}^j(\boldsymbol{k}) + \imath\mu \frac{c}{v} \left[T_{\mathrm{RL}}^j(\boldsymbol{k}) + Q_{\mathrm{RL}}^j(\boldsymbol{k}) \right] \right)$$

$$+ e^{-2\imath\psi} \left(\mu^2 P_{\mathrm{LR}}^j(\boldsymbol{k}) - \frac{c^2}{v^2} R_{\mathrm{LR}}^j(\boldsymbol{k}) - \imath\mu \frac{c}{v} \left[T_{\mathrm{LR}}^j(\boldsymbol{k}) + Q_{\mathrm{LR}}^j(\boldsymbol{k}) \right] \right) \right]$$

$$+ \frac{J_n(W)}{\sqrt{2(1-\mu^2)}} \left[J_{n+1}(W) \left(e^{-\imath\psi} \left[\mu P_{\|\mathrm{R}}^j(\boldsymbol{k}) - \frac{\imath c}{v} T_{\|\mathrm{R}}^j(\boldsymbol{k}) \right] \right. \right.$$

$$\left. + e^{\imath\psi} \left[\mu P_{\mathrm{R}\|}^j(\boldsymbol{k}) + \frac{\imath c}{v} Q_{\mathrm{R}\|}^j(\boldsymbol{k}) \right] \right) + J_{n-1}(W)$$

$$\left. \left. \times \left(e^{-\imath\psi} \left[\mu P_{\mathrm{L}\|}^j(\boldsymbol{k}) - \frac{\imath c}{v} Q_{\mathrm{L}\|}^j(\boldsymbol{k}) \right] + e^{\imath\psi} \left[\mu P_{\|\mathrm{L}}^j(\boldsymbol{k}) + \frac{\imath c}{v} T_{\|\mathrm{L}}^j(\boldsymbol{k}) \right] \right) \right] \right\} \ . \quad (12.2.18)$$

12.3 The Diffusion Approximation

The Fokker-Planck equation (12.1.26) with its 25 Fokker-Planck coefficients of the type (12.2.11)–(12.2.17) is very complicated and cannot be solved in most cases. From the illustrative Fokker-Planck coefficients given, and also the others, one notices, however, that in comparison to

$$\frac{\partial}{\partial \mu} D_{\mu\mu} \frac{\partial F_a}{\partial \mu} + \frac{\partial}{\partial \mu} D_{\mu\phi} \frac{\partial F_a}{\partial \phi} + \frac{\partial}{\partial \phi} D_{\phi\mu} \frac{\partial F_a}{\partial \mu} + \frac{\partial}{\partial \phi} D_{\phi\phi} \frac{\partial F_a}{\partial \phi} \quad (12.3.1)$$

(apart from the injection function S_a) all other terms on the right-hand side of (12.1.26) are of the order (V_{ph}/v), $(V_{\mathrm{ph}}/v)^2$, $(V_{\mathrm{ph}}/v)(R_{\mathrm{L}}/R)$ or $(R_{\mathrm{L}}/R)^2$, where $V_{\mathrm{ph}} = \omega_{\mathrm{R}}/k$ is the plasma waves' phase speed, $R_{\mathrm{L}} = v/\Omega$ is the gyroradius of a particle and R a typical length scale for the variation of F_a in X and Y. Therefore, if we consider only particle distributions which are weakly variable in X and Y, i.e. $R_{\mathrm{L}} \ll R$, we can say that the fastest particle-wave interaction processes are diffusion in gyrophase ϕ and pitch angle μ (described in (12.3.1)), since for low-frequency magnetohydrodynamical waves the phase speed V_{ph} (see Chap. 9) is much less than the individual cosmic ray particle speeds, i.e. $V_{\mathrm{ph}}/v \ll 1$. We can then follow the analysis of Jokipii (1966 [244]), Hasselmann and Wibberenz (1968 [218]), and Skilling (1975 [498]), and split the particle distribution function into its average in pitch angle and an anisotropic part:

12.3 The Diffusion Approximation

$$F_a(\boldsymbol{x}, p, \mu, \phi, t) = M_a(\boldsymbol{x}, p, t) + G_a(\boldsymbol{x}, p, \mu, \phi, t), \qquad (12.3.2)$$

where

$$M_a(\boldsymbol{x}, p, t) = \frac{1}{4\pi} \int_0^{2\pi} d\phi \int_{-1}^1 d\mu \, F_a(\boldsymbol{x}, p, \mu, \phi, t), \qquad (12.3.3)$$

$$\frac{1}{4\pi} \int_0^{2\pi} d\phi \int_{-1}^1 d\mu \, G_a(\boldsymbol{x}, p, \mu, \phi, t) = 0. \qquad (12.3.4)$$

Substituting (12.3.2) in the Fokker-Planck equation (12.1.26) and averaging over μ and ϕ we obtain using (12.3.3)–(12.3.4)

$$\begin{aligned}
&\frac{\partial M_a}{\partial t} + \frac{v}{4\pi} \frac{\partial}{\partial Z} \int_0^{2\pi} d\phi \int_{-1}^1 d\mu\mu \, G_a - S_a(\boldsymbol{x}, \boldsymbol{p}, t) \\
&= \frac{1}{p^2} \frac{\partial}{\partial p} \left[\frac{1}{4\pi} \int_0^{2\pi} d\phi \int_{-1}^1 d\mu \, p^2 \left(D_{pp} \frac{\partial M_a}{\partial p} + D_{pX} \frac{\partial M_a}{\partial X} + D_{pY} \frac{\partial M_a}{\partial Y} \right) \right. \\
&\quad \times \frac{\partial}{\partial X} \left[\frac{1}{4\pi} \int_0^{2\pi} d\phi \int_{-1}^1 d\mu \, \left(D_{Xp} \frac{\partial M_a}{\partial p} + D_{XX} \frac{\partial M_a}{\partial X} + D_{XY} \frac{\partial M_a}{\partial Y} \right) \right. \\
&\quad + \frac{\partial}{\partial Y} \left[\frac{1}{4\pi} \int_0^{2\pi} d\phi \int_{-1}^1 d\mu \, \left(D_{Yp} \frac{\partial M_a}{\partial p} + D_{YX} \frac{\partial M_a}{\partial X} + D_{YY} \frac{\partial M_a}{\partial Y} \right) \right. \\
&\quad + \left. \frac{1}{p^2} \frac{\partial}{\partial x_i}_{i=(p,X,Y)} \left[\frac{1}{4\pi} \int_0^{2\pi} d\phi \int_{-1}^1 d\mu \, p^2 \left(D_{ij} \frac{\partial G_a}{\partial x_j} \right)_{j=(p,\mu,\phi,X,Y)} \right) \right],
\end{aligned}$$
$$(12.3.5)$$

where we have used $D_{\mu x_i} = 0$ for $|\mu| = 1$ and the periodicity $G_a(\phi) = G_a(\phi + 2\pi)$. The diffusion approximation applies if the particle densities are slowly varying in time and space, i.e.

$$\frac{\partial M_a}{\partial t} = o\left(\frac{M_a}{T}\right), \quad \frac{\partial G_a}{\partial t} = o\left(\frac{G_a}{T}\right), \qquad (12.3.6a)$$

$$\frac{\partial M_a}{\partial Z} = o\left(\frac{M_a}{L}\right), \quad \frac{\partial G_a}{\partial Z} = o\left(\frac{G_a}{L}\right), \qquad (12.3.6b)$$

where $T \gg \tau \simeq o(1/D_{\mu\mu})$ and $L \gg v\tau$, where τ denotes the pitch angle relaxation time. In this case, the cosmic ray particles have time to adjust locally to a near-isotropic equilibrium, so that $G_a \ll M_a$. Subtracting the averaged Fokker-Planck equation (12.3.5) from the full Fokker-Planck equation (12.1.26) and keeping only terms which are at most of first-order in the small quantities (τ/T), $(v\tau/L)$, (V_{ph}/v), (G_a/M_a) and (R_L/R), we obtain to lowest order an approximation for the anisotropy G_a in terms of the isotropic distribution function M_a:

$$\left[\epsilon \Omega + \frac{\partial}{\partial \mu} D_{\mu\phi}\right] \frac{\partial G_a}{\partial \phi} + \frac{\partial}{\partial \phi} \left(D_{\phi\phi} \frac{\partial G_a}{\partial \phi}\right) + \frac{\partial}{\partial \mu} \left(D_{\mu\mu} \frac{\partial G_a}{\partial \mu}\right)$$

$$\simeq v\mu \frac{\partial M_a}{\partial Z} - \frac{\partial}{\partial \mu}\left(D_{\mu p}\frac{\partial M_a}{\partial p}\right) - \frac{\partial}{\partial \mu}\left(D_{\mu X}\frac{\partial M_a}{\partial X}\right)$$
$$- \frac{\partial}{\partial \mu}\left(D_{\mu Y}\frac{\partial M_a}{\partial Y}\right) - \frac{\partial D_{\phi X}}{\partial \phi}\frac{\partial M_a}{\partial X} - \frac{\partial D_{\phi Y}}{\partial \phi}\frac{\partial M_a}{\partial Y} . \tag{12.3.7}$$

Since G_a is periodic in the gyrophase we can express it as a Fourier series

$$G_a(\boldsymbol{x},p,\mu,\phi) = \sum_{s=-\infty}^{\infty} G_s(\boldsymbol{x},p,\mu) \exp(\imath s\phi) . \tag{12.3.8}$$

For our purposes only G_{-1}, G_0 and G_1 are of interest since all other terms do not contribute to the averaged Fokker-Planck equation (12.3.5) because all Fokker-Planck coefficients have a periodicity in gyrophase with a maximal period of 2π. Inserting (12.3.8) into (12.3.7) and comparing coefficients we derive the two equations

$$\frac{\partial}{\partial \mu}\left[D_{\mu\mu}\frac{\partial G_0}{\partial \mu} + D_{\mu p}\frac{\partial M_a}{\partial p}\right] = v\mu\frac{\partial M_a}{\partial Z} , \tag{12.3.9}$$

and

$$[\imath\epsilon\Omega - D_{\phi\phi}] G_1 + \frac{\partial}{\partial \mu}D_{\mu\mu}\frac{\partial G_1}{\partial \mu} = \Lambda , \tag{12.3.10}$$

where

$$\Lambda = -\left[\frac{\partial D_{\mu X}}{\partial \mu} + \frac{\partial D_{\phi X}}{\partial \phi}\right]\frac{\partial M_a}{\partial X} - \left[\frac{\partial D_{\mu Y}}{\partial \mu} + \frac{\partial D_{\phi Y}}{\partial \phi}\right]\frac{\partial M_a}{\partial Y} , \tag{12.3.11}$$

and where we neglected $D_{\mu\phi}$ in comparison to Ω. The equation for G_{-1} is not necessary since the anisotropy G_a is real and therefore the complex conjugate $G_1^* = G_{-1}$. Since $\Omega/D_{\mu\mu} \simeq o(B_0^2/\delta B^2)$ we may neglect the pitch angle diffusion term in (12.3.10) to obtain

$$G_1 \simeq \frac{\Lambda}{\imath\epsilon\Omega - D_{\phi\phi}} \simeq \frac{\Lambda}{\imath\epsilon\Omega} , \tag{12.3.12}$$

since, again, gyrorotation is faster than gyrophase diffusion. Combining (12.3.11) and (12.3.12) we note that

$$G_1 \simeq o\left(\frac{(\delta B)^2}{B_0^2}M_a\right) \tag{12.3.13}$$

would be of second-order in the small parameter $(\delta B/B_0)^2 \ll 1$. Reinserting G_1 and G_{-1} into the averaged Fokker-Planck equation (12.3.5) would then yield terms of the order $(\delta B/B_0)^4$. However, the Fokker-Planck equation was derived by a truncation of the order $(\delta B/B_0)^2$. For consistency, G_1 has to be neglected and we approximate (12.3.8) by

$$G_a \simeq G_0 , \tag{12.3.14}$$

where G_0 obeys (12.3.9).

12.3.1 Cosmic Ray Anisotropy

Integrating (12.3.9) over μ we obtain

$$D_{\mu\mu}\frac{\partial G_a}{\partial \mu} + D_{\mu p}\frac{\partial M_a}{\partial p} = c_1 + \frac{v\mu^2}{2}\frac{\partial M_a}{\partial Z}, \qquad (12.3.15)$$

where the integration constant c_1 is determined from the requirement that the left-hand side of (12.3.15) vanishes for $\mu = \pm 1$, yielding

$$c_1 = -\frac{v}{2}\frac{\partial M_a}{\partial Z}. \qquad (12.3.16)$$

By using (12.3.16) in (12.3.15) we derive

$$\frac{\partial G_a}{\partial \mu} = -\frac{1-\mu^2}{2D_{\mu\mu}}v\frac{\partial M_a}{\partial Z} - \frac{D_{\mu p}}{D_{\mu\mu}}\frac{\partial M_a}{\partial p}. \qquad (12.3.17)$$

Integrating this equation gives for the anisotropy

$$G_a(\mu) = c_2 - \frac{v}{2}\frac{\partial M_a}{\partial Z}\int_{-1}^{\mu} d\nu \frac{1-\nu^2}{D_{\mu\mu}(\nu)} - \frac{\partial M_a}{\partial p}\int_{-1}^{\mu} d\nu \frac{D_{\mu p}(\nu)}{D_{\mu\mu}(\nu)}, \qquad (12.3.18)$$

where the integration constant c_2 is determined by condition (12.3.4). We then obtain for the anisotropy

$$G_a(\mu) = \frac{v}{4}\frac{\partial M_a}{\partial Z}\left[\int_{-1}^{1} d\mu \frac{(1-\mu^2)(1-\mu)}{D_{\mu\mu}(\mu)} - 2\int_{-1}^{\mu} d\nu \frac{1-\nu^2}{D_{\mu\mu}(\nu)}\right]$$
$$+ \frac{1}{2}\frac{\partial M_a}{\partial p}\left[\int_{-1}^{1} d\mu\,(1-\mu)\frac{D_{\mu p}(\mu)}{D_{\mu\mu}(\mu)} - 2\int_{-1}^{\mu} d\nu \frac{D_{\mu p}(\nu)}{D_{\mu\mu}(\nu)}\right]. \qquad (12.3.19)$$

Apparently the anisotropy (12.3.19) is made up of two components. The first is related to pitch angle scattering and the spatial gradient of the isotropic distribution M. This term will provide the spatial diffusion of particles. Evidently, pitch angle diffusion produces spatial effects in a plasma if a density gradient along the ordered magnetic field, $\partial M_a/\partial Z$, is present. The second contribution to the anisotropy (12.3.18) stems from the momentum gradient of M_a and is related to the Compton-Getting effect (Compton and Getting 1935 [103]).

Expanding the anisotropy (12.3.19) into orthonormal Legendre polynomials $P_l(\mu)$, i.e.

$$G_a(\mu, Z, p) = \sum_{l=0}^{\infty} A_l(Z,p) P_l(\mu), \qquad (12.3.20)$$

defines the harmonics $A_l(Z,p)$ of the anisotropy.

12.3.2 The Diffusion-Convection Transport Equation

We insert (12.3.17) and (12.3.18) into the averaged Fokker-Planck equation (12.3.5). For the two μ-integrals we obtain

$$\int_{-1}^{1} d\mu \mu G_a(\mu) = -\frac{v}{2}\frac{\partial M_a}{\partial Z}\int_{-1}^{1} d\mu \mu \int_{-1}^{\mu} d\nu \frac{1-\nu^2}{D_{\mu\mu}(\nu)}$$

$$-\frac{\partial M_a}{\partial p}\int_{-1}^{1} d\mu \mu \int_{-1}^{\mu} d\nu \frac{D_{\mu p}(\nu)}{D_{\mu\mu}(\nu)}$$

$$= -\frac{v}{4}\frac{\partial M_a}{\partial Z}\int_{-1}^{1} d\mu \frac{(1-\mu^2)^2}{D_{\mu\mu}(\mu)}$$

$$-\frac{1}{2}\frac{\partial M_a}{\partial p}\int_{-1}^{1} d\mu \frac{(1-\mu^2)D_{\mu p}(\mu)}{D_{\mu\mu}(\mu)}, \qquad (12.3.21)$$

where we partially integrated the right-hand side, and

$$\int_{-1}^{1} d\mu D_{\mu p}\frac{\partial G_a}{\partial \mu} = -\frac{v}{2}\frac{\partial M_a}{\partial Z}\int_{-1}^{1} d\mu \frac{(1-\mu^2)D_{\mu p}(\mu)}{D_{\mu\mu}}$$

$$-\frac{\partial M_a}{\partial p}\int_{-1}^{1} d\mu \frac{D_{\mu p}^2(\mu)}{D_{\mu\mu}(\mu)}. \qquad (12.3.22)$$

Collecting terms in (12.3.5) we obtain

$$\frac{\partial M_a}{\partial t} - S_a(\boldsymbol{x},\boldsymbol{p},t) = \frac{\partial}{\partial z}\kappa_{zz}\frac{\partial M_a}{\partial z} - \frac{1}{4p^2}\frac{\partial(p^2 v A_1)}{\partial p}\frac{\partial M_a}{\partial z}$$

$$+\frac{\partial}{\partial X}\left[\kappa_{XX}\frac{\partial M_a}{\partial X} + \kappa_{XY}\frac{\partial M_a}{\partial Y}\right]$$

$$+\frac{\partial}{\partial Y}\left[\kappa_{YY}\frac{\partial M_a}{\partial Y} + \kappa_{YX}\frac{\partial M_a}{\partial X}\right]$$

$$+\frac{1}{p^2}\frac{\partial}{\partial p}\left(p^2 A_2 \frac{\partial M_a}{\partial p}\right) + \frac{v}{4}\frac{\partial A_1}{\partial z}\frac{\partial M_a}{\partial p}, \qquad (12.3.23)$$

where the components of the spatial diffusion tensor κ_{ij}, the rate of adiabatic deceleration A_1 and the momentum diffusion coefficient A_2 are determined by the pitch angle averages of Fokker-Planck coefficients as

$$\kappa_{zz} \equiv v\lambda/3 = \frac{v^2}{8}\int_{-1}^{1} d\mu \frac{(1-\mu^2)^2}{D_{\mu\mu}(\mu)}, \qquad (12.3.24)$$

$$\kappa_{XX} = \frac{1}{2}\int_{-1}^{1} d\mu\, D_{XX}(\mu), \qquad (12.3.25)$$

$$\kappa_{XY} = \frac{1}{2}\int_{-1}^{1} d\mu\, D_{XY}(\mu), \qquad (12.3.26)$$

$$\kappa_{YY} = \frac{1}{2}\int_{-1}^{1} d\mu \, D_{YY}(\mu) , \qquad (12.3.27)$$

$$\kappa_{YX} = \frac{1}{2}\int_{-1}^{1} d\mu \, D_{YX}(\mu) , \qquad (12.3.28)$$

$$A_1 = \int_{-1}^{1} d\mu \frac{(1-\mu^2) \, D_{\mu p}(\mu)}{D_{\mu\mu}} , \qquad (12.3.29)$$

and

$$A_2 = \frac{1}{2}\int_{-1}^{1} d\mu \left[D_{pp}(\mu) - \frac{D_{\mu p}^2(\mu)}{D_{\mu\mu}(\mu)} \right] . \qquad (12.3.30)$$

Equation (12.3.23) is commonly referred to as the *diffusion-convection equation* for the isotropic pitch-angle averaged particle distribution function $M_a(\mathbf{x}, p, t)$. In its most general form it contains spatial diffusion and convection terms, describing the propagation of cosmic rays in space, as well as momentum diffusion and convection terms, describing the acceleration of cosmic rays, i.e. the transport in momentum space. Whether in a specific physical situation all seven cosmic ray transport parameters (12.3.24)–(12.3.30) arise and are different from zero, depends solely on the nature and the statistical properties of the plasma turbulence and the background medium. In (12.3.24) we also defined the spatial diffusion coefficient along the ordered magnetic field κ_{zz} in terms of the mean free path for scattering λ.

12.3.2.1 Moving Background Medium. Very often the situation arises that the background plasma, supporting the plasma waves, moves with respect to the observer with the bulk speed U along the ordered magnetic field; e.g. the solar wind moves with respect to the Earth with velocity $\simeq 400$ km/s. However, the diffusion-convection equation (12.3.23) has been derived in the comoving reference frame, i.e. the rest system of the moving plasma. In this case it is convenient to apply the diffusion approximation in the mixed comoving coordinate system (see Kirk et al. 1988 [262]), in which the space coordinates ($\mathbf{x}$) are measured in the laboratory system and the particle's momentum coordinates (p^*, μ^*) are measured in the rest frame of the streaming plasma. This is particularly important in relativistic flows (see the discussion in Kirk et al. (1988 [262])).

For non-relativistic bulk speed $U \ll c$ we can apply a simple Galilean transformation to (12.3.23) and the diffusion-convection equation in the mixed comoving coordinate system becomes with $M_a(\mathbf{x}, p*)$ denoting the corresponding particle distribution function

$$\frac{\partial M_a}{\partial t} - S_a(\mathbf{x}, p, t) = \frac{\partial}{\partial z}\kappa_{zz}\frac{\partial M_a}{\partial z} - \left[U + \frac{1}{4p^{*2}}\frac{\partial(p^{*2}v^*A_1)}{\partial p^*} \right]\frac{\partial M_a}{\partial z}$$
$$+ \frac{\partial}{\partial X}\left[\kappa_{XX}\frac{\partial M_a}{\partial X} + \kappa_{XY}\frac{\partial M_a}{\partial Y} \right] + \frac{\partial}{\partial Y}\left[\kappa_{YY}\frac{\partial M_a}{\partial Y} + \kappa_{YX}\frac{\partial M_a}{\partial X} \right]$$

$$+ \frac{1}{p^{*2}} \frac{\partial}{\partial p^*} \left(p^{*2} A_2 \frac{\partial M_a}{\partial p^*} \right) + \left[\frac{p^*}{3} \frac{\partial U}{\partial z} + \frac{v^*}{4} \frac{\partial A_1}{\partial z} \right] \frac{\partial M_a}{\partial p^*} ,$$
(12.3.31)

where the transport parameters (12.3.24)–(12.3.30) remain unchanged but have to be calculated with the appropriate transformed momentum variables p^* and μ^*. Schlickeiser (1989c [459]) has shown that for non-relativistic flows (12.3.31) can also be derived from the untransformed (12.2.23) by incorporating the background plasma flow in the dispersion relation of the plasma waves.

13. Test Particle Approach 2. Calculation of Transport Parameters

In the preceding chapter we have developed the quasilinear theory for calculating the cosmic ray transport parameters for arbitrary plasma wave turbulence. As we have learned from the streaming instability studies in Sect. 11.2 in particular low-frequency Alfvén waves and magnetosonic waves, described by the dispersion relations (9.2.63), have short growth times in cosmic plasmas. We therefore calculate the cosmic ray transport parameters (12.3.24)–(12.3.30) for these two plasma modes. The plasma wave properties enter the Fokker-Planck coefficients through the electric and magnetic field correlation functions (12.2.6) which for individual plasma modes are related because of Maxwell's induction law (12.2.9).

13.1 Linearly Polarized Alfvén Waves

In the cold (electron temperature $T_e \to 0$) plasma limit, which is equivalent to the low beta limit since $\beta = 2c_e^2/V_A^2 = 8\pi n k_B T_e/B_0^2 \to 0$, the Alfvén waves have the dispersion relation at small enough wavenumber (see (9.2.63b))

$$\omega_{R,j}^2 = V_A^2 k_\|^2, \quad \omega_{R,j} = j V_A k_\|, \quad j = \pm 1 \tag{13.1.1}$$

at frequencies well below the proton gyrofrequency $\Omega_{0,p}$. As $|\omega_R|$ approaches $\Omega_{0,p}$, the simple relation (13.1.1) is no longer valid, and if the wave vector values $|\boldsymbol{k}|$ approaches infinity these waves are referred to as electromagnetic ion cyclotron waves, but they are on the same branch. For propagation along the ordered magnetic field the Alfvén wave corresponds to the left-hand polarized mode, discussed in detail in Sect. 9.2.3.2.

Without loss of generality we may choose the particular coordinate system where the y-component of the wave vector $k_y = 0$ vanishes, implying $\psi = \cot^{-1}(k_x/k_y) = 0$ in all Fokker-Planck coefficients (12.2.11)–(12.2.18). As mentioned already in Sect. 9.2.5.1 in this coordinate system

$$\boldsymbol{k} = (k_\perp, 0, k_\|) = k(\sin\theta, 0, \cos\theta) \tag{13.1.2}$$

the x-component is the only non-vanishing electric field component of the Alfvén waves

$$\delta \boldsymbol{E}_{\mathrm{A}}^{j} = (E_x, 0, 0) \,. \tag{13.1.3}$$

Since $\delta\boldsymbol{E}_{\mathrm{A}} \cdot \boldsymbol{k} = kE_x \sin\theta$ they are purely transversal waves for $\theta = 0$ but become more longitudinal as the propagation angle increases. Faraday's law (12.2.9) shows that the y-component is the only non-vanishing magnetic field component of the Alfvén waves

$$\delta \boldsymbol{B}_{\mathrm{A}}^{j} = \frac{ck}{\omega_{\mathrm{R},j}} E_x(0, \cos\theta, 0) = j\frac{c}{V_{\mathrm{A}}} E_x(0, \cos\theta, 0) \,. \tag{13.1.4}$$

According to (13.1.3)–(13.1.4) we find for the respective (see (12.1.10) and (12.2.3)) right- and left-handed polarized and the parallel components for Alfvén waves

$$E_{\mathrm{L}} = E_{\mathrm{R}}\,, \qquad E_{\|} = 0\,, \tag{13.1.5}$$
$$B_{\mathrm{L}} = -B_{\mathrm{R}}\,, \qquad B_{\|} = 0\,, \tag{13.1.6}$$

implying for the components of correlation tensors (12.2.6) that

$$-P_{\mathrm{LR}} = -P_{\mathrm{RL}} = P_{\mathrm{LL}} = P_{\mathrm{RR}}\,, \tag{13.1.7a}$$
$$Q_{\mathrm{LR}} = -Q_{\mathrm{RL}} = -Q_{\mathrm{LL}} = Q_{\mathrm{RR}}\,, \tag{13.1.7b}$$
$$-T_{\mathrm{LR}} = T_{\mathrm{RL}} = -T_{\mathrm{LL}} = T_{\mathrm{RR}}\,, \tag{13.1.7c}$$
$$R_{\mathrm{LR}} = R_{\mathrm{RL}} = R_{\mathrm{LL}} = R_{\mathrm{RR}}\,, \tag{13.1.7d}$$
$$R_{\|\|\|} = R_{\|\mathrm{R}} = R_{\|\mathrm{L}} = T_{\mathrm{L}\|} = T_{\mathrm{R}\|} = T_{\|\mathrm{L}} = T_{\|\mathrm{R}} = P_{\|\mathrm{L}} = P_{\|\mathrm{R}}$$
$$= P_{\|\|\|} = P_{\mathrm{L}\|} = P_{\mathrm{R}\|} = Q_{\|\mathrm{R}} = Q_{\|\mathrm{L}} = Q_{\mathrm{L}\|} = Q_{\mathrm{R}\|} = 0\,. \tag{13.1.7e}$$

Likewise (13.1.4) implies

$$E_x = \frac{\omega_{\mathrm{R},j}}{ck_\|} B_y\,, \tag{13.1.8a}$$

yielding for the two electric field components

$$E_{\mathrm{L}} = E_{\mathrm{R}} = \frac{\imath\omega_j}{ck_\|} B_{\mathrm{R}} = \frac{-\imath\omega_j}{ck_\|} B_{\mathrm{L}} = \frac{\imath\omega_j}{2ck_\|} [B_{\mathrm{R}} - B_{\mathrm{L}}]\,, \tag{13.1.8b}$$

which allows us to express the tensors $R_{\alpha\beta}^{j}(\boldsymbol{k})$, $Q_{\alpha\beta}^{j}(\boldsymbol{k})$ and $T_{\alpha\beta}^{j}(\boldsymbol{k})$ in terms of the magnetic field fluctuation tensor $P_{\alpha\beta}^{j}(\boldsymbol{k})$. In addition to (13.1.7) we obtain

$$R_{\mathrm{RR}} = \frac{\omega_j^2}{c^2 k_\|^2} P_{\mathrm{RR}} = \frac{V_{\mathrm{A}}^2}{c^2} P_{\mathrm{RR}}\,, \tag{13.1.9a}$$

$$Q_{\mathrm{RR}} = \frac{\imath\omega_j}{ck_\|} P_{\mathrm{RR}} = \frac{\imath j V_{\mathrm{A}}}{c} P_{\mathrm{RR}} \tag{13.1.9b}$$

and

$$T_{\mathrm{RR}} = -\frac{\imath\omega_j}{ck_\|} P_{\mathrm{RR}} = -\frac{\imath j V_{\mathrm{A}}}{c} P_{\mathrm{RR}}\,. \tag{13.1.9c}$$

13.1.1 Alfvén Wave Fokker-Planck Coefficients

Inserting the relations (13.1.7) and (13.1.9) into (12.2.11)–(12.2.18) we obtain for the Fokker-Planck coefficients of Alfvén waves after straightforward but tedious algebra

$$D_{\mu\mu} = \frac{2\Omega^2 \left(1-\mu^2\right)}{B_0^2} \sum_{j=\pm 1} \sum_{n=-\infty}^{\infty} \int d^3k \mathcal{R}_j \left(\boldsymbol{k}, \omega_j\right)$$

$$\times \left[1 - \frac{\mu \omega_j}{k_\| v}\right]^2 \left(\frac{n J_n(W)}{W}\right)^2 P_{\mathrm{RR}}^j(\boldsymbol{k}), \qquad (13.1.10)$$

$$D_{\mu p} = \frac{2\Omega^2 p \left(1-\mu^2\right)}{v B_0^2} \sum_{j=\pm 1} \sum_{n=-\infty}^{\infty} \int d^3k \mathcal{R}_j \left(\boldsymbol{k}, \omega_j\right) \frac{\omega_j}{k_\|}$$

$$\times \left[1 - \frac{\mu \omega_j}{k_\| v}\right] \left(\frac{n J_n(W)}{W}\right)^2 P_{\mathrm{RR}}^j(\boldsymbol{k}), \qquad (13.1.11)$$

$$D_{pp} = \frac{2\Omega^2 p^2 \left(1-\mu^2\right)}{v^2 B_0^2} \sum_{j=\pm 1} \sum_{n=-\infty}^{\infty} \int d^3k \mathcal{R}_j(\boldsymbol{k}, \omega_j)$$

$$\times \left(\frac{\omega_j}{k_\|}\right)^2 \left(\frac{n J_n(W)}{W}\right)^2 P_{\mathrm{RR}}^j(\boldsymbol{k}), \qquad (13.1.12)$$

$$D_{XX} = D_{XY} = D_{YX} = 0, \qquad (13.1.13)$$

and

$$D_{YY} = \frac{2}{B_0^2} \sum_{j=\pm 1} \sum_{n=-\infty}^{\infty} \int d^3k \mathcal{R}_j(\boldsymbol{k}, \omega_j) \left[\frac{\omega_j}{k_\|} - \mu v\right]^2 J_n^2(W) P_{\mathrm{RR}}^j(\boldsymbol{k}), \qquad (13.1.14)$$

where we recall from (12.2.2b) that

$$W = \frac{kv\sqrt{1-\mu^2}\sin\theta}{\Omega}. \qquad (13.1.15)$$

Equations (13.1.10)–(13.1.14) represent the exact Fokker-Planck coefficients for linearly polarized Alfvén waves and are expressed in terms of the (measurable) components of the magnetic fluctuation correlation tensor P_{RR}^j. For the quantitative analysis of the Fokker-Planck coefficients we have to specify the magnetic turbulence tensor $P_{lm}^j(\boldsymbol{k})$.

13.1.2 Magnetic Turbulence Tensors

We start to derive the general properties of the magnetic fluctuation tensor

$$P_{lm}^j(\boldsymbol{k}) = \left\langle B_l^j(\boldsymbol{k}) B_m^{j*}(\boldsymbol{k}) \right\rangle, \qquad (13.1.16)$$

where $B_{l,m}$ denote the Cartesian components of the fluctuating magnetic field $\boldsymbol{\delta B}$, making physical assumptions about the allowed directions of the plasma wave vectors.

13.1.2.1 Slab Turbulence.
Guided by the results of the calculations of the relevant growth rates (see Sect. 11.2) that waves propagating along the ordered magnetic field have the largest growth rate, we first assume that the wave vectors of the fluctuations are all either parallel or antiparallel to the background magnetic field. Then the magnetic fluctuation tensor has the form

$$P_{lm}^j(\mathbf{k}) = \begin{cases} g_s^j(k_\parallel)(\delta(k_\perp)/k_\perp)\left[\delta_{lm} + i\sigma^j(k_\parallel)\epsilon_{lm3}\right] & \text{for } l,m = 1,2 \\ 0 & \text{for } l,m = 3 \end{cases}, \tag{13.1.17}$$

where δ_{lm} is the Kronecker δ-symbol and ϵ_{ijl} is the total antisymmetric tensor of rank 3. The purely real quantity $\sigma^j(k_\parallel)$ can be identified with the magnetic helicity, and for linearly polarized waves has to be set to $\sigma = 0$ throughout. The function $g_s^j(k_\parallel)$ is determined by the normalization requirement that the magnetic energy density in wave component j is given by

$$(\delta B_j)^2 = \int d^3k \left(P_{11}^j(\mathbf{k}) + P_{22}^j(\mathbf{k}) + P_{33}^j(\mathbf{k})\right) = 4\pi \int_{-\infty}^{\infty} dk_\parallel \, g_s^j(k_\parallel). \tag{13.1.18}$$

The tensor (13.1.17) is commonly referred to as the *slab turbulence* model.

13.1.2.2 Isotropic Turbulence.
In the isotropic turbulence model it is demanded that the magnetic fluctuations are isotropic. Mathematically, this can be defined by the requirement that the nine tensor components $P_{lm}^j(\mathbf{k})$ transform like a rank-2 tensor in wavenumber space if $\mathbf{k}$ is rotated. There are only three independent rank-2 tensors that can be constructed by the components of the wave vector $\mathbf{k}$ alone, namely $a_{ij} = k_i k_j$, $b_{ij} = \delta_{ij}$ and $c_{ij} = \epsilon_{ijl} k_l$. The magnetic fluctuation tensor then must be a linear combination of these:

$$P_{lm}^j(\mathbf{k}) = A^j(k)\delta_{lm} + C^j(k)k_l k_m + H^j(k)\epsilon_{lmn}k_n, \tag{13.1.19}$$

where the functions A^j, C^j, H^j depend only on $k = |\mathbf{k}|$. Obviously the magnetic fluctuations fulfill $\nabla \cdot \delta \mathbf{B}(\mathbf{x}) = 0$ yielding

$$\mathbf{k} \cdot \mathbf{B}^j(\mathbf{k}) = 0 \tag{13.1.20a}$$

for their Fourier transforms. Relation (13.1.20a) implies

$$\left\langle (\mathbf{k} \cdot \mathbf{B}^j(\mathbf{k}))(\mathbf{k} \cdot \mathbf{B}^j(\mathbf{k}))^* \right\rangle = 0, \tag{13.1.20b}$$

from which we obtain with (13.1.19) that

$$A^j(k) = -C^j(k)k^2, \quad \Re H^j(k) = 0. \tag{13.1.21}$$

Thus the fluctuation tensor (13.1.19) is of the form

$$P^j_{lm}(\boldsymbol{k}) = \frac{g^j(k)}{k^2}\left[\delta_{lm} - \frac{k_l k_m}{k^2} + i\sigma^j(k)\epsilon_{lmn}\frac{k_n}{k}\right]. \qquad (13.1.22)$$

The purely real quantity $\sigma^j(k)$ can be identified with the magnetic helicity. For linearly polarized waves we therefore have to set $\sigma = 0$ throughout, so that (13.1.22) reduces to

$$\begin{aligned}P^j_{lm}(\boldsymbol{k}) &= \frac{g^j_i(k)}{8\pi k^2}\left[\delta_{lm} - \frac{k_l k_m}{k^2}\right] \\ &= \frac{g^j_i(k)}{8\pi k^2}\begin{pmatrix} \cos^2\theta & 0 & -\sin\theta\cos\theta \\ 0 & 1 & 0 \\ -\sin\theta\cos\theta & 0 & \sin^2\theta \end{pmatrix}, \end{aligned} \qquad (13.1.23)$$

where in the last step we used the wavenumber representation (13.1.2). The function $g^j_i(k)$ can be determined again from the normalization (13.1.18) which in this case yields

$$\begin{aligned}(\delta B_j)^2 &= \int d^3k \left(P^j_{11}(\boldsymbol{k}) + P^j_{22}(\boldsymbol{k}) + P^j_{33}(\boldsymbol{k})\right) \\ &= \int d^3k \frac{g^j(k)}{4\pi k^2} = \int_0^\infty dk\, g^j(k). \end{aligned} \qquad (13.1.24)$$

More general magnetic turbulence models can be generated, see the discussion in Jaekel and Schlickeiser (1992 [236]) and Jaekel et al. (1994 [237]).

13.1.2.3 Kolmogorov-Type Turbulence Spectra. When illustrating our results we will often adopt a Kolmogorov-type power law dependence (index $q > 1$) of g^j_s and g^j_i above some minimum wavenumber $k_{\|,\min}$ and $k_{\min}$, respectively, i.e.

$$g^j_s(k_\|) = g^j_{s0}k_\|^{-q} \quad \text{for } k_\| > k_{\|,\min}, \qquad (13.1.25)$$

and

$$g^j_i(k) = g^j_{i0}k^{-q} \quad \text{for } k > k_{\min}, \qquad (13.1.26)$$

since this form is suggested by in-situ observations in the interplanetary medium (compare with Fig. 15.3). The normalizations (13.1.18) and (13.1.24) then imply

$$g^j_{s0} = \frac{q-1}{4\pi}(\delta B_j)^2 k_{\|,\min}^{q-1}, \qquad (13.1.27)$$

and

$$g^j_{i0} = (q-1)(\delta B_j)^2 k_{\min}^{q-1}, \qquad (13.1.28)$$

respectively.

13.1.3 Slab Linearly Polarized Alfvén Turbulence

From the slab turbulence tensor (13.1.17) we obtain for

$$P^j_{RR} = \frac{1}{2}\left[P^j_{11} + P^j_{22} + \imath\left(P^j_{12} - P^j_{21}\right)\right] = g^j_s(k_\parallel)\frac{\delta(k_\perp)}{k_\perp}, \qquad (13.1.29)$$

and with (12.2.12) the Fokker-Planck coefficients (13.1.10)–(13.1.14) reduce to

$$D_{\mu\mu} = \frac{\pi\Omega^2(1-\mu^2)}{B_0^2}\sum_{j=\pm 1}\int_{-\infty}^{\infty}dk_\parallel g^j_s(k_\parallel)\left[1 - \frac{\mu\omega_j}{k_\parallel v}\right]^2$$
$$\times \left[\frac{\Gamma_j}{\Gamma_j^2 + \left(v\mu k_\parallel - \omega_{R,j} + \Omega\right)^2} + \frac{\Gamma_j}{\Gamma_j^2 + \left(v\mu k_\parallel - \omega_{R,j} - \Omega\right)^2}\right], (13.1.30)$$

$$D_{\mu p} = \frac{\pi\Omega^2 p(1-\mu^2)}{vB_0^2}\sum_{j=\pm 1}\int_{-\infty}^{\infty}dk_\parallel g^j_s(k_\parallel)\frac{\omega_j}{k_\parallel}\left[1 - \frac{\mu\omega_j}{k_\parallel v}\right]$$
$$\times \left[\frac{\Gamma_j}{\Gamma_j^2 + \left(v\mu k_\parallel - \omega_{R,j} + \Omega\right)^2} + \frac{\Gamma_j}{\Gamma_j^2 + \left(v\mu k_\parallel - \omega_{R,j} - \Omega\right)^2}\right], (13.1.31)$$

$$D_{pp} = \frac{\pi\Omega^2 p^2(1-\mu^2)}{v^2 B_0^2}\sum_{j=\pm 1}\int_{-\infty}^{\infty}dk_\parallel g^j_s(k_\parallel)\left(\frac{\omega_j}{k_\parallel}\right)^2$$
$$\times \left[\frac{\Gamma_j}{\Gamma_j^2 + \left(v\mu k_\parallel - \omega_{R,j} + \Omega\right)^2} + \frac{\Gamma_j}{\Gamma_j^2 + \left(v\mu k_\parallel - \omega_{R,j} - \Omega\right)^2}\right] \quad (13.1.32)$$

and

$$D_{YY} = \frac{4\pi}{B_0^2}\sum_{j=\pm 1}\int_{-\infty}^{\infty}dk_\parallel\frac{g^j_s(k_\parallel)}{k_\parallel^2}\left[\omega_j - \mu v k_\parallel\right]^2\frac{\Gamma_j}{\Gamma_j^2 + \left(v\mu k_\parallel - \omega_{R,j}\right)^2}. \qquad (13.1.33)$$

13.1.3.1 Undamped Waves.
In the case of negligible damping ($\Gamma_j = 0$) the use of (12.2.13) simplifies the Fokker-Planck coefficients further:

$$D_{\mu\mu} = \sum_{j=\pm 1} D(j), \qquad (13.1.34)$$

$$D_{\mu p} = \sum_{j=\pm 1} \frac{j\epsilon p}{1 - j\epsilon\mu} D(j), \qquad (13.1.35)$$

$$D_{pp} = \sum_{j=\pm 1} \frac{\epsilon^2 p^2}{(1 - j\epsilon\mu)^2} D(j), \qquad (13.1.36)$$

$$D_{YY} = 0, \qquad (13.1.37)$$

where $\epsilon = V_A/v$ and

$$D(j) \equiv \frac{\pi \Omega^2 (1-\mu^2)}{v B_0^2} \frac{(1-j\epsilon\mu)^2}{|\mu - j\epsilon|} \left[g_s^j(k_r^j) + g_s^j(-k_r^j) \right] , \qquad (13.1.38)$$

where

$$k_r^j = \frac{\Omega/v}{\mu - j\epsilon} = \frac{R_{\rm L}^{-1}}{\mu - j\epsilon} \qquad (13.1.39)$$

is the resonant parallel wavenumber in terms of the cosmic ray particles gyroradius $R_{\rm L}$. We defer the detailed discussion of these Fokker-Planck coefficients and the implied cosmic ray transport parameters to Sect. 13.2.2 where we include non-zero magnetic helicity values for slab waves.

13.1.4 Isotropic Linearly Polarized Alfvén Waves

From the isotropic turbulence tensor (13.1.23) we find

$$P_{\rm RR}^j = \frac{1}{2}\left[P_{11}^j + P_{22}^j + \imath\left(P_{12}^j - P_{21}^j\right)\right] = \frac{g^j(k)}{16\pi k^2}\left[1+\eta^2\right] , \qquad (13.1.40)$$

with

$$\eta = \cos\theta . \qquad (13.1.41)$$

For the Fokker-Planck coefficients (13.1.10)–(13.1.14) in this case we obtain

$$\begin{Bmatrix} D_{\mu\mu} \\ D_{\mu p} \\ D_{pp} \end{Bmatrix} = \frac{\Omega^2(1-\mu^2)}{4B_0^2} \sum_{n=-\infty}^{\infty} \sum_{j=\pm 1} \int_{-1}^{1} d\eta (1+\eta^2) \int_0^{\infty} dk\, g^j(k)$$

$$\times \frac{\Gamma_j}{\Gamma_j^2 + (vk\mu\eta - \omega_{{\rm R},j} + n\Omega)^2} \left(\frac{nJ_n(W)}{W}\right)^2 \begin{Bmatrix} \left[1 - \frac{\mu\omega_j}{\eta k v}\right]^2 \\ \frac{p}{v}\frac{\omega_j}{\eta k}\left[1 - \frac{\mu\omega_j}{\eta k v}\right] \\ \left(\frac{p}{v}\frac{\omega_j}{\eta k}\right)^2 \end{Bmatrix} \qquad (13.1.42)$$

and

$$D_{YY} = \frac{1}{4B_0^2} \sum_{n=-\infty}^{\infty} \sum_{j=\pm 1} \int_{-1}^{1} d\eta \frac{1+\eta^2}{\eta^2} \int_0^{\infty} dk\, g^j(k) k^{-2} J_n^2(W)$$

$$\times [\omega_j - \mu\eta k v]^2 \frac{\Gamma_j}{\Gamma_j^2 + (vk\mu\eta - \omega_{{\rm R},j} + n\Omega)^2} , \qquad (13.1.43)$$

where

$$W = kR_{\rm L}\sqrt{(1-\mu^2)(1-\eta^2)} . \qquad (13.1.44)$$

13.1.4.1 Undamped Waves. In the case of negligible damping ($\Gamma_j = 0$) the use of (12.2.13) and (13.1.1) simplifies these Fokker-Planck coefficients further:

$$\begin{Bmatrix} D_{\mu\mu} \\ D_{\mu p} \\ D_{pp} \end{Bmatrix} = \frac{\pi \Omega^2 (1 - \mu^2)}{4 B_0^2} \sum_{n=-\infty}^{\infty} \sum_{j=\pm 1} \begin{Bmatrix} [1 - j\mu\epsilon]^2 \\ j\epsilon p [1 - j\mu\epsilon] \\ (\epsilon p)^2 \end{Bmatrix}$$

$$\times \int_{-1}^{1} d\eta (1 + \eta^2) \int_{0}^{\infty} dk g^j(k) \delta \left([v\mu - jV_A] \eta k + n\Omega \right) \left(\frac{n J_n(W)}{W} \right)^2 \quad (13.1.45)$$

and

$$D_{YY} = \frac{\pi V_A^2}{4 B_0^2} \sum_{j=\pm 1} (1 - j\mu\epsilon)^2 \int_{-1}^{1} d\eta (1 + \eta^2) \sum_{n=-\infty}^{\infty}$$

$$\times \int_{0}^{\infty} dk g^j(k) J_n^2(W) \delta \left([v\mu - jV_A] \eta k + n\Omega \right), \quad (13.1.46)$$

where we used (13.1.37). The η-integration can be performed readily with the help of the δ-function, and we obtain after straightforward algebra

$$\begin{Bmatrix} D_{\mu\mu} \\ D_{\mu p} \\ D_{pp} \end{Bmatrix} = \frac{\pi \Omega^2}{2 v B_0^2} \sum_{j=\pm 1} |\mu - j\epsilon| \begin{Bmatrix} [1 - j\mu\epsilon]^2 \\ j\epsilon p [1 - j\mu\epsilon] \\ (\epsilon p)^2 \end{Bmatrix}$$

$$\times \sum_{n=1}^{\infty} \int_{1}^{\infty} dt t^{-3} \frac{t^2 + 1}{t^2 - 1} g^j \left(\frac{nt}{R_L |\mu - j\epsilon|} \right) J_n^2 \left(\frac{n\sqrt{1 - \mu^2}}{|\mu - j\epsilon|} \sqrt{t^2 - 1} \right) \quad (13.1.47)$$

and

$$D_{YY} = \frac{\pi V_A^2}{4 v B_0^2} \sum_{j=\pm 1} \frac{(1 - j\mu\epsilon)^2}{|\mu - j\epsilon|} \left[\int_{0}^{\infty} dk \frac{g^j(k)}{k} J_0^2 \left(k R_L \sqrt{1 - \mu^2} \right) \right.$$

$$\left. + 2 \sum_{n=1}^{\infty} \int_{1}^{\infty} dt \frac{t^2 + 1}{t^3} g^j \left(\frac{nt}{R_L |\mu - j\epsilon|} \right) J_n^2 \left(\frac{n\sqrt{1 - \mu^2}}{|\mu - j\epsilon|} \sqrt{t^2 - 1} \right) \right], \quad (13.1.48)$$

for general turbulence spectra $g^j(k)$.

13.2 Parallel Propagating Magnetohydrodynamic Waves

The dispersion relation (13.1.1) is only asymptotically correct at frequencies well below the proton gyrofrequency. In the case of parallel propagation we have discussed the exact dispersion relation of magnetohydrodynamic waves in Sect. 9.2.3.2 which included the left-handed polarized Alfvén-ion

13.2 Parallel Propagating Magnetohydrodynamic Waves

cyclotron branch and the right-handed polarized Alfvén-whistler-electron cyclotron branch. It is straightforward to modify the derivation of the Fokker-Planck coefficients presented in the preceding Sect. 13.1 for these special waves, using the slab turbulence tensor (13.1.17) for non-vanishing magnetic helicity ($\sigma^j(k_\parallel) \neq 0$) (see e.g. Achatz et al. 1991 [6]) to derive

$$\begin{Bmatrix} D_{\mu\mu} \\ D_{\mu p} \\ D_{pp} \end{Bmatrix} = \frac{\Omega^2(1-\mu^2)}{2B_0^2} \sum_{j=\pm 1} \sum_{n=1}^{\infty} \int d^3k \begin{Bmatrix} \left[1 - \frac{\mu\omega_j}{k_\parallel v}\right]^2 \\ \frac{p}{v}\frac{\omega_j}{k_\parallel}\left[1 - \frac{\mu\omega_j}{k_\parallel v}\right] \\ \left(\frac{p}{v}\frac{\omega_j}{k_\parallel}\right)^2 \end{Bmatrix}$$

$$\times \mathcal{R}_j(\mathbf{k},\omega_j) \left[J_{n+1}^2(W)\, P_{\mathrm{RR}}^j(\mathbf{k}) + J_{n-1}^2(W)\, P_{\mathrm{LL}}^j(\mathbf{k})\right] \qquad (13.2.1)$$

and

$$\begin{Bmatrix} D_{XX} \\ D_{YY} \\ D_{XY} \\ D_{YX} \end{Bmatrix} = \frac{1}{B_0^2} \sum_{j=\pm 1} \sum_{n=-\infty}^{\infty} \int d^3k\, \mathcal{R}_j(\mathbf{k},\omega_j) \left[\frac{\omega_j}{k_\parallel} - \mu v\right]^2 J_n^2(W)$$

$$\times \begin{Bmatrix} P_{\mathrm{RR}}^j - P_{\mathrm{RL}}^j - P_{\mathrm{LR}}^j + P_{\mathrm{LL}}^j \\ P_{\mathrm{RR}}^j + P_{\mathrm{RL}}^j + P_{\mathrm{LR}}^j + P_{\mathrm{LL}}^j \\ P_{\mathrm{RR}}^j - P_{\mathrm{RL}}^j - P_{\mathrm{LR}}^j + P_{\mathrm{LL}}^j \\ P_{\mathrm{RR}}^j + P_{\mathrm{RL}}^j - P_{\mathrm{LR}}^j - P_{\mathrm{LL}}^j \end{Bmatrix}. \qquad (13.2.2)$$

From the full slab turbulence tensor (13.1.17) we obtain

$$P_{\mathrm{RR}}^j = \frac{1}{2}\left[P_{11}^j + P_{22}^j + \imath\left(P_{12}^j - P_{21}^j\right)\right] = \left[1 - \sigma^j(k_\parallel)\right] g^j(k_\parallel) \frac{\delta(k_\perp)}{k_\perp}, \qquad (13.2.3a)$$

$$P_{\mathrm{LL}}^j = \frac{1}{2}\left[P_{11}^j + P_{22}^j - \imath\left(P_{12}^j - P_{21}^j\right)\right] = \left[1 + \sigma^j(k_\parallel)\right] g^j(k_\parallel) \frac{\delta(k_\perp)}{k_\perp}, \qquad (13.2.3b)$$

$$P_{\mathrm{RL,LR}}^j = \frac{1}{2}\left[P_{11}^j - P_{22}^j \mp \imath\left(P_{12}^j + P_{21}^j\right)\right] = 0, \qquad (13.2.3c)$$

so that with (12.2.12) the Fokker-Planck coefficients (13.2.1)–(13.2.2) become

$$\begin{Bmatrix} D_{\mu\mu} \\ D_{\mu p} \\ D_{pp} \end{Bmatrix} = \frac{\pi\Omega^2(1-\mu^2)}{B_0^2} \sum_{j=\pm 1} \int_{-\infty}^{\infty} dk_\parallel g_s^j(k_\parallel) \begin{Bmatrix} \left[1 - \frac{\mu\omega_j}{k_\parallel v}\right]^2 \\ \frac{p}{v}\frac{\omega_j}{k_\parallel}\left[1 - \frac{\mu\omega_j}{k_\parallel v}\right] \\ \left(\frac{p}{v}\frac{\omega_j}{k_\parallel}\right)^2 \end{Bmatrix}$$

$$\times \left[\frac{\Gamma_{j,\mathrm{R}}\left(1 - \sigma^j(k_\parallel)\right)}{\Gamma_{j,\mathrm{R}}^2 + \left(v\mu k_\parallel - \omega_{\mathrm{R}j,\mathrm{R}} - \Omega\right)^2} + \frac{\Gamma_{j,\mathrm{L}}\left(1 + \sigma^j(k_\parallel)\right)}{\Gamma_{j,\mathrm{L}}^2 + \left(v\mu k_\parallel - \omega_{\mathrm{R}j,\mathrm{L}} + \Omega\right)^2}\right] \qquad (13.2.4)$$

and

$$\begin{Bmatrix} D_{XX} \\ D_{YY} \\ D_{XY} \\ D_{YX} \end{Bmatrix} = \frac{2\pi}{B_0^2} \sum_{j=\pm 1} \sum_{n=-\infty}^{\infty} \int_{-\infty}^{\infty} dk_\parallel g_s^j(k_\parallel) \frac{\Gamma_j}{\Gamma_j^2 + \left(v\mu k_\parallel - \omega_{R,j}\right)^2}$$

$$\times \left[\frac{\omega_j}{k_\parallel} - \mu v\right]^2 J_n^2(W) \begin{Bmatrix} 1 \\ 1 \\ -\sigma^j \\ -\sigma^j \end{Bmatrix}. \qquad (13.2.5)$$

13.2.1 Undamped Waves

In the case of negligible damping ($\Gamma_j = 0$) the use of (12.2.13) yields readily

$$D_{XX} = D_{YY} = D_{XY} = D_{YX} = 0, \qquad (13.2.6)$$

and

$$\begin{Bmatrix} D_{\mu\mu} \\ D_{\mu p} \\ D_{pp} \end{Bmatrix} = \frac{\pi^2 \Omega^2 (1-\mu^2)}{B_0^2} \sum_{j=\pm 1} \int_{-\infty}^{\infty} dk_\parallel g_s^j(k_\parallel)$$

$$\times \left[\begin{Bmatrix} \left[1 - \frac{\mu\omega_{Rj,R}}{k_\parallel v}\right]^2 \\ \frac{p}{v}\frac{\omega_{Rj,R}}{k_\parallel}\left[1 - \frac{\mu\omega_{Rj,R}}{k_\parallel v}\right] \\ \left(\frac{p}{v}\frac{\omega_{Rj,R}}{k_\parallel}\right)^2 \end{Bmatrix} \left(1 - \sigma^j(k_\parallel)\right) \delta\left(v\mu k_\parallel - \omega_{Rj,R}(k_\parallel) - \Omega\right) \right.$$

$$\left. + \begin{Bmatrix} \left[1 - \frac{\mu\omega_{Rj,L}}{k_\parallel v}\right]^2 \\ \frac{p}{v}\frac{\omega_{Rj,L}}{k_\parallel}\left[1 - \frac{\mu\omega_{Rj,L}}{k_\parallel v}\right] \\ \left(\frac{p}{v}\frac{\omega_{Rj,L}}{k_\parallel}\right)^2 \end{Bmatrix} \left(1 + \sigma^j(k_\parallel)\right) \delta\left(v\mu k_\parallel - \omega_{Rj,L}(k_\parallel) + \Omega\right) \right].$$

$$(13.2.7)$$

In general the dispersion relation of undamped right-handed ($\omega_{Rj,R}$) and left-handed ($\omega_{Rj,L}$) circularly polarized waves is given by (9.2.35), which has to be inserted into (13.2.7).

13.2.2 Undamped Non-dispersive Alfvén Waves

In the Alfvénic part of the dispersion relation (9.2.35), i.e. at wavenumbers well below $|k_\parallel| \ll k_c = 2\Omega_p/V_A$, the approximation (9.2.38) holds:

$$\omega_{Rj,R} = \omega_{Rj,L} \simeq jV_A k_\parallel. \qquad (13.2.8)$$

In this case the Fokker-Planck coefficients (13.2.7) readily reduce to

13.2 Parallel Propagating Magnetohydrodynamic Waves

$$\left\{ \begin{matrix} D_{\mu\mu} \\ D_{\mu p} \\ D_{pp} \end{matrix} \right\} = \sum_{j=\pm 1} \left\{ \begin{matrix} 1 \\ j\epsilon p/1 - j\epsilon\mu \\ \epsilon^2 p^2/(1-j\epsilon\mu)^2 \end{matrix} \right\} D(j) \,, \qquad (13.2.9)$$

where

$$D(j) \equiv \frac{\pi^2 \Omega^2 (1-\mu^2)}{v B_0^2} \frac{(1-j\epsilon\mu)^2}{|\mu - j\epsilon|}$$
$$\times \left[\left(1 - \sigma^j(k_r^j)\right) g_s^j(k_r^j) + \left(1 + \sigma^j(-k_r^j)\right) g_s^j(-k_r^j) \right] \,, \quad (13.2.10)$$

where the resonant parallel wavenumber k_r^j is given by (13.1.39). Note that for vanishing magnetic helicity values ($\sigma^j = 0$), i.e. linearly polarized waves, (13.2.9)–(13.2.10) agree exactly with (13.1.34)–(13.1.38).

13.2.2.1 Influence of Magnetic and Cross Helicity. The ratio of the intensities of forward ($j = 1$) to backward ($j = -1$) propagating waves is referred to as the *normalized cross helicity state* H_c of the Alfvén waves (Dung and Schlickeiser 1990a [147])

$$H_c = \frac{g^{j=+1} - g^{j=-1}}{g^{j=+1} + g^{j=-1}} = 2\frac{g^{j=+1}}{g_{\text{tot}}} - 1 \,, \qquad (13.2.11)$$

where we also introduced the total wave intensity

$$g_{\text{tot}}(k_\|) = g^{j=1}(k_\|) + g^{j=-1}(k_\|) \,. \qquad (13.2.12)$$

The values of H_c vary between -1 and $+1$, attaining -1 if no forward moving waves are present, $+1$ if no backward moving waves are present.

Schlickeiser (1989b [458]), as well as Dung and Schlickeiser (1990a [147], b [148]), have investigated in detail the influence of cross and magnetic helicities both on the cosmic ray Fokker-Planck coefficients and their pitch angle averaged transport parameters, under the assumption that the helicity values are constants independent of the parallel wavenumber "isospectral turbulence"). In Fig. 13.1 we show the pitch angle Fokker-Planck coefficient $D_{\mu\mu}$ of energetic protons of velocity $v = 10V_A$ if only forward moving waves ($H_c = 1$) are present consisting of right- and left-handed waves of equal intensity ($\sigma_f = 0$). At one point in pitch angle $\mu = \epsilon = V_A/v = 0.1$ all three Fokker-Planck coefficients $D_{\mu\mu}(\mu = \epsilon) = D_{\mu p}(\mu = \epsilon) = D_{pp}(\mu = \epsilon) = 0$ become zero. Omitting for the discussion here small but finite scattering contributions from additional terms neglected so far, such a point is referred to as a resonance gap.

In Fig. 13.2 we show the pitch angle Fokker-Planck coefficient for the same protons now in a wave field consisting of only right-handed polarized waves ($\sigma_f = \sigma_b = \sigma - 1$) streaming with equal intensity into forward and backward directions, while Fig. 13.3 shows the result in a wave field consisting of left-handed polarized waves ($\sigma_f = \sigma_b = 1$) streaming with equal intensity ($H_c = 0$) into forward and backward directions. In both cases resonance

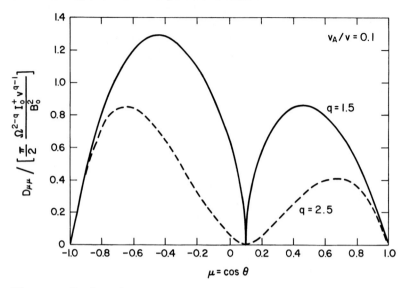

Fig. 13.1. Pitch angle Fokker-Planck coefficient $D_{\mu\mu}$ for energetic protons in a linearly polarized ($\sigma_f = 0$) Alfvén wave field of forward moving waves only ($H_c = 1$) assuming the Kolmogorov spectrum (13.1.25) and two values of the spectral index q

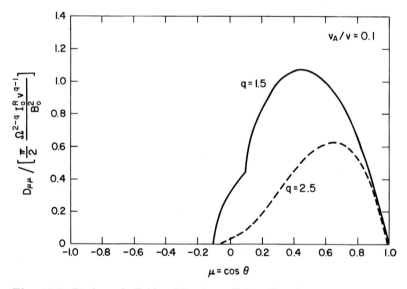

Fig. 13.2. Pitch angle Fokker-Planck coefficient $D_{\mu\mu}$ for energetic protons due to right-handed polarized ($\sigma_f = \sigma_b = -1$) Alfvén waves streaming with equal intensity into forward and backward direction ($H_c = 0$). The Kolmogorov spectrum (13.1.25) of magnetic irregularities is adopted with two different values of the spectral index q

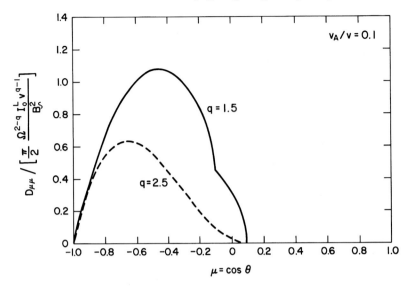

Fig. 13.3. Pitch angle Fokker-Planck coefficient $D_{\mu\mu}$ for energetic protons due to left-handed polarized ($\sigma_f = \sigma_b = 1$) Alfvén waves streaming with equal intensity into forward and backward direction ($H_c = 0$). The Kolmogorov spectrum (13.1.25) of magnetic irregularities is adopted with two different values of the spectral index q

gaps occur over the wide pitch angle range $-1 \leq \mu \leq -\epsilon$ for $\sigma = -1$ and $\epsilon \leq \mu \leq 1$ for left-handed waves. Particles at pitch angles within these gaps find no waves to interact with and therefore are not scattered, if no other (so far neglected) terms provide finite scattering.

Intermediate cases of pitch angle Fokker-Planck coefficients are shown in Fig. 13.4 where we assumed a linearly polarized Alfvén wave field with cross helicity values $H_c = \pm 1/3$ resulting in asymmetric pitch angle Fokker-Planck coefficients with respect to $\mu = 0$. Moreover, no resonance gap occurs in this case.

In Figs. 13.5 and 13.6 we plot the Fokker-Planck coefficients $D_{\mu p}$ and D_{pp} for energetic protons in a linearly polarized Alfvén wave field with zero cross helicity value $H_c = 0$. In this case the two coefficients are antisymmetric ($D_{\mu p}$) and symmetric (D_{pp}) with respect to $\mu = 0$.

The calculations reveal that resonance gaps do not occur if waves of both polarization state (i.e. $|\sigma_{f,b}| \neq 1$) traveling in both directions (i.e. $|H_c| \neq 1$) are present. With Alfvén waves of only one handedness propagating in one or both directions, a whole resonance gap interval occurs (see Figs. 13.2 and 13.3). With waves of both polarization states traveling in only one direction (Fig. 13.1), the resonance gap occurs at one point in μ-space.

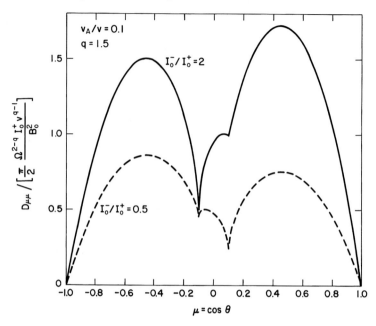

Fig. 13.4. Pitch angle Fokker-Planck coefficient $D_{\mu\mu}$ for energetic protons in a linearly polarized Alfvén wave field and cross helicity values $H_c = \pm 1/3$ calculated with the Kolmogorov spectrum (13.1.25) and two values of the spectral index q

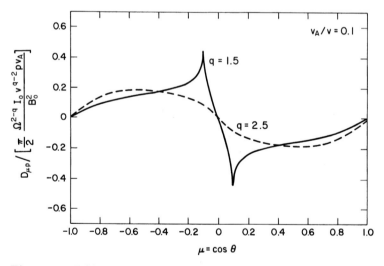

Fig. 13.5. Fokker-Planck coefficient $D_{\mu p}$ for energetic protons in a linearly polarized Alfvén wave field and zero cross helicity value $H_c = 0$ calculated with the Kolmogorov spectrum (13.1.25) and two values of the spectral index q

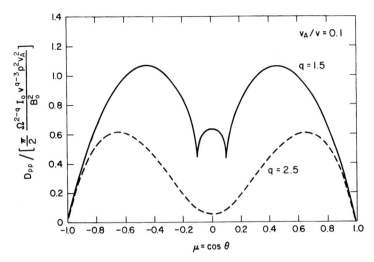

Fig. 13.6. Fokker-Planck coefficient D_{pp} for energetic protons in a linearly polarized Alfvén wave field and zero cross helicity value $H_c = 0$ calculated with the Kolmogorov spectrum (13.1.25) and two values of the spectral index q

13.2.2.2 Cosmic Ray Mean Free Path for Isospectral Kolmogorov Turbulence.
In the isospectral turbulence case where the helicities are independent of wavenumber $k_\parallel$ and for the Kolmogorov turbulence spectrum (13.1.25) the cosmic ray mean free path (12.3.24) can be expressed as

$$\lambda = \lambda_0 F(Z, H_c, \sigma_{f,b}) , \qquad (13.2.13)$$

where

$$\lambda_0 = \frac{3}{2\pi(q-1)} \left(\frac{B_0}{\delta B}\right)^2 (R_L k_{\min})^{1-q} R_L \qquad (13.2.14)$$

and the integral

$$F(Z, H_c, \sigma_{f,b}) \equiv \int_{-1}^{1} d\mu \frac{1-\mu^2}{N(\mu)} , \qquad (13.2.15)$$

with

$$\begin{aligned}N(\mu) = & \; r(1-\mu\epsilon)^2 |\mu - \epsilon|^{q-1} \\ & \times \left[(1+\sigma_f)\,\bar{\theta}\,[Z(\epsilon - \mu)] + (1-\sigma_f)\,\bar{\theta}\,[Z(\mu - \epsilon)]\right] \\ & + (1-r)(1+\mu\epsilon)^2 |\mu + \epsilon|^{q-1} \\ & \times \left[(1+\sigma_b)\bar{\theta}\,[Z(-\epsilon - \mu)] + (1-\sigma_b)\,\bar{\theta}\,[Z(\mu + \epsilon)]\right] . \quad (13.2.16)\end{aligned}$$

$Z = Q/|Q|$ accounts for positively and negatively charged particles, and $\bar{\theta}[x]$ denotes the Heaviside function.

$$r = \frac{1 + H_c}{2} \qquad (13.2.17)$$

denotes the fractional abundance of forward moving waves.

For energetic particles the ratio $\epsilon = V_A/v \ll 1$ is small compared to unity, and the integral (13.2.15) can be evaluated to lowest order in ϵ (Dung and Schlickeiser 1990a [147]) in the relevant range of turbulence spectral indices $1 < q < 3$:

$$F(Z, H_c, \sigma_{f,b}) \simeq \frac{2}{(2-q)(4-q)} \left[\frac{1}{a} + \frac{1}{\alpha}\right]$$
$$+ \epsilon^{2-q} \left[\frac{2}{d} - \frac{b}{a} - \frac{\beta}{\alpha} + \frac{1}{q-2}\left(\frac{1}{a} + \frac{1}{\alpha}\right)\right], \qquad (13.2.18)$$

with

$$a = r\,(1 - Z\sigma_f) + (1-r)\,(1 - Z\sigma_b), \qquad (13.2.19a)$$
$$ab = (1-r)\,(1 - Z\sigma_b) - r\,(1 - Z\sigma_f), \qquad (13.2.19b)$$
$$\alpha = r\,(1 + Z\sigma_f) + (1-r)\,(1 + Z\sigma_b), \qquad (13.2.19c)$$
$$\alpha\beta = r\,(1 + Z\sigma_f) - (1-r)\,(1 + Z\sigma_b), \qquad (13.2.19d)$$
$$d = r\,(1 + Z\sigma_f) + (1-r)\,(1 - Z\sigma_b). \qquad (13.2.19e)$$

Equations (13.2.18)–(13.2.19) are generalizations of the earlier results of Hasselmann and Wibberenz (1968 [218]) and Goldstein and Matthaeus (1991 [191]), who discussed the influence of the magnetic helicity on cosmic ray transport of Alfvén waves propagating with one phase speed only ($r = 1$ or $r = 0$), and Schlickeiser (1989b [458]) who specialized to linearly polarized Alfvén waves with vanishing magnetic helicity ($\sigma_f = \sigma_b = 0$) and equal intensities of forward and backward moving waves ($r = 1/2$).

For flat turbulence spectra $q < 2$ we may neglect the term proportional to $\epsilon^{2-q} \ll 1$ with respect to the first term in (13.2.18), and the variation of the function F (apart from the charge sign Z) is determined only by helicity parameters which are independent of the energetic particle properties. In Fig. 13.7 we show the function F for positively charged cosmic ray particles as a function of the cross helicity $H_c = 2r - 1$ for different magnetic helicities. The function F is constant and independent of r for equal magnetic helicities $\sigma_f = \sigma_b$. For non-equal magnetic helicities $\sigma_f \neq \sigma_b$ the function F attains its minimum value at $F_{\min}(q < 2) = 4/[(2-q)(4-q)]$ at $r_{\min} = [1 - (\sigma_f/\sigma_b)]^{-1}$. The maximum values occur for either $r = 0$ and $r = 1$ with $F(q < 2, r = 0) = [1 - \sigma_b^2]^{-1} F_{\min}$ and $F(q < 2, r = 1) = [1 - \sigma_f^2]^{-1} F_{\min}$.

For turbulence spectra with $q = 2$ we apply L'Hopital's rule to

$$\lim_{q \to 2} \left[\frac{2}{(2-q)(4-q)} + \frac{\epsilon^{2-q}}{q-2}\right] = \lim_{x \to 0} \frac{2 - (x+2)\exp[x\ln(\epsilon)]}{x(x+2)} = -\frac{1}{2} + \ln\epsilon^{-1},$$
$$(13.2.20)$$

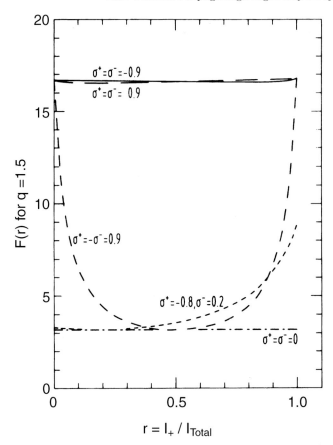

Fig. 13.7. Function F from (13.2.18) for positively charged particles with $\epsilon = 2 \times 10^{-4}$ as a function of r and different magnetic helicities in isospectral Kolmogorov turbulence with a flat spectrum $q = 1.5$

so that (13.2.18) is well approximated by

$$F(Z=1, q=2) \simeq \frac{2 \ln \epsilon^{-1}}{1 - [\sigma_b + r(\sigma_f - \sigma_b)]^2}. \qquad (13.2.21)$$

For equal magnetic helicities (13.2.21) is a constant with respect to r, in agreement with the variation shown in Fig. 13.8. For non-equal magnetic helicities (13.2.21) attains its minimum value $F_{\min}(q=2) = 2 \ln \epsilon^{-1}$ at $r_{\min} = [1 - (\sigma_f/\sigma_b)]^{-1}$.

For steep turbulence spectra $q > 2$ the term proportional to $\epsilon^{2-q} \gg 1$ dominates the first term in (13.2.18), and the values of the function F become very large due to its proportionality $\propto (v/V_A)^{q-2} \gg 1$ as can be seen also in Fig. 13.9.

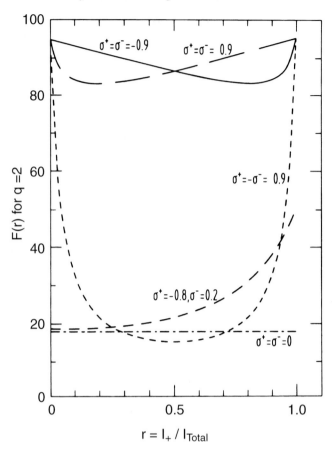

Fig. 13.8. Function F for positively charged particles with $\epsilon = 2 \times 10^{-4}$ as a function of r and different magnetic helicities in isospectral Kolmogorov turbulence with spectrum $q = 2.0$. The function F for $q = 2$ diverges at $r = 0$ and $r = 1$, and the calculated values should be regarded as the result at $r = 0 + \eta$ and $r = 1 - \eta$ with $\eta \ll 1$ small but finite

Finally, in Fig. 13.10 we show the resulting variation of the normalized mean free path $\lambda/\lambda_0(q)$ as a function of particle kinetic energy and total particle momentum for different values of the power spectrum spectral index q in a linearly polarized ($\sigma_f = \sigma_b = 0$) Alfvén wave field of zero cross helicity ($r = 1/2$). $\lambda_0(q)$ is related to (13.2.14) as

$$\lambda_0 = \lambda_0(q)(\gamma^2 - 1)^{(2-q)/2} \ . \tag{13.2.22}$$

It can be seen that the behavior is quite different for flat ($q < 2$) and steep ($q > 2$) turbulence spectra. For $q < 2$ we find that $\lambda(q < 2) \propto E_{\text{kin}}^{(2-q)/2}$ at non-relativistic particle energies and $\lambda(q < 2) \propto E_{\text{kin}}^{2-q}$ at relativistic energies.

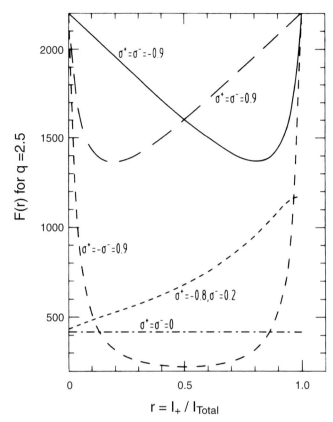

Fig. 13.9. Function F for positively charged particles with $\epsilon = 2 \times 10^{-4}$ as a function of r and different magnetic helicities in isospectral Kolmogorov turbulence with the steep spectrum $q = 2.5$. The function F for $q > 2$ diverges at $r = 0$ and $r = 1$, and the calculated values should be regarded as the result at $r = 0+\eta$ and $r = 1-\eta$ with $\eta \ll 1$ small but finite

However, for $q > 2$ the mean free path is constant at non-relativistic energies, but decreases $\lambda(q > 2) \propto E_{\text{kin}}^{-(q-2)}$ at relativistic energies.

Regarding its dependence on cosmic ray particle properties we note from (13.2.13)–(13.2.14) for $q < 2$, that the value of the mean free path (apart from all helicity values) is mainly determined by the Larmor radius of the cosmic ray particle, the ratio of energy densities in ordered and turbulent magnetic fields, and the ratio of the maximum Alfvén wavelength $l_{\text{max}} = 2\pi k_{\text{min}}^{-1}$ to the Larmor radius:

$$\lambda \simeq R_L \left(\frac{B_0}{\delta B}\right)^2 \left(\frac{l_{\text{max}}}{R_L}\right)^{q-1}. \tag{13.2.23}$$

The maximum Alfvén wave wavelength l_{max} is determined by the physics of turbulent input at large eddies. It certainly has to be smaller than the size L, of the astrophysical system considered. So using $l_{\text{max}} \leq L$ we obtain

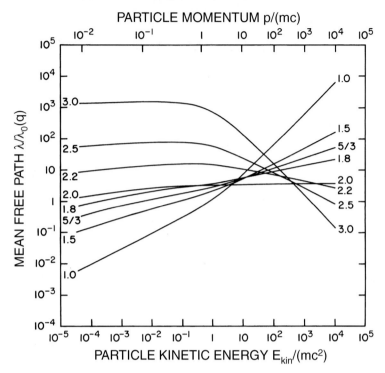

Fig. 13.10. Normalized mean free path of cosmic rays $\lambda/\lambda_0(q)$ as a function of particle kinetic energy and total particle momentum for different values of the power spectrum spectral index q in a linearly polarized Alfvén wave field of zero cross helicity

$$\lambda \simeq R_{\rm L} \left(\frac{B_0}{\delta B}\right)^2 \left(\frac{L}{R_{\rm L}}\right)^{q-1}. \qquad (13.2.24)$$

If applied to the case of the Galaxy with $B_0 = 3$ µG, $\delta B = 0.9$ µG, $L = 10^{21}$ cm, and $q = 1.5$, we find for protons a value

$$\lambda \simeq 3 \times 10^{17} \text{ cm} \left(\frac{L}{300 \text{ pc}}\right)^{0.5} \left(\frac{p}{1 \text{ GeV}/c}\right)^{0.5}, \qquad (13.2.25)$$

implying a spatial diffusion coefficient along the ordered field of

$$\kappa_{zz} \simeq 3 \times 10^{27} \text{ cm} \left(\frac{L}{300 \text{ pc}}\right)^{0.5} \beta \left(\frac{p}{1 \text{ GeV}/c}\right)^{0.5}, \qquad (13.2.26)$$

with $\beta = v/c$.

13.2.2.3 Rate of Adiabatic Deceleration for Isospectral Kolmogorov Turbulence.
In the isospectral Kolmogorov turbulence case the rate of adiabatic deceleration (12.3.29) can be expressed as

$$A_1 = \frac{4}{3}\epsilon p H(Z, H_c, \sigma_{f,b}), \tag{13.2.27}$$

with the integral

$$H(Z, H_c, \sigma_{f,b}) \equiv \frac{3}{4\epsilon} \int_{-1}^{1} d\mu (1-\mu^2) \frac{rl_+ + (1-r)l_-}{rh_+ + (1-r)h_-}, \tag{13.2.28}$$

where we used the same notation as above (13.2.17), and

$$l_\pm = \pm\epsilon\,(1\mp\mu\epsilon)\,|\mu\mp\epsilon|^{q-1}\,\bar{\theta}_\pm, \tag{13.2.29}$$
$$h_\pm = (1\mp\mu\epsilon)^2|\mu\mp\epsilon|^{q-1}\bar{\theta}_\pm, \tag{13.2.30}$$

and

$$\bar{\theta}_\pm = (1+\sigma_{f,b})\,\bar{\theta}\,[-Z(\mu\mp\epsilon)] \;+\; (1-\sigma_{f,b})\,\bar{\theta}\,[Z(\mu\mp\epsilon)]\,. \tag{13.2.31}$$

Again for energetic particles and turbulence spectral indices between $1 < q < 3$ we can evaluate the integral (13.2.28) to lowest order in $\epsilon = V_A/v \ll 1$ to obtain

$$H(r,\sigma_{f,b}) \simeq \frac{r^2(1-\sigma_f^2) - (1-r)^2(1-\sigma_b^2)}{r^2(1-\sigma_f^2) + (1-r)^2(1-\sigma_b^2) + 2r(1-r)(1-\sigma_f\sigma_b)}, \tag{13.2.32}$$

which is independent of the sign of the particle's charge (Z) and the turbulence spectral index (q).

In Fig. 13.11 we calculate the function $H(H_c, \sigma_{f,b})$ for various values of the helicities r, σ_f, and σ_b. It follows immediately from (13.2.32) that the function H takes its minimum values ($= -1$) at $r = 0$, and its maximum value ($= +1$) at $r = 1$. For every combination of magnetic helicities there exists a cross helicity r_0 where H vanishes, and where

$$r_0 = \frac{1}{\sigma_b^2 - \sigma_f^2}\left[\sigma_b^2 - 1 + \sqrt{(1-\sigma_b^2)(1-\sigma_f^2)}\right], \tag{13.2.33}$$

if $\sigma_f \neq \sigma_b$. For equal magnetic helicities ($\sigma_f = \sigma_b$) the value of $r_0 = 1/2$ and (13.2.28) reduces to the cross helicity value

$$H(r, \sigma_f = \sigma_b) = (2r - 1) = H_c\,. \tag{13.2.34}$$

13.2.2.4 Momentum Diffusion Coefficient for Isospectral Kolmogorov Turbulence.

In the isospectral Kolmogorov turbulence case the momentum diffusion coefficient (12.3.30) can be expressed as

$$A_2 = \frac{\epsilon^2 p^2}{\tau(p,q)} G(Z, H_c, \sigma_{f,b}), \tag{13.2.35}$$

with the time scale τ given by

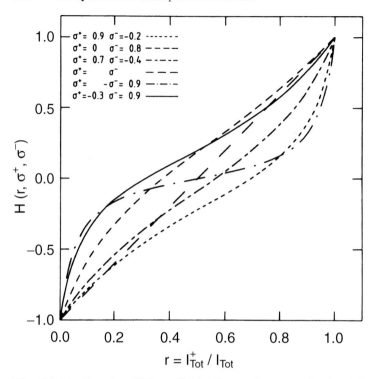

Fig. 13.11. Function H from (13.2.32) as a function of r for different magnetic helicities in the case of isospectral turbulence

$$\tau(p,q) = \frac{q(q+2)}{4(q-1)\pi|\Omega|} \left(\frac{B_0}{\delta B}\right)^2 (R_L k_{\min})^{1-q}, \qquad (13.2.36)$$

and the integral

$$G(Z, H_c, \sigma_{f,b}) \equiv \frac{q(q+2)}{16\epsilon^2} \int_{-1}^{1} d\mu \, (1-\mu^2)$$
$$\times \left(rk_+ + (1-r)k_- - \frac{[rl_+ + (1-r)l_-]^2}{rh_+ + (1-r)h_-}\right), \quad (13.2.37)$$

with the same notation as above ((13.2.29) ff.), and

$$k_\pm = \epsilon^2 |\mu \mp \epsilon|^{q-1} \, \bar{\theta}_\pm \,. \qquad (13.2.38)$$

Again to lowest order in $\epsilon \ll 1$ the integral (13.2.37) becomes

$$G(H_c, \sigma_{f,b}) \simeq r\,(1-r)$$
$$\times \frac{r(1-\sigma_f^2) - (1-r)(1-\sigma_b^2)}{r^2(1-\sigma_f^2) + (1-r)^2(1-\sigma_b^2) + 2r(1-r)(1-\sigma_f \sigma_b)},$$
$$(13.2.39)$$

which also is independent of the sign of the particle's charge (Z) and the turbulence spectral index (q).

In Fig. 13.12 we calculate the function $G(H_c, \sigma_{f,b})$ for various values of the helicities r, σ_f, and σ_b. From (13.2.39) we obtain immediately that the function G and thus the momentum diffusion coefficient A_2 are equal to zero if only forward ($r = 1$) or only backward ($r = 0$) propagating waves are present, which is also clearly visible in Fig. 13.12, i.e.

$$A_2(H_c = \pm 1) = 0 \,. \tag{13.2.40}$$

Momentum diffusion from parallel propagating non-dispersive Alfvén waves requires the presence of waves with at least two different phase speeds, which is equivalent to a non-degenerate cross helicity state $H_c \neq \pm 1$.

For non-degenerate cross helicity values $H_c \neq \pm 1$, the function G is finite for all magnetic helicity values. For equal magnetic helicities ($\sigma_f = \sigma_b$) (13.2.39) reduces to

$$G(r, \sigma_f = \sigma_b) = r(1-r) \,, \tag{13.2.41}$$

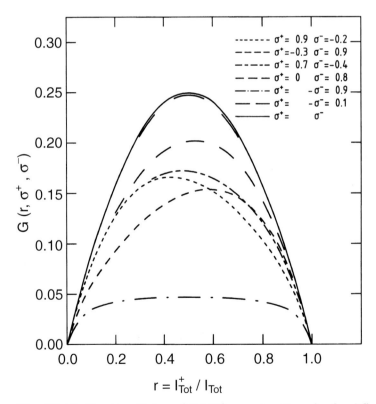

Fig. 13.12. Function G from (13.2.39) as a function of r for different magnetic helicities in the case of isospectral turbulence

which takes its maximum at $r_m = 1/2$, i.e. for equal intensity of forward and backward moving waves.

13.2.2.5 A Useful Relation Between Spatial and Momentum Diffusion Coefficient for Isospectral Kolmogorov Slab Alfvén Turbulence.

If we calculate the product between the momentum diffusion coefficient (13.2.35) and the spatial diffusion coefficient $\kappa_{zz} = v\lambda/3$ obtained from (13.2.13) we obtain

$$\kappa_{zz} \cdot A_2 = \frac{v\lambda_0 \epsilon^2}{3\tau(p,q)} p^2 F(Z, H_c, \sigma_{f,b}) G(H_c, \sigma_{f,b})$$
$$= \frac{2 F(Z, H_c, \sigma_{f,b}, q, \epsilon) G(H_c, \sigma_{f,b})}{q(q+2)} V_A^2 p^2 , \quad (13.2.42)$$

which for known cross helicity states allows us to express one diffusion coefficient in terms of the other.

If one introduces the typical time scale $t_D = L^2/\kappa_{zz}$ to diffuse a distance L along the ordered magnetic field, and the typical acceleration time scale $t_F = p^2/A_2$ due to momentum diffusion, the relation (13.2.42) is equivalent to

$$t_D(p,z) \cdot t_F(p,z) = \left(\frac{L}{V_A}\right)^2 \frac{q(q+2)}{2F(Z, H_c, \sigma_{f,b}, q, \epsilon) G(H_c, \sigma_{f,b})} , \quad (13.2.43)$$

which is a constant for relativistic particles at all momenta, being determined mainly by the macroscopic properties of the physical system (as typical size L, Alfvén speed V_A) but not by the actual level of slab Alfvén turbulence intensity.

13.2.2.6 The Effect of Dissipation Turbulence Spectra.

Measurements of the interplanetary magnetic fluctuation spectra (see also Chap. 15) with high time resolution (Denskat et al. 1983 [120]; Smith et al. 1990 [500]) clearly indicate a significant steepening or cutoff of the power spectra at frequencies of approximately $\omega_{R,c} \simeq 1$ Hz to a few tens of Hz, suggesting that the Kolmogorov-type turbulence spectrum (13.1.25) should be modified as

$$g_s^j(k_\parallel) = g_{s0}^j k_\parallel^{-q} Q(k_\parallel) \, \bar{\theta} \left[k_\parallel - k_{\parallel,\min}\right] , \quad (13.2.44)$$

where $Q(k_\parallel)$ denotes the cutoff function of the form

$$Q(k_\parallel) = \exp\left(-l_d k_\parallel\right) , \quad \text{or} \quad Q(k_\parallel) = \exp\left(-l_d^2 k_\parallel^2\right) , \quad \text{or}$$
$$Q(k_\parallel) = \left[1 + l_d^2 k_\parallel^2\right]^{-\nu} , \quad (13.2.45)$$

with the characteristic scale $l_d \simeq \omega_{R,c}/V_A$.

Schlickeiser et al. (1991 [473]) have calculated the modifications on the quasilinear cosmic ray mean free path resulting from this dissipation of parallel propagating Alfvénic plasma turbulence at large wavenumbers. In their

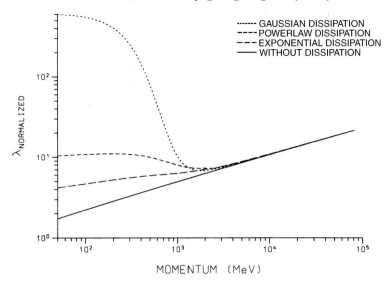

Fig. 13.13. Normalized cosmic ray proton mean free path as a function of particle momentum for the three dissipation functions (13.2.45) in comparison to the case of no dissipation $Q = 1$. Calculations assume $q = 5/3$, zero cross and magnetic helicity, $V_A = 50$ km/s and the same dissipation scale $l_d = 3 \times 10^7$ cm

calculation they kept the simple linear Alfvénic frequency-wavenumber relation $\omega_R \propto k_\parallel$ at all wavenumbers. The cutoff of the plasma wave intensity at large wavenumbers implied by (13.2.45) leads to a reduction of the scattering rate near pitch angles with $\mu \simeq 0$ and thus to an enhancement of the value of the mean free path at small particle momenta. This is clearly evident in Fig. 13.13 the normalized cosmic ray proton scattering length for the three dissipation functions (13.2.45) in comparison to the straight power law variation in the case of no dissipation.

Achatz et al. (1993 [7]) pointed out that the linear Alfvénic frequency-wavenumber-relation does not hold at large wavenumbers, and repeated the analysis of Schlickeiser et al. (1991 [473]) using the correct dispersion relation (9.2.35) of circularly polarized parallel propagating magnetohydrodynamic waves. Although the analysis becomes more complex they also conclude that the wave dissipation leads to drastically enhanced values of the mean free path at small particle momenta.

13.2.3 Influence of Damping and Dissipation

The steepening of the magnetic power spectra at high wavenumbers (see Sect. 13.2.2.6) most probably results from the collisionless cyclotron damping (see Sect. 10.4.1) of the transverse Alfvén waves at wavenumbers near $\simeq \Omega_{0,p}/V_A$. It has been emphasized by Schlickeiser and Achatz (1993a [462]) that this damping enters twice into the calculation of quasilinear cosmic ray transport parameters:

(1) the plasma wave intensity at long wavelength is drastically reduced, which is reflected by the cutoff function (13.2.45);
(2) as we have discussed already in Sect. 12.3, wave damping also modifies the resonance functions in the Fokker-Planck coefficients from sharp delta-function resonances (see (12.2.13)) to broadened Breit-Wigner resonance functions (see (12.2.12)).

Achatz et al. (1993 [7]) demonstrated that considering only the first effect leads to quasilinear mean free paths of cosmic rays drastically larger than those measured, in obvious contradiction to the observational evidence. For a consistent description Schlickeiser and Achatz (1993a [462], 1993b [463]) correctly accounted also for the second effect of wave damping and included the resulting wave damping. In the case of cosmic ray protons they demonstrated that the scattering of particles in pitch angle is markedly different at small and large particle momenta. At large relativistic momenta at all pitch angles the resonant interaction with undamped waves controls the scattering $D_{\mu\mu}$. However, at non-relativistic energies there exists a small pitch-angle interval $|\mu| \leq V_A/c$ where the undamped left- and right-handed polarized Alfvén waves do not contribute. In this interval the scattering relies entirely on the small but finite resonance-broadened contribution from damped right-handed polarized waves. Because, according to (12.3.24) the mean free path is sensitively determined by the minimum value of $D_{\mu\mu}$ one finds a quite different behavior of the mean free path at small and large particle momenta. The mean free path shown in Fig. 13.14 consists of two terms: the first describes the resonant scattering from undamped waves and exhibits the familiar power law dependence in rigidity if the turbulence spectrum is of the Kolmogorov-type in the inertial range; the second term reflects the contribution from scattering by the damped waves and only occurs below some limiting rigidity, above which particles at all pitch angles interact with undamped waves. Towards small energies this second contribution approaches a constant independent of particle rigidity and dominates over the resonant scattering contribution.

The mean free paths of different cosmic ray particle species for parallel propagating magnetohydrodynamic waves have been calculated by Schlickeiser et al. (1997 [474]). As Fig. 13.15 indicates, at particle rigidities above several 10^9 V/c cosmic ray particles of the same rigidity have the same mean free path, whereas at particle rigidities below 10^9 V/c the mean free paths are significantly different for cosmic ray protons, negatrons and positrons. In particular, the cosmic ray positrons have a much larger mean free path, i.e. they are less often scattered, than cosmic ray negatrons and protons of the same rigidity. The differences between negatrons and protons at low rigidities is due to their opposite handedness. At rigidities above several 10^9 V/c, corresponding to relativistic energies of protons (Lorentz factor > 3) and ul-

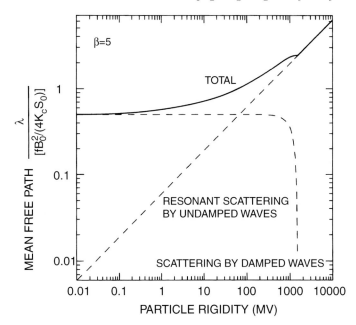

Fig. 13.14. Normalized cosmic ray proton mean free path as a function of rigidity calculated from parallel propagating magnetohydrodynamic waves with Kolmogorov-type turbulence spectrum in the inertial range and an exponential dissipation function. Calculations assume $q = 3/2$, zero magnetic helicity, cross helicity $H_c = 0.5$ and a background plasma with plasma beta $\beta = 5$

trarelativistic (Lorentz factor > 1836) energies of negatrons and positrons, all particles interact with the weakly damped Alfvén waves, and their mean free paths become essentially equal since we assumed vanishing cross and magnetic helicities for the interplanetary turbulence. While a cosmic ray proton mainly interacts resonantly with the left-handed polarized weakly damped Alfvén and strongly damped ion-cyclotron waves, and a cosmic ray negatron interacts with the right-handed polarized weakly damped whistler waves and the strongly-damped electron-cyclotron waves, a cosmic ray positron in the rigidity range less than 10^9 V/c finds no weakly damped plasma waves to interact with, since thermal positrons are absent in the solar wind background plasma. At nearly all pitch angles cosmic ray positrons are then only slightly scattered by non-resonant interplanetary plasma waves due to the thermal resonance broadening effect that produces small, but finite, tails of the Breit-Wigner resonance functions.

This markedly different behavior of positron and negatron mean free paths at energies below 1 GeV is a characteristic prediction of cosmic ray particle transport in parallel propagating plasma wave turbulence.

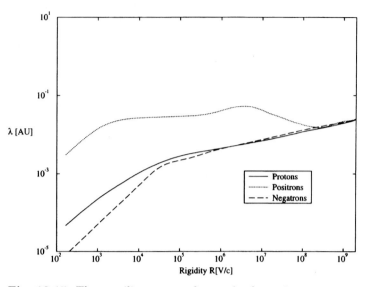

Fig. 13.15. The quasilinear mean free path of cosmic ray protons, negatrons and positrons with rigidity values between 10^2 and 10^{12} V/c in the interplanetary medium

13.2.4 Numerical Test of Quasilinear Theory

The quasilinear values for the spatial diffusion coefficient along the ordered magnetic field obtained from (13.2.13), the momentum diffusion coefficient (13.2.35), and the product relation (13.2.42) for the two, have been compared with the results of Monte Carlo simulations of the transport of relativistic particles in slab Alfvén turbulence by Michalek and Ostrowski (1996 [351]). They calculated numerically the trajectories in phase space of relativistic cosmic ray particles in a uniform field with superposed plane parallel propagating Alfvén waves with vanishing magnetic and cross helicity. By averaging over these trajectories the values of the transport parameters are derived. The calculations were performed for various wave amplitudes ranging from $\delta B/B_0$ from 0.15 to 2.0. for a Kolmogorov turbulence spectral index value $q = 1$. Figure 13.16 shows the comparison of their simulation results with the quasilinear values. Over the whole range of wave amplitudes the agreement between simulated and quasilinear values of the diffusion coefficients is remarkably good, and within factors of order 2 even at large amplitudes.

These calculations demonstrate that, for small-amplitude $((\delta B)^2/B_0^2 \leq 4)$ plasma wave fields, quasilinear theory provides an accurate description of cosmic ray particle transport.

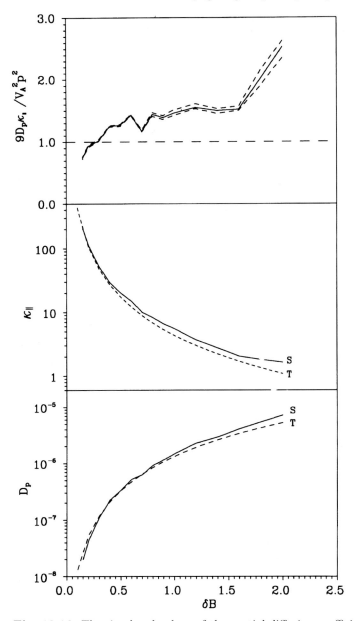

Fig. 13.16. The simulated values of the spatial diffusion coefficient ($\kappa_\parallel$), the momentum diffusion coefficient (D_p) and their product in comparison with the quasi-linear values versus the wave amplitude δB (in units of B_0) for linearly polarized plane Alfvén waves. From Michalek and Ostrowski (1996 [351])

13.3 Magnetosonic Waves

In the cold (electron temperature $T_e \to 0$) plasma limit the fast and slow magnetosonic waves in the fluid plasma (e.g. Thompson 1962 [543]) merge to the fast mode waves with dispersion relation at small enough wavenumber (see (9.2.63a))

$$\omega_{R,j} = jV_A k, \quad j = \pm 1 \tag{13.3.1}$$

at frequencies well below the proton gyrofrequency $\Omega_{0,p}$. The fast mode branch extends from low frequencies past the proton gyrofrequency $\Omega_{0,p}$, where they enter the whistler regime, up to the electron cyclotron frequency. For propagation along the ordered magnetic field the fast mode wave corresponds to the right-hand polarized mode, discussed in detail in Sect. 9.2.3.2.

As before we choose the particular coordinate system with the wave vector representation (13.1.2). As mentioned in Sect. 9.2.5.1 in this coordinate system the y-component is the only non-vanishing electric field component of the fast mode wave,

$$\boldsymbol{e}_F \equiv \frac{\boldsymbol{E}_F}{|\boldsymbol{E}_F|} = (0, 1, 0), \tag{13.3.2}$$

so that the fast mode wave is a transverse wave $\boldsymbol{e}_F \cdot \boldsymbol{k} = 0$. Faraday's law (12.2.8) yields for the fast wave's normalized magnetic field component

$$\boldsymbol{b}_F \equiv \frac{\boldsymbol{B}_F}{|\boldsymbol{B}_F|} = (-\cos\theta, 0, \sin\theta). \tag{13.3.3}$$

For purely transverse waves Faraday's law can also be expressed as

$$\boldsymbol{E}^j(\boldsymbol{k}) = -\frac{\omega_j}{ck^2} \boldsymbol{k} \times \boldsymbol{B}^j(\boldsymbol{k}). \tag{13.3.4}$$

Consequently, low-frequency fast mode waves have a simple polarization. If a wave is parallel, then its electric field $\delta \boldsymbol{E}_w$ is transverse to the ambient magnetic field $\boldsymbol{B}_0$ and right-hand circularly polarized; if is oblique, then the wave is linearly polarized, and $\delta \boldsymbol{E}_w$ is in the direction of $\boldsymbol{B}_0 \times \boldsymbol{k}$. According to (13.3.3) the wave magnetic field $\delta \boldsymbol{B}_w$ has both a compressive (along $\boldsymbol{B}_0$) and a linearly polarized transverse component.

From (13.3.2)–(13.3.4) we obtain for the respective (see (12.1.10) and (12.2.3)) right- and left-handed polarized and the parallel components for the fast mode waves

$$E_L = -E_R, \quad E_\parallel = 0, \tag{13.3.5}$$

and

$$B_L = B_R, \quad B_\parallel \neq 0, \tag{13.3.6}$$

implying for the components of correlation tensors (12.2.6) in this case

$$P_{\text{LR}} = P_{\text{RL}} = P_{\text{LL}} = P_{\text{RR}} , \qquad (13.3.7a)$$
$$Q_{\text{LR}} = -Q_{\text{RL}} = Q_{\text{LL}} = -Q_{\text{RR}} , \qquad (13.3.7b)$$
$$T_{\text{LR}} = -T_{\text{RL}} = -T_{\text{LL}} = T_{\text{RR}} , \qquad (13.3.7c)$$
$$R_{\text{LR}} = R_{\text{RL}} = -R_{\text{LL}} = -R_{\text{RR}} , \qquad (13.3.7d)$$
$$R_{\|\|} = R_{\|\text{R}} = R_{\|\text{L}} = T_{\text{L}\|} = T_{\text{R}\|} = Q_{\|\text{R}} = Q_{\|\text{L}} = 0 , \qquad (13.3.7e)$$
$$T_{\text{L}\|} = -T_{\|\text{R}} , \qquad (13.3.7f)$$
$$Q_{\text{L}\|} = -Q_{\text{R}\|} , \qquad (13.3.7g)$$
$$P_{\text{L}\|} = P_{\text{R}\|} , \qquad P_{\|\text{L}} = P_{\|\text{R}} . \qquad (13.3.7h)$$

Likewise (13.3.4) implies the two electric field components

$$E_{\text{L}} = -\frac{\imath\omega_j}{ck^2}\left[k_\| B_{\text{L}} - \frac{k_\perp}{\sqrt{2}} B_\|\right] , \qquad (13.3.8a)$$

$$E_{\text{R}} = -\frac{\imath\omega_j}{ck^2}\left[-k_\| B_{\text{R}} + \frac{k_\perp}{\sqrt{2}} B_\|\right] , \qquad (13.3.8b)$$

allowing one to express the tensors $R^j_{\alpha\beta}(\boldsymbol{k})$, $Q^j_{\alpha\beta}(\boldsymbol{k})$ and $T^j_{\alpha\beta}(\boldsymbol{k})$ in terms of the magnetic field fluctuation tensor $P^j_{\alpha\beta}(\boldsymbol{k})$. In addition to (13.3.7) we obtain

$$\begin{aligned}R_{\text{RR}} &= \frac{\omega_j^2}{c^2 k^4}\left[k_\|^2 P_{\text{RR}} + \frac{k_\perp^2}{2}P_{\|\|} - \frac{k_\perp k_\|}{\sqrt{2}}(P_{\text{R}\|} + P_{\|\text{R}})\right] \\ &= \frac{V_{\text{A}}^2}{c^2}\left[\cos^2\Theta P_{\text{RR}} + \frac{\sin^2\Theta}{2}P_{\|\|} - \frac{\sin\Theta\cos\Theta}{\sqrt{2}}(P_{\text{R}\|} + P_{\|\text{R}})\right] ,\end{aligned}$$
$$(13.3.9a)$$

$$Q_{\text{RR}} = \frac{\imath\omega_j}{ck^2}\left[k_\| P_{\text{RR}} - \frac{k_\perp}{\sqrt{2}}P_{\|\text{R}}\right] = \frac{\imath j V_{\text{A}}}{c}\left[\cos(\Theta)P_{\text{RR}} - \frac{\sin\Theta}{\sqrt{2}}P_{\|\text{R}}\right] ,$$
$$(13.3.9b)$$

$$T_{\text{RR}} = \frac{\imath\omega_j}{ck^2}\left[-k_\| P_{\text{RR}} + \frac{k_\perp}{\sqrt{2}}P_{\text{R}\|}\right] = \frac{\imath j V_{\text{A}}}{c}\left[-\cos(\Theta)P_{\text{RR}} + \frac{\sin\Theta}{\sqrt{2}}P_{\text{R}\|}\right] ,$$
$$(13.3.9c)$$

$$T_{\|\text{R}} = \frac{\imath\omega_j}{ck^2}\left[-k_\| P_{\|\text{R}} + \frac{k_\perp}{\sqrt{2}}P_{\|\|}\right] = \frac{\imath j V_{\text{A}}}{c}\left[-\cos(\Theta)P_{\|\text{R}} + \frac{\sin\Theta}{\sqrt{2}}P_{\|\|}\right] ,$$
$$(13.3.9d)$$

$$Q_{\text{R}\|} = \frac{\imath\omega_j}{ck^2}\left[k_\| P_{\text{R}\|} - \frac{k_\perp}{\sqrt{2}}P_{\|\|}\right] = \frac{\imath j V_{\text{A}}}{c}\left[\cos(\Theta)P_{\text{R}\|} - \frac{\sin\Theta}{\sqrt{2}}P_{\|\|}\right] .$$
$$(13.3.9e)$$

13.3.1 Fast Mode Wave Fokker-Planck Coefficients

Inserting the relations (13.3.7) and (13.3.9) into (12.2.11)–(12.2.18) one obtains for the Fokker-Planck coefficients of fast mode waves after straightforward but tedious algebra

$$D_{\rm pp} = \frac{2\Omega^2 p^2 c^2 (1-\mu^2)}{B_0^2 v^2} \sum_{j=\pm 1} \sum_{n=-\infty}^{\infty} \int {\rm d}^3 k \ R_{\rm RR}^j({\bm k}) \mathcal{R}_j({\bm k},\omega_j) \left[J_n'(Z)\right]^2$$

$$= \frac{2\Omega^2 p^2 V_{\rm A}^2 (1-\mu^2)}{B_0^2 v^2} \sum_{j=\pm 1} \sum_{n=-\infty}^{\infty} \int {\rm d}^3 k \ \mathcal{R}_j({\bm k},\omega_j)$$

$$\times \left[\cos^2\Theta P_{\rm RR}^j + \frac{\sin^2\Theta}{2} P_{\|\|}^j - \frac{\sin\Theta\cos\Theta}{\sqrt{2}}\left(P_{\rm R\|}^j + P_{\|\rm R}^j\right)\right], \quad (13.3.10)$$

$$D_{\mu p} = \frac{2\Omega^2 (1-\mu^2) pc}{v B_0^2} \Re \sum_{j=\pm 1} \sum_{n=-\infty}^{\infty} \int {\rm d}^3 k \mathcal{R}_j \cdot \imath \left(J_n'(Z)\right)^2$$

$$\times \left[T_{\rm RR}^j({\bm k}) + \imath\mu\frac{c}{v} R_{\rm RR}^j({\bm k})\right]$$

$$= \frac{2\Omega^2 (1-\mu^2) p V_{\rm A}}{v B_0^2} \Re \sum_{j=\pm 1} j \sum_{n=-\infty}^{\infty} \int {\rm d}^3 k \ \mathcal{R}_j \left(J_n'(Z)\right)^2$$

$$\times \left[P_{\rm RR}^j \cos\Theta - \frac{\sin\Theta}{\sqrt{2}} P_{\rm R\|}^j - \frac{j\mu V_{\rm A}}{v}\left(\cos^2\Theta P_{\rm RR}^j \right.\right.$$

$$\left.\left. + \frac{\sin^2\Theta}{2} P_{\|\|}^j - \frac{\sin\Theta\cos\Theta}{\sqrt{2}}\left(P_{\rm R\|}^j + P_{\|\rm R}^j\right)\right)\right] \quad (13.3.11)$$

and

$$D_{\mu\mu} = \frac{\Omega^2 (1-\mu^2)}{2 B_0^2} \sum_{j=\pm 1} \sum_{n=-\infty}^{\infty} \int {\rm d}^3 k \mathcal{R}_j({\bm k},\omega_j) \left\{\left[J_{n+1}^2(Z) + J_{n-1}^2(Z)\right]\right.$$

$$\times \left(P_{\rm RR}^j({\bm k}) + \mu^2 \frac{c^2}{v^2} R_{\rm RR}^j({\bm k}) + \imath\mu\frac{c}{v}\left[T_{\rm RR}^j({\bm k}) - Q_{\rm RR}^j({\bm k})\right]\right)$$

$$- 2 J_{n-1}(Z) J_{n+1}(Z)$$

$$\left. \times \left[P_{\rm RR}^j({\bm k}) + \mu^2 \frac{c^2}{v^2} R_{\rm RR}^j({\bm k}) - \imath\mu\frac{c}{v}\left[T_{\rm RR}^j({\bm k}) - Q_{\rm RR}^j({\bm k})\right]\right]\right\}$$

$$= \frac{2\Omega^2 (1-\mu^2)}{B_0^2} \sum_{j=\pm 1} \sum_{n=-\infty}^{\infty} \int {\rm d}^3 k \mathcal{R}_j({\bm k},\omega_j) \left[J_n'(Z)\right]^2$$

$$\times \left[\left(P_{\rm RR}^j({\bm k}) + \mu^2 \frac{c^2}{v^2} R_{\rm RR}^j({\bm k})\right) + \imath\mu\frac{c}{v}\left[Q_{\rm RR}^j({\bm k}) - T_{\rm RR}^j({\bm k})\right]\right], \quad (13.3.12)$$

where

$$Z \equiv \frac{k_\perp v_\perp}{|\Omega|}, \quad J_n(Z)' \equiv \frac{{\rm d}J_n(Z)}{{\rm d}Z}. \quad (13.3.13)$$

Equations (13.3.10)–(13.3.12) represent the exact Fokker-Planck coefficients for fast mode waves and they have been expressed in terms of the (measurable) components of the magnetic fluctuation correlation tensor.

For very relativistic particles $v \gg V_A$ we may expand the general expressions $D_{\mu\mu}$ and $D_{\mu p}$ to first order in the small parameter $V_A/v \ll 1$ to obtain approximately

$$D_{\mu p}(v \gg V_A) \simeq \frac{2\Omega^2(1-\mu^2)pV_A}{vB_0^2} \sum_{j=\pm 1} j \sum_{n=-\infty}^{\infty} \int d^3k \, \mathcal{R}_j(\boldsymbol{k},\omega_j) \left(J_n'(Z)\right)^2$$
$$\times \left[\cos(\Theta)P_{RR}^j(\boldsymbol{k}) - \frac{\sin\Theta}{\sqrt{2}}P_{R\|}^j(\boldsymbol{k})\right], \qquad (13.3.14)$$

and

$$D_{\mu\mu}(v \gg V_A) \simeq \frac{2\Omega^2(1-\mu^2)}{B_0^2}$$
$$\times \sum_{j=\pm 1} \sum_{n=-\infty}^{\infty} \int d^3k \, \mathcal{R}_j(\boldsymbol{k},\omega_j) \left[J_n'(Z)\right]^2 P_{RR}^j(\boldsymbol{k}). \qquad (13.3.15)$$

13.3.2 Isotropic Fast Mode Waves

From the isotropic turbulence tensor (13.1.23) we obtain

$$P_{RR}^j = \frac{(1+\cos^2\Theta)g^j(k)}{16\pi k^2}, \qquad (13.3.16a)$$
$$P_{\|\|}^j = \frac{\sin^2\Theta \, g^j(k)}{8\pi k^2}, \qquad (13.3.16b)$$
$$P_{\|R}^j = P_{R\|}^j = -\frac{\sin\Theta\cos\Theta \, g^j(k)}{8\sqrt{2}\pi k^2}. \qquad (13.3.16c)$$

We again adopt the Kolmogorov-type turbulence spectrum (13.1.26) and (13.1.28), and consider undamped waves using (12.2.13) for the resonance function $\mathcal{R}_j = \pi\delta(v\mu k_\| - \omega_{R,j} + n\Omega)$.

Moreover, we limit our analysis to the important cases of (a) very fast ($v \gg V_A$) cosmic ray particles, so that we can use (13.3.14)–(13.3.15), and (b) vanishing cross helicity $H_{c,\mathrm{fmw}} = 0$ of the fast mode waves, i.e. equal intensity of forward ($j = 1$ in (13.3.1)) and backward ($j = -1$) moving waves. With (13.3.16) and with $\eta = \cos\Theta$ we obtain

$$D_{\mu\mu} = \frac{\pi\Omega^2(1-\mu^2)}{4B_0^2} \sum_{j=\pm 1} \sum_{n=-\infty}^{\infty} \int_{-1}^{1} d\eta \int_0^\infty dk \, g^j(k)\left(1+\eta^2\right)$$
$$\times \delta(kv\mu\eta - jV_Ak + n\Omega) \left[J_n'\left(\frac{kv_\perp\sqrt{1-\eta^2}}{|\Omega|}\right)\right]^2 \qquad (13.3.17)$$

and

$$D_{pp} = \frac{\pi\Omega^2 p^2 V_A^2 (1-\mu^2)}{4B_0^2 v^2} \sum_{j=\pm 1} \sum_{n=-\infty}^{\infty} \int_{-1}^{1} d\eta \int_{0}^{\infty} dk \, g^j(k) \left(1+\eta^2\right)$$
$$\times \delta(kv\mu\eta - jV_A k + n\Omega) \left[J_n' \left(\frac{kv_\perp \sqrt{1-\eta^2}}{|\Omega|} \right) \right]^2 = \frac{p^2 V_A^2}{v^2} D_{\mu\mu} \,.$$
(13.3.18)

It can be shown that in the case of zero cross helicity the third Fokker-Planck coefficients (13.3.14) vanishes, yielding for the rate of adiabatic deceleration (12.3.29)

$$A_1 = 0 \,. \tag{13.3.19}$$

Also, the two Fokker-Planck coefficient (13.3.17) and (13.3.18) are symmetric functions in μ in this case. Consequently, the spatial diffusion coefficient (12.3.24) and the momentum diffusion coefficient (12.3.30) simplify to

$$\kappa_{zz} = v\lambda/3 = \frac{v^2}{4} \int_0^1 d\mu \, \frac{(1-\mu^2)^2}{D_{\mu\mu}(\mu)} \tag{13.3.20}$$

and

$$A_2 = \int_{-1}^{1} d\mu \, D_{pp}(\mu) = \frac{p^2 V_A^2}{v^2} \int_{-1}^{1} d\mu \, D_{\mu\mu}(\mu) \,, \tag{13.3.21}$$

respectively. Obviously we can express (13.3.17) as

$$D_{\mu\mu} = \frac{\pi\Omega^2(1-\mu^2)}{4B_0^2} \sum_{j=\pm 1} \left[D_T^j + D_G^j \right] \,, \tag{13.3.22}$$

where

$$D_T^j \equiv \int_{-1}^{1} d\eta \int_{0}^{\infty} dk \, g^j(k) \left(1+\eta^2\right) \delta$$
$$\times \left(k\left(v\mu\eta - jV_A\right)\right) \left[J_0' \left(\frac{kv_\perp \sqrt{1-\eta^2}}{|\Omega|} \right) \right]^2$$
$$= \frac{1}{|v\mu|} \int_{0}^{\infty} dk \, g^j(k) \, k^{-1} \int_{-1}^{1} d\eta \, (1+\eta^2) \, \delta$$
$$\times \left(\eta - \frac{jV_A}{v\mu} \right) J_1^2 \left(\frac{kv_\perp \sqrt{1-\eta^2}}{|\Omega|} \right)$$
$$= \frac{H\left[|\mu| - (V_A/v)\right] \left(1 + (V_A^2/v^2\mu^2)\right)}{|v\mu|}$$
$$\times \int_{0}^{\infty} dk \, g^j(k) k^{-1} J_1^2 \left(\frac{kv_\perp \sqrt{1-(V_A^2/v^2\mu^2)}}{|\Omega|} \right) \tag{13.3.23}$$

represents the contribution from $n = 0$ resonances, and

$$D_{\text{G}}^j \equiv \sum_{n=-\infty,\, n\neq 0}^{\infty} \int_{-1}^{1} d\eta \int_0^\infty dk \ g^j(k) \left(1 + \eta^2\right)$$

$$\times \delta\left(k\left[v\mu\eta - jV_{\text{A}}\right] + n\Omega\right) \left[J_n'\left(\frac{kv_\perp \sqrt{1-\eta^2}}{|\Omega|}\right)\right]^2$$

$$= \sum_{n=1}^{\infty} \int_{-1}^{1} d\eta \frac{1 + \eta^2}{|v\mu\eta - jV_{\text{A}}|}$$

$$\times \int_0^\infty dk \ g^j(k) \left[J_n'\left(\frac{kv_\perp \sqrt{1-\eta^2}}{|\Omega|}\right)\right]^2$$

$$\times \left[\delta\left(k + \frac{n\Omega}{v\mu\eta - jV_{\text{A}}}\right) + \delta\left(k - \frac{n\Omega}{v\mu\eta - jV_{\text{A}}}\right)\right] \quad (13.3.24)$$

represents the contribution from gyroresonant interactions by resonances with $n \geq 1$.

The $n = 0$ contribution (13.3.23) is referred to as "transit-time damping". According to the general resonance condition (12.2.13), energetic charged particles of velocity v and Larmor frequency Ω resonantly interact with undamped waves of frequency ω if the gyroresonance interaction condition,

$$\omega_{\text{R}} - k_\| v_\| = n\Omega \quad (13.3.25)$$

with entire n running between $-\infty \leq n \leq \infty$, is fulfilled, where $v_\| = v\mu$ and $k_\|$ are the parallel velocity component and the parallel wavenumber, respectively. Unlike for purely parallel propagating waves in slab turbulence, the presence of the compressive magnetic field component of δB_w for fast mode waves allows the cosmic ray particles to resonantly interact with these waves through the $n = 0$ resonance, if the particle's parallel velocity matches exactly the parallel component of the wave's phase velocity $\omega_{\text{R}}/k\eta$. This effect is called transit-time damping (Fisk 1976 [171]; Achterberg 1981 [8]; Stix 1992 [527], p. 273) and is the magnetic analog of Landau damping (which involves the $n = 0$ resonance and a parallel electric field). The resonance condition for transit-time damping $V_{\text{ph}}/\eta = v\mu$ implies that only super-Alfvénic cosmic ray particles with pitch angles outside the interval $|\mu| \geq V_{\text{A}}/v = \epsilon$ undergo this interaction since $\eta \leq 1$.

Adopting the Kolmogorov-type turbulence spectrum (13.1.26) we obtain for (13.3.23)

$$\sum_j D_{\text{T}}^j(\mu) = g_{i0} \frac{H\left[|\mu| - \epsilon\right]\left(1 + (\epsilon^2/\mu^2)\right)}{|v\mu|} \int_{k_{\min}}^{\infty} dk \ k^{-1-q}$$

$$\times J_1^2\left[kR_{\text{L}}\sqrt{(1-\mu^2)\left(1 - \frac{\epsilon^2}{\mu^2}\right)}\right]$$

$$= (q-1)(\delta B)^2 \frac{R_L}{v} (R_L k_{\min})^{q-1} f_T(\mu), \qquad (13.3.26)$$

where we used the normalization (13.1.28) and introduced the transit-time damping function

$$f_T(\mu) \equiv H\left[|\mu| - \epsilon\right] \frac{(1+(\epsilon^2/\mu^2))}{|\mu|} \left[(1-\mu^2)\left(1 - \frac{\epsilon^2}{\mu^2}\right)\right]^{q/2}$$
$$\times \int_U^\infty ds\, s^{-(1+q)} J_1^2(s) \qquad (13.3.27)$$

where the lower integration boundary is

$$U = k_{\min} R_L \sqrt{(1-\mu^2)(1-\epsilon^2 \mu^{-2})}, \qquad (13.3.28)$$

and where we introduced

$$R_L = \frac{v}{|\Omega|}, \quad \epsilon = \frac{V_A}{v} \ll 1. \qquad (13.3.29)$$

Equation (13.3.24) can be cast into the form

$$\sum_j D_G^j = \frac{1}{v} \sum_{n=1}^\infty \int_0^1 d\eta\, (1+\eta^2) \left[G_1(|\mu\eta - \epsilon|) + G_1(|\mu\eta + \epsilon|)\right], \qquad (13.3.30)$$

with

$$G_1(x) \equiv x^{-1} \sum_j g^j \left(\frac{n}{R_L x}\right) \left[J_n'\left(\frac{n\sqrt{(1-\mu^2)(1-\eta^2)}}{x}\right)\right]^2$$
$$\times H\left[1 - R_L x k_{\min}\right]$$
$$= g_{i0}\, n^{-q} R_L^q\, x^{q-1} \left[J_n'\left(\frac{n\sqrt{(1-\mu^2)(1-\eta^2)}}{x}\right)\right]^2$$
$$\times H\left[1 - R_L x k_{\min}\right]. \qquad (13.3.31)$$

Throughout this analysis we will neglect cutoff effects, i.e.

$$k_{\min} R_L \ll 1. \qquad (13.3.32)$$

For (13.3.30) we then obtain

$$\sum_j D_G^j = \frac{g_{i0}}{v} R_l^q f_G(\mu) = (q-1)(\delta B)^2 \frac{R_L}{v} (R_L k_{\min})^{q-1} f_G(\mu), \qquad (13.3.33)$$

in terms of the gyroresonance function

$$f_G(\mu) \equiv \sum_{n=1}^{\infty} n^{-q} \int_0^1 d\eta \, (1+\eta^2)$$
$$\times \left(|\mu\eta - \epsilon|^{q-1} \left[J'_n \left(\frac{n\sqrt{(1-\mu^2)(1-\eta^2)}}{|\mu\eta - \epsilon|} \right) \right]^2 \right.$$
$$\left. + |\mu\eta + \epsilon|^{q-1} \left[J'_n \left(\frac{n\sqrt{(1-\mu^2)(1-\eta^2)}}{|\mu\eta + \epsilon|} \right) \right]^2 \right). \quad (13.3.34)$$

Collecting terms, we obtain for the Fokker-Planck coefficients

$$D_{\mu\mu} = \frac{\pi |\Omega|(q-1)(1-\mu^2)}{4} \frac{(\delta B)^2}{B_0^2} (k_{\min} R_L)^{q-1} [f_T(\mu) + f_G(\mu)] \quad (13.3.35a)$$

and

$$D_{pp} = p^2 \epsilon^2 D_{\mu\mu} . \quad (13.3.35b)$$

According to (13.3.20)–(13.3.21) we obtain for the corresponding mean free path and momentum diffusion coefficient from fast mode waves

$$\lambda = \frac{3}{\pi(q-1)} R_L (R_L k_{\min})^{1-q} \left(\frac{B_0}{\delta B} \right)^2 I_\lambda , \quad (13.3.36)$$

and

$$A_2 = \int_0^1 d\mu \, D_{pp}(\mu) = \frac{\pi(q-1)}{4} |\Omega| (R_L k_{\min})^{q-1} \left(\frac{\delta B}{B_0} \right)^2 \frac{V_A^2 p^2}{v^2} I_2 , \quad (13.3.37)$$

involving the two integrals

$$I_\lambda = \int_0^1 d\mu \, \frac{1-\mu^2}{f_T(\mu) + f_G(\mu)} , \quad (13.3.38)$$

and

$$I_2 = \int_0^1 d\mu \, (1-\mu^2) [f_T(\mu) + f_G(\mu)] . \quad (13.3.39)$$

We find for the product of the spatial and momentum diffusion coefficients, which is related to the product of the related diffusive escape and acceleration time scales (see Sect. 13.2.2.5),

$$\kappa \cdot A_2 = \left(\frac{V_A}{2} \right)^2 \cdot p^2 \cdot I_\lambda \cdot I_2 . \quad (13.3.40)$$

Since the transit-time damping function f_T vanishes for $\mu \leq \epsilon$ (see (13.3.27)) the two integrals obviously can be separated as

$$I_\lambda = \int_0^\epsilon d\mu \, \frac{1-\mu^2}{f_G(\mu)} + \int_\epsilon^1 d\mu \, \frac{1-\mu^2}{f_T(\mu) + f_G(\mu)}, \qquad (13.3.41)$$

and

$$I_2 = \int_0^1 d\mu \, (1-\mu^2) f_G(\mu) + \int_\epsilon^1 d\mu \, (1-\mu^2) f_T(\mu). \qquad (13.3.42)$$

Schlickeiser and Miller (1998 [469]) have investigated in detail the variation of the functions $f_T(\mu)$ and $f_G(\mu)$ in the two pitch angle intervals $0 \leq \mu \leq \epsilon$ and $\epsilon < \mu \leq 1$. The transit-time damping function f_T is non-zero only in the interval $\mu > \epsilon$ where it can be approximated as

$$f_T\left(1 \leq x \leq \epsilon^{-2}, q \leq 2\right) = \frac{c_1(q)}{\epsilon} \frac{1+x}{x^{(3+q)/2}} \left[\left(1+\epsilon^2\right) x - 1 - \epsilon^2 x^2\right]^{q/2}, \qquad (13.3.43)$$

for flat turbulence spectra, where $x = (\mu/\epsilon)^2$, $c_1(2) = 3/4$ and

$$c_1(q < 2) = 2^{1-q} \frac{q}{4-q^2} \frac{\Gamma[q]\Gamma[2-(q/2)]}{\Gamma^3[1+(q/2)]}. \qquad (13.3.44)$$

For steeper turbulence spectra

$$f_T\left(q > 2, k_{\min} R_L \ll 1\right) = c_4(q) \left(R_L k_{\min}\right)^{2-q}$$
$$\times H\left[|\mu| - \epsilon\right] \frac{\left(1-\mu^2\right)\left(1-\left(\epsilon^4/\mu^4\right)\right)}{|\mu|}, \qquad (13.3.45)$$

where

$$c_4(q) = \frac{2q^2 - 3q + 4}{4q(2q-3)}. \qquad (13.3.46)$$

For the gyroresonance function $f_G(\mu)$ one obtains for all values of q the approximation

$$f_G(\mu) \simeq \begin{cases} \frac{3\zeta(q+1)\epsilon^q}{2\sqrt{1-\mu^2}} & \text{for } 0 \leq \mu \leq \sqrt{2}\epsilon \\ \frac{3\zeta(q+1)\epsilon^q}{2\sqrt{1-\mu^2}} \left[1 + \frac{4(3+2q)}{3(1+q)} (\mu/\epsilon)^q\right] & \text{for } \sqrt{2}\epsilon < \mu \leq 2^{-1/2} \\ \frac{\zeta(q)\mu^{q-1}}{2q} & \text{for } 2^{-1/2} < \mu \leq 1 \end{cases}, \qquad (13.3.47)$$

where $\zeta(q)$ denotes Riemann's zeta function.

In Figs. 13.17 and 13.18 we calculate the transit-time damping function f_T (dashed curve) for the value $\epsilon = 0.01$ for the turbulence spectral index $q = 5/3$. Whereas Fig. 13.17 shows the variation over the whole $0 \leq \mu \leq 1$

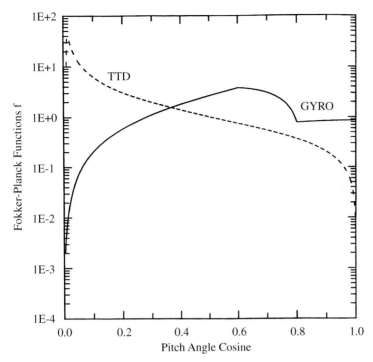

Fig. 13.17. Variation of the Fokker-Planck gyroresonance function f_G (**solid curve**) transit-time damping function f_T (**dashed curve**) as a function of the cosine of pitch angle $0 \leq \mu \leq 1$ calculated for values of $\epsilon = V_A/v = 0.01$ and $q = 5/3$

interval, Fig. 13.18 highlights the variation for small $\mu \leq 5\epsilon$. First, we note that the maximum value of f_G is much less than the maximum value of f_T, in agreement with (13.3.43) and (13.3.47). Secondly, Fig. 13.18 indicates that in the interval $|\mu| < \epsilon$, where no transit-time damping occurs, the gyroresonant interactions f_G provide a small (of order ϵ^q, see (13.3.47a)), but finite contribution to the scattering rate of particles. These properties have profound consequences for the behavior of the two integrals (13.3.38) and (13.3.39) that we discuss next.

13.3.2.1 The Integral I_λ for Flat Turbulence Spectra $q \leq 2$. By using (13.3.47a) we obtain for the first integral in (13.3.41)

$$\int_0^\epsilon d\mu \frac{1-\mu^2}{f_G(\mu)} \simeq \frac{2}{3\zeta(q+1)} \epsilon^{1-q} , \qquad (13.3.48)$$

which is much larger than unity since $\epsilon \ll 1$ and $q > 1$.

The second integral in (13.3.41) can be approximated as

$$\int_\epsilon^1 d\mu \frac{1-\mu^2}{f_T(\mu)+f_G(\mu)} \simeq \int_\epsilon^{\mu_T} d\mu \frac{1-\mu^2}{f_T(\mu)} + \int_{\mu_T}^1 d\mu \frac{1-\mu^2}{f_G(\mu)} , \qquad (13.3.49)$$

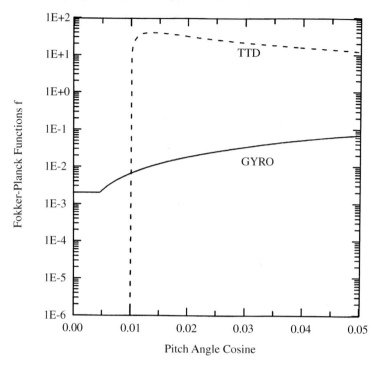

Fig. 13.18. Variation of the Fokker-Planck gyroresonance function f_G (**solid curve**) transit-time damping function f_T (**dashed curve**) as a function in the cosine of pitch angle interval $0 \leq \mu \leq 0.05$ calculated for values of $\epsilon = V_A/v = 0.01$ and $q = 1.5$. One clearly notices that transit-time damping does not contribute to the scattering of particles with pitch angles $|\mu| < \epsilon = 0.01$

where μ_T is that pitch angle cosine where $f_T(\mu_T) = f_G(\mu_T)$. Inspection of Figs. 13.16 and 13.17 shows, μ_T is much larger than ϵ, and in particular much larger than μ_M where f_T has its maximum. With the approximation (13.3.43) and (13.3.47) for f_T and f_G, respectively, for values of $q \leq 2$ and $s \gg s_M$, one determines

$$\mu_T(q) \simeq \sqrt{\frac{c_5(q)}{1 + c_5(q)}}, \qquad (13.3.50)$$

where

$$c_5(q) = \left[\frac{c_1(q)(1+q)}{2(3+2q)\zeta(q+1)}\right]^{2/(q+1)}, \qquad (13.3.51)$$

which is solely determined by the value of q and independent of the value of ϵ.

For the first integral in (13.3.49) we derive

$$I_T \equiv \int_\epsilon^{\mu_T} d\mu \frac{1-\mu^2}{f_T(\mu)}$$

13.3 Magnetosonic Waves

$$\begin{aligned}
&= \frac{1}{c_1(q)} \int_\epsilon^{\mu_T} d\mu \frac{\mu \left(1-\mu^2\right)^{(2-q)/2}}{\left(1+\epsilon^2\mu^{-2}\right)\left(1-\epsilon^2\mu^{-2}\right)^{q/2}} \\
&= \frac{2\epsilon^2}{c_1(q)} \int_1^{(\mu_T/\epsilon)^2} dx \frac{x^{(2+q)/2}\left(1-\epsilon^2 x\right)^{(2-q)/2}}{(x+1)(x-1)^{q/2}},
\end{aligned} \quad (13.3.52)$$

after obvious substitution. An upper bound to this integral is given by

$$\begin{aligned}
I_T < I_0 &= \frac{2\epsilon^2}{c_1(q)} \int_1^{(\mu_T/\epsilon)^2} dx \frac{x^{(2+q)/2}}{(x+1)(x-1)^{q/2}} \\
&\simeq \frac{2\epsilon^2}{c_1(q)} \left[\left(\frac{\mu_T}{\epsilon}\right)^2 - 4\ln(\mu_T/\epsilon) + \ln 9 + \frac{2q-3}{2-q} \right] \\
&\simeq \frac{2\mu_T^2}{c_1(q)} = \frac{2c_5(q)}{c_1(q)\left[1+c_5(q)\right]},
\end{aligned} \quad (13.3.53)$$

which is of order unity.

For the second integral in (13.3.49) we derive, with approximations (13.3.50)

$$\begin{aligned}
I_G &\equiv \int_{\mu_T}^1 d\mu \frac{1-\mu^2}{f_G(\mu)} \\
&\simeq \frac{1+q}{2(3+2q)\zeta(q+1)} \int_{\mu_t}^{2^{-1/2}} d\mu\, \mu^{-q}\left(1-\mu^2\right)^{3/2} \\
&\quad + \frac{2q}{\zeta(q)} \int_{2^{-1/2}}^1 d\mu\, \mu^{1-q}\left(1-\mu^2\right) \\
&\simeq \frac{1+q}{2(q-1)(3+2q)\zeta(q+1)} \left[\mu_T^{1-q} - 2^{(q-1)/2}\right] \\
&\quad + \frac{4q}{(2-q)(4-q)\zeta(q)} \left[1-(6-q)2^{(6+q)/2}\right],
\end{aligned} \quad (13.3.54)$$

which also is of order unity because it is only determined by the value of q.

Collecting terms in the integral (13.3.41) we find that both contributions (13.3.53) and (13.3.54) are negligibly small compared to contribution (13.3.48), so that the integral (13.3.38) for flat turbulence spectra $q \leq 2$ becomes

$$I_\lambda(q \leq 2) \simeq \frac{2}{3\zeta(q+1)} \epsilon^{1-q}. \quad (13.3.55)$$

We thus have shown that for the quasilinear interaction with fast mode waves with flat power law turbulence spectrum ($q \leq 2$) the mean free path λ and the associated spatial diffusion coefficient κ, which are determined by the integral (13.3.38) (see (13.3.36)), are solely determined by the $|\mu| < \epsilon$-interval, where transit-time damping does not contribute and the pitch-angle scattering from gyroresonant interactions with $n \geq 1$ is the only contributor. The small gyroresonance scattering in this interval is crucial to have quasilinear scattering through $\mu = 0$ in this case, (and thus establish an isotropic particle distribution function,) and therefore determines the quantitative value of the spatial diffusion coefficient.

13.3.2.2 The Integral I_λ for Steep Turbulence Spectra $2 < q < 6$.

Repeating the analysis for steep power spectra we have to use (13.3.45) instead of (13.3.43) for the transit-time damping function. Except near the end-points $\mu \simeq \epsilon$ and $\mu \simeq 1$, the transit-time damping function (13.3.45) is much larger than the gyroresonance function (13.3.47), so that we obtain for the second integral in (13.3.41)

$$\int_\epsilon^1 d\mu \frac{1-\mu^2}{f_T(\mu) + f_G(\mu)} \simeq \int_{\epsilon(1+x_1)}^{1-x_2} d\mu \frac{1-\mu^2}{f_T(\mu)}$$
$$+ \int_\epsilon^{\epsilon(1+x_1)} d\mu \frac{1-\mu^2}{f_G(\mu)} + \int_{1-x_2}^1 d\mu \frac{1-\mu^2}{f_G(\mu)}, \quad (13.3.56)$$

where

$$x_1 \simeq \frac{3}{8} \frac{\zeta(q+1)}{c_4(q)} \epsilon^{q+1} (R_L k_{\min})^{q-2} \ll 1$$

and

$$x_2 \simeq \frac{\zeta(q)}{4qc_4(q)} (R_L k_{\min})^{q-2} \ll 1$$

are determined by the conditions

$$f_T(\epsilon(1+x_1)) = f_G(\epsilon(1+x_1)), \quad f_T(1-x_2) = f_G(1-x_2).$$

Performing the integrals (13.3.56) becomes

$$\int_\epsilon^1 d\mu \frac{1-\mu^2}{f_T(\mu)+f_G(\mu)} \simeq \frac{(R_L k_{\min})^{q-2}}{c_4(q)} \left(\frac{1}{2}\left[(1-x_2)^2 - \epsilon^2(1+x_1)^2\right] \right.$$
$$+ \frac{\epsilon^4}{4}\left[\ln\left(\frac{(1-x_2)^2 - \epsilon^2}{(1-x_2)^2 + \epsilon^2}\right) - \ln\left(\frac{(1+x_1)^2 - 1}{(1+x_1)^2 + 1}\right)\right]$$
$$\left. + \frac{\epsilon^2}{4c_4(q)} + \frac{\zeta(q)}{8qc_4(q)} (R_L k_{\min})^{q-2} \right) \simeq \frac{(R_L k_{\min})^{q-2}}{2c_4(q)},$$
$$(13.3.57)$$

where the last approximation holds to lowest order in $\epsilon \ll 1$.

Compared with the contribution (13.3.48), which also holds in the case of steep turbulence spectra, we find that (13.3.57) is negligibly small, and we reproduce (13.3.55) also for steep turbulence spectra,

$$I_\lambda(q > 2) \simeq \frac{2}{3\zeta(q+1)} \epsilon^{1-q}. \quad (13.3.58)$$

13.3.2.3 The Integral I_2. The integral (13.3.42) is easier to evaluate since the contributions simply add. With (13.3.47) we obtain for the gyroresonance contribution

$$H_G \equiv \int_0^1 d\mu \, (1-\mu^2) \, f_G(\mu)$$
$$\simeq \frac{3(\pi+2)}{16} \zeta(q+1)\epsilon^q + \frac{\zeta(q)}{q^2(2+q)} \left[1 - (q+4)2^{-(4+q)/2}\right]$$
$$+ \frac{2(3+2q)\zeta(q+1)}{1+q} H_0 \,, \qquad (13.3.59)$$

where

$$H_0 \equiv \int_0^{2^{-1/2}} d\mu \mu^q \sqrt{1-\mu^2} < \int_0^{2^{-1/2}} d\mu \mu^q = \frac{2^{-(1+q)/2}}{1+q} \,. \qquad (13.3.60)$$

The first term on the right hand side of (13.3.59) is much smaller than the other two because $\epsilon \ll 1$ and, with (13.3.60) used as an estimate for H_0, we obtain for the gyroresonance contribution

$$H_G \leq \frac{\zeta(q)}{q^2(2+q)} \left[1 - (q+4)2^{-(4+q)/2}\right] + \frac{2^{(1-q)/2}(3+2q)\zeta(q+1)}{(1+q)^2} \,, \qquad (13.3.61)$$

which is of order unity and solely determined by the value of q.

Expressing (13.3.43) in terms of $\mu = \epsilon x^{1/2}$ the transit-time damping contribution to the integral (13.3.42) for flat turbulence spectra is

$$H_T(q \leq 2) \equiv \int_\epsilon^1 d\mu \, (1-\mu^2) \, f_T(\mu, q \leq 2)$$
$$= c_1(q) \int_\epsilon^1 d\mu \, \frac{1-\mu^2}{\mu} \left(1 + \frac{\epsilon^2}{\mu^2}\right) \left[(1-\mu^2)\left(1 - \frac{\epsilon^2}{\mu^2}\right)\right]^{q/2} \,. \qquad (13.3.62)$$

Substituting

$$y = \sqrt{(1-\mu^2)\left(1 - \frac{\epsilon^2}{\mu^2}\right)}$$

the integral (13.3.62) can be cast in the form

$$H_T(q \leq 2) = c_1(q)(1-\epsilon)^{q+2} D(q, \alpha) \,, \qquad (13.3.63)$$

with

$$\alpha = \left(\frac{1-\epsilon}{1+\epsilon}\right)^2 \qquad (13.3.64)$$

and

$$D(q,\alpha) \equiv \int_0^1 dy\; y^{q/2}(1-y)^{-1/2}(1-\alpha y)^{-1/2}$$
$$= \frac{\pi^{1/2}\Gamma[(2+q)/2]}{\Gamma[(3+q)/2]}\; F\left(\frac{2+q}{2},\frac{1}{2};\frac{3+q}{2};\alpha\right) \quad (13.3.65)$$

expressed in terms of the hypergeometric function. Using the quadratic transformation formula (15.4.14) of Abramowitz and Stegun (1972 [3]) the hypergeometric function can be expressed as an associated Legendre function of the second kind $Q_{q/2}$ of zeroth order and degree $(q/2)$, and (13.3.65) reduces to

$$D(q,\alpha) = 2^{1/2}\frac{(\chi^2-1)^{1/4}Q_{q/2}(\chi)}{\alpha^{(1+q)/4}(1-\alpha)^{1/2}}, \quad (13.3.66)$$

where

$$\chi = \frac{1+\alpha}{2\alpha^{1/2}} = \frac{1+\epsilon^2}{1-\epsilon^2}. \quad (13.3.67)$$

Equation (13.3.66) then becomes

$$D(q,\alpha) = \left(\frac{1+\epsilon}{1-\epsilon}\right)^{(q+2)/2} Q_{q/2}\left(\frac{1+\epsilon^2}{1-\epsilon^2}\right). \quad (13.3.68)$$

Inserting (13.3.68) into (13.3.63) we find for the transit-time contribution

$$H_T(q \leq 2) = c_1(q)(1-\epsilon^2)^{(q+2)/2}\, Q_{q/2}\left(\frac{1+\epsilon^2}{1-\epsilon^2}\right). \quad (13.3.69)$$

For arguments $\chi \to 1$ approaching unity from above, the Legendre function Q_ν logarithmically diverges (Magnus et al. 1966 [313], p. 196) with the leading term

$$\lim_{\chi\to 1^+} Q_\nu(\chi) \simeq -\gamma - \psi[(2+q)/2] - \frac{1}{2}\ln\left[\frac{\chi-1}{2}\right], \quad (13.3.70)$$

in terms of Euler's constant $\gamma = 0.5772$ and the Digamma function ψ. The limit $\chi \to 1^+$ corresponds to the limit $\epsilon \to 0$. Applying this limit to (13.3.69) we obtain

$$\lim_{\epsilon\to 0} H_T(q \leq 2) \to c_1(q)\left(1-\epsilon^2\right)^{(q+2)/2}\left(-\frac{1}{2}\ln\frac{\epsilon^2}{1-\epsilon^2} - \gamma - \psi[(2+q)/2]\right). \quad (13.3.71)$$

To lowest order in $\epsilon \ll 1$ we thus obtain for the transit-time damping contribution in the case of flat turbulence spectra

$$H_T(\epsilon \ll 1, q \leq 2) \simeq c_1(q)\, \ln\epsilon^{-1}, \quad (13.3.72)$$

which, at least for relativistic particles, is about one order of magnitude larger than unity because of the term $\ln \epsilon^{-1} = \ln v/V_A \sim O(10)$.

Comparing the transit-time contribution (13.3.72) with the gyroresonance contribution (13.3.61) we find that for relativistic particles transit-time damping provides the dominant contribution. Collecting terms in (13.3.42) we find in this case for the integral

$$I_2(q \leq 2) \simeq c_1(q)\left[\ln \epsilon^{-1} + c_6(q)\right] \simeq c_1(q)\ln \frac{v}{V_A}, \qquad (13.3.73)$$

where

$$c_6(q) \equiv \left(\frac{\zeta(q)}{q^2(2+q)}\left[1-(q+4)2^{-(4+q)/2}\right]\right.$$
$$\left.+ \frac{2^{(1-q)/2}(3+2q)\zeta(q+1)}{(1+q)^2}\right)/c_1(q), \qquad (13.3.74)$$

as indicated for fast particles $v \gg V_A$ can be neglected in (13.3.73).

For steep power spectra we use (13.3.45) to obtain, after simple integrations

$$H_T(q>2) \equiv \int_\epsilon^1 d\mu\, (1-\mu^2)\, f_T(\mu, q>2)$$
$$= c_4(q)(R_L k_{\min})^{2-q} \int_\epsilon^1 d\mu\, \mu^{-1}(1-\mu^2)^2(1-\epsilon^4\mu^{-4})$$
$$= c_4(q)(R_L k_{\min})^{2-q}(1-\epsilon^2)\left[(1+\epsilon^2)\ln \epsilon^{-1} - 1 + \epsilon^2\right]$$
$$\simeq c_4(q)(R_L k_{\min})^{2-q}\ln \epsilon^{-1}, \qquad (13.3.75)$$

which is much larger than the gyroresonance contribution (13.3.61) since $R_L k_{\min} \ll 1$. Here we obtain

$$I_2(q>2) \simeq c_4(q)(R_L k_{\min})^{2-q}\ln \frac{v}{V_A}. \qquad (13.3.76)$$

Inspecting (13.3.73) and (13.3.76) we find, quite to the contrary to our results on the spatial transport of cosmic ray particles, that transit-time damping provides the bulk of the contribution to the stochastic acceleration of fast cosmic ray particles, whereas the contribution from gyroresonant acceleration is negligibly small.

13.3.3 Cosmic Ray Mean Free Path from Fast Mode Waves

With the results (13.3.55) and (13.3.58) for the integral I_λ we obtain for the cosmic ray mean free path (13.3.36) resulting from interactions with fast mode waves of total intensity $(\delta B)_F^2$, spectral index q_F and minimum wavenumber $k_{\min,F}$,

13. Quasilinear Transport Parameters

$$\lambda_F (R_L k_{min,F} \ll 1) = \lambda_{F0}\beta \left[\left|\frac{\Omega_{p,0}}{\Omega_0}\right| \gamma \right]^{2-q_F}, \qquad (13.3.77)$$

where $\beta = v/c$ and the constant

$$\lambda_{F0} = \frac{2c}{\pi(q_F - 1)\zeta(q_F + 1)V_A} \frac{B_0^2}{(\delta B)_F^2} k_{min,F}^{-1}(k_{min,F}c/\omega_{p,i})^{2-q_F}. \qquad (13.3.78)$$

Apart from numerical factors (of order unity) the physical dependence is

$$\lambda_F \simeq R_L \left(\frac{B_0}{\delta B_F}\right)^2 (k_{min,F} R_L)^{1-q_F} \epsilon^{1-q_F} \qquad (13.3.79)$$

for all values of $q_F > 1$ provided the Larmor radius of the cosmic ray particles $R_L \ll k_{min,F}^{-1}$ is smaller than the outer scale of the fast mode turbulence. In Fig. 13.19 we calculate the mean free path (13.3.77) for different cosmic ray particles and the value $q_F = 5/3$. At non-relativistic energies the mean free path resulting from the interaction with fast mode waves varies linearly with the particle velocity independently of the value of q_F, while at relativistic energies it is proportional to γ^{2-q_F}.

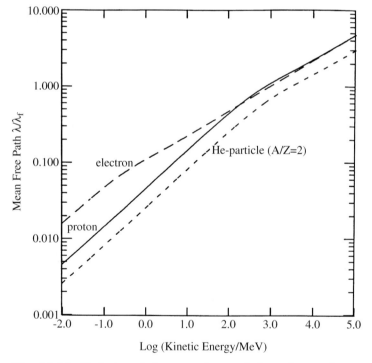

Fig. 13.19. Variation of the mean free path from fast mode waves (13.3.77) as a function of kinetic energy for three different cosmic ray particles calculated for a power spectral index $q_F = 5/3$ and neglecting any cutoff effects

For the associated spatial diffusion coefficient we obtain

$$\kappa_F\left(R_L k_{\min,F} \ll 1\right) = \frac{v}{3}\lambda_F = \frac{2}{3\pi(q_F - 1)\zeta(q_F + 1)}\left(\frac{B_0}{\delta B_F}\right)^2$$
$$\times \left[\left|\frac{\Omega_{p,0}}{\Omega_0}\right|\frac{ck_{\min,F}}{\omega_{p,i}}\right]^{2-q_F} \gamma^{2-q_F} \frac{v^2}{V_A k_{\min,F}}, \quad (13.3.80)$$

which is proportional to v^2 or $E_{\rm kin}$ at non-relativistic energies, and proportional to γ^{2-q_F} at relativistic energies.

The time scale for particles to diffuse a distance L by pitch angle scattering along the ordered magnetic field can be calculated from the spatial diffusion coefficient as

$$T_D = \frac{L^2}{\kappa_F} = \frac{3\pi(q_F - 1)\zeta(q_F + 1)}{2}\left(\frac{\delta B_F}{B_0}\right)^2 \frac{L^2(R_L k_{\min,F}\epsilon)^{q_F - 1}}{vR_L}. \quad (13.3.81)$$

13.3.4 Cosmic Ray Momentum Diffusion Coefficient from Fast Mode Waves

In the case of the momentum diffusion coefficient (13.3.37) we have to consider the two cases of flat ($1 < q_F \leq 2$) and steep ($q_F > 2$) turbulence power spectra. Using (13.3.73) and (13.3.76), respectively, we obtain

$$A_{2,F}(1 < q_F \leq 2) \simeq \frac{\pi(q_F - 1)c_1(q_F)}{4}\left(\frac{\delta B_F}{B_0}\right)^2 |\Omega|\left(R_L k_{\min,F}\right)^{q_F - 1}$$
$$\times \frac{V_A^2 p^2}{v^2}\ln\frac{v}{V_A}, \quad (13.3.82)$$

and

$$A_{2,F}(2 < q_F < 6) \simeq \frac{\pi(q_F - 1)c_4(q_F)}{4}\left(\frac{\delta B_F}{B_0}\right)^2 |\Omega| R_L k_{\min,F} \frac{V_A^2 p^2}{v^2}\ln\frac{v}{V_A}$$
$$= \frac{\pi(q_F - 1)c_4(q_F)}{4}\left(\frac{\delta B_F}{B_0}\right)^2 \frac{V_A^2 k_{\min,F}\, p^2}{v}\ln\frac{v}{V_A}, \quad (13.3.83)$$

respectively. The momentum diffusion coefficient (13.3.82) is about a factor $\ln(v/V_A)$ larger than the momentum diffusion coefficient (13.2.35) in the case of pure slab Alfvén wave turbulence.

For the associated acceleration time scale we derive

$$T_A(1 < q_F \leq 2) \equiv \frac{p^2}{A_{2,F}} = T_0 \frac{vc}{V_A^2 \ln(v/V_A)}\left[\left|\frac{\Omega_{p,0}}{\Omega_0}\right|\beta\gamma\right]^{2-q_F}, \quad (13.3.84)$$

with the constant

$$T_0 = \frac{4}{\pi(q_F - 1)c_1(q_F)} \left(\frac{B_0}{\delta B_F}\right)^2 [k_{\min}c]^{-1} \left(\frac{k_{\min}c}{\Omega_{p,0}}\right)^{2-q_F} \quad (13.3.85)$$

and

$$T_A(2 < q_F < 6) = \frac{4}{\pi(q_F - 1)c_1(q_F)} \left(\frac{B_0}{\delta B_F}\right)^2 \frac{vk_{\min,F}^{-1}}{V_A^2 \ln(v/V_A)} . \quad (13.3.86)$$

In Fig. 13.20 we calculate the acceleration time scale (13.3.84) as a function of kinetic energy assuming a value of $q_F = 5/3$ for several cosmic ray particle species. At non-relativistic energies the acceleration time scales increases proportional to $T_A(\beta \ll 1) \propto \beta^{3-q_F}/\ln(\beta/\beta_A) \simeq E_{\rm kin}^{(3-q_F)/2}$ while at relativistic energies it increases, $T_A(\gamma \gg 1) \propto \gamma^{2-q_F} \propto E_{\rm kin}^{2-q_F}$.

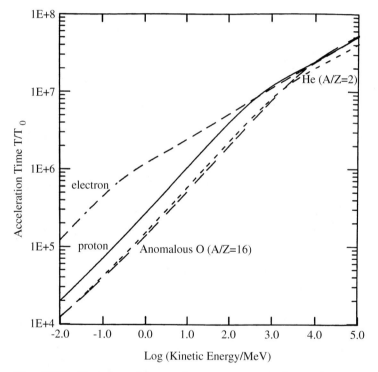

Fig. 13.20. Variation of the acceleration time scale (13.3.84) resulting from transit-time damping of fast mode waves as a function of kinetic energy for four different cosmic ray particles calculated for a power spectral index $q_F = 5/3$ and neglecting any cutoff effects

13.3.5 Fast Mode Time Scale Relation

Inspecting (13.3.81) and (13.3.84)–(13.3.86) we obtain for the product of acceleration and diffusion time scales due to fast mode wave interactions

$$[T_A(p) \cdot T_D(p)]_F \, (1 < q_F \leq 2) = \frac{6\zeta(q+1)}{c_1(q)} \left(\frac{L}{V_A}\right)^2 \left(\frac{V_A}{v}\right)^{q-1} \frac{1}{\ln(v/V_A)} \quad (13.3.87)$$

for flat turbulence spectra, which is a constant for relativistic particles ($v = c$) at all momenta. The value of the constant is independent of microscopic quantities (e.g. as the turbulence level $(\delta B)^2$) and solely given by the macroscopic properties (typical size L, Alfvén speed V_A) of the physical system considered.

In the case of steep power spectra we obtain

$$[T_A(p) \cdot T_D(p)]_F \, (2 < q_F < 6)$$
$$= \frac{6\zeta(q+1)}{c_4(q)} \left(\frac{L}{V_A}\right)^2 [R_L k_{\min,F}]^{q_F-2} \left(\frac{V_A}{v}\right)^{q-1} \frac{1}{\ln(v/V_A)} \, , (13.3.88)$$

which is about a factor $(R_L k_{\min,F})^{q_F-2} \ll 1$ smaller than (13.3.87).

13.4 Cosmic Transport Parameters from Fast Mode Waves and Slab Alfvén Waves

We will now calculate the quasilinear cosmic ray transport parameters for plasma turbulence consisting of a mixture of isotropic fast mode waves and slab Alfvén waves by combining our results from Sect. 13.3 and 13.2, respectively.

13.4.1 Modifications Due to Slab Alfvén Waves

Assuming vanishing magnetic and cross helicity of the slab Alfvén waves and adopting the power law turbulence spectrum (13.1.25) and (13.1.27) with total Alfvén wave intensity $(\delta B_A)^2$, spectral index q_A and outer scale $k_{\min,A}$, we use (13.2.9)–(13.2.10) for the two Fokker-Planck coefficients $D_{\mu\mu}$ and D_{pp} from Alfvén waves. With the same notation (13.3.53) as before, the slab Alfvén waves give rise to the additional cyclotron-resonance function

$$f_A = \frac{1}{2}\left[(1-\mu\epsilon)^2 |\mu-\epsilon|^{q_A-1} + (1+\mu\epsilon)^2 |\mu+\epsilon|^{q_A-1}\right]$$
$$\simeq \frac{1}{2}\left[\,|\mu-\epsilon|^{q_A-1} + |\mu+\epsilon|^{q_A-1}\right] \, , \quad (13.4.1)$$

which additionally contributes to the two integrals (13.3.38) and (13.3.39). Now these integrals become

$$I_\lambda = \int_0^1 d\mu \, \frac{1-\mu^2}{f_A(\mu) + f_T(\mu) + f_G(\mu)}, \tag{13.4.2}$$

and

$$I_2 = \int_0^1 d\mu \, (1-\mu^2)[f_A(\mu) + f_T(\mu) + f_G(\mu)]. \tag{13.4.3}$$

In the interval $\epsilon < \mu \leq 1$ the function

$$f_A(\epsilon < \mu \leq 1) \simeq \mu^{q_A - 1} \tag{13.4.4}$$

is comparable to the gyroresonance function (13.3.47) from fast mode waves, but is clearly smaller than the transit time damping function, see (13.3.43) and (13.3.45). In this interval we may neglect the contribution of the Alfvén wave function f_A to the integrals (13.4.2) and (13.4.3).

However, in the interval $0 \leq \mu < \epsilon$ the function

$$f_A(0 \leq \mu \leq \epsilon) \simeq \epsilon^{q_A - 1} \tag{13.4.5}$$

is larger than the gyroresonance contribution (13.3.47a), provided the power spectrum spectral indices of Alfvén (q_A) and fast mode (q_F) waves fulfill

$$q_A < q_F + 1, \tag{13.4.6}$$

where we tacitly assume that the total wave intensities $(\delta B_A)^2$ and $(\delta B_F)^2$ are of comparable orders of magnitude.

As a consequence the integral (13.4.3) remains unchanged as compared to the fast mode values, i.e. (13.3.73) and (13.3.76) still hold, so that the momentum diffusion is still dominated by transit-time damping of the fast mode waves. However, the integral (13.4.2) instead of (13.3.55) and (13.3.58) now becomes

$$I_\lambda \simeq \int_0^\epsilon d\mu \, \frac{1-\mu^2}{f_A(\mu)} \simeq \epsilon^{2-q_A}. \tag{13.4.7}$$

13.4.2 Transport Parameters in the Case of Admixture of Slab Alfvén Waves to Fast Mode Waves

Generalizing the result (13.3.77) in the presence of slab Alfvén waves and fast mode waves of arbitrary outer scale ($k_{\min,F}, k_{\min,A}$), total intensities ($\delta B_F^2, \delta B_A^2$) and power law shape ($q_F, q_A$) the cosmic ray mean free path (13.3.36) is

$$\lambda = \frac{3}{\pi(q_A - 1)} \frac{B_0^2}{(\delta B_A)^2} k_{\min,A}^{-1} (k_{\min,A} R_L \epsilon)^{2-q_A}$$

$$\times \left[1 + \frac{3\zeta(q_F + 1)(q_F - 1)}{2(q_A - 1)} \frac{(\delta B_F)^2}{(\delta B_A)^2} \epsilon \, k_{\min,A}^{1-q_A} k_{\min,F}^{q_F - 1} (R_L \epsilon)^{q_F - q_A}\right]^{-1}$$

$$= \frac{3}{\pi(q_A-1)} \frac{B_0^2}{(\delta B_A)^2} k_{\min,A}^{-1} \left(k_{\min,A} V_A / |\Omega_0|\right)^{2-q_A} \gamma^{2-q}$$

$$\times \left[1 + \frac{3\zeta(q_F+1)(q_F-1)}{2(q_A-1)} \frac{(\delta B_F)^2}{(\delta B_A)^2} \epsilon k_{\min,A}^{1-q_A} k_{\min,F}^{q_F-1} \left(\frac{V_A \gamma}{|\Omega_0|}\right)^{q_F-q_A}\right]^{-1},$$

(13.4.8)

whereas the momentum diffusion coefficients (13.3.82) and (13.3.83) remain unchanged.

Obviously, the detailed variation of the mean free path (13.4.8) depends sensitively on the assumptions concerning the intensities $(\delta B_F^2, \delta B_A^2)$, shape (q_F, q_A) and outer scale $(k_{\min,F}, k_{\min,A})$ of the two wave fields.

13.4.2.1 Equal Intensity, Shape and Outer Scale. We will consider here only the most simple possibility that Alfvén and fast mode waves have equal intensities, identical shapes and outer scales, i.e.

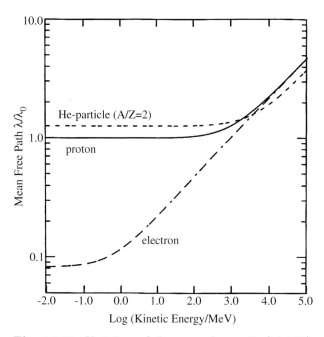

Fig. 13.21. Variation of the mean free path (13.4.10) as a function of kinetic energy for three different cosmic ray particles in the case of admixture of slab Alfvén waves to fast mode waves assuming equal intensities, shapes and outer scales and neglecting any cutoff effects. A turbulence power spectral index of $q = 5/3$ is assumed. The neglect of cutoff effects is particularly problematic for electrons with energies below 10^3 MeV, and probably leads to a severe underestimation of the electron mean free path

$$\delta B_{\rm F}^2 = \delta B_{\rm A}^2 = \delta B_{\rm t}^2/2\,, \quad q_{\rm F} = q_{\rm A} = q\,, \quad k_{\rm min,F} = k_{\rm min,A} = k_{\rm min}\,. \quad (13.4.9)$$

The examination of more general cases is left as an exercise for the interested reader.

The mean free path in this case reduces to

$$\begin{aligned}
\lambda &= \frac{6}{\pi(q-1)} \frac{B_0^2}{(\delta B_{\rm t})^2} k_{\rm min}^{-1} (k_{\rm min} R_{\rm L} \epsilon)^{2-q} \left[1 + \frac{3\zeta(q+1)}{2}\epsilon\right] \\
&\simeq \frac{6}{\pi(q-1)} \frac{B_0^2}{(\delta B_{\rm t})^2} k_{\rm min}^{-1} (k_{\rm min} V_{\rm A}/|\Omega_0|)^{2-q} \gamma^{2-q} = \lambda_0 \left[\left|\frac{\Omega_{\rm p,0}}{\Omega_0}\right| \gamma\right]^{2-q},
\end{aligned}$$
(13.4.10)

where

$$\lambda_0 = \frac{6}{\pi(q-1)} \frac{B_0^2}{(\delta B_{\rm t})^2} k_{\rm min}^{-1} (k_{\rm min} c/\omega_{\rm p,i})^{2-q}\,. \quad (13.4.11)$$

In Fig. 13.21 we show the mean free path (13.4.10) for different cosmic ray species. As can be seen the mean free path is proportional to $(\gamma/|\Omega_0|)^{2-q} \propto (E_{\rm tot}/Z)^{2-q}$ which is a constant determined by $|\Omega_0|^{q-2} = (|Z|e/m)^{q-2}$ at non-relativistic energies and varies $\propto (E_{\rm tot}/Z)^{2-q}$ at relativistic energies.

14. Acceleration and Transport Processes of Cosmic Rays

Essential to an explanation of most phenomena of high-energy astrophysics is the understanding of the physics of the acceleration and transport of the energetic charged particles, i.e. the cosmic rays. The in-situ study of these transport and acceleration processes in interplanetary space offers a unique opportunity to test our conceptual understanding of cosmic ray dynamics, and thus provides an increased confidence in the applicability of the developed concepts. The transport and acceleration of cosmic rays in interplanetary space are quite similar to those in other cosmic objects. Quite generally, the cosmic ray particles have to propagate in a collisionless, high-conductive, magnetized and tenuous background plasma consisting mainly of protons and electrons. Very often the energy density of cosmic ray particles is comparable to that of the background medium, the magnetic field and the convective motion of the medium. As a consequence, the electromagnetic fields in the system are severely influenced by the cosmic ray particles, and the description of cosmic ray transport and acceleration is more complex than solving the equation of motion of charged cosmic ray test particles in a fixed and given electromagnetic field (see the discussion in Chap. 8).

14.1 Structure of the Cosmic Ray Transport Equation

The most important and generally used approximation to cosmic ray dynamics is the diffusion approximation, which we have developed in the last two chapters. This is based chiefly on the observation that energetic particles in many important cases are observed to have a very nearly isotropic pitch angle distribution relative to the local plasma. This isotropy is a consequence of the fast scattering of cosmic ray particles by the low-frequency magnetohydrodynamic turbulence as, for example, Alfvén waves and fast magnetosonic waves, whose magnetic field is much larger than their electric field components ($|\delta \boldsymbol{B}| = (c/V_\mathrm{A})|\delta \boldsymbol{E}|$, Alfvén velocity $V_\mathrm{A} \ll c$). In the mixed comoving coordinate system (Sect. 12.3.2.1) the isotropic part of the particle's distribution function $M_a(\boldsymbol{x}, p, t)$ obeys the diffusion-convection equation (12.3.31) for non-relativistic bulk speeds of the background medium, which may be written with the various physical effects noted next to the corresponding terms

$$\frac{\partial M_a}{\partial t} - S_a(\boldsymbol{x}, \boldsymbol{p}^*, t)$$
$$= \frac{\partial}{\partial z} \kappa_{zz} \frac{\partial M_a}{\partial z} + \frac{\partial}{\partial X}\left[\kappa_{XX}\frac{\partial M_a}{\partial X} + \kappa_{XY}\frac{\partial M_a}{\partial Y}\right]$$
$$+ \frac{\partial}{\partial Y}\left[\kappa_{YY}\frac{\partial M_a}{\partial Y} + \kappa_{YX}\frac{\partial M_a}{\partial X}\right] \quad \text{(spatial diffusion)}$$
$$- \left[U + \frac{1}{4p^{*2}}\frac{\partial(p^{*2}v^* A_1)}{\partial p^*}\right]\frac{\partial M_a}{\partial z} \quad \text{(spatial convection)}$$
$$+ \frac{1}{p^{*2}}\frac{\partial}{\partial p^*}\left(p^{*2} A_2 \frac{\partial M_a}{\partial p^*}\right) \quad \text{(momentum diffusion)}$$
$$+ \left[\frac{p^*}{3}\frac{\partial U}{\partial z} + \frac{v^*}{4}\frac{\partial A_1}{\partial z}\right]\frac{\partial M_a}{\partial p^*} \quad \text{(momentum convection)}. \tag{14.1.1}$$

In the preceding chapter we showed that the presence or absence of the seven transport parameters depends solely on the nature and the statistical properties of the plasma turbulence in the background medium. In particular, perpendicular spatial diffusion does not occur for slab Alfvénic turbulence (see (13.2.6)) which controls the spatial transport even in the presence of fast mode waves (Sect. 13.4). Omitting perpendicular diffusion reduces (14.1.1) to

$$\frac{\partial M_a}{\partial t} - S_a(\boldsymbol{x}, p^*, t) = \frac{\partial}{\partial z}\kappa_{zz}\frac{\partial M_a}{\partial z} - V\frac{\partial M_a}{\partial z}$$
$$+ \frac{1}{p^{*2}}\frac{\partial}{\partial p^*}\left(p^{*2} A_2 \frac{\partial M_a}{\partial p^*}\right) + \frac{p^*}{3}\frac{\partial V}{\partial z}\frac{\partial M_a}{\partial p^*}, \tag{14.1.2}$$

where we introduced the effective cosmic ray bulk speed

$$V \equiv U + \frac{1}{4p^{*2}}\frac{\partial(p^{*2}v^* A_1)}{\partial p^*}. \tag{14.1.3}$$

This introduction is possible since the rate of adiabatic deceleration (12.3.28) in general scales as

$$A_1(z, p) = A_1^0(z)\, p/v. \tag{14.1.4}$$

Equation (14.1.2) includes both spatial parallel diffusion and convection terms ("transport of cosmic rays"), as well as momentum diffusion and convection terms ("acceleration of cosmic rays"). Since the pioneering work of Fermi (1949 [162], 1954 [164]) it has become customary to refer to the latter two as Fermi acceleration of the second- and first-order, respectively. Note that the momentum convection term can describe both acceleration and cooling of particles depending on whether the velocity gradient dV/dz is negative or positive, respectively.

14.1.1 Inclusion of Momentum Losses

The diffusion-convection equation (14.1.2) adequately describes the effects of particle-wave interactions but is incomplete with respect to the additional momentum loss processes that we discussed in Chaps. 4 and 5, resulting from interactions with ambient target photon and matter fields, as Coulomb interactions, ionization, pion production, bremsstrahlung, spallation, synchrotron radiation and inverse Compton interactions. If these particle interaction processes operate on time scales of the same order of magnitude as the diffusion and convection time scales, (14.1.2) has to be complemented by additional loss terms. The loss processes have to be divided into two groups: continuous and catastrophic losses. Continuous loss processes (such as synchrotron radiation) are those through which the interacting particle loses momentum but which conserve the total number of particles. In contrast catastrophic loss processes (such as spallation or pair annihilation) refer to those where the total number of particles changes.

Continuous loss processes are accounted for by the additional term

$$-\frac{1}{p^{*2}}\frac{\partial}{\partial p^*}\left(p^{*2}\dot{p}^* M_a\right)$$

in the transport equation (14.1.2) with the appropriate negatively counted loss rate $\dot{p}^*$ calculated in Chaps. 4 and 5. Catastrophic loss processes are accounted for by the additional term

$$-\frac{M_a}{T_c(p^*, z)}$$

in the transport equation (14.1.2) with the appropriate loss time $T_c(p, z)$ calculated in Chaps. 4 and 5.

The diffusion-convection transport equation then becomes

$$\frac{\partial M_a}{\partial t} - S_a(\boldsymbol{x}, p, t) = \frac{\partial}{\partial z}\kappa_{zz}\frac{\partial M_a}{\partial z} - V\frac{\partial M_a}{\partial z}$$
$$+ \frac{1}{p^2}\frac{\partial}{\partial p}\left(p^2 A_2 \frac{\partial M_a}{\partial p} - p^2 \dot{p} M_a\right)$$
$$- \frac{M_a}{T_c(z, p)} + \frac{p}{3}\frac{\partial V}{\partial z}\frac{\partial M_a}{\partial p}, \qquad (14.1.5)$$

where for brevity we omit the index $*$, i.e. we set $p = p^*$ with the understanding that the transport equation (14.1.5) holds in the mixed comoving coordinate system.

Noting that

$$-\frac{\partial}{\partial z}[VM_a] + \frac{1}{p^2}\frac{\partial}{\partial p}\left[\frac{p^3}{3}\frac{\partial V}{\partial z}M_a\right] = \frac{p}{3}\frac{\partial V}{\partial z}\frac{\partial M_a}{\partial p} - V\frac{\partial M_a}{\partial z},$$

we can cast (14.1.5) in the alternative form

$$\frac{\partial M_a}{\partial t} - S_a(\boldsymbol{x}, p, t)$$
$$= \frac{\partial}{\partial z}\left[\kappa_{zz}\frac{\partial M_a}{\partial z} - V M_a\right]$$
$$+ \frac{1}{p^2}\frac{\partial}{\partial p}\left(p^2 A_2 \frac{\partial M_a}{\partial p} + \frac{p^3}{3}\frac{\partial V}{\partial z} M_a - p^2 \dot{p} M_a\right) - \frac{M_a}{T_c(z,p)}. \quad (14.1.6)$$

In this way the generalized convection-diffusion equation (14.1.6)–hereafter referred to as transport equation–contains, as a special case, nearly all of the cosmic ray acceleration and transport mechanisms discussed in the literature, and will be the basis of most of our applications in this and the following chapters. Of course, in many cases, some of the terms in this equation may be omitted to simplify the analysis. In addition, sometimes the transport equation (14.1.6) is still not sufficiently accurate and has to be augmented by additional terms.

Nearly all models of cosmic ray transport and acceleration in astrophysics make use of this transport equation, or some simplification of it. As we will discuss in the following chapters, the propagation of cosmic rays in galaxies (Chap. 17) follows (14.1.6), and the basic theory of particle acceleration at hydrodynamic shock waves (Sect. 16.4) also uses (14.1.6). In view of the wide-reaching importance of this equation, the fact that it applies also to cosmic ray transport in the solar wind, which may be observed in much more detail than more-distant regions, allows us to subject it to more stringent tests (Chap. 15) than might otherwise be possible (Jokipii 1983 [245]). Before we get to these tests and particular applications we discuss some necessary formal aspects related to the basic transport equation (14.1.6).

14.1.2 Continuous and Catastrophic Momentum Losses

The different representation of continuous and catastrophic losses in (14.1.6) becomes obvious if we multiply this equation with $4\pi p^2 dp$ and integrate over all momenta from $p = 0$ to $p = \infty$. Denoting with

$$N_a(\boldsymbol{x}, t) = 4\pi \int_0^\infty dp\, p^2 M_a(\boldsymbol{x}, p, t) \quad (14.1.7)$$

and

$$Q_a(\boldsymbol{x}, t) = 4\pi \int_0^\infty dp\, p^2 S_a(\boldsymbol{x}, p, t) \quad (14.1.8)$$

the total number density cosmic ray particles of type a at position $\boldsymbol{x}$ and the corresponding total injection rate, respectively, we then find from (14.1.6) the relation

14.1 Structure of the Cosmic Ray Transport Equation

$$\frac{\partial N_a}{\partial t} - Q_a(\boldsymbol{x}, t) = \frac{\partial}{\partial z}\left[\bar{\kappa}_{zz}\frac{\partial N_a}{\partial z} - VN_a\right] - \frac{N_a}{T_c(z)} + F(0, \infty), \quad (14.1.9)$$

where

$$\bar{\kappa}_{zz} \equiv \frac{4\pi \int_0^\infty p^2 \kappa_{zz} M_a}{N_a}, \quad (14.1.10)$$

$$T_c(z) \equiv \frac{N_a}{4\pi \int_0^\infty \mathrm{d}p\, p^2 M_a/T_c(z, p)} \quad (14.1.11)$$

denote the momentum-averaged spatial diffusion coefficient and catastrophic loss time, respectively.

The expression

$$F(0, \infty) \equiv 4\pi \left[p^2 \left(A_2 \frac{\partial M_a}{\partial p} - \dot{p}M_a + \frac{p}{3}\frac{\partial V}{\partial z}M_a \right) \right]_{p=0}^{p=\infty} = 0, \quad (14.1.12)$$

represents the flux of particles in momentum space. It seems to make little physical sense for particles to traverse into a region of negative absolute momentum, or to cross the boundary at $p = \infty$ to "higher" momentum values, so that the "zero-energy-flux" boundary condition (14.1.12) is well justified. Integrating the balance equation (14.1.9) for the total particle density over the whole volume of the considered cosmic ray confinement region V, using Gauss' theorem to convert the volume integral into a surface integral, we obtain for the balance of the total number of particles

$$\mathcal{N}_a(t) = \int_V \mathrm{d}^3 x\, N_a(\boldsymbol{x}, t), \quad \mathcal{Q}_a(t) = \int_V \mathrm{d}^3 x\, Q_a(\boldsymbol{x}, t), \quad (14.1.13)$$

that

$$\frac{\partial \mathcal{N}_a}{\partial t} = \mathcal{Q}_a(t) + \mathcal{G}_a - \frac{\mathcal{N}_a}{\mathcal{T}_c}, \quad (14.1.14)$$

where

$$\mathcal{G}_a = \int_S \mathrm{d}\boldsymbol{\sigma} \cdot \left[\bar{\kappa}_{zz}\frac{\partial N_a}{\partial z} - VN_a\right] \quad (14.1.15)$$

denotes the number flux of cosmic rays escaping through the surface (S) of the cosmic ray confinement region.

The balance equation (14.1.14) has a very simple interpretation: the number of cosmic ray particles in the confinement region can only change by

1. particles being injected from sources with rate $\mathcal{Q}_a$,
2. particles being destroyed in catastrophic spallation and/or annihilation interactions characterized by the volume-averaged loss time $\mathcal{T}_c$, and
3. by particles escaping from the confinement region through the boundaries.

Continuous loss and acceleration processes only move the particles around in momentum space but conserve the total number of particles, and therefore do not appear in the balance equations (14.1.9) and (14.1.14).

As an aside we note that it is possible to calculate also higher moments of the cosmic ray transport equation (14.1.6) in the same way, and thus derive balance equations for the cosmic ray pressure by multiplying with $4\pi p^2 v$ before integrating over p (see e.g. Schlickeiser and Lerche 1985 [466]).

14.2 Steady-State and Time-Dependent Transport Equations

Generalizing the fundamental cosmic ray transport equation (14.1.6) back to 3-dimensional spatial transport and introducing the spatial operator

$$\mathcal{L}_x M_a \equiv \nabla \cdot [\kappa(\boldsymbol{x}, p)\nabla M_a - \boldsymbol{V} M_a] \tag{14.2.1}$$

and the momentum operator

$$\mathcal{L}_p M_a \equiv \frac{1}{p^2}\frac{\partial}{\partial p}\left(p^2 A_2 \frac{\partial M_a}{\partial p} + \frac{p^3}{3}\nabla \cdot \boldsymbol{V} M_a - p^2 \dot{p} M_a\right) - \frac{M_a}{T_c(z,p)}, \tag{14.2.2}$$

we may briefly write for the full time-dependent transport equation (14.1.6)

$$\frac{\partial M_a}{\partial t} - S_a(\boldsymbol{x}, p, t) = \mathcal{L}_x M_a + \mathcal{L}_p M_a . \tag{14.2.3}$$

Applying the Laplace transform to (14.2.3), i.e. expressing

$$M_a(\boldsymbol{x}, p, t) = \int_0^\infty ds\, F_a(\boldsymbol{x}, p, s)\, e^{-st}, \tag{14.2.4}$$

and

$$S_a(\boldsymbol{x}, p, t) = \int_0^\infty ds\, Q_a(\boldsymbol{x}, p, s)\, e^{-st}, \tag{14.2.5}$$

we obtain for the transformed distribution function the equation

$$\mathcal{L}_x F_a + \mathcal{L}_p F_a + s F_a = -Q_a(\boldsymbol{x}, p, s) - M_a(\boldsymbol{x}, p, t=0) . \tag{14.2.6}$$

On the other hand the steady-state version of (14.2.3) is

$$\mathcal{L}_x M_a(\boldsymbol{x}, p) + \mathcal{L}_p M_a(\boldsymbol{x}, p) = -S_a(\boldsymbol{x}, p) , \tag{14.2.7}$$

whose mathematical structure is almost identical to (14.2.6): the only differences occur in the source term, with the additional appearance of the initial value of M_a at time $t = 0$ in (14.2.6), and with respect to the catastrophic loss time, being simply M_a/T_c in the steady-state case (14.2.7) but F_a/T_s with

$$T_s^{-1} = T_c^{-1} + s \tag{14.2.8}$$

in the time-dependent case (14.2.6). It is therefore convenient to solve first the steady-state transport equation (14.2.7). With the replacement (14.2.8) we can use these also as solutions for the Laplace-transformed time-dependent distribution function (14.2.4). By Laplace inversion the time-dependent solution $M_a(\boldsymbol{x}, p, t)$ is obtained.

14.3 Scattering Time Method: Separation of Spatial and Momentum Problem

An important class of exact analytical solutions of the transport equation (14.2.7) can be obtained (Wang and Schlickeiser 1987 [553], Lerche and Schlickeiser 1988 [298]) if the spatial and momentum operators can be separated as

$$\mathcal{L}_{\boldsymbol{x}} = g(p)\mathcal{O}_{\boldsymbol{x}}, \quad \mathcal{L}_p = h(\boldsymbol{x})\mathcal{O}_p \tag{14.3.1}$$

where $\mathcal{O}_{\boldsymbol{x}}$ is a coordinate-space operator only and $\mathcal{O}_p$ is a momentum-space dependent operator only, and if we take the source term $S_a(\boldsymbol{x}, p)$ to be a single product of two separable functions, i.e.

$$S_a(\boldsymbol{x}, p) = Q_1(\boldsymbol{x})Q_2(p) . \tag{14.3.2}$$

The condition (14.3.2) is taken for ease of exposition, and can easily be relaxed to more general source representations (see Lerche and Schlickeiser 1988 [298], Appendix A).

14.3.1 Formal Mathematical Solution

Under these two conditions (14.2.7) is solved by the convolution

$$M_a(\boldsymbol{x}, p) = \int_0^\infty du\, G(p, u) P(\boldsymbol{x}, u) , \tag{14.3.3}$$

where $G(p, u)$ satisfies given momentum boundary conditions and

$$\frac{\partial G}{\partial u} = \frac{1}{g(p)} \mathcal{O}_p G \tag{14.3.4}$$

with

$$G(p, u = 0) = Q_2(p)/g(p) \tag{14.3.5}$$
$$G(p, u = \infty) = 0 , \tag{14.3.6}$$

and where $P(\boldsymbol{x}, u)$ satisfies given spatial boundary conditions and

$$\frac{\partial P}{\partial u} = \frac{1}{h(\boldsymbol{x})}\mathcal{O}_{\boldsymbol{x}}P \qquad (14.3.7)$$

with

$$P(\boldsymbol{x}, u = 0) = Q_1(\boldsymbol{x})/h(\boldsymbol{x}) \qquad (14.3.8)$$
$$P(\boldsymbol{x}, u = \infty) = 0. \qquad (14.3.9)$$

$P(\boldsymbol{x}, u)$, if normalized as

$$\int_0^\infty du\, P(\boldsymbol{x}, u) = 1, \qquad (14.3.10)$$

is commonly referred to as the *age distribution* of cosmic rays at position $\boldsymbol{x}$, or as the *cosmic ray propagator*. After choosing the appropriate geometry and spatial and momentum boundary conditions, we can calculate the steady-state particle phase space density at any position by solving the two partial differential equations (14.3.4) and (14.3.7) and convolving according to (14.3.3).

The convolution (14.3.3) can also be used if more general (than diffusion and convection) spatial transport modes (such as compound diffusion, subdiffusion and/or superdiffusion (Duffy et al. 1995 [145], Ragot and Kirk 1997 [414])) are considered. However, here we limit our discussion to spatial diffusion and convection only. Instead of solving the two partial differential equations (14.3.4) and (14.3.7), in this case a simpler approach is possible.

14.3.2 Leaky-Box Equations

In the case of diffusion and convection the spatial operator $\mathcal{O}_{\boldsymbol{x}}$ (see (14.2.1) and (14.3.1a)) is of Sturm-Liouville type (e.g. Arfken 1970 [17]) and thus has a complete eigenfunction system $H_i(\boldsymbol{x})$. The age distribution $P(\boldsymbol{x}, u)$ then can be expanded in this orthonormal system as

$$P(\boldsymbol{x}, u) = \sum_i A_i(\boldsymbol{x})\, e^{-\lambda_i^2 u}, \qquad (14.3.11)$$

with

$$A_i(\boldsymbol{x}) = c_i H_i(\boldsymbol{x}), \qquad (14.3.12)$$

where λ_i^2 denote the eigenvalues of the spatial operator, being determined by the spatial boundary conditions, and where the expansion coefficients c_i have to be calculated from the orthonormality relations of H_i and the normalization (14.3.10).

Inserting the expansion (14.3.11) into the convolution (14.3.3) leads to

$$M_a(\boldsymbol{x}, p) = \sum_i A_i(\boldsymbol{x}) R_i(p), \qquad (14.3.13)$$

where each of the functions

$$R_i(p) \equiv \int_0^\infty du\, G(p,u)\, e^{-\lambda_i^2 u} \tag{14.3.14}$$

obeys the ordinary differential equation

$$\mathcal{O}_p R_i(p) - \lambda_i^2 g(p) R_i(p) = -Q_2(p), \tag{14.3.15}$$

commonly referred to as the "leaky-box equation".

Equation (14.3.15) is most easily derived by multiplying (14.3.4) by $P(\boldsymbol{x}, u)$, as given in (14.3.11), and integrating over u from 0 to ∞ to obtain

$$\sum_i A_i(\boldsymbol{x})\mathcal{O}_p\, R_i(p) = g(p) \sum_i A_i(\boldsymbol{x}) \int_0^\infty du\, \frac{\partial G}{\partial u} e^{-\lambda_i^2 u}$$

$$= g(p) \sum_i A_i(\boldsymbol{x}) \left[-G(p, u=0) + \lambda_i^2 R_i(p) \right], \tag{14.3.16}$$

where the definition (14.3.14) has been used and the right-hand side has been integrated by parts using (14.3.6). After substituting the initial condition (14.3.5) and transposing, this becomes

$$\sum_i A_i(\boldsymbol{x}) \left[\mathcal{O}_p R_i(p) - \lambda_i^2 g(p) R_i(p) + Q_2(p) \right] = 0, \tag{14.3.17}$$

from which (14.3.15) follows.

We see from (14.3.13) that the steady-state phase space density $M_a(\boldsymbol{x}, p)$ can be expressed as an (infinite) sum of products of functions $R_i(p)$ and $A_i(\boldsymbol{x})$. The spatial functions $A_i(\boldsymbol{x})$ are determined by the cosmic ray age distribution at location $\boldsymbol{x}$ and depend on the geometry of the considered cosmic ray confinement region, the spatial boundary conditions and the spatial source distribution. Each momentum function $R_i(p)$ satisfies a leaky-box equation (14.3.15) where the corresponding spatial eigenvalues λ_i^2 enter in the form of an inverse catastrophic loss time.

14.3.3 Illustrative Example: Galactic Cosmic Ray Diffusion 1

We illustrate the formal solution of the steady-state transport equation developed above by modeling the propagation of cosmic ray particles in the Galaxy at large particle momenta $p > 10$ GeV/c where parallel spatial diffusion dominates convection and adiabatic deceleration. Because of the geometry of our Galaxy as a highly flattened cylinder we adopt a model, in which the cosmic ray sources are distributed in a uniform ($Q_1(\boldsymbol{x}) = q_1 = $ const.), axisymmetric, cylindrical disk of radius R_0 and half-thickness z_0, and the confinement region is a concentric disk with radius $R_1 > R_0$ and half-thickness $z_1 > z_0$. At the edges of the confinement region, we apply free-escape boundary conditions in the form

$$P(R, z = \pm z_1, u) = P(R = R_1, z, u) = 0 \,. \tag{14.3.18}$$

We assume that the spatial diffusion coefficient is momentum dependent but independent of position, i.e.

$$\kappa_{zz}(\boldsymbol{x}, p) = K_0 \kappa(p) \,, \tag{14.3.19}$$

with $\kappa(p)$ a dimensionless function. Moreover, we assume that the momentum operator (14.3.1b) is independent of position (i.e. $h(\boldsymbol{x}) = 1$). The cosmic ray age distribution can then be derived from the solutions of (14.3.7)–(14.3.8) which in cylindrical coordinates with no azimuthal dependence read

$$\frac{\partial P}{\partial u} = K_0 \left[\frac{1}{R} \left(R \frac{\partial P}{\partial R} \right) + \frac{\partial^2 P}{\partial z^2} \right] \,, \tag{14.3.20}$$

$$P(R, z, u = 0) = q_1 H\left[R_0 - R\right] H\left[z_0 - |z|\right] \,, \tag{14.3.21}$$

where $H[x]$ is the Heaviside step function. Demanding that the solution is finite at $R = 0$ and symmetric with respect to the galactic plane $z = 0$ the solution of (14.3.20) is (Le Guet and Stanton 1974 [288])

$$P(R, z, u) = \sum_{l=0}^{\infty} \sum_{m=1}^{\infty} A_{lm}(R, z) \, \exp\left(-\lambda_{lm}^2 u\right) \,, \tag{14.3.22}$$

where

$$A_{lm}(R, z) = c_{lm} \cos\left[\frac{(2l+1)\pi z}{2z_1}\right] J_0\left(\frac{y_m R}{R_1}\right) \,, \tag{14.3.23}$$

$$\lambda_{lm}^2 = K_0 \left(\frac{y_m^2}{R_1^2} + \frac{(2l+1)^2 \pi^2}{4 z_1^2} \right) \,. \tag{14.3.24}$$

Here, $J_0(y)$ is the Bessel function of the first kind and order zero; y_m denotes the mth zero of $J_0(y)$ and is tabulated in Abramowitz and Stegun (1972 [3]). Solution (14.3.22) indeed is of the form (14.3.11)–(14.3.12).

The expansion coefficients c_{lm} in (14.3.23) are determined by applying the initial condition (14.3.21) and making use of the orthonormality relation for Bessel functions

$$\int_0^{R_1} dR\, R\, J_0\left(y_m \frac{R}{R_1}\right) J_0\left(y_n \frac{R}{R_1}\right) = \frac{1}{2} R_1^2 J_1^2(y_m) \delta_{mn} \,. \tag{14.3.25}$$

One thus obtains

$$c_{lm} = \frac{q_1}{z_1 R_1^2 J_1^2(y_m)} \int_0^{R_0} dR\, R\, J_0\left(y_m \frac{R}{R_1}\right) \int_{-z_0}^{z_0} dz\, \cos\left[\frac{(2l+1)\pi z}{2z_1}\right] \,. \tag{14.3.26}$$

This completes the derivation of the cosmic ray age distribution.

Following (14.3.13)ff. the full solution of the galactic cosmic rays obviously is given by

$$M_a(R, z, p) = \sum_{l=0}^{\infty} \sum_{m=1}^{\infty} A_{lm}(R, z) N_{lm}(p) ,\qquad (14.3.27)$$

where each function $N_{lm}(p)$ obeys the leaky-box equation

$$\mathcal{O}_p N_{lm}(p) - \lambda_{lm}^2 \kappa(p) N_{lm}(p) = -Q_2(p) ,\qquad (14.3.28)$$

which indeed is of the form (14.3.15). We will continue solving this example in Sect. 17.1.

14.3.4 Momentum Problem

The use of the scattering time method has left us with leaky-box type equations of the general form (14.3.15). Spatial inhomogeneities, geometries and source distributions do not enter these equations but, of course, are incorporated correctly both in terms of the spatial eigenfunctions and expansion coefficients (14.3.12) and the spatial eigenvalues λ_i^2. With the explicit form of the momentum operator (14.2.2) and (14.3.1) the leaky-box equation (14.3.15) reads

$$\frac{1}{p^2} \frac{d}{dp} \left(p^2 A_2(p) \frac{dR_i}{dp} + V_1 p^3 R_i - p^2 \dot{p}(p) R_i \right)$$
$$- \left[\lambda_i^2 g(p) + T_c^{-1}(p) \right] R_i(p) = - Q_2(p) ,\qquad (14.3.29)$$

where we introduced the constant

$$V_1 \equiv \nabla \cdot \mathbf{V}/3 . \qquad (14.3.30)$$

We recall that (14.3.29) is exact when all terms of the momentum operator have the same spatial dependence, which requires that V_1 has to be constant, and that $g(p)$ denotes the momentum dependence of the spatial operator. Equation (14.3.29) can also be written in self-adjoint form as

$$\frac{d}{dp} \left[p^2 A_2(p) s(p) \frac{dR_i}{dp} \right] - r_i(p) s(p) R_i(p) = -p^2 s(p) Q_2(p) ,\qquad (14.3.31)$$

where

$$s(p) \equiv \exp \left[\int^p dt \frac{V_1 t - \dot{p}(t)}{A_2(t)} \right] ,\qquad (14.3.32)$$

and

$$r_i(p) = \frac{d}{dp} \left[p^2 \dot{p}(p) \right] + p^2 \left[\lambda_i^2 g(p) + T_c^{-1}(p) - 3V_1 \right] .\qquad (14.3.33)$$

We solve the self-adjoint equation (14.3.31) for general source distributions $Q_2(p)$ in terms of the Green's function $G_i(p, p_0)$ as

$$R_i(p) = \int_0^\infty dp_0\, p_0^2 s(p_0) Q_2(p_0)\, G_i(p, p_0)\,, \tag{14.3.34}$$

where the Green's function satisfies the equation

$$\frac{d}{dp}\left[p^2 A_2(p) s(p) \frac{dG_i}{dp}\right] - r_i(p) s(p) G_i(p) = -\delta(p - p_0)\,, \tag{14.3.35}$$

and is constructed from two linear independent solutions $U_{1,2}(p)$ of the corresponding homogenous (i.e. $Q_2(p) = 0$) equation to (14.3.31) by the standard procedure.

14.3.4.1 Momentum Boundary Conditions. Equation (14.3.35) is an ordinary singular differential equation of second order in momentum and in order to solve it, two boundary conditions at $p = 0$ and $p = \infty$ have to be specified. From a purely mathematical point of view, Park and Petrosian (1995 [386]) have emphasized that proper care of these boundaries has to be taken since the relevant momentum interval $0 \le p \le \infty$ is not finite. The spectral theory for second-order differential equations, explained in detail in Stakgold (1967 [513]), is an extension of the familiar Sturm-Liouville eigenfunction expansion theory for regular second-order differential equations, which refers to finite variable intervals, and is fairly well developed. Park and Petrosian (1995 [386]) also noted that the mathematical apparatus to handle singular momentum problems is at hand, and can be employed to treat these problems correctly.

However, the choice of the correct momentum boundary conditions certainly is more complex than this mathematical aspect. Recalling that all the expansions were done based on the smallness of $V_{\rm ph}/v \ll 1$ of the plasma wave phase velocities as compared to the individual cosmic ray particle velocities, made particularly in Sect. 12.3 to derive the diffusion-convection transport equation, our basic transport equation (14.1.6) is invalid at small particle momentum values. Moreover, the representation of elastic Coulomb scatterings between the ions and electrons (e.g. Rosenbluth et al. 1957 [440]) as simple continuous Coulomb loss terms of individual cosmic ray test particles is appropriate at large particle momentum, but is inadequate at suprathermal particle kinetic energies near the mean thermal energy of the plasma. An improved approach for the dynamics of suprathermal but non-relativistic particles has been developed by Parker and Tidman (1958a [390],1958b [391]) (see also the discussion in Schlickeiser and Steinacker (1989 [470], Appendix)). They argue that a more (than (14.1.6)) complete particle transport equation may be written as

$$\frac{\partial M_a}{\partial t} - S_a(\boldsymbol{x}, p, t) = \left(\frac{\partial M_a}{\partial t}\right)_C + \left(\frac{\partial M_a}{\partial t}\right)_F, \tag{14.3.36}$$

where $(\partial M_a/\partial t)_\mathrm{F}$ stands for the right-hand side of (14.1.6) and represents the effects of particle-wave interactions and continuous and catastrophic momentum loss processes, and $(\partial M_a/\partial t)_\mathrm{C}$ is the contribution of Coulomb interactions between the particles which, in the case of isotropic distributions, has been calculated by MacDonald et al. (1957 [311]). We are particularly interested in the transfer of particles on the high-velocity suprathermal tail of the distribution function. Parker and Tidman (1958a [390]) have noted that in this velocity range the Coulomb interaction term reduces to

$$\left(\frac{\partial M_a}{\partial t}\right)_\mathrm{C} \simeq \frac{\alpha}{v^2}\frac{\partial}{\partial v}\left[\frac{1}{v}\frac{\partial M_a}{\partial v} + \frac{m_a}{k_\mathrm{B}T}M_a\right], \qquad (14.3.37)$$

where $\alpha = 4\pi N_a k_\mathrm{b} T q_a^4 \ln \Lambda / m_a^3$ with $\Lambda = 1.34 \times 10^4 T^{3/2} N_a^{-1/2}$ in cgs units. Obviously, the steady-state solution of (14.3.37) yields a Maxwellian distribution.

From simplified forms of the particle-wave interaction term $(\partial M_a/\partial t)_\mathrm{F}$ Parker and Tidman (1958a [390]) demonstrated that below a critical velocity v_0 the solution of the full transport equation (14.3.36) yields a Maxwellian distribution while at velocities $v > v_0$ the particle-wave interaction terms shape the distribution function. This suggests in general for the solution of the full transport equation (14.3.36), to split the range of velocity at the critical point v_0, supposing that we have only Coulomb interactions at lower velocities, yielding as a solution the Maxwellian $M_1(p \leq p_0)$, and only particle-wave interactions plus momentum loss processes at higher velocities. The two solutions, the Maxwellian $M_1(p \leq p_0)$ and $M_2(p \geq p_0)$ from solving (14.1.6), are then matched continuously at $p_0 = m_a v_0 \gamma_0$ by requiring both continuity of the respective distribution functions

$$M_1(p_0) = M_2(p_0), \qquad (14.3.38)$$

and continuity of the particle flux in momentum space at p_0, similarly to (14.1.12), i.e.

$$\left[p^2\left(A_2\frac{\partial M_2}{\partial p} - \dot{p}M_2 + \frac{p}{3}\frac{\partial V}{\partial z}M_2\right)\right]_{p=p_0} = \alpha\left[\frac{1}{v}\frac{\partial M_1}{\partial v} + \frac{m_a}{k_\mathrm{B}T}M_1\right]_{p=p_0}. \qquad (14.3.39)$$

Obviously by (14.3.39) we adjust the rate of transport across momentum at the critical momentum p_0. It then remains to determine the critical momentum p_0 or velocity v_0 where particle-wave and Coulomb interactions balance each other.

At infinitely large momentum values we would use the zero-momentum flux boundary condition (14.1.12).

14.4 Solutions Without Momentum Diffusion

It is not possible in all cases to separate the spatial and momentum operators as in (14.3.1) which is the basis for applying the scattering time method. Another important class of exact analytical solutions of the transport equation (14.2.7) can be obtained (Lerche and Schlickeiser 1981 [294]) if the momentum diffusion term in (14.2.7) is omitted. The momentum operator (14.2.2) then is a first-order differential operator in momentum

$$\mathcal{L}_p M_a \simeq \frac{1}{p^2} \frac{\partial}{\partial p} \left(\frac{p^3}{3} \nabla \cdot \boldsymbol{V} M_a - p^2 \dot{p} M_a \right) - \frac{M_a}{T_c(z,p)}, \qquad (14.4.1)$$

whereas the spatial operator (14.2.1) remains unchanged,

$$\mathcal{L}_{\boldsymbol{x}} M_a = \nabla \cdot \left[\kappa(\boldsymbol{x},p) \nabla M_a - \boldsymbol{V} M_a \right]. \qquad (14.4.2)$$

If we introduce the associated isotropic differential number density of cosmic ray particles, N_a, and the related isotropic differential number density source term, s_a, i.e.

$$N_a(\boldsymbol{x},p,t) = 4\pi p^2 M_a(\boldsymbol{x},p,t), \quad s_a(\boldsymbol{x},p,t) = 4\pi p^2 S_a(\boldsymbol{x},p,t), \qquad (14.4.3)$$

the full time-dependent transport equation (14.4.3) in this case reduces to

$$\frac{\partial N_a}{\partial t} - \nabla \cdot \left[\kappa(\boldsymbol{x},p) \nabla N_a - \boldsymbol{V} N_a \right] - \frac{\partial}{\partial p} \left(\frac{p}{3} \nabla \cdot \boldsymbol{V} N_a - \dot{p} N_a \right)$$
$$+ \frac{N_a}{T_c(z,p)} = -s_a(\boldsymbol{x},p,t). \qquad (14.4.4)$$

The most complete analytical treatment of (14.4.4) has adopted the one-dimensional Cartesian form of the spatial diffusion-convection operator (Lerche and Schlickeiser 1981 [294]), since the results were applied to the cosmic ray transport in our Galaxy which has a highly flattened cylindrical shape (see Chap. 2). This justifies the neglect of any gradients in the (r, ϕ)-galactic plane with respect to perpendicular gradients. However, it is straightforward, although tedious, to generalize the arguments for more general geometries. A variation of the source term s_a in the plane can be treated parametrically by this approach. With this one-dimensional approach (14.4.4) reduces to

$$\frac{\partial N}{\partial t} - \frac{\partial}{\partial z} \left[\kappa(z,p) \frac{\partial N}{\partial z} - V(z) N \right] - \frac{\partial}{\partial p} \left(\frac{p^3}{3} \frac{dV}{dz} N - p^2 \dot{p} N \right)$$
$$+ \frac{N}{T_c(z,p)} = -s(\boldsymbol{x},p,t), \qquad (14.4.5)$$

where, for ease of exposition, the index a is omitted.

Special analytic solutions of (14.4.5), which incorporate the spatial and momentum dependence of the spatial diffusion coefficient $\kappa(z,p)$ and the momentum loss terms $\dot{p}(z,p)$ and $T_c(z,p)$, have been given in the literature for the case of a static cosmic ray transport volume with no convective motion $V = 0$ (e.g. Syrovatskii 1959 [536]; Jokipii and Meyer 1968 [247]; Webster 1970 [561]; Bulanov and Dogiel 1974 [79], 1975 [80]). Lerche and Schlickeiser (1981 [294]) presented analytic solutions to the steady-state and time-dependent transport equation (14.4.5) which allow for finite and spatially varying convection velocities, and therefore include in particular energy gain (by first-order Fermi acceleration) effects. We restrict our discussion here to the steady-state case and refer the reader to the original publication (Lerche and Schlickeiser 1981 [294]) for time-dependent solutions.

Again the analytic solutions are based on the assumption of separability of the spatial diffusion coefficient $\kappa(z,p)$ and the momentum loss terms $\dot{p}(z,p)$ and $T_c(z,p)$ in the variables z and p, i.e. we assume that

$$\kappa(p,z) = K(p)k(z) , \quad (14.4.6)$$

with $k(z=0) = 1$ and dimensionless, $K(p)$ arbitrary but prescribed;

$$\dot{p}(p,z) = -B(p)b(z) , \quad (14.4.7)$$

where $B(p)$ is arbitrary but prescribed and may be either positive or negative depending on whether momentum losses or momentum gains dominate, respectively;

$$T_c(z,p) = T(p)t(z) , \quad (14.4.8)$$

with $T(p)$ arbitrary but prescribed.

Moreover, we assume that the spatial variation of the diffusion coefficient $k(z)$ is proportional to the spatial variation of the catastrophic loss term $t(z)$ and inversely proportional to the spatial variation of the continuous momentum loss term $b(z)$ so that

$$k(z) = t(z) = \frac{\alpha}{b(z)} , \quad (14.4.9)$$

where α is a constant.

The separability assumptions (14.4.6)–(14.4.8) are conventionally made in any endeavor to glean information from the transport equation. Assumptions (14.4.9) require some physical justification in particular physical situations. For example, in the case of cosmic ray nucleon propagation in the Galaxy the catastrophic fragmentation time T_c is inversely proportional to the interstellar gas density and will therefore increase with galactic height z since the gas density decreases. On the other hand, the continuous ionization, Coulomb and pion production losses $\dot{p}(p,z)$ are proportional to the gas density and therefore decrease with galactic height, so that the second proportionality in (14.4.9) is exactly fulfilled in this case. We also expect that the spatial diffusion coefficient increases with galactic height, although here the detailed dependence on galactic parameters is less transparent (cf. Chap. 13). The situation $k(z) \propto [b(z)]^{-1}$ captures the essential physical behavior of these two effects while at the same time allowing complete integrability of the transport equation. It therefore provides a vehicle for illustrating the basic physical dependences with a precise mathematical solution.

On assumptions (14.4.6)–(14.4.9) the steady-state form of the transport equation (14.4.5) can be written after multiplying with $k(z)$

$$k(z)\frac{\partial}{\partial z}\left[K(p)k(z)\frac{\partial N}{\partial z} - V(z)N\right]$$
$$+ \frac{\partial}{\partial p}\left(\left[\frac{1}{3}k(z)\frac{dV}{dz}p + \alpha B(p)\right]N\right)$$
$$- \frac{N}{T(p)} + k(z)s(r,\phi,z,p) = 0\,. \qquad (14.4.10)$$

Introduce a new independent variable x in place of z through

$$dx = dz/k(z), \quad \text{corresponding to } x(z) = \int_0^z \frac{dz'}{k(z')}\,, \qquad (14.4.11)$$

with $x = 0$ on $z = 0$, (14.4.10) takes on the form

$$\frac{\partial}{\partial x}\left[K(p)\frac{\partial N}{\partial x} - V(x)N\right]$$
$$+ \frac{\partial}{\partial p}\left(\left[\frac{1}{3}\frac{dV}{dx}p + \alpha B(p)\right]N\right)$$
$$- \frac{N}{T(p)} + k(x)s(r,\phi,x,p) = 0\,. \qquad (14.4.12)$$

In order to illustrate as simply as possible the effect of the adiabatic deceleration or acceleration term on the convective and diffusive transport of cosmic rays we shall assume that the convective velocity is linear in the implicit variable x so that

$$V(x) = V_0 + 3V_1 x\,, \qquad (14.4.13)$$

with $(1/3)(dV/dx) = V_1$. If we consider a problem symmetric in galactic height z it suffices to consider only positive values of $z \geq 0$ and $x \geq 0$. With this assumption (14.4.13) made the solution of (14.4.12) in $x \geq 0$ can be expressed as

$$N(x,p) = \int_0^\infty dp_0 \int_0^\infty dx_0 G(x,x_0,p,p_0)k(x_0)s(r,\phi,x_0,p_0)\,, \qquad (14.4.14)$$

in terms of the Green's function $G(x,x_0,p,p_0)$ that satisfies the equation

$$\frac{\partial}{\partial x}\left[K(p)\frac{\partial G}{\partial x} - (V_0 + 3V_1 x)\,G\right]$$
$$+ \frac{\partial}{\partial p}\left(\left[V_1 p + \alpha B(p)\right] G\right) - \frac{G}{T(p)} = -\delta(x-x_0)\delta(p-p_0)\,. \quad (14.4.15)$$

For propagation in an infinitely extended medium and adiabatic deceleration ($V_1 > 0$) the solution of (14.4.15) is (slightly generalizing the derivation of Lerche and Schlickeiser (1981 [294]))

$$G(x,x_0,p,p_0) = \frac{H[p_0 - p]}{2\sqrt{\pi}\,[V_1 p + \alpha B(p)]}\left[\frac{\Delta(p,p_0)}{\Psi(p,p_0) + \Phi^2(p,p_0)}\right]^{1/2}$$
$$\times\, \mathcal{P}(p,p_0)\exp\left[-\frac{\Delta(p,p_0)}{4\left[\Psi(p,p_0) + \Phi^2(p,p_0)\right]}\right.$$
$$\left.\times\left(x - x_0 + \frac{V_0}{3V_1}\left(1 - \frac{A(p_0)}{A(p)}\right)\right)^2\right], \quad (14.4.16)$$

with the propagation function

$$\Delta(p,p_0) = A^{-2}(p)\int_{\tau(p)}^{\tau(p_0)} d\tau(u)\, K(u)\left[A(u) - A(p)\right]^2, \quad (14.4.17)$$

$$\Psi(p,p_0) = \left[2A^2(p)\right]^{-1}\int_{\tau(p)}^{\tau(p_0)} d\tau(u)$$
$$\times \int_{\tau(p)}^{\tau(p_0)} d\tau(v)\, K(u)K(v)\left[A(u) - A(v)\right]^2, \quad (14.4.18)$$

$$\Phi(p,p_0) = \left[A^{-2}(p)\right]\int_{\tau(p)}^{\tau(p_0)} d\tau(u)\, K(u)A(u)\left[A(E) - A(u)\right], \quad (14.4.19)$$

where

$$A(u) = \exp\left[3V_1 \tau(u)\right], \quad (14.4.20)$$

and

$$d\tau = \frac{du}{V_1 u + \alpha B(u)}. \quad (14.4.21)$$

$H[y]$ denotes Heaviside's step function and

$$\mathcal{P}(p,p_0) = \exp\left[-\int_p^{p_0} \frac{du}{T(u)\left[V_1 u + \alpha B(u)\right]}\right] \quad (14.4.22)$$

the cosmic ray particle survival probability against catastrophic losses.

Note that the solution (14.4.14)–(14.4.22) is valid for *any* momentum dependence of the spatial diffusion coefficient, the catastrophic loss time and the continuous gain/loss term, and also for *any* given source distribution. The propagation functions (14.4.17)–(14.4.20) have been explicitly worked out for cosmic ray electron and nucleon propagation in the Galaxy by Lerche and Schlickeiser (1982a [295],1982b [296]); the influence of spatially extended source distributions has been analyzed by Pohl and Schlickeiser (1990 [403]).

15. Interplanetary Transport of Cosmic Ray Particles

We have demonstrated in Chaps. 12 and 13 that the quasilinear cosmic ray transport parameters in a given astrophysical system are determined by the statistical properties of the electromagnetic fluctuations in this system, expressed through the correlation tensors of the electric and magnetic field fluctuations (12.2.6). If these electromagnetic fluctuations can be identified with normal low-frequency plasma wave modes and/or exhibit certain symmetry properties with respect to the uniform background magnetic field, the calculation of the cosmic ray transport parameters simplifies enormously under the diffusion approximation, as shown in Chap. 13.

In-situ observations carried out in the interplanetary medium offer a unique opportunity to test the quasilinear theory of cosmic ray transport. Here we can directly measure both the cosmic ray particle properties and the fluctuation tensors of the ambient electromagnetic fluctuations. Using the latter as input quantities for the quasilinear calculation of the cosmic ray transport parameters, the comparison of the calculated quasilinear transport parameters with the empirically derived transport parameters from the particle observations provides a definite test for quasilinear theory.

15.1 Solar Flare Events

The first quantitative application of the fundamental cosmic transport equation (14.1.6) to explain observations was to the time-intensity profile of relativistic particles during the February 23, 1956, solar particle event by Meyer et al. (1956 [349]). For these particles the spatial diffusion term is the only important one on the right hand side of (14.1.6), and it was established that a diffusive profile with a reasonable value for the diffusion coefficient $\kappa_{zz} = v\lambda/3$ provided a compelling interpretation of the event. A more recent comparison of theory and observation is shown in Fig. 15.1.

15.1.1 Observations

A summary of near-Earth mean free paths $\lambda_{\rm fit}$ versus particle rigidity, originally compiled by Palmer (1982 [385]), and supplemented with some new

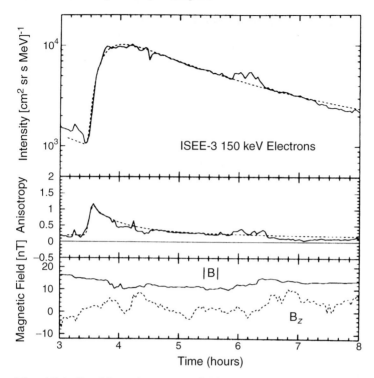

Fig. 15.1. Particle and magnetic field observations of the 1980 June 7 event observed on ISEE-3. Electron data (**solid lines**) can be fitted (**dashed lines**) in intensity (**upper panel**) and anisotropy (**middle panel**) assuming a parallel mean free path of 0.4 AU. Simultaneous measurements of the magnetic field (**lower panel**) are utilized to monitor the propagation conditions and to determine the fluctuation spectra of the magnetic field component perpendicular to the average field (B_z). From Dröge (1993 [139])

observations, is shown in Fig. 15.2. The dashed line is an old quasilinear theory prediction assuming the existence of steady solar wind turbulence conditions which obviously disagrees with the data.

The first point to note is that any comparison between theory and experiment should be made on an event-by-event basis, i.e. values of λ_{th} derived from fluctuation spectra show as large a variation as values of λ_{fit}. It has been demonstrated by Dröge et al. (1993 [139]) for three individual solar particle events, for which simultaneous measurements of the fluctuation tensors existed, that theoretical and fitted mean free paths are in good agreement, when the actual fluctuation measurements at the time of the solar particle event are used as input for the quasilinear calculation.

The observed mean free paths of non-relativistic cosmic ray ions with rigidities between 30 and 1000 MV are practically constant with respect to particle rigidity. Electrons with rigidities below 30 MV also exhibit a flat

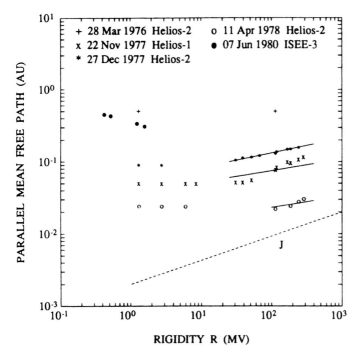

Fig. 15.2. Parallel mean free path as a function of particle rigidity for selected solar particle events. The **dashed line** with label "J" represents the prediction of standard quasilinear theory based on typical solar wind conditions (Jokipii and Coleman 1968 [246]). **Solid lines** are the quasilinear predictions of Dröge et al. (1993 [139]) derived from the magnetic field measurements at the times of events. From Dröge (1993 [139])

rigidity variation. Obviously the rather strong rigidity variation predicted by the old quasilinear calculations is not in accord with the "flatness" of the rigidity dependence of the observed mean free paths.

The quasilinear analysis in Chap. 13 has demonstrated that the scattering rate of particles in pitch angle $D_{\mu\mu}$, and the closely related quasilinear mean free path, e.g. (13.3.20), are remarkably sensitive to the detailed properties of the electromagnetic fluctuations. Determination of the nature of the electromagnetic fluctuations, particularly at high frequencies, in the interplanetary space during a particular solar particle event, is the key issue to reach agreement of the quasilinear and phenomenologically determined mean free paths.

A representative power spectrum (Denskat et al. 1983 [120]) of magnetic field fluctuations in the solar wind perpendicular to the average magnetic field is shown in Fig. 15.3. In the frequency range below 0.1 Hz the spectra typically exhibit a moderately steep power law of the form (13.1.25) or (13.1.26), assuming a non-dispersive dispersion relation as, for example, the

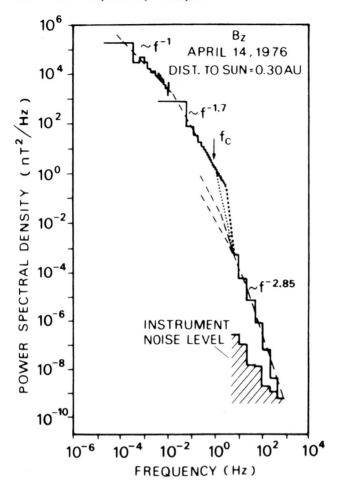

Fig. 15.3. Power spectral density estimate of the transverse magnetic field component observed on Helios 2. From Denskat et al. (1983 [120])

Alfvén wave dispersion relation $\omega = \pm V_A k_\parallel$, with the exponent q varying from 1.5 to 1.9, whereas the high frequencies above 1 Hz display a considerable steepening of of the turbulence spectrum with values of q of approximately 3.

One of the fundamental observational limitations of turbulence measurements is that in order to determine the cosmic ray particle scattering rate the full knowledge of the *spatial* correlation tensor of the fluctuations is required, i.e. the three-dimensional correlation tensor with respect to the wave vector component k. However, magnetometer and plasma wave experiments on single spacecraft measure, at best, a *time* series of the magnetic field or solar wind vector, respectively, along one single line yielding the power density of fluctuations in one direction. From such measurements alone the exact nature of the correlation tensor cannot be deduced, and additional assumptions on the three-dimensional structure of the turbulence have to be made. As we explained in Sect. 12.2.2, one key assumption is to model the electromagnetic fluctuations as superpositions of small-amplitude plasma waves whose dispersion relations reflect the physical state of the plasma wave-carrying background medium. And in this case it is of interest to know exactly the cross helicity and magnetic helicity states of all relevant plasma waves, because these quantities have a profound effect on the quantitative value of the cosmic ray transport parameters, as our computations in Chap. 13 have shown. Attempts to extract all these important turbulence properties from single spacecraft observations have only been started recently (Schmidt and Dröge 1997 [477]). On the other hand, quasilinear calculations developed fully to include all turbulence helicity effects yield remarkably good agreement with the observations from solar particle events (e.g. Dröge et al. 1993 [139]).

15.1.2 Comparison with Quasilinear Theory

For parallel propagating waves (referred to as the "slab model") the calculation of quasilinear Fokker-Planck coefficients and transport parameters of the diffusion-convection equation is well developed. At large particle energies the influence of particle interactions with damped ion-cyclotron and electron-cyclotron waves is negligibly small, and the Fokker-Planck coefficients can be calculated in the resonant diffusion limit. Schlickeiser (1992 [460]) has reviewed the status of cosmic ray transport parameter calculations in this limit. The most noteworthy results are that

(i) the famous discrepancy between observed (λ_{fit}) and quasilinear (λ_{ql}) spatial mean free paths for solar flare particles can be resolved once the magnetostatic approximation is discarded and the influence of cross (i.e. the fraction of forward and backward moving waves) and magnetic helicities are properly taken into account,
(ii) dissipation range spectra yield finite values of κ_{ql}.

At small, especially non-relativistic, particle energies the situation is more complex. On the one hand, damping of the cyclotron waves reduces the intensity of waves necessary to scatter non-relativistic particles through pitch angles of 90°. Smith et al. (1990 [500]) and Achatz et al. (1993 [7]) have pointed out that this wave dissipation would lead to unreasonably large spatial diffusion coefficients of these particles if calculated in the resonant diffusion limit. However, as explained in Sect. 13.2.3, as a consequence of wave damping resonance broadening of the wave-particle interaction sets in, so that the pitch angle Fokker-Planck coefficients have to be calculated with the full Breit-Wigner resonance function. In Fig. 13.14 we calculated the cosmic ray mean free path in such a case which consists of two terms: the first describes the resonant scattering from undamped waves and exhibits the well-known power law behavior in particle rigidity if the turbulence spectrum is of Kolmogorov-type in the inertial range; the second term describes the contribution from scattering by the damped waves and only occurs below some limiting kinetic energy E_H above which particles at all pitch angles interact with undamped waves. Towards small energies this contribution approaches a constant, independent of particle kinetic energy, which dominates the resonant scattering contribution and for interplanetary plasma parameters agrees remarkably well with the consensus value of Fig. 15.2.

Moreover, extending the quasilinear theory to obliquely propagating fast magnetosonic waves (as was done in Sect. 13.3) also provides an alternative explanation for the flatness problem of solar flare event observations. In Sect. 13.4 we calculated the quasilinear Fokker-Planck coefficients in turbulent electromagnetic fields consisting of a mixture of undamped slab Alfvén waves and isotropic fast magnetosonic waves in a low-β plasma, assuming a vanishing cross helicity state of the plasma waves, and Kolmogorov-like turbulence spectra. The resulting mean free path (13.4.10), shown in Fig. 13.21 for different cosmic ray species, is $\propto (\gamma/|\Omega_0|)^{2-q}$, which is a constant determined by $|\Omega_0|^{q-2} = |m/Ze|^{2-q}$ at non-relativistic energies, and thus can account for the flatness problem. Bearing in mind that the mean free path (13.4.10) has been calculated under the hypothesis that Alfvén and fast mode waves have equal intensities, identical shapes and outer scales, it is clear that the generalization (see (13.4.8)) to the case of slab Alfvén waves and fast mode waves of arbitrary outer scale ($k_{min,F}, k_{min,A}$), total intensities (δB_F^2, δB_A^2) and power law shape (q_F, q_A) allows enough flexibility also to account for the event-to-event variability of solar flare event mean free paths, i.e. the magnitude problem. If interplanetary turbulence consists of a mixture of slab Alfvén waves and isotropic fast mode waves, both the magnitude and flatness problems can be resolved within quasilinear theory.

15.2 Solar Modulation of Galactic Cosmic Rays

Possibly the most complicated and detailed application of the cosmic ray transport equation (14.1.6) has been to the modulation of galactic cosmic rays by the Sun (see Sect. 3.2). Here the solar system is regarded as residing in a constant, isotropic bath of galactic cosmic rays, and the electromagnetic fluctuations carried by the outflowing solar wind act to partially exclude these particles from the inner solar system in phase with the solar activity. In this case, all the terms in the transport equation (14.1.6) except the momentum diffusion (A_2) term play important roles (Jokipii 1983 [245]). Indeed, such a form of the cosmic ray transport equation was first written in response to the challenges presented by the measured solar modulation (e.g. Parker 1965 [389], Axford and Gleeson 1967 [21], Jokipii and Parker 1969 [248]). Then the modulation may be regarded as a balance between the inward random walk or diffusion, the outward convection by the solar wind, gradient and curvature drifts caused by the large-scale structure of the interplanetary magnetic field, and the adiabatic cooling due to the radial expansion of the solar wind.

Sophisticated fully three-dimensional numerical solutions of the transport equation (Le Roux and Potgieter 1990 [299]; Kota and Jokipii 1983 [269]) have been developed incorporating the interplanetary distribution of the solar wind and its entrained magnetic field. Despite the uncertainties in the values of the ambient magnetic field and flow velocity outside the equatorial regions, the calculated properties of the model are in good agreement with the observed properties of cosmic rays. The modulation of cosmic rays by the solar wind thus provides a reasonably complete verification of the basic transport equation with the exception of the stochastic momentum diffusion term which, at these particle energies, plays only a minor role.

16. Acceleration of Cosmic Ray Particles at Shock Waves

According to the fundamental cosmic ray transport equation (14.1.6) energetic particles get accelerated both

(1) by momentum diffusion (A_2) due to cyclotron and/or transit-time damping of the electric fields of the ambient electromagnetic fluctuations, and
(2) by momentum convection due to the compression, i.e. $(dV/dz) < 0$, of the cosmic ray scattering centers.

Both acceleration processes occur near cosmic shock waves. Acceleration near shocks is the only applicable particle acceleration mechanism which has been directly observed in space (see Fig. 3.12). Since shocks are common in astrophysics, and since they have considerable free energy, they are regarded as an important site of cosmic ray acceleration.

16.1 Astrophysical Shock Waves

Cosmic shock waves are a direct consequence of violent dynamical phenomena in the universe, as high-velocity stellar winds and supernova envelopes impinging on interplanetary and interstellar plasmas with small signal speeds (speed of sound $c_s \simeq 10\text{--}100$ km s^{-1} and Alfvén speed $V_A \simeq 30\text{--}50$ km s^{-1}. We mention three prominent examples:

(a) *Interplanetary shock waves* result when a fast ($V_s \geq 500$ km s^{-1}) solar wind stream overtakes a slow ($V_s \sim 300$ km s^{-1}) one. Both velocities are larger than the interplanetary speed of sound and the Alfvén speed (~ 50 km s^{-1}), so that these shocks most often are super-Alfvénic.
(b) Spectroscopic studies of supernovae show that they involve initial expansion motions in excess of 5000 km s^{-1} which, at later phases, decrease to several hundred km s^{-1}, still much larger than the interstellar Alfvén speed and speed of sound (≤ 100 km s^{-1}). *Supernova shock waves* for most of their lifetime are fast and super-Alfvénic shock waves.
(c) The hypothesis of relativistic ($V > c/\sqrt{2}$) bulk outflow motion from active galactic nuclei currently represents the basic model to explain the origin of the luminous broadband nonthermal radiation emitted from these objects (see the discussion in Sect. 6.3.1), since it avoids gamma

ray transparency problems and provides a natural explanation for superluminal component motion, rapid flux density variations and efficient transfer of energy over kpc distances. Intergalactic signal speeds are of comparable value to interstellar signal speeds, implying again the formation of super-Alfvénic shocks at the point of termination.

16.2 Magnetohydrodynamic Shock Discontinuities

The equations of motion (compare Sect. 8.1.3) for an ideal fluid (having zero viscosity, thermal conductivity and electric resistance) admit discontinuous flows as in ordinary fluid dynamics (e.g. Landau and Lifshitz 1960 [281], Chap. 8). The relationships between the thermodynamic state, velocity and magnetic field far upstream and far downstream of a discontinuity (the Rankine-Hugoniot equations) are found by application of the laws of conservation of mass, momentum, and energy, and from the requirements of Maxwell's equations that the tangential component of the electric field and the normal component of the magnetic field are continuous across a discontinuity (for a detailed derivation of these conditions we refer the reader to the monograph of Cabannes (1970 [86])). Here we follow mainly the monograph of Anderson (1963 [14]).

It is possible to chose a coordinate system such that the gas velocity vectors $\boldsymbol{u}$ and the magnetic field vectors $\boldsymbol{B}$ on both sides of a magnetohydrodynamic discontinuity lie in the same plane (de Hoffmann and Teller 1950 [229]). We take this plane to be the $(x-y)$-plane, so that

$$\boldsymbol{u} = (u_x, u_y, 0), \quad \boldsymbol{B} = (B_x, B_y, 0), \quad (16.2.1)$$

and assume the discontinuity to lie in the $(y-z)$-plane. Obviously, u_x and B_x are the normal components of the gas velocity and magnetic field vectors, while u_y and B_y are the respective transverse components. The relationships between the upstream and downstream physical quantities follow from the continuity conditions:

(1) *continuity of the mass flux*

$$[\rho u] \equiv \rho_1 u_1 - \rho_2 u_2 = 0, \quad (16.2.2)$$

where throughout this chapter we use the notation that the braces [ ] indicate the difference of the enclosed quantity on the sides of the discontinuity. The index "1" and "2" refer to the upstream and downstream values, respectively;

(2) *continuity of the normal magnetic field component*

$$[B_x] = 0. \quad (16.2.3)$$

The momentum flux must also be continuous. The momentum flux density tensor is the sum of the hydrodynamic contribution and Maxwell's stress tensor and reads

$$\Pi_{ik} = \rho u_i u_k + P\delta_{ik} - \frac{1}{4\pi}\left(B_i B_k - \frac{1}{2}B^2 \delta_{ik}\right), \qquad (16.2.4)$$

where P denotes the scalar gas pressure. The continuity condition is

$$[\Pi_{ik} n_k] = 0, \qquad (16.2.5)$$

where $\boldsymbol{n}$ denotes the unit vector normal to the shock surface which, for our coordinate system, is simply $n_k = (1,0,0)$. Equation (16.2.5) then reduces to

$$\left[\rho u_i u_x + P\delta_{ix} - \frac{1}{4\pi}\left(B_i B_x - \frac{1}{2}B^2 \delta_{ix}\right)\right] = 0, \qquad (16.2.6)$$

with $i = x, y, z$.
For $i = x$ (16.2.6) becomes

$$\left[P + \rho u_x^2 + \frac{1}{8\pi}(B_y^2 - B_x^2)\right] = 0. \qquad (16.2.7)$$

Using the continuity (16.2.3) of the normal magnetic field component, (16.2.7) yields

(3) *continuity of the normal-momentum flux*

$$\left[P + \rho u_x^2 + \frac{B_y^2}{8\pi}\right] = 0. \qquad (16.2.8)$$

For $i \neq x$ (16.2.6) leads to

(4) *continuity of the transverse-momentum flux*

$$\left[\rho u_x u_y - \frac{1}{4\pi} B_x B_y\right] = 0. \qquad (16.2.9)$$

In a medium of vanishing electric resistance or equivalently very large electrical conductivity, the electric field is frozen into the medium and given by

$$\boldsymbol{E} = -\,\boldsymbol{u}\times\boldsymbol{B}/c, \qquad (16.2.10)$$

so that

(5) *continuity of the transverse electric field* is

$$c\,[E_z] = [u_x B_y - u_y B_x] = 0. \qquad (16.2.11)$$

Finally, the normal component of the energy flux must be continuous. The energy flux q is the sum of the hydrodynamic contribution and the Poynting flux,

$$q = \rho u \left(w + \frac{u^2}{2}\right) + \frac{c}{4\pi} E \times B ,\qquad(16.2.12)$$

where the enthalpy per unit fluid mass

$$w = e + \frac{P}{\rho} \qquad(16.2.13)$$

is related to the internal energy e. With (16.2.10) we derive

(6) *continuity of the energy flux*

$$[q_x] = \left[\rho u_x \left(w + \frac{u_x^2 + u_y^2}{2}\right) + \frac{1}{4\pi}\left(u_x B_y^2 - u_y B_x B_y\right)\right] = 0 . \qquad(16.2.14)$$

Now it is convenient to use the specific volume of the fluid $V = 1/\rho$ in place of its density ρ, so that the mass flux density through the discontinuity becomes

$$G \equiv \rho u_x = u_x / V . \qquad(16.2.15)$$

Because G, according to (16.2.2), and B_x, according to (16.2.3), are continuous, the remaining boundary conditions (16.2.8), (16.2.9), (16.2.11) and (16.2.14) can be written in the form

$$[P] + G^2 [V] + [B_y^2]/8\pi = 0 , \qquad(16.2.16a)$$
$$G[u_y] = B_x \lfloor B_y \rfloor /4\pi , \qquad(16.2.16b)$$
$$B_x [u_y] = G [V B_y] , \qquad(16.2.16c)$$
$$G\left[w + \frac{1}{2}G^2 V^2 + \frac{1}{2}u_y^2 + V B_y^2/4\pi\right] = B_x [u_y B_y]/4\pi . \qquad(16.2.16d)$$

This is the fundamental system of equations of discontinuities in magnetic fluid dynamics, equivalent to (53.1)–(53.4) in Landau and Lifshitz (1960 [281]).

Two general classes of discontinuities can be distinguished (Burlaga 1971 [82]):

(1) Stationary discontinuities ($G = 0$),
(2) Shocks ($G \neq 0$).

We consider both in turn.

16.2.1 Stationary Discontinuities

Since by definition $G = 0$ for a stationary discontinuity, $u_{x,1} = u_{x,2} = 0$, which implies that the surface of the discontinuity does not move relative to the plasma. There are two types of such discontinuities, corresponding to the normal magnetic field component $B_x = 0$ and $B_x \neq 0$. These are called *tangential discontinuity* and *contact discontinuity*, respectively.

16.2.1.1 Tangential Discontinuity ($G = 0$, $B_x = 0$). At such a discontinuity the velocity and the magnetic field on both sides of the discontinuity surface are tangential and can have any discontinuity in both magnitude and direction (see Fig. 16.1); hence the shock normal vector is

$$\boldsymbol{n} = \boldsymbol{B}_1 \times \boldsymbol{B}_2 / |\boldsymbol{B}_1 \times \boldsymbol{B}_2| = \boldsymbol{u}_1 \times \boldsymbol{u}_2 / |\boldsymbol{u}_1 \times \boldsymbol{u}_2| \ . \tag{16.2.17}$$

The equation for the normal momentum flux (16.2.16a) gives

$$\left[P + \frac{B_y^2}{8\pi} \right] = 0 \ , \tag{16.2.18}$$

which says that the pressure discontinuity is related to that of the magnetic field. The equations for the transverse momentum flux, (16.2.16b), and the energy flux, (16.2.16d), are identically satisfied. To identify positively a tangential discontinuity in in-situ interplanetary observations, one must show

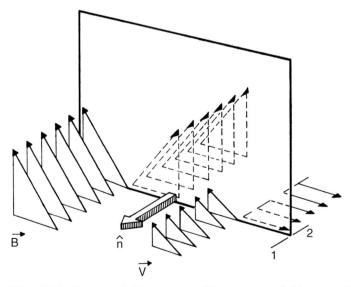

Fig. 16.1. Tangential discontinuity: The magnetic field vector $\boldsymbol{B}$ is parallel to the surface of the discontinuity, but its direction may change across the surface. From Burlaga (1971 [82])

that all of the above conditions are satisfied, which requires simultaneous magnetic field data ($\boldsymbol{B}(t)$) and plasma data (density $\rho(t)$, temperature $T(t)$, to infer the gas pressure $P(t) = k_{\rm B}\rho(t)T(t)/m_p$, and $\boldsymbol{u}$ for protons and electrons) and the arrival times at 4 spacecrafts (to obtain the speed and orientation). Such a definite measurement of a tangential discontinuity has still not been performed, but Burlaga (1971 [82]) discusses the available evidence from less complete observations that are consistent with these conditions, and leaves little doubt about the existence of tangential discontinuities in the solar wind.

16.2.1.2 Contact Discontinuity ($G = 0$, $B_x \neq 0$).
Equation (16.2.16c) shows that at such a discontinuity one must have

$$u_{y,1} = u_{y,2} \,, \tag{16.2.19}$$

i.e. there can be no relative motions along the discontinuity surface. The conservation of transverse momentum (16.2.16b) requires that

$$B_{y,1} = B_{y,2} \,, \tag{16.2.20}$$

and the conservation of parallel momentum (16.2.16a) requires the pressure to be continuous

$$P_1 = P_2 \,. \tag{16.2.21}$$

So there is no change in either the magnetic field vector $\boldsymbol{B}$ or the velocity vector $\boldsymbol{u}$ across a contact discontinuity: no change in the magnitude, no change in the direction. The density (and therefore the temperature in agreement with condition (16.2.21)) may have any discontinuity. A contact discontinuity is simply the boundary between two media at rest which, at equal pressure, have different densities and temperatures (see Fig. 16.2).

16.2.1.3 Rotational Discontinuities.
For completeness we mention here rotational continuities for which

$$G \neq 0, \quad [V] = [u_x] \,. \tag{16.2.22}$$

On the right-hand side of (16.2.16c) we can take V outside the brackets and divide this equation by (16.2.16b), obtaining

$$G = B_x/\sqrt{4\pi V}, \quad [u_y] = \sqrt{V/4\pi}\,[B_y] \,. \tag{16.2.23}$$

In the energy flux equation (16.2.16d) we use (16.2.13) again, take V outside the brackets and find, with (16.2.23a), that

$$G\,[e] + GV\left[P + \frac{B_y^2}{8\pi}\right] + \frac{1}{2}\left[\left(u_y - \sqrt{\frac{V}{4\pi}}B_y\right)^2\right] = 0 \,. \tag{16.2.24}$$

The second term is zero by (16.2.16a), and the third term is zero by (16.2.23b), so that (16.2.24) reduces to $[e] = 0$, i.e. the internal energy is

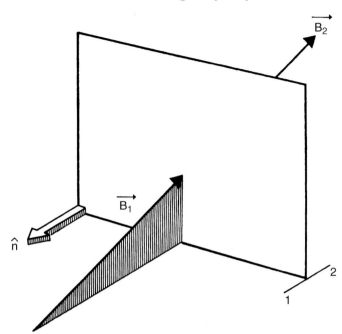

Fig. 16.2. Contact discontinuity. From Burlaga (1971 [82])

also continuous, Hence all other thermodynamic quantities, including the pressure $[p] = 0$, are continuous, implying from (16.2.16a) that the magnetic field strength $[B_y^2] = 0$ is continuous too.

At a rotational discontinuity all thermodynamic gas properties are continuous, the magnitude of the magnetic field is continuous, but the magnetic field direction rotates through an angle about the normal. The tangential magnetic field vector and, by (16.2.23b), the tangential velocity component are discontinuous, while the normal velocity component

$$u_x = GV = B_x\sqrt{V/4\pi} = B_x/\sqrt{4\pi\rho} \qquad (16.2.25)$$

is continuous.

16.2.2 Shock Waves

A discontinuity is called a shock wave if $G \neq 0$. In hydrodynamics there is basically only one kind of shock, and it corresponds to the one type of wave in an unmagnetized gas, the sound wave. Gas arrives at a hydrodynamic shock at a speed greater than the sound speed and leaves at a speed less than the sound speed. The addition of an ordered magnetic field $\boldsymbol{B}$ in magnetohydrodynamics implies the existence of additional wave modes, the magnetosonic waves and the shear Alfvén waves (see Chap. 9), which lead to a greater variety of shocks.

We shall consider here only the simple cases of shocks moving parallel to $\boldsymbol{B}$ or perpendicular to $\boldsymbol{B}$. A detailed analytical theory of hydromagnetic shocks can be found in the monograph of Jeffrey and Taniuti (1964 [241]).

In principle, shocks in the solar wind (or in stellar winds) can move either away from the Sun (from the central star) or toward it. Shocks moving away from the Sun are called forward shocks, while shocks moving toward the Sun are called reverse shocks. Physically there is no difference between a forward shock and a reverse shock.

16.2.2.1 Parallel Shock Wave. A parallel shock is one whose normal is parallel to $\boldsymbol{B}$ (see Fig. 16.3), i.e.

$$B_{y,1} = B_{y,2} = 0, \quad B_{x,1} = B_{x,2}, \tag{16.2.26}$$

where we used (16.2.3). Equation (16.2.16b) then implies

$$u_{y,1} = u_{y,2}, \tag{16.2.27}$$

but there is no requirement that $\boldsymbol{u_1}$, $\boldsymbol{u_2}$, and $\boldsymbol{n}$ be coplanar. The magnetic field drops out of the Rankine-Hugoniot equations (16.2.16a) and (16.2.16d), which become identical to those of hydrodynamic shocks,

$$\left[P + G^2 V\right] = \left[P + G u_x\right] = 0, \tag{16.2.28}$$

and

$$\left[w + \frac{1}{2}G^2 V^2\right] = \left[w + \frac{1}{2}u_x^2\right] = 0, \tag{16.2.29}$$

where we used (16.2.15). Combining these two equations (16.2.28)–(16.2.29) readily yields

$$w_1 - w_2 + \frac{1}{2}(V_1 + V_2)(P_2 - P_1) = 0. \tag{16.2.30}$$

In ideal gases the enthalpy is given by

$$w = \frac{\gamma}{\gamma - 1} PV, \tag{16.2.31}$$

where γ denotes the ratio of specific heats.

Using (16.2.31) in (16.2.30) immediately yields

$$\frac{V_2}{V_1} = \frac{(\gamma + 1) P_1 + (\gamma - 1) P_2}{(\gamma - 1) P_1 + (\gamma + 1) P_2}, \tag{16.2.32}$$

and with the ideal gas law $PV \propto T$ also

$$\frac{T_2}{T_1} = \frac{P_2}{P_1} \frac{(\gamma + 1) P_1 + (\gamma - 1) P_2}{(\gamma - 1) P_1 + (\gamma + 1) P_2}. \tag{16.2.33}$$

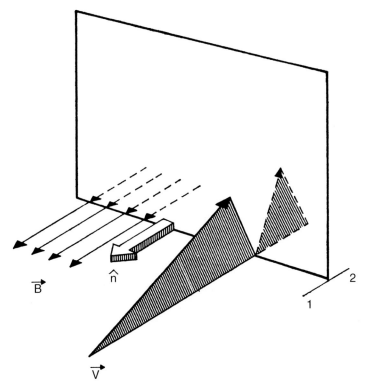

Fig. 16.3. Parallel shock wave. The magnetic field vector $\boldsymbol{B}$ is parallel to the shock normal $\boldsymbol{n}$. From Burlaga (1971 [82])

We introduce the compression ratio r of up- and downstream quantities as

$$r \equiv \frac{\rho_2}{\rho_1} = \frac{u_{x,1}}{u_{x,2}} = \frac{GV_1}{GV_2} = \frac{(\gamma-1)P_1 + (\gamma+1)P_2}{(\gamma+1)P_1 + (\gamma-1)P_2}, \quad (16.2.34)$$

where we used (16.2.32). Equation (16.2.28) readily gives

$$G^2 = \frac{P_1 - P_2}{V_2 - V_1}. \quad (16.2.35)$$

At this point it is convenient to introduce the upstream Alfvénic Mach number of the flow,

$$M_{A,1} \equiv u_{x,1}/V_{A,1}, \quad (16.2.36)$$

that measures the upstream gas speed in units of the upstream Alfvén speed, and the upstream sound Mach number

$$M_{s,1} \equiv u_{x,1}/c_{s,1}, \quad (16.2.37)$$

that gives the upstream gas speed in units of the local speed of sound $c_{s,1} = \sqrt{\gamma P_1/\rho_1}$. Obviously, the two Mach numbers are related

$$\frac{M_{A,1}}{M_{s,1}} = \frac{c_{s,1}}{V_{A,1}} = \sqrt{\beta} \tag{16.2.38}$$

by the upstream plasma beta

$$\beta = 4\pi\gamma P_1/B_x^2 . \tag{16.2.39}$$

From the definition (16.2.37) we obtain with (16.2.35)

$$M_{s,1}^2 = \frac{\rho_1 u_{x,1}^2}{\gamma P_1} = \frac{G^2 V_1}{\gamma P_1} = \frac{1-(P_2/P_1)}{\gamma\left[(V_2/V_1) - 1\right]} . \tag{16.2.40}$$

Inserting (V_2/V_1) from (16.2.32) yields

$$\frac{P_2}{P_1} = \frac{2\gamma M_{s,1}^2}{\gamma+1} - \frac{\gamma-1}{\gamma+1} . \tag{16.2.41}$$

According to (16.2.34) we also have

$$\frac{P_2}{P_1} = \frac{r(\gamma+1) - (\gamma-1)}{\gamma+1 - r(\gamma-1)} . \tag{16.2.42}$$

Equating the last two equations leads to

$$r = \frac{(\gamma+1)M_{s,1}^2}{(\gamma-1)M_{s,1}^2 + 2} = \frac{(\gamma+1)M_{A,1}^2}{(\gamma-1)M_{A,1}^2 + 2\beta} , \tag{16.2.43}$$

where we used (16.2.38). The relation (16.2.43) shows that for strong shocks $M_{A,1} \gg 1$ the compression ration approaches the limit

$$r(M_{A,1} \gg 1) \to \frac{\gamma+1}{\gamma-1} . \tag{16.2.44}$$

With (16.2.33) we obtain for the temperature ratio

$$\frac{T_2}{T_1} = \frac{\left[2\gamma M_{s,1}^2 - (\gamma-1)\right]\left[(\gamma-1)M_{s,1}^2 + 2\right]}{(\gamma+1)^2 M_{s,1}^2} . \tag{16.2.45}$$

16.2.2.2 Perpendicular Shock Wave. A shock is called a perpendicular shock if the shock normal is perpendicular to $\boldsymbol{B}$. In this case the normal magnetic field component $B_x = 0$ and the magnetic field is parallel to the shock surface on both sides of the shock, as shown in Fig. 16.4.

The only magnetohydrodynamic wave that propagates perpendicular to $\boldsymbol{B}$ is the fast magnetosonic wave, whose speed is $V_M^2 = V_A^2 + c_s^2$. The flow enters

16.2 Magnetohydrodynamic Shock Discontinuities

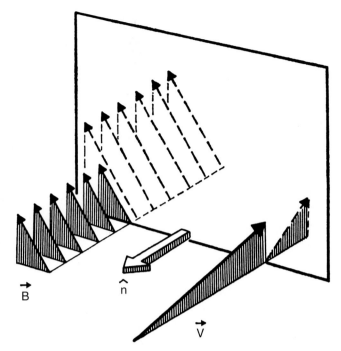

Fig. 16.4. Perpendicular shock wave. The magnetic field vector $\boldsymbol{B}$ is perpendicular to the shock normal $\boldsymbol{n}$. From Burlaga (1971 [82])

a perpendicular shock at a speed greater than V_M and leaves at a speed less than V_M. The shock propagates super-magnetosonically.

For a perpendicular shock the mass flux (16.2.2) and frozen field equation (16.2.16c) give

$$\frac{\boldsymbol{B}_1}{\rho_1} = \frac{\boldsymbol{B}_2}{\rho_2}, \qquad (16.2.46)$$

which states that

(a) the magnetic field direction does not change across the shock, and
(b) its intensity changes in the same ratio as the density, $B_1/B_2 = \rho_1/\rho_2$.

Equation (16.2.16b) shows that there is no change in the transverse momentum across a shock when $B_x = 0$, so $u_{y,1} = u_{y,2}$. This implies that $\boldsymbol{u}_2 - \boldsymbol{u}_1 = (u_{x,2} - u_{x,1})\boldsymbol{n}$, so that the shock normal $\boldsymbol{n}$ is parallel to $\boldsymbol{u}_2 - \boldsymbol{u}_1$ for a perpendicular shock.

The normal momentum flux equation (16.2.16a) gives

$$[\rho u_x^2] = \left[P + \frac{B^2}{8\pi}\right], \qquad (16.2.47)$$

i.e. the change in the normal momentum flux is balanced by a change in the total pressure that now is the sum of the particle pressure and the magnetic field pressure.

16.2.2.3 Classification of Shocks. We note that the special types of shocks discussed above can be viewed as different special solutions of (16.2.16b). The frozen field condition (16.2.11) in addition with the normal field continuity (16.2.3) gives

$$[u_y] = \frac{1}{B_x}[u_x B_y] \,. \tag{16.2.48}$$

Inserting (16.2.48) into (16.2.16b) we obtain

$$G[u_x B_y] = \frac{B_x^2}{4\pi}[B_y] \,, \tag{16.2.49}$$

that can be written as

$$B_{y,1}\rho_1 V_{A,1}^2 \left(M_{A,1}^2 - 1\right) = B_{y,2}\rho_2 V_{A,2}^2 \left(M_{A,2}^2 - 1\right) \,. \tag{16.2.50}$$

Obviously, the cases discussed above correspond to the following special solutions of (16.2.50):

Parallel shock: $B_{y,1} = B_{y,2} = 0$
Perpendicular shock: $V_{A,1}^2 = V_{A,2}^2 = 0$.

Other solutions are also possible:

Alfvén shock: $M_{A,1} = 1$; $M_{A,2} = 1$,
Fast shock: $M_{A,1} > 1$; $M_{A,2} > 1$,
Slow shock: $M_{A,1} < 1$; $M_{A,2} < 1$,
Switch-off shock: $M_{A,1} = 1$; $M_{A,2} \neq 1$,
Switch-on shock: $M_{A,1} \neq 1$; $M_{A,2} = 1$.

Here, we will consider only particle acceleration at parallel fast shocks.

16.3 Alfvén Wave Transmission Through a Parallel Fast Shock

As we have shown in Chaps. 12–14, in general, the dynamics of cosmic rays is controlled by their interaction with the magnetohydrodynamic plasma waves. Also in this special application here, the shock structure continuity equations (16.2.16) allow us to relate the properties of the downstream turbulent electromagnetic fields to those of the prescribed upstream turbulent electromagnetic fields, as we shall demonstrate next.

16.3.1 Basic Equations

We derive formulae for the transmission and reflection coefficients for Alfvén waves striking a plane, parallel, hydromagnetic fast but non-relativistic shock following the treatment of McKenzie and Westphal (1969 [336]) and Scholer and Belcher (1971 [479]). In extension of this original work we keep particular track of the change of the wave helicity state while going from upstream to downstream (Campeanu and Schlickeiser 1992 [89], Vainio and Schlickeiser 1998 [547]). We will also correct two mistakes in the treatment of Campeanu and Schlickeiser (1992 [89] – in this section referred to as CS).

The shock configuration in a frame at rest with respect to the shock interface is sketched in Fig. 16.5. The plane $y = 0$ contains the flow velocity $\boldsymbol{U} = (-U, 0, 0)$, the magnetic field $\boldsymbol{B}_0 = (-B_0, 0, 0)$, and the shock is located in the plane $x = 0$. The indices 1 and 2 refer to the corresponding quantities in the upstream and downstream regions, respectively. Keeping the minus sign in the x-component of the ordered magnetic field vector $\boldsymbol{B}_0$ corrects for the first mistake in the CS-treatment, who adopted $\boldsymbol{B}_0^{\text{CS}} = (+B_0, 0, 0)$, which has led to a misidentification of forward moving (i.e. parallel gas velocity and Alfvén velocity) and backward moving (i.e. antiparallel gas and Alfvén velocity) Alfvén waves.

The one-dimensional inviscid cold flows ahead of and behind the shock are subject to perturbations supposed as a result of non-dispersive Alfvén waves propagating along $\boldsymbol{B}_0$. Such a configuration can support four types of Alfvén waves (see Sects. 9.2.3 and 13.2): forward (f), i.e. moving in the same direction as the fluid, (phase velocity $|U| + |V_A|$) and backward (b),

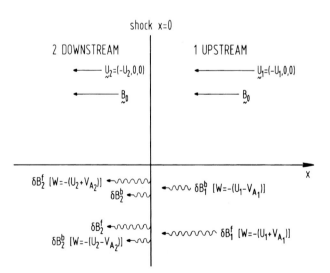

Fig. 16.5. Schematic representation of the plane parallel shock and the incident and transmitted Alfvén waves

i.e. moving in the opposite direction as the fluid (phase velocity $|U| - |V_A|$) moving waves, each being right (R)- and left (L)-handed circularly polarized. For the perturbations in flow velocity, magnetic and electric field in the up- and downstream region we obtain

$$\delta \boldsymbol{U}_{1,2} = \delta \boldsymbol{U}^{\mathrm{f}}_{1,2} + \delta \boldsymbol{U}^{\mathrm{b}}_{1,2} , \qquad (16.3.1\mathrm{a})$$

$$\delta \boldsymbol{B}_{1,2} = \delta \boldsymbol{B}^{\mathrm{f}}_{1,2} + \delta \boldsymbol{B}^{\mathrm{b}}_{1,2}$$
$$= \delta B^{\mathrm{f,L}}_{1,2}(0,1,-i) + \delta B^{\mathrm{f,R}}_{1,2}(0,1,i)$$
$$+ \delta B^{\mathrm{b,L}}_{1,2}(0,1,-i) + \delta B^{\mathrm{b,R}}_{1,2}(0,1,i) . \qquad (16.3.1\mathrm{b})$$

The velocity and magnetic field perturbations are related as (e.g. Chandrasekhar 1961 [97], p. 156)

$$\delta \boldsymbol{U}^{\mathrm{f}}_{1,2} = -\delta \boldsymbol{B}^{\mathrm{f}}_{1,2} (4\pi\rho_{1,2})^{-1/2} , \qquad (16.3.2\mathrm{a})$$

$$\delta \boldsymbol{U}^{\mathrm{b}}_{1,2} = +\delta \boldsymbol{B}^{\mathrm{b}}_{1,2} (4\pi\rho_{1,2})^{-1/2} . \qquad (16.3.2\mathrm{b})$$

The appropriate boundary conditions to be considered are again the continuity of transverse momentum (16.2.9), which in this notation reads

$$\left[\rho U_\mathrm{n} \boldsymbol{U}_\mathrm{t} - B_\mathrm{n} \frac{\boldsymbol{B}_\mathrm{t}}{4\pi} \right] = 0 , \qquad (16.3.3\mathrm{a})$$

and the continuity of tangential electric field (16.2.11)

$$[U_\mathrm{n} \boldsymbol{B}_\mathrm{t} - B_\mathrm{n} \boldsymbol{U}_\mathrm{t}] = 0 , \qquad (16.3.3\mathrm{b})$$

where ρ refers to the density of the flow, using the continuity of the mass flux (16.2.2)

$$\rho_1 U_1 = \rho_2 U_2 . \qquad (16.3.3\mathrm{c})$$

The indices n and t on a vector $\boldsymbol{a}$ refer to the normal $a_\mathrm{n} = a_x$ and tangential $\boldsymbol{a}_\mathrm{t} = (0, a_y, a_z)$ components.

In previous work by McKenzie and Westphal (1969 [336]), Scholer and Belcher (1971 [479]), Achterberg and Blandford (1986 [9]) the Alfvén waves were treated as purely magnetic waves and as solely linearly polarized. Our treatment relaxes these assumptions. Because of their prevailing astrophysical interest (see discussion in Sect. 16.1) we restrict our analysis to fast but non-relativistic shocks, i.e. we require that the shock wave is super-Alfvénic both downstream and upstream

$$|U_2| > V_{A_2}, \quad |U_1| > V_{A_1} , \qquad (16.3.3\mathrm{d})$$

where

$$V_{A_{1,2}} = B_0 (4\pi\rho_{1,2})^{-1/2} \qquad (16.3.3\mathrm{e})$$

are the respective Alfvén speeds. In terms of the compression ratio

$$r = U_1/U_2 = \rho_2/\rho_1 , \qquad (16.3.3f)$$

the upstream Alfvénic Mach number

$$M = U_1/V_{A_1} , \qquad (16.3.3g)$$

and the upstream plasma beta (16.2.39), the conditions (16.3.3d) imply

$$M > r^{1/2} . \qquad (16.3.3h)$$

The shock requirement $|U_1| > c_{s,1}$ yields

$$M > \beta^{1/2} . \qquad (16.3.3i)$$

16.3.2 Reflection and Transmission Coefficients

The boundary conditions (16.3.3a–16.3.3b) allow us to calculate the transmission coefficients for left- and right-hand polarized waves

$$\begin{array}{ll} T_{\rm f,L} \equiv \delta B_2^{\rm f,L}/\delta B_1^{\rm f,L} , & T_{\rm f,R} \equiv \delta B_2^{\rm f,R}/\delta B_1^{\rm f,R} \\ T_{\rm b,L} \equiv \delta B_2^{\rm b,L}/\delta B_1^{\rm f,L} , & T_{\rm b,R} \equiv \delta B_2^{\rm b,R}/\delta B_1^{\rm b,R} \end{array} \qquad (16.3.4)$$

and the reflection coefficients for left- and right-hand polarized waves

$$\begin{array}{ll} R_{\rm f,L} \equiv \delta B_2^{\rm b,L}/\delta B_1^{\rm f,L} , & R_{\rm f,R} \equiv \delta B_2^{\rm b,R}/\delta B_1^{\rm f,R} \\ R_{\rm b,L} \equiv \delta B_2^{\rm f,L}/\delta B_1^{\rm b,L} , & R_{\rm b,R} \equiv \delta B_2^{\rm f,R}/\delta B_1^{\rm b,R} \end{array} \qquad (16.3.5)$$

each for forward and backward moving waves, once the upstream wave helicity state ($\delta B_1^{\rm f,L}$, $\delta B_1^{\rm f,R}$, $\delta B_1^{\rm b,L}$, $\delta B_1^{\rm b,R}$) has been specified.

For reasons discussed in Sect. 16.3.7 we consider two special cases of the upstream wave helicity state:

(a) only forward moving waves exist upstream, i.e. $\delta B_1^{\rm b,L} = \delta B_1^{\rm b,R} = 0$,
(b) only backward moving waves exist upstream, i.e. $\delta B_1^{\rm f,L} = \delta B_1^{\rm f,R} = 0$.

Obviously, results for more general cases can be obtained by combining these two special cases. We consider each case in turn.

16.3.3 Only Forward Moving Waves Upstream

Setting

$$\delta B_1^{\rm f,L} = \alpha \delta B_1^{\rm f} , \qquad \delta B_1^{\rm f,R} = (1-\alpha) \delta B_1^{\rm f} \qquad (16.3.6a)$$

$$\delta B_2^{\rm f,L} = \epsilon \delta B_2^{\rm f} , \qquad \delta B_2^{\rm f,R} = (1-\epsilon) \delta B_2^{\rm f} \qquad (16.3.6b)$$

$$\delta B_2^{\rm b,L} = \eta \delta B_2^{\rm b} , \qquad \delta B_2^{\rm b,R} = (1-\eta) \delta B_2^{\rm b} \qquad (16.3.6c)$$

Equations (16.3.2)–(16.3.3) then yield the four relations

$$M + 1 = \left(Mr^{-1/2} + 1\right) T_f - \left(Mr^{-1/2} - 1\right) R_f \quad (16.3.7a)$$

$$(1 - 2\alpha)(M + 1) = (1 - 2\epsilon)\left(Mr^{-1/2} + 1\right) T_f$$
$$- (1 - 2\eta)\left(Mr^{-1/2} - 1\right) R_f \quad (16.3.7b)$$

$$(M + 1)r^{1/2} = \left(Mr^{-1/2} + 1\right) T_f + \left(Mr^{-1/2} - 1\right) R_f \quad (16.3.7c)$$

$$(1 - 2\alpha)(M + 1)r^{1/2} = (1 - 2\epsilon)\left(Mr^{-1/2} + 1\right) T_f$$
$$+ (1 - 2\eta)\left(Mr^{-1/2} - 1\right) R_f \quad (16.3.7d)$$

with

$$T_f \equiv \delta B_2^f / \delta B_1^f, \quad R_f \equiv \delta B_2^b / \delta B_1^f . \quad (16.3.7e)$$

In an adiabatic medium with the ratio of specific heats, γ, and the plasma beta, β, the shock's compression ratio is given according to (16.2.43),

$$r = \frac{\gamma + 1}{\gamma - 1 + (2\beta/M^2)} . \quad (16.3.7f)$$

Combining (16.3.7a) and (16.3.7c) readily yields

$$T_f = \frac{r^{1/2} \left(r^{1/2} + 1\right)}{2} \frac{(M + 1)}{M + r^{1/2}} \quad (16.3.8a)$$

$$R_f = \frac{r^{1/2} \left(r^{1/2} - 1\right)}{2} \frac{(M + 1)}{M - r^{1/2}} . \quad (16.3.8b)$$

Inserting these results in (16.3.7b) and (16.3.7d) gives

$$\alpha = \eta = \epsilon , \quad (16.3.8c)$$

indicating that the transmission and reflection coefficients are the same for each polarization state:

$$T_{f,R} = T_{f,L} = T_f, \quad R_{f,R} = R_{f,L} = R_f , \quad (16.3.8d)$$

whereas the remaining coefficients $T_{b,R}, T_{b,L}, R_{b,R}, R_{b,L}$ do not occur in this case.

With relation (16.3.7f) we obtain for a non-relativistic adiabatic plasma ($\gamma = 5/3$)

$$T_f(r, \beta) = \frac{\sqrt{r + 1}}{2} \frac{\sqrt{3\beta r} + \sqrt{4 - r}}{\sqrt{3\beta} + \sqrt{4 - r}} , \quad (16.3.9a)$$

$$R_f(r, \beta) = \frac{\sqrt{r - 1}}{2} \frac{\sqrt{3\beta r} + \sqrt{4 - r}}{\sqrt{3\beta} - \sqrt{4 - r}} . \quad (16.3.9b)$$

16.3 Alfvén Wave Transmission Through a Parallel Fast Shock

Note that because of restrictions (16.3.3h) and (16.3.3i)

$$r > r_{\min} = \max(1, 4 - 3\beta), \qquad (16.3.9c)$$

so that T_f remains finite.

In Figs. 16.6 and 16.7 we show the forward transmission and reflection coefficients as a function of r for several values of the plasma beta. One can see that the transmission coefficient attains values larger than unity, indicating that the interaction with the shock wave leads to an amplification of the instreaming upstream waves. Moreover, the non-vanishing reflection coefficient indicates that, through the interaction with the shock wave, backward moving waves are generated in the downstream medium.

As a result of this we have a strong modification of the normalized cross helicity state H_c of the waves, defined in (13.2.11). The latter specifies the fraction of forward moving wave intensity of the total wave intensity

$$H_c = \frac{(\delta B_f)^2 - (\delta B_b)^2}{(\delta B_f)^2 + (\delta B_b)^2}. \qquad (16.3.10)$$

The value of H_c varies between -1 and $+1$, attaining -1 if no forward waves are present, $+1$ if no backward waves are present, and 0 for equal intensity of forward and backward moving waves. Clearly in the upstream region by assumption we have

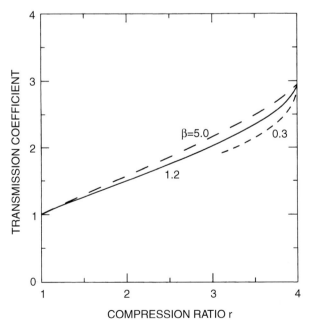

Fig. 16.6. Transmission coefficient of forward waves as a function of the gas compression ratio r for three values of the upstream plasma beta β

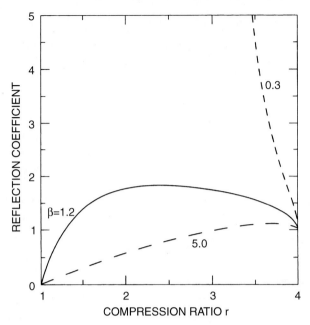

Fig. 16.7. Reflection coefficient of forward waves as a function of the gas compression ratio r for three values of the upstream plasma beta β

$$H_{c,1} = +1 . \qquad (16.3.11a)$$

In the downstream region we can calculate the cross helicity value from the transmission and reflection coefficients (16.3.9)

$$H_{c,2}^f = \frac{T_f^2 - R_f^2}{T_f^2 + R_f^2} = \frac{1 - \psi_f^2}{1 + \psi_f^2}, \qquad (16.3.11b)$$

where

$$\psi_f = \frac{R_f}{T_f} = \frac{r^{1/2} - 1}{r^{1/2} + 1} \frac{M + r^{1/2}}{M - r^{1/2}} = \frac{\sqrt{r-1}}{\sqrt{r+1}} \frac{\sqrt{3\beta} + \sqrt{4-r}}{\sqrt{3\beta} - \sqrt{4-r}} . \qquad (16.3.11c)$$

Moreover, the total wave intensity changes by the factor

$$W_f = \frac{(\delta B_{f,2})^2 + (\delta B_{b,2})^2}{(\delta B_{f,1})^2} = T_f^2 + R_f^2 = T_f^2 \left(1 + \psi_f^2\right)$$

$$= \frac{r}{4} \left(r^{1/2} + 1\right)^2 \left(\frac{M+1}{M + r^{1/2}}\right)^2 \left(1 + \psi_f^2\right) . \qquad (16.3.12)$$

In contrast to the drastic modification of the wave cross helicity state across the shock, the wave magnetic helicity values σ_f, σ_b remain unchanged. These measure the net polarization state of the forward and backward waves (see (13.1.17)):

$$\sigma_{\rm f} \equiv \frac{(\delta B_{\rm f,L})^2 - (\delta B_{\rm f,R})^2}{(\delta B_{\rm f})^2}, \qquad (16.3.13{\rm a})$$

$$\sigma_{\rm b} \equiv \frac{(\delta B_{\rm b,L})^2 - (\delta B_{\rm b,R})^2}{(\delta B_{\rm b})^2}. \qquad (16.3.13{\rm b})$$

With the definitions (16.3.6) we have in the upstream region in our case

$$\sigma_{\rm f,1} = 2\alpha - 1, \qquad (16.3.14{\rm a})$$

whereas $\sigma_{b,1}$ does not occur, and in the downstream region, because of (16.3.8c),

$$\sigma_{\rm f,2} = (2\epsilon - 1) = \sigma_{\rm f,1}, \qquad (16.3.14{\rm b})$$
$$\sigma_{\rm b,2} = (2\gamma - 1) = \sigma_{\rm f,1} = \sigma_{\rm f,2}. \qquad (16.3.14{\rm c})$$

The magnetic helicity state of the waves is not modified in the interaction process.

16.3.4 Only Backward Moving Waves Downstream

In the opposite case of having no forward moving upstream waves ($\delta B_1^{\rm f,L} = \delta B_1^{\rm f,R} = 0$) we obtain

$$T_{\rm b} \equiv \frac{\delta B_2^{\rm b}}{\delta B_1^{\rm b}} = \frac{r^{1/2}\left(r^{1/2}+1\right)}{2}\frac{M-1}{M-r^{1/2}}, \qquad (16.3.15{\rm a})$$

$$R_{\rm b} \equiv \frac{\delta B_2^{\rm f}}{\delta B_1^{\rm b}} = \frac{r^{1/2}\left(r^{1/2}-1\right)}{2}\frac{M-1}{M+r^{1/2}}. \qquad (16.3.15{\rm b})$$

The total wave intensity increases by the factor

$$W_{\rm b} \equiv \frac{\left(\delta B_2^{\rm f}\right)^2 + \left(\delta B_2^{\rm b}\right)^2}{\left(\delta B_1^{\rm b}\right)^2} = T_{\rm b}^2 + R_{\rm b}^2$$
$$= \frac{r(M-1)^2}{4}\left[\left(\frac{r^{1/2}+1}{M-r^{1/2}}\right)^2 + \left(\frac{r^{1/2}-1}{M+r^{1/2}}\right)^2\right], \qquad (16.3.16)$$

and the cross helicity changes from $H_{c_1} = -1$ through the shock to

$$H_{c_2}^{\rm b} = \frac{R_{\rm b}^2 - T_{\rm b}^2}{T_{\rm b}^2 + R_{\rm b}^2} = \frac{\psi_{\rm b}^2 - 1}{\psi_{\rm b}^2 + 1}, \qquad (16.3.17)$$

where

$$\psi_{\rm b} \equiv \frac{R_{\rm b}}{T_{\rm b}} = \frac{r^{1/2}-1}{r^{1/2}+1}\frac{M-r^{1/2}}{M+r^{1/2}}. \qquad (16.3.18)$$

As before, the magnetic helicity state of the waves is not modified in the interaction process.

16.3.5 Alternative Notation

Evidently, the transmission and reflection coefficients (16.3.9) and (16.3.15) can be conveniently cast into the form

$$T \equiv \frac{\delta B_2'}{\delta B_1} = \frac{r^{1/2}\left(r^{1/2}+1\right)}{2} \frac{M+H_{c_1}}{M+r^{1/2}H_{c_1}} = \frac{r^{1/2}+1}{2r^{1/2}} \frac{V_1}{V_2'} \qquad (16.3.19)$$

and

$$R \equiv \frac{\delta B_2''}{\delta B_1} = \frac{r^{1/2}\left(r^{1/2}-1\right)}{2} \frac{M+H_{c_1}}{M-r^{1/2}H_{c_1}} = \frac{r^{1/2}-1}{2r^{1/2}} \frac{V_1}{V_2''}, \qquad (16.3.20)$$

where V_i denote the shock frame phase velocities of the relevant upstream ($i=1$) and downstream ($i=2$) wave modes, and primes and double primes are used to denote the transmitted and reflected downstream waves, respectively; i.e. $V_1 = u_1 + H_{c_1}V_{A_1}$, $V_2' = u_2 + H_{c_1}V_{A_2}$, and $V_2'' = u_2 - H_{c_1}V_{A_2}$. With this representation the wave amplification factor and the downstream cross helicity may be written in compact form

$$W = T^2\left(1+\psi^2\right), \qquad (16.3.21)$$

$$H_{c_2} = H_{c_1}\frac{1-\psi^2}{1+\psi^2}, \qquad (16.3.22)$$

where

$$\psi \equiv \frac{R}{T} = \frac{r^{1/2}-1}{r^{1/2}+1} \frac{M+r^{1/2}H_{c_1}}{M-r^{1/2}H_{c_1}} = \frac{r^{1/2}-1}{r^{1/2}+1} \frac{V_2'}{V_2''}. \qquad (16.3.23)$$

16.3.6 Shock Effect on Wavenumbers

CS assumed the conservation of wavenumber during the interaction of the Alfvén waves with the shock. In this work, we demand the conservation of wave frequency, $\omega_1 = \omega_2$. For circularly polarized waves this is a direct consequence of the coplanarity theorem, which states that the downstream magnetic field has to be in the plane defined by the upstream magnetic field and the shock normal. We will ignore magnetic helicity since it has no effect on the transmission of the waves, and the wave spectra are taken to be of Kolmogorov form

$$I_1(k) = \mathcal{H}(k-k_1)\mathcal{I}_1 k^{-q}, \qquad (16.3.24a)$$
$$I_1(k) = \mathcal{H}(k-k_1)\mathcal{I}_1 k^{-q}, \qquad (16.3.24b)$$
$$I_2''(k) = \mathcal{H}(k-k_2'')\mathcal{I}_2'' k^{-q}, \qquad (16.3.24c)$$

where k is the wavenumber, $\mathcal{H}(k)$ is the step function, k_1, k_2', and k_2'' are the minimum wavenumbers of the incident, transmitted and reflected

16.3 Alfvén Wave Transmission Through a Parallel Fast Shock

waves, respectively. We consider values $1 < q < 2$ for the spectral index. The normalization coefficients are given by $\mathcal{I}_1 = (q-1)(\delta B_1)^2 k_1^{q-1}$, $\mathcal{I}_2' = (q-1)(\delta B_2')^2 k_2'^{q-1}$, and $\mathcal{I}_2'' = (q-1)(\delta B_2'')^2 k_2''^{q-1}$. From the frequency conservation condition, we have

$$k_2' = k_1 \frac{V_1}{V_2'} = k_1 r \frac{M + H_{c_1}}{M + r^{1/2} H_{c_1}}, \quad (16.3.25a)$$

$$k_2'' = k_1 \frac{V_1}{V_2''} = k_1 r \frac{M + H_{c_1}}{M - r^{1/2} H_{c_1}}. \quad (16.3.25b)$$

Thus, we may write the relations

$$\frac{\mathcal{I}_2'}{\mathcal{I}_1} = \left(\frac{\delta B_2'}{\delta B_1}\right)^2 \left(\frac{k_2'}{k_1}\right)^{q-1} = \left(\frac{\delta B_2'}{\delta B_1}\right)^2 \left(\frac{V_1}{V_2'}\right)^{q-1},$$

$$\frac{\mathcal{I}_2''}{\mathcal{I}_1} = \left(\frac{\delta B_2''}{\delta B_1}\right)^2 \left(\frac{k_2''}{k_1}\right)^{q-1} = \left(\frac{\delta B_2'}{\delta B_1}\right)^2 \left(\frac{V_1}{V_2''}\right)^{q-1}.$$

These equations give the wave transmission coefficients at *constant wavenumber*. Using (16.3.19)–(16.3.20) we obtain

$$T_k \equiv \left(\frac{I_2'(k)}{I_1(k)}\right)^{1/2} = \frac{r^{1/2}+1}{2r^{1/2}} \left(\frac{V_1}{V_2'}\right)^{(q+1)/2}$$

$$= \frac{r^{1/2}+1}{2r^{1/2}} \left(r \frac{M + H_{c_1}}{M + r^{1/2} H_{c_1}}\right)^{(q+1)/2}, \quad (16.3.26)$$

$$R_k \equiv \left(\frac{I_2''(k)}{I_1(k)}\right)^{1/2} = \frac{r^{1/2}-1}{2r^{1/2}} \left(\frac{V_1}{V_2''}\right)^{(q+1)/2}$$

$$= \frac{r^{1/2}-1}{2r^{1/2}} \left(r \frac{M + H_{c_1}}{M - r^{1/2} H_{c_1}}\right)^{(q+1)/2}, \quad (16.3.27)$$

where k has to be larger than the largest of the cutoff wavenumbers, $k > \max(k_i)$. We can also define the total wave amplification factor and downstream cross helicity at constant wavenumber $k > \max(k_i)$,

$$W_k \equiv \frac{I_2'(k) + I_2''(k)}{I_1(k)} = T_k^2 \left(1 + \psi_k^2\right), \quad (16.3.28)$$

$$H_{c_2}^k \equiv H_{c_1} \frac{I_2'(k) - I_2''(k)}{I_2'(k) + I_2''(k)} = H_{c_1} \frac{1 - \psi_k^2}{1 + \psi_k^2}, \quad (16.3.29)$$

where

$$\psi_k \equiv \frac{R_k}{T_k} = \frac{r^{1/2}-1}{r^{1/2}+1} \left(\frac{M + r^{1/2} H_{c_1}}{M - r^{1/2} H_{c_1}}\right)^{(q+1)/2}$$

$$= \frac{r^{1/2}-1}{r^{1/2}+1} \left(\frac{V_2'}{V_2''}\right)^{(q+1)/2}. \quad (16.3.30)$$

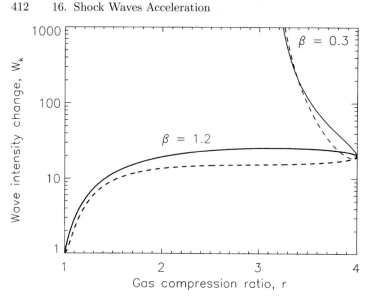

Fig. 16.8. Total wave amplification factor at constant wavenumber for a shock with a constant upstream plasma beta. **Dashed** and **solid lines** give the results for $H_{c_1} = +1$ and -1, respectively

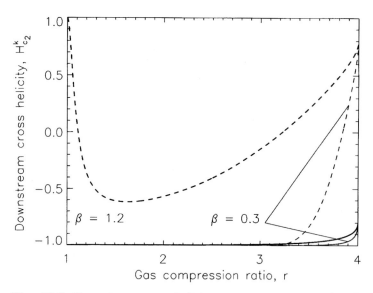

Fig. 16.9. Downstream cross helicity at constant wavenumber for a shock with a constant upstream plasma beta. **Dashed** and **solid lines** give the results for $H_{c_1} = +1$ and -1, respectively

Note that these representations will be changed if we consider wavenumbers smaller than the largest of the cutoff values. It can be seen that the transmission coefficients at constant wavenumber agree with those calculated for the integrated spectra only if $q \to 1$. We have plotted W_k and $H_{c_2}^k$ for some sets of shock parameters in Figs. 16.8 and 16.9, respectively.

The total wave amplification plots are qualitatively similar to those of CS. However, because of their misidentification of the wave propagation direction (their forward wave, in fact, should be the backward wave and vice versa), the drastic changes in the downstream cross helicity state for low Mach number shocks occur for forward, and not for backward, upstream waves. When the critical limit, $M \to r^{1/2}$ is approached, the downstream helicity state always approaches $H_{c_2}^k \to -1$, not $+1$ as was concluded by CS.

16.3.7 Precursor Streaming Instability

According to the following section (see Sect. 16.5.2) the cosmic ray particle distribution upstream of the shock is given by the precursor distribution

$$F_1(x,p) = F_0(p)\, \exp\left[-(u_1 + H_{c_1} V_{A,1}) \int_0^x \frac{dx^*}{\kappa_1(x^*)}\right], \qquad (16.3.31)$$

so that the spatial gradient is negative

$$\frac{\partial F_1}{\partial x} = -\frac{u_1 + H_{c_1} V_{A,1}}{\kappa_1} F_1(x,p) < 0. \qquad (16.3.32)$$

In (11.2.51) and (11.2.52) we calculated the damping and/or growth rate of left-handed (L) and right-handed (R) polarized, forward (f) and backward (b) moving Alfvén waves, respectively. For Alfvén speeds $V_A \ll c$, corresponding to index of refraction $|N| \gg 1$, we can simplify both equations to (notation as in Sect. 11.2.1.2)

$$\Gamma_{L,R}^{f,b} \simeq \pm \frac{\pi^2 V_A}{2c^2 k_\|} \sum_r \frac{\omega_{p,a}^2 N_r (m_r c)^3}{n_a}$$

$$\times \int_{\sqrt{1+x_r^2}}^\infty dE \; \frac{E^2 - 1 - x_r^2}{\sqrt{E^2-1}} \frac{\partial f_r}{\partial \mu} \delta(\mu - \mu_{L,R}), \qquad (16.3.33)$$

where we used $y = \mu\sqrt{E^2-1}$ and where

$$\mu_{L,R} \equiv \frac{\mp x_r}{\sqrt{E^2-1}}. \qquad (16.3.34)$$

According to (12.3.2) and (12.3.17) we can express the cosmic ray particle anisotropy appearing in (16.3.33) in terms of the spatial and momentum gradients of the isotropic part of the distribution function $M_a(p,z) = F_1(p,x)$, as

16. Shock Waves Acceleration

$$\frac{\partial f_r}{\partial \mu} = \frac{\partial G_a}{\partial \mu} = -\left[\frac{1-\mu^2}{2D_{\mu\mu}} v \frac{\partial M_a}{\partial z} + \frac{D_{\mu p}}{D_{\mu\mu}} \frac{\partial M_a}{\partial p}\right]$$
$$\simeq -\frac{1-\mu^2}{2D_{\mu\mu}} c \frac{\sqrt{E^2-1}}{E} \frac{\partial F_1}{\partial x}, \qquad (16.3.35)$$

where we neglected the momentum gradient because for parallel propagating Alfvén waves, $D_{\mu p}$ is much smaller than $D_{\mu\mu}$, see (13.2.9). Combining (16.3.34) and (16.3.35) we obtain

$$\Gamma_{L,R}^{f,b} = \frac{\partial \gamma_{L,R}^{f,b}}{\partial x}, \qquad \gamma_{L,R}^{f,b} = \mp \gamma_{L,R}^0, \qquad (16.3.36)$$

where

$$\gamma_{L,R}^0(x) = \frac{\pi^2 V_A}{4ck_\parallel} \sum_r \frac{\omega_{p,a}^2 N_r (m_r c)^3}{n_a}$$
$$\times \int_{\sqrt{1+x_r^2}}^{\infty} dE \frac{E^2 - 1 - x_r^2}{E} \frac{1-\mu_{L,R}^2}{D_{\mu\mu}(\mu_{R,L})} F_1(x,p) \qquad (16.3.37)$$

is determined by the spatial distribution of the precursor F_1.

In quasi-equilibrium the differential intensities of forward and backward moving Alfvén waves satisfy the wave kinetic equation (11.3.6). If we neglect spontaneous emission, Landau damping and cascading effects, but include wave propagation, these wave equations are

$$(u_1 \pm V_{A,1}) \frac{\partial W_{f,b}}{\partial x} = \Gamma^{f,b} W_{f,b}. \qquad (16.3.38)$$

By using (16.3.36) we derive

$$\frac{d}{dx}\left[(u_1 \pm V_{A,1}) \ln W_{f,b} - \gamma_{L,R}^{f,b}\right] = 0, \qquad (16.3.39)$$

leading to

$$W_{f,b}(x) \exp\left[-\frac{\gamma_{L,R}^{f,b}(x)}{u_1 \pm V_{A,1}}\right] = W_{f,b}(x=\infty) \exp\left[-\frac{\gamma_{L,R}^{f,b}(x=\infty)}{u_1 \pm V_{A,1}}\right]. \qquad (16.3.40)$$

At $x \to \infty$, i.e. far ahead of the shock, the precursor distribution (16.3.31) vanishes, implying according to (16.3.36)–(16.3.37) that $\gamma_{L,R}^{f,b}(\infty) = 0$. Equation (16.3.40) then relates the wave intensities at any upstream position, $W_{f,b}(x)$, to the ambient wave intensity far ahead of the shock, $W_{f,b}(\infty)$,

$$W_{f,b}(x) = W_{f,b}(\infty) \exp\left[\mp \frac{\gamma_{L,R}^0(x)}{u_1 \pm V_{A,1}}\right]. \qquad (16.3.41)$$

Obviously the intensity of forward moving waves is exponentially *reduced* with respect to the ambient level of forward moving waves far ahead of the shock, while the intensity of backward moving waves is exponentially *amplified* with respect to the ambient level of backward moving waves far ahead of the shock. As a consequence we can expect that in the upstream medium the intensity of backward moving Alfvén waves is much larger than the intensity of forward moving Alfvén waves. A posteriori, this justifies the assumptions made in Sect. 16.3.2. It also indicates that the upstream Alfvén waves are predominantly backward traveling.

16.4 Cosmic Ray Transport and Acceleration Parameters

After having determined the wave helicity state we are now in a position to determine the transport and acceleration parameters of cosmic ray particles. The isotropic part of the gyrophase-averaged cosmic ray particle phase space density $F(x,p,t)$ evolves in the quasilinear approximation of the wave-particle interaction according to the diffusion-convection equation (see (12.3.31) and (14.1.6)), which in the shock wave rest system and under steady state conditions reads

$$\frac{\partial}{\partial x} \kappa \frac{\partial F}{\partial x} - \left(u_\text{n} + \frac{1}{4p^2} \frac{\partial}{\partial p} (p^2 vA)\right) \frac{\partial F}{\partial x}$$
$$+ \left(\frac{p}{3} \frac{\partial u_\text{n}}{\partial x} + \frac{v}{4} \frac{\partial A}{\partial x}\right) \frac{\partial F}{\partial p}$$
$$+ \frac{1}{p^2} \frac{\partial}{\partial p} p^2 \theta \frac{\partial F}{\partial p} = -Q(x,p) , \qquad (16.4.1)$$

where x is distance from the shock along the shock normal, p is particle momentum, v is particle velocity, $Q(x,p)$ represents sources and sinks, and the general forms of the acceleration and transport parameters, spatial diffusion coefficient $\kappa(x,p)$, rate of adiabatic deceleration $A(x,p) = A_1(x,p)$, and momentum diffusion coefficient $\theta(x,p) = A_2(x,p)$, may be found according to (13.2.23)–(13.3.29).

The transport and acceleration parameters for undamped slab Alfvénic turbulence and Kolmorov-turbulence (see (16.3.24)) follow from the Fokker-Planck coefficients $D_{\mu\mu}, D_{\mu p}, D_{pp}$ given in (13.2.9), in which the momentum coordinates p and μ are calculated in the local fluid rest system.

For the spatial diffusion coefficient we obtain, according to (13.2.13)

$$\kappa(x,p) = \frac{v}{2\pi(q-1)} \left(\frac{B_0}{\delta B}\right)^2 (R_\text{L} k_\text{min})^{1-q} R_\text{L} \, S\left(H_\text{c}, \sigma_\text{f}, \sigma_\text{b}, q\right) , \qquad (16.4.2)$$

where R_L is the cosmic ray proton gyroradius and

$$S = \frac{2}{(2-q)(4-q)} \left[\left\{ (\sigma_f - \sigma_b) \frac{1+H_c}{2} + (1+\sigma_b) \right\}^{-1} \right.$$
$$\left. + \left\{ (1-\sigma_b) - (\sigma_f - \sigma_b) \frac{1+H_c}{2} \right\}^{-1} \right]. \quad (16.4.3)$$

All terms in (16.4.2) have to be evaluated in the local fluid rest system, i.e. $u_1 = 0$ at $x > 0$ and $u_2 = 0$ at $x < 0$.

For the rate of adiabatic deceleration and the momentum diffusion coefficient we find according to (13.2.27) and (13.2.35), respectively,

$$A(x,p) = -\frac{4}{3} \frac{V_A}{v} p J(H_c, \sigma_f, \sigma_b), \quad (16.4.4)$$

with

$$J(H_c, \sigma_f, \sigma_b)$$
$$= \frac{(1+H_c)^2 (1-\sigma_f^2) - (1-H_c)^2 (1-\sigma_b^2)}{(1+H_c)^2 (1-\sigma_f^2) + (1-H_c)^2 (1-\sigma_b^2) + 2(1-H_c^2)(1-\sigma_f \sigma_b)},$$
$$(16.4.5)$$

and

$$\theta(x,p) = \frac{2}{\pi^2 q(q+2)} G(H_c, \sigma_b, \sigma_f) S(H_c, \sigma_b, \sigma_f, q) \frac{V_A^2 p^2}{\kappa(x,p)}, \quad (16.4.6)$$

with

$$G(H_c, \sigma_b, \sigma_f) = \frac{1}{2} (1 - H_c^2)$$
$$\times \frac{(1+H_c)(1-\sigma_f^2) + (1-H_c)(1-\sigma_b^2)}{(1+H_c)^2 (1-\sigma_f^2) + (1-H_c)^2 (1-\sigma_b^2) + 2(1-H_c^2)(1-\sigma_f \sigma_b)}.$$
$$(16.4.7)$$

By using the respective helicity states in the upstream and downstream region calculated in Sect. 16.3 the cosmic ray transport parameters in both regions follow immediately.

16.4.1 Upstream Cosmic Ray Transport Equation

According to the analysis of the precursor streaming instability (Sect. 16.3.7) it is reasonable to start from the idealized picture that in the upstream region the cross helicity of the Alfvénic turbulence is either $H_{c,1} = -1$ or $H_{c,1} = +1$, in which case $G(\pm 1, \sigma_b, \sigma_f) = 0$ according to (16.4.7), and upstream momentum diffusion is absent. We obtain for the upstream transport parameters

$$\kappa_1(x,p) = \frac{v R_L S_1}{2\pi(q-1)} \left(\frac{B_0}{\delta B_1}\right)^2 (k_1 R_L)^{q-1} , \qquad (16.4.8)$$

$$A_1(x,p) = -H_{c_1} \frac{4}{3} \frac{V_{A_1}}{v} p , \qquad (16.4.9a)$$

$$\theta_1(x,p) = 0 , \qquad (16.4.9b)$$

where

$$S_1 = \frac{4}{(2-q)(4-q)} \frac{1}{1-\sigma^2} , \qquad (16.4.10)$$

and σ is the magnetic helicity state of the upstream waves. Hence, the transport equation for the upstream ($x > 0$) phase space density $F_1(x,p)$ reads

$$\frac{\partial}{\partial x} \kappa_1 \frac{\partial F_1}{\partial x} + (u_1 + H_{c_1} V_{A_1}) \frac{\partial F_1}{\partial x} - \frac{p}{3} \frac{\partial}{\partial x}(u_1 + H_{c_1} V_{A_1}) \frac{\partial F_1}{\partial p} = -Q_1(x,p) . \qquad (16.4.11)$$

16.4.2 Downstream Cosmic Ray Transport Equation

In the downstream region, we obtain with the transmission and reflection coefficients ((16.3.26)–(16.3.30)) the transport parameters

$$\kappa_2(x,p) = \kappa_1(x,p)/W_k , \qquad (16.4.12)$$

$$A_2(x,p) = r^{-1/2} \frac{1-\psi_k^2}{1+\psi_k^2} A_1(x,p) , \qquad (16.4.13)$$

and

$$\theta_2(x,p) = \vartheta \frac{W_k V_{A_1}^2 p^2}{r \kappa_1} , \qquad (16.4.14)$$

where

$$\vartheta \equiv \frac{2\psi_k^2 S_1}{q(q+2)(1+\psi_k^2)^2} , \qquad (16.4.15)$$

Note that these transport parameters are valid only for particles with $R_L < \max(k_i)^{-1}$, which we shall assume. Thus, the steady-state transport equation for the downstream ($x < 0$) cosmic ray phase space density $F_2(x,p)$ reads

$$\frac{\partial}{\partial x} \frac{\kappa_1}{W_k} \frac{\partial F_2}{\partial x} + \frac{1}{r}\left(u_1 + r^{1/2} H_{c_1} \frac{1-\psi_k^2}{1+\psi_k^2} V_{A_1}\right) \frac{\partial F_2}{\partial x}$$
$$- \frac{p}{3r} \frac{\partial}{\partial x}\left(u_1 + r^{1/2} H_{c_1} \frac{1-\psi_k^2}{1+\psi_k^2} V_{A_1}\right) \frac{\partial F_2}{\partial p}$$
$$+ \frac{\vartheta}{p^2} \frac{\partial}{\partial p} \frac{W_k V_{A_1}^2 p^4}{r \kappa_1} \frac{\partial F_2}{\partial p} = -Q_2(x,p) , \qquad (16.4.16)$$

which inevitably contains the momentum diffusion term.

16.4.3 Scattering Center Compression Ratio

Comparing the upstream (16.4.11) and downstream (16.4.16) cosmic ray transport equations, we notice that at $x = 0$ the cosmic ray bulk speed changes abruptly from $-(u_1 + H_{c_1} V_{A,1})$ upstream to $-(u_1 + r^{1/2} H_{c_1} 1 - \psi_k^2 / 1 + \psi_k^2 V_{A,1})/r$. We therefore define the scattering center compression ratio

$$r_k \equiv \frac{V_1}{V_2} = \frac{u_1 + H_{c_1} V_{A_1}}{u_2 + H_{c_2}^k V_{A_2}} = r \frac{M + H_{c_1}}{M + r^{1/2} H_{c_2}^k}, \qquad (16.4.17)$$

which, in general, is different from the gas compression ratio (r). Figure 16.10 shows the variation of r_k for various plasma betas.

It has been recognized in a series of papers by Axford et al. (1977 [22]), Krymsky (1977 [276]), Bell (1978 [41]) and Blandford and Ostriker (1978 [58]) that this compression (16.4.17) of the scattering centers at the shock leads to efficient acceleration of cosmic ray particles at the shock, by the first-order Fermi mechanism (see also Sect. 16.6.1). As discussed in Sect. 14.1, the compression (16.4.17) corresponds to a negative velocity gradient ($dV/dz < 0$) which accelerates particles by positive momentum convection.

The principal difference between the gas compression ratio and the scattering center compression ratio, being equivalent to the difference between the effective wave velocity and the gas velocity, has been already noted by Bell (1978 [41]), see his equations (11) and (12) (although he did no quan-

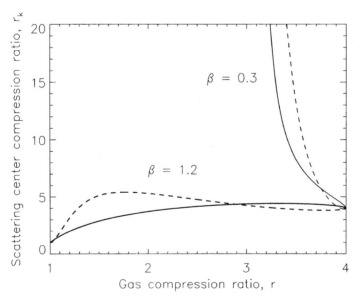

Fig. 16.10. Scattering center compression ratio r_k for a shock with a constant upstream plasma beta. **Dashed** and **solid lines** give the results for $H_{c_1} = +1$ and -1, respectively

titative calculations of this effect). By calculating the correct transmission coefficients of Alfvén waves through the shock we obtain the correct relation between r_k and r for these types of shock waves.

16.4.4 Importance of Downstream Momentum Diffusion

To assess the importance of downstream momentum diffusion we compare the respective acceleration time scales of the first-order adiabatic compression acceleration associated with the scattering center compression at the shock and second-order acceleration by momentum diffusion, i.e. the inverse logarithmic momentum gain rates $\tau = (\mathrm{d}\log p/\mathrm{d}t)^{-1}$, of the two acceleration processes. For the first-order process we have (e.g. Forman and Webb 1985 [175])

$$\tau_\mathrm{s} \equiv \frac{3}{\Delta V}\left(\frac{\kappa_1}{V_1}+\frac{\kappa_2}{V_2}\right) = \frac{\kappa_1}{(u_1+H_{c_1}V_{A_1})^2}\frac{3r_k}{r_k-1}\left(1+\frac{r_k}{W_k}\right). \qquad (16.4.18)$$

The systematic part of the momentum diffusion has the inverse rate $\tau_\mathrm{F} = p^3/\left[\partial(p^2\theta_2)/\partial p\right]$. Hence, the ratio of the two time scales is given by

$$\frac{\tau_\mathrm{F}/\tau_\mathrm{s}}{(1-\sigma^2)B(q)\alpha(q,v)} = \frac{(M+H_{c_1})^2(r_k-1)}{1+r_k/(T_k^2+R_k^2)}\frac{3r}{5r_k}\left(\frac{1}{T_k^2}+\frac{1}{R_k^2}\right), \qquad (16.4.19)$$

where $B(q) = 5q(q+2)(2-q)(4-q)/24$ is a slightly varying function of q, $B(1.5) \approx 1.4$, $B(5/3) \approx 1.0$, and $B(1.8) \approx 0.63$, and $\alpha(q,v) = \partial \log(p^4\kappa^{-1})\partial p/3$, which equals $(1+q)/3$ at non-relativistic particle velocities and $(2+q)/3$ at relativistic velocities, i.e. $2/3 < \alpha < 4/3$.

The ratio of acceleration time scales (16.4.19) for different values of the shock parameters is calculated in Fig. 16.11. For high Mach number shocks, the acceleration is dominated by the first order mechanism, if the magnetic helicity of the waves is not large, $\sigma^2 \ll 1$. If the upstream waves are dominantly backward waves and the magnetic helicity is not large, the acceleration is always dominated by the first-order Fermi process regardless of the Mach number. Note, however, that if the magnetic helicity of the waves is close to $\sigma = \pm 1$, the dominant acceleration process is the downstream stochastic acceleration. This occurs because the degenerate magnetic helicity states lead to large regions in velocity space where particles do not see any waves at all. This, in turn, makes it impossible for the particles to get reflected back to the shock. However, in that part of the velocity space where the particles do see the waves, they see waves traveling in both directions in the downstream region. Thus, the second-order mechanism still accelerates particles. Stochastic acceleration is also important in the case of forward upstream waves if the Mach number of the shock and plasma beta are not large. However, since one expects the upstream waves to be predominantly backward traveling (see Sect. 16.3.7) and not to have degenerate magnetic helicity values ($\sigma \neq 1$), we expect the first-order process to provide the dominant part to the acceleration in realistic physical environments.

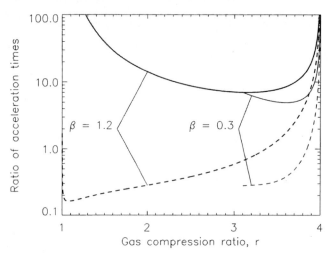

Fig. 16.11. The ratio of acceleration times for a shock with a constant upstream plasma beta. **Dashed** and **solid lines** give the results for $H_{c_1} = +1$ and -1, respectively

16.4.5 Interlude

The present calculations of the transmissions of the Alfvén waves through the shock indicate a qualitatively different behavior from that found by CS. It was concluded by CS that a forward wave field experiences little changes in its cross helicity state as it is convected through the shock, and that the downstream wave field with large intensity would be the forward traveling one. In the case of backward upstream waves, the wave field was calculated to be effectively reversed by the shock, if the compression ratio was moderately strong ($r \simeq 2-3$) and the plasma beta was small. This implied that the scattering center compression ratio was small, even smaller than unity for certain conditions, and that the first-order Fermi acceleration by the shock crossings was inefficient in such shocks or could even turn to deceleration. Noting that the physically more plausible case is the one of backward upstream waves, CS concluded that stochastic second-order Fermi acceleration in the downstream region would generally be more efficient than first-order Fermi acceleration by multiple shock crossings. We have shown here that these results by CS are due to their misidentification of the wave propagation direction, and we have corrected their calculations for this mistake.

Based on the corrected Alfvén wave transmission coefficients we have shown that it is always the backward downstream wave field that seems to have an infinite amplitude in a $\beta < 1$ plasma at the critical limit, $M \to r^{1/2}$. This is also qualitatively correct, because in this limit the downstream backward waves are propagating with small speed in the shock frame; as the upstream waves are continuously flowing in to the shock, the part of their energy that is converted to the downstream backward waves is slowly transported away from the shock. This results in large intensities of the downstream backward waves regardless of the cross helicity state of the upstream medium. Our study showed that the downstream stochastic acceleration works slower than the acceleration by multiple shock crossings under physically plausible conditions. However, we confirmed the conclusion of CS that in no case can stochastic acceleration be totally absent from the downstream region, and that this second-order mechanism would work even in the extreme case of degenerate magnetic helicity state $\sigma = \pm 1$, where the first-order mechanism ceases to operate. Second-order acceleration would also be important in the case of forward upstream waves if the Alfvénic Mach number of the shock was small.

For a given value of plasma beta, (16.3.7f) gives a unique relationship between the gas compression ratio and the Alfvénic Mach number of the shock. For $\beta < 1$, there is a singularity in the backward wave intensity at the critical Mach number, which can also be given in the form $M \to (4 - 3\beta)^{1/2}$, when (16.3.7f) holds. However, this singularity occurs only because the wave pressure has not been taken into account in (16.3.7f). In a situation where the critical limit is approached, e.g. if the shock is slowing down in a uniform medium or the shock is propagating towards a decreasing plasma beta, the gas compression ratio will adjust to a lower value due to the additional downstream pressure produced by the waves. In this way the shock will avoid the singularity and continue to propagate as a super-Alfvénic shock. An interesting subject for the future is to determine the upper limits for the wave amplification factor and the scattering center compression ratio including the wave pressure effects in (16.3.7f), i.e. to see how close the Mach number can get to $r^{1/2}$.

The final assessment of the importance and influence of the momentum diffusion term on the resulting particle spectra from diffusive shock wave acceleration should be based on a fully self-consistent treatment solving simultaneously for the particle's spectrum and the Alfvén wave turbulence spectra including their mutual interaction and finite spatial boundaries beyond which the acceleration is turned off. In the next subsection we will address the solution of the full test-particle cosmic ray transport equations (16.4.11) and (16.4.16) with the unavoidable downstream momentum diffusion term to follow the evolution of the particle distribution function in more detail.

16.5 Solution of the Transport Equations

The full system of (16.4.11) and (16.4.16) together with the appropriate boundary conditions has been solved analytically by Schlickeiser et al. (1993 [472]; hereafter referred to as SCL) for the case that the power law spectral index of the Kolmogorov turbulence (16.4.24) equals $q = 2$. In this case the spatial diffusion coefficients (16.4.8) and (16.4.12) for relativistic particles are independent of momentum (see the discussion in Sect. 13.2.2.2). Earlier numerical (Krülls 1992 [275], Ostrowski and Schlickeiser 1993 [380]) and analytical (Webb 1983 [557], Schlickeiser 1989c [459]) investigations of second-order Fermi acceleration at shock waves suffer from the fact that there was no link of the momentum diffusion acceleration rate to the shock wave properties, or that they did not consider the full spatial problem (Dröge and Schlickeiser 1986 [136]).

As we shall see, an important result of the work of SCL is that stochastic acceleration by momentum diffusion can be very efficient, implying that wave damping by cyclotron acceleration will be significant. The SCL-test-particle solution does not take this wave damping self-consistently into account, which would require the solution of a system of three coupled differential equations describing the quasilinear evolution of energetic particle spectra and Alfvén wave intensities, including their mutual nonlinear interaction in the downstream region. The work of Bogdan et al. (1991 [71]), Ko (1992 [265]) as well as Jiang et al. (1996 [243]) for much simpler homogeneous (i.e. non-shock) physical systems, and, in the case of the latter two papers, performed on a non-kinetic equation level, has indicated how complex and rich in phenomena such an investigation can be. Therefore the test-particle solution of this work has to be regarded as a necessary and important, but not sufficient, step forward towards a complete understanding of the diffusive shock wave acceleration process, which in future work has to be extended to the inclusion of the self-consistent wave damping. But it serves one important purpose: it demonstrates that the diffusive shock wave acceleration process can only be fully understood if the mutual quasilinear wave-particle interaction in the downstream region is explored, whereas in the earlier original treatments of this acceleration process it had been claimed that the resulting particle spectra are insensitive to the actual microscopic physical conditions in the downstream region.

16.5.1 Basic Equations

Here we consider the following physical situation: in the upstream region of a fast, but non-relativistic, super-Alfvénic parallel shock wave ratio only backward or forward moving right- and left-handed polarized Alfvén waves are present with respective turbulence power spectra (16.4.24) with $q = 2$. The steady-state upstream cosmic ray transport equation (16.4.11) for relativistic cosmic ray particles then reads

$$\frac{\partial}{\partial x}\left(\kappa_1 \frac{\partial F_1}{\partial x}\right) + V_1 \frac{\partial F_1}{\partial x} = -Q_1(x,p) \quad (x > 0) , \qquad (16.5.1)$$

where $V_1 = u_1 + H_{c_1} V_{A,1}$. The spatial diffusion coefficient κ_1 is independent of momentum for relativistic particles.

The steady-state downstream cosmic ray transport equation (16.4.16) is

$$\frac{\partial}{\partial x}\left(\kappa_2(x)\frac{\partial F_2}{\partial x}\right) + V_2\frac{\partial F_2}{\partial x} + \frac{\vartheta}{p^2}\frac{\partial}{\partial p}\left[\frac{V_{A,1}^2 p^4}{r\kappa_2}\frac{\partial F_2}{\partial p}\right] = -Q_2(x,p) \ (x<0),$$
(16.5.2)

with the constant downstream cosmic ray bulk speed $V_2 \equiv u_2 + H_{C_2}^k V_{A,2}$. In contrast to the upstream region, the downstream transport equation (16.5.2) contains the momentum diffusion term whose strength is controlled by the parameter ϑ given in (16.4.15). The downstream and upstream spatial diffusion coefficients are related by (16.4.12).

To solve the transport equations (16.5.1) and (16.5.2) we impose the following boundary and continuity conditions:

(1) far upstream ($x \to \infty$) we require no accelerated particles, i.e.

$$F_1(+\infty, p) = 0 ;$$
(16.5.3)

(2) because, in the test-particle approach, no damping of the Alfvén waves is included, the background waves act as an infinite source of energy and particles can always continue to gain energy from the waves if the downstream region is treated as infinitely extended. To avoid this unphysical situation we introduce a finite downstream free-escape boundary condition, beyond which the particles do not return to the shock front and are lost from the acceleration process, i.e. we require

$$F_2(x = -L, p) = 0 .$$
(16.5.4)

Physically, the scale L coincides with the extent of enhanced wave intensity on the downstream side of the shock. We want to emphasize that this approach of simply adding a spatial boundary condition to the problem does not correspond to a solution of the wave damping treatment.

(3) we restrict our discussion to particle injection at the position of the shock only, at the rate $q_0(p)$. We therefore adopt

$$Q_1 = Q_2 = 0$$
(16.5.5)

in (16.5.1) and (16.5.2), respectively.

(4) at the position of the shock wave, $x = 0$, we require the continuity of the cosmic ray particles phase space density,

$$F_1(0, p) = F_2(0, p) ,$$
(16.5.6)

and the cosmic ray current density,

$$\left[\kappa_1\frac{\partial F_1(x,p)}{\partial x} + V_1 F_1(x,p)\right]_{(x \to 0+)} - \left[\kappa_2\frac{\partial F_2(x,p)}{\partial x} + V_2 F_2(x,p)\right]_{(x \to 0-)}$$
$$+ \frac{V_1 - V_2}{3p^2}\frac{\partial}{\partial p}\left[p^3 F_1(0,p)\right] + q_0(p) = 0 .$$
(16.5.7)

The condition (16.5.7) can be derived by (1) spatially integrating the cosmic ray transport equation (16.4.1) across the discontinuity at $x = 0$ as $\int_{-\epsilon}^{+\epsilon}$, and (2) then taking the limit $\epsilon \to 0$ (see e.g. Toptygin 1980 [544], Drury 1983a [141]).

16.5.2 Upstream Solution

We multiply (16.5.1) with $\kappa_1(x)$ and then introduce the new variable

$$\mu_1(x) = \int_0^x \frac{\mathrm{d}x^*}{\kappa_1(x^*)} \,. \tag{16.5.8}$$

The solution in the upstream region readily follows from (16.5.1) and condition (16.5.3) as

$$F_1(x > 0, p) = F_0(p) \exp(-V_1 \mu_1) \,, \tag{16.5.9}$$

and has the "precursor" behavior discussed in Sect. 16.3.7.

16.5.3 Downstream Solution

We multiply (16.5.2) with $\kappa_2(x)$ and introduce the new variable

$$\mu(x) = \int_0^x \frac{\mathrm{d}x^*}{\kappa_2(x^*)} \,, \tag{16.5.10}$$

and the abbreviation

$$A \equiv \frac{\vartheta V_{A,1}^2}{r} \,, \tag{16.5.11}$$

to obtain

$$\frac{\partial^2 F_2}{\partial \mu^2} + V_2 \frac{\partial F_2}{\partial \mu} + \frac{A}{p^2} \frac{\partial}{\partial p}\left[p^4 \frac{\partial F_2}{\partial p} \right] = 0 \,. \tag{16.5.12}$$

16.5.3.1 Analytical Solution.
Equation (16.5.12) can be solved by separation of variables (yielding with (16.5.4)) the general solution

$$F_2(\mu, p) = \int_C \mathrm{d}\lambda \left[J_1(\lambda) p^{\omega_1'} + J_2(\lambda) p^{\omega_2'} \right] \left[e^{\eta(\gamma_1 - \gamma_2) + \gamma_1 \mu} - e^{\gamma_2 \mu} \right] \,, \tag{16.5.13}$$

where we have introduced

$$\eta \equiv -\mu(-L) \,, \tag{16.5.14}$$

$$\omega_{1,2}' = \frac{1}{2}\left[-3 \pm \sqrt{9 - (4\lambda/A)} \right] \,, \tag{16.5.15}$$

$$\gamma_{1,2} = \frac{1}{2}\left[-V_2 \pm \sqrt{V_2^2 + 4\lambda} \right] \,. \tag{16.5.16}$$

The functions $J_1(\lambda)$ and $J_2(\lambda)$ are, as yet, unknown functions of the eigenvalues λ. These functions, the proper integration contour C, and the eigenvalues

16.5 Solution of the Transport Equations

must, of course, be chosen so that $F_2(\mu,p)$ is real, positive and well-behaved as $p \to 0$ and $p \to \infty$. The behaviors of J_1 and J_2 will be determined by the continuity conditions (16.5.6) and (16.5.7) and the source term $q_0(p)$.

The first continuity condition (16.5.6) yields, with (16.5.9) and (16.5.13), the relation

$$F_0(p) = \int_C d\lambda \left[J_1(\lambda) p^{\omega'_1} + J_2(\lambda) p^{\omega'_2} \right] \left[e^{\eta(\gamma_1 - \gamma_2)} - 1 \right]. \qquad (16.5.17)$$

Combining (16.5.17) with the second continuity condition (16.5.7) we obtain after straightforward manipulations

$$\int_C d\lambda J_1(\lambda) p^{\omega'_1} \left[\left(e^{\eta(\gamma_1 - \gamma_2)} - 1 \right) \left(1 + (R-1)\left(1 + \frac{\omega'_1}{3} \right) \right) \right.$$
$$\left. + \frac{1}{V_2} \left(\gamma_1 e^{\eta(\gamma_1 - \gamma_2)} - \gamma_2 \right) \right]$$
$$+ \int_C d\lambda J_2(\lambda) p^{\omega'_2} \left[\left(e^{\eta(\gamma_1 - \gamma_2)} - 1 \right) \left(1 + (R-1)\left(1 + \frac{\omega'_2}{3} \right) \right) \right.$$
$$\left. + \frac{1}{V_2} \left(\gamma_1 e^{\eta(\gamma_1 - \gamma_2)} - \gamma_2 \right) \right] = \frac{q_0(p)}{V_2}, \qquad (16.5.18)$$

where we have introduced the scattering center compression ratio (16.4.17), i.e. $R = r_k = V_1/V_2$. Transforming the first integral in (16.5.18) from the variable λ to ω'_1 and the second integral from the variable λ to ω'_2 using definition (16.5.15) we find that both integrals are identical in structure, yielding the relation

$$\int_C d\omega J\left[\omega(3+\omega)\right]\left(\omega + \frac{3}{2}\right) p^\omega D_0(\omega) = -\frac{q_0(p)}{2AV_2}, \qquad (16.5.19)$$

with $J = J_1$ and

$$\omega = \omega'_1 = \frac{1}{2}\left[-3 + \sqrt{9 - (4\lambda/A)}\right], \qquad (16.5.20)$$

where we have introduced the function

$$D_0(\omega) \equiv 2\left[\left(e^{\eta(\gamma_1 - \gamma_2)} - 1 \right) \left(1 + (R-1)\left(1 + \frac{\omega}{3} \right) \right) \right.$$
$$\left. + \frac{1}{V_2}\left(\gamma_1 e^{\eta(\gamma_1 - \gamma_2)} - \gamma_2 \right) \right]. \qquad (16.5.21)$$

From (16.5.19) it is evident that the downstream solution (16.5.13) for a general injection rate $q_0(p)$ can be expressed as

$$F_2(\mu,p) = \int_0^\infty dt\, q_0(t) G(\mu,p,t), \qquad (16.5.22)$$

in terms of the Green's function

$$G(\mu, p, t) = -4A \int_C d\omega J_0 \left[\omega(3+\omega)\right] \left(\omega + \frac{3}{2}\right) p^\omega$$
$$\times \left[e^{\eta(\gamma_1 - \gamma_2) + \gamma_1 \mu} - e^{\gamma_2 \mu}\right]. \quad (16.5.23)$$

The function J_0 is obtained from the relation

$$\int_C d\omega J_0 \left[\omega(3+\omega)\right] \left(\omega + \frac{3}{2}\right) p^\omega D_0(\omega) = -\frac{\delta(p-t)}{2AV_2}. \quad (16.5.24)$$

With the δ-function representation

$$\delta(p-t) = \frac{1}{t}\delta\left(\ln\frac{p}{t}\right) = \frac{-i}{2\pi t}\int_{c_0 - i\infty}^{c_0 + i\infty} d\omega \left(\frac{p}{t}\right)^\omega, \quad (16.5.25)$$

we readily obtain from relation (16.5.24) that

$$\left(\omega + \frac{3}{2}\right) J_0(\omega) = \frac{it^{-(\omega+1)}}{4\pi A V_2 D_0(\omega)}. \quad (16.5.26)$$

Inserting this result into (16.5.23) yields the Green's function

$$G(\mu, p, t) = \frac{-i}{\pi t V_2} \int_{c_0 - i\infty}^{c_0 + i\infty} d\omega (p/t)^\omega \frac{\left[e^{\eta(\gamma_1 - \gamma_2) + \gamma_1 \mu} - e^{\gamma_2 \mu}\right]}{D_0(\omega)}$$
$$= \frac{-ie^{-V_2 \mu/2}}{\pi t V_2} \int_{c_0 - i\infty}^{c_0 + i\infty} d\omega \frac{(p/t)^\omega \sin\left[(S/2)(1 + (\mu/\eta))\right]}{D(\omega)}, \quad (16.5.27)$$

where

$$S = S(\omega) = i\Phi_0 \eta V_2 Y_0(\omega), \quad (16.5.28)$$

with

$$\Phi_0 \equiv \frac{2A^{1/2}}{V_2} = 2\vartheta^{1/2}\left(Mr^{-1/2} + H_{c_2}^k\right)^{-1} > 0, \quad (16.5.29)$$

$$Y_0(\omega) \equiv \sqrt{(\omega_1 - \omega)(\omega - \omega_2)}, \quad (16.5.30)$$

$$\omega_1 \equiv -(\omega_2 + 3) < -3, \quad (16.5.31)$$

$$\omega_2 \equiv \sqrt{\frac{9}{4} + \Phi^{-2}} - \frac{3}{2} > 0. \quad (16.5.32)$$

The function $D(\omega)$ in (16.5.27) is

$$D(\omega) = \frac{S}{\eta V_2}\cos\frac{S}{2} + \left[(2R - 1) + \frac{2}{3}(R-1)\omega\right]\sin\frac{S}{2}. \quad (16.5.33)$$

The integrand in (16.5.27) is analytic everywhere in the finite ω-plane with the exception of the zeros of the function $D(\omega)$,

$$D(\omega_m) = 0, \quad m = 1, 2, 3, \ldots, \tag{16.5.34}$$

which are poles of first order. Employing the residue theorem we then obtain

$$G(\mu, p, t) = \frac{2e^{-V_2\mu/2}}{tV_2} \sum_{m=1}^{\infty} (p/t)^{\omega_m} \left[\frac{\partial D(\omega)}{\partial \omega}\right]^{-1}_{\omega=\omega_m}$$

$$\times \sin\left[\frac{S_m}{2}\left(1 + \frac{\mu}{\eta}\right)\right], \tag{16.5.35}$$

where $S_m = S(\omega = \omega_m)$. Of course the path c_0 is chosen differently in the two cases $p \leq t$ and $p \geq t$ to avoid divergences of the resulting Green's function at zero and infinite momentum, respectively.

Equations (16.5.22) and (16.5.35) represent the full analytical solution of the problem. It remains to determine the eigenvalues according to (16.5.34). First we discuss the behavior of the function (16.5.29).

16.5.3.2 Behavior of the Function Φ_0 for $q = 2$.

The function Φ_0, defined in (16.5.29), determines the strength of momentum diffusion of the particles in the downstream region of the shock. As shown in (13.2.21), for turbulence power spectral indices $q = 2$ and upstream backward moving waves only (i.e. $H_{c_1} = -1$), the constant (16.4.10) approaches

$$S_1(q = 2) = 2\ln\left(\frac{c}{V_A}\right) / (1 - \sigma^2), \tag{16.5.36}$$

so that (16.5.29) becomes with (16.3.29) and (16.4.15),

$$\Phi_0 = \sqrt{\ln(c/V_A)/(1-\sigma^2)}\Phi, \tag{16.5.37}$$

with

$$\Phi \equiv \frac{\sqrt{1 - \left(H_{c_2}^k\right)^2}}{\sqrt{2}\left(a + H_{c_2}^k\right)}, \tag{16.5.38}$$

and

$$a \equiv \frac{M}{\sqrt{r}} = \sqrt{\frac{3\beta}{4 - r}}. \tag{16.5.39}$$

Figure 16.12 shows the variation of the function Φ as a function of the downstream cross helicity and the parameter a.

The parameter a is always greater than or equal to 1, and it attains this critical value when the shock is just slightly super-Alfvénic in the downstream region. As Fig. 16.12 exhibits in this case a singularity occurs, as $H_c \to -1$, as discussed already at the end of Sect. 16.3.6. Since in this limit the backward moving waves are slowly propagating in the downstream region, large wave intensities result, leading to efficient downstream momentum diffusion of particles.

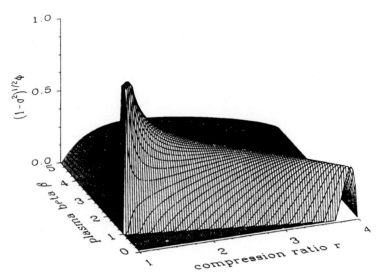

Fig. 16.12. Plot of the function Φ describing momentum diffusion dependence on the plasma beta β and the gas compression ration r

All curves $a > 1$ lie beneath the curve $a = 1$ and reveal a maximum $\Phi_{\max} = 1/[\sqrt{2(a^2-1)}]$ at $H_{c_2} = -(1/a)$. Curves flatten when a increases, i.e. with increasing plasma beta or gas compression, which is understandable since momentum diffusion becomes smaller if the plasma beta rises because of the decreasing Alfvèn velocity and, on the other hand, if the gas compression becomes greater, since the cross helicity tends to $+1$. In the case $\beta < 1$ there will be the restriction $r > 4 - 3\beta$, following from the requirement of super-Alfvènic flow on both sides of the shock, which leads to a smooth flattening of the maximum of the curves with decreasing β.

Finally, net polarization of the upstream waves ($\sigma \neq 0$ in (16.5.37)) increases the value of $\Phi_0 \propto (1 - \sigma^2)^{-1/2}$.

16.6 Qualitative Results

Three physical parameters control the behavior of the distribution function (16.5.35) of accelerated particles:

(i) the scattering center compression ratio, (16.4.17), which determines the strength of first-order Fermi acceleration of particles at the shock;
(ii) the Peclet number,

$$N = \eta V_2/2 = -\mu(-L)V_2/2, \qquad (16.6.1)$$

which measures the extent of the downstream Alfvén wave field that accelerates the particles by momentum diffusion;

(iii) the parameter Φ_0, defined in (16.5.37) that indicates the strength of momentum diffusion.

Note that the only parameter freely available in a test particle calculation is the Peclet number N. For given gas compression, upstream plasma beta and upstream wave intensity, the other two parameters follow according to our calculations in Sects. 16.4 and 16.5. It is instructive to look first at extreme end-member cases of the downstream spectrum of accelerated particles.

16.6.1 Zero Momentum Diffusion

The case of zero momentum diffusion, $A = 0$ and δ-function injection $q_0(p) = \delta(p - p_0)$ yields the downstream particle spectrum

$$F_2(\mu, p) = \frac{3(p/p_0)^{-s}}{p_0 V_2(R-1)} \frac{\exp(-V_2\mu/2)}{(\exp(2N)-1)}$$
$$\times \left[\exp N\left(2 + \frac{\mu}{\eta}\right) - \exp\left(-\frac{N\mu}{\eta}\right) \right], \quad (16.6.2)$$

where

$$-\omega = s = 3\left[R + (\exp(2N) - 1)^{-1}\right]/(R-1), \quad (16.6.3)$$

is the only non-vanishing eigenvalue.

At the position of the shock front ($\mu = 0$) this spectrum becomes

$$F_2(0, p) = \frac{3(p/p_0)^{-s}}{p_0 V_2(R-1)}. \quad (16.6.4)$$

In the case of large Peclet numbers $N \gg 1$, the spectral index approaches the classical result (e.g. Axford et al. 1977 [22], Krymsky 1977 [276], Bell 1978 [41], Blandford and Ostriker 1978 [58])

$$s(N \gg 1) = \frac{3R}{R-1}, \quad (16.6.5a)$$

whereas in the opposite case, $N \ll 1$, we find a strong reduction in the number density of accelerated particles due to the much steeper power law spectral index,

$$s(N \ll 1) = \frac{3}{2N(R-1)}. \quad (16.6.5b)$$

The case (16.6.5a) of large Peclet numbers deserves special attention. Here the spectral index of the power law of accelerated particles is controlled exclusively by the scattering center compression ratio $R = r_k$ given in (16.4.17). For the differential intensity of cosmic ray particles this gives

$$dJ/dE \propto p^2 F_2(0, p) \propto p^{-\Gamma}, \quad (16.6.6)$$

with
$$\Gamma = s - 2 = \frac{r_k + 2}{r_k - 1}. \tag{16.6.7}$$

Because r_k is a *ratio* of two characteristic velocities up- and downstream from the shock wave, it can account for the observational fact that in a great variety of different cosmic sources rather similar particle spectral indices are measured or inferred from nonthermal radiation information.

If the scattering center compression ratio equals the gas compression ratio, as at the infinite Mach number limit in (16.4.17), the spectral index (16.6.7) is $\Gamma_{\text{gas}} = (r+2)/(r-1) \geq 2$, since $1 < r \leq 4$ for an adiabatic shock with $\gamma = 5/3$. Using our results on the scattering center compression ratio shown in Fig. 16.10, we have calculated the resulting spectral index (16.6.7) for different shock parameter values in Fig. 16.13. Especially in a low beta plasma, the scattering center compression ratio attains large values, and consequently the resulting particle spectral indices Γ approach the infinite compression ratio limit, $\Gamma \to 1$. Quite interestingly, if the plasma beta is close to unity we obtain a spectral index between 1.5 and 3, almost independent of the gas compression ratio ($\Gamma_{\text{gas}} = 8$ corresponds to $r = 10/7 \approx 1.43$).

The spectra resulting from the first-order mechanism were generally harder than predicted by conventional theory, i.e. s_{gas}. If the plasma beta is close to unity the spectral index seems to vary slowly with the gas compression ratio and is between 1.8 and 2.8 for forward upstream waves in a wide

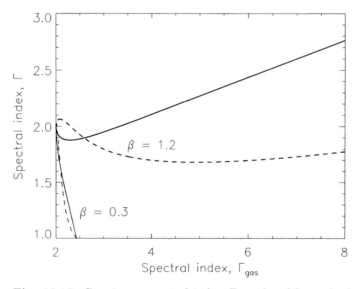

Fig. 16.13. Cosmic ray spectral index Γ produced by a shock with a constant upstream plasma beta neglecting stochastic acceleration in the downstream region and the final extent of the downstream region. Note that $\Gamma_{\text{gas}} = (r+2)/(r-1)$. **Dashed** and **solid lines** give the results for $H_{c_1} = +1$ and -1, respectively

range of $1.4 < r \leq 4$. The most interesting result is the possibility of generating extremely hard power-law spectra in $\beta < 1$ plasmas by the first-order mechanism, if the critical Mach number limit, $M \to r^{1/2}$, is approached. These kinds of conditions can be met, for example, in the solar corona, where the Alfvén speeds are of the same order as the shock speeds and much larger than the sound speed, and also near supernova shocks propagating into an interstellar medium with a low plasma-beta. This possibility, to generate particle energy power law spectra harder than the originally limiting value $\Gamma = 2$, removes the legendary (e.g. Lerche 1980 [293], Drury 1983b [142], Dröge et al. 1987 [137]) discrepancy of the original shock wave acceleration theory with the explanation of flat particle spectra in shell-type supernova remnants and bright spiral galaxies, and with the independence of spectral indices from evolutionary effects in these sources (see the discussion in Sect. 6.1).

16.6.2 Finite Momentum Diffusion

For finite momentum diffusion $\Phi_0 > 0$, the solutions are mathematically much more complex (see Sect. 16.5). According to our analysis (see (16.5.35)) the downstream solution is given as an infinite sum of power laws whose spectral indices follow from a transcendental eigenvalue equation. Each individual power law component is weighted by expansion coefficients that depend on the actual downstream position. Both the eigenvalues and the expansion coefficients are controlled by the scattering center compression ratio r_k, the shock wave Peclet number N defined in (16.6.1), and Φ_0 defined in (16.5.37).

From existing self-consistent treatments (Bogdan et al. 1991 [71], Ko 1992 [265]) of second-order Fermi acceleration and Alfvén wave evolution in homogeneous (i.e. non-shock) physical systems, it can be expected that a fully self-consistent analysis of our shock wave system, which treats simultaneously the acceleration of particles by wave damping and the evolution of the Alfvén wave intensities and helicities in the downstream region, will ultimately yield a close relationship between the level of momentum diffusion Φ_0 and the extent of the downstream region determining the Peclet number N.

The downstream solution (16.5.35) allows for

(i) any momentum dependence of the particle injection at the shock,
(ii) any spatial dependence of the upstream spatial diffusion coefficient, and
(iii) any spatial dependence of the downstream spatial diffusion coefficient.

The importance of the solutions is that they are simple and exact within the assumed, but not unreasonable, framework. They allow us to illustrate the basic effects of competing first and second-order Fermi mechanism in forming the momentum spectrum of the accelerated particles. According to the detailed analysis of Schlickeiser et al. (1993 [472]) the solutions exhibit the following main properties:

(1) In the most general case of finite N and Φ_0, the downstream solution is a superposition of an infinite sum of power laws; the shape of the downstream distribution function is concavely curved. This is clearly visible in Fig. 16.14 where we calculated the momentum spectrum of accelerated particles at the shock for strong momentum diffusion $\Phi_0 = 3$.

At large momenta $p \gg p_0$, far from the injection momentum p_0, the largest negative eigenvalue dominates the spectrum, in contrast to momenta close to injection where many eigenvalues contribute to the total spectrum. The convergence at large momenta of the spectral index towards the largest negative eigenvalue ω_1 is also evident from Fig. 16.15.

(2) In the formal limit of vanishing momentum diffusion $\Phi_0 \to 0$, and infinite extent of the downstream region $N \to \infty$, the general solution approaches the classical result of Axford et al. (1977 [22]), Krymsky (1977 [276]), Bell (1978 [41]), and Blandford and Ostriker (1978 [58]).

(3) For small values of the Peclet number $N < \min(1, \Phi_0^{-1})$ diffusive shock wave acceleration becomes very inefficient. At large momenta the power law spectral index is proportional to N^{-1}.

(4) For efficient momentum diffusion of particles $\Phi_0 \gg \max(1, N^{-1})$ the acceleration is dominated by the second-order Fermi mechanism. At large momenta the spectral index of the resulting power law approaches the limiting value -3 below (e.g. see Fig. 16.15). This is an additional mechanism (to that discussed in Sect. 16.6.1) to produce hard power law spectra.

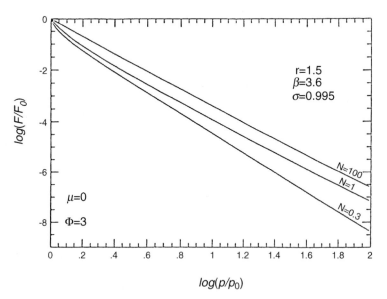

Fig. 16.14. Normalized momentum spectrum of cosmic ray protons at the shock for $\Phi_0 = 3$ and different values of the Peclet number N

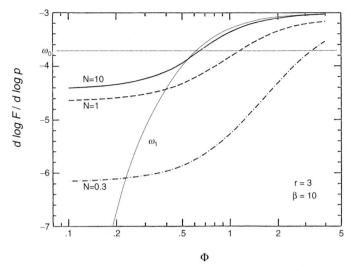

Fig. 16.15. Spectral index of accelerated cosmic ray protons at the shock at momentum $p = 10^3 p_0$ as a function of Φ_0 and N for a shock with gas compression $r = 3$ and plasma beta $\beta = 10$

16.7 Maximum Particle Energy

An underlying assumption made in calculating the cosmic ray transport parameters, not only in Sect. 16.4, is that the ratio of the cosmic ray gyroradius $R_{\rm L}$ to the longest wavelength $\lambda_{\rm max}$ of the plasma waves is less than unity,

$$k_{\rm min} R_{\rm L} = R_{\rm L}/\lambda_{\rm max} \leq 1 \; . \tag{16.7.1}$$

An upper bound to the longest wavelength available in a physical system is certainly given by the characteristic size L of the system, so that (16.7.1) becomes $R_{\rm L} \leq L$, which is equivalent to a limit on the maximum particle rigidity R:

$$R = \frac{p}{Z} \leq eB_0 L \; . \tag{16.7.2}$$

An alternative way to express the condition (16.7.2) is

$$(B_0/\mu{\rm G})(L/{\rm pc}) \geq E_{15}/Z \; , \tag{16.7.3}$$

where E_{15} denotes the cosmic ray particle energy in units of 10^{15} eV.

In Fig. 16.16 are plotted many cosmic sites where particle acceleration may occur, with sizes from kilometers to megaparsecs. Sites lying below the diagonal line fail to satisfy conditions (16.7.2) and (16.7.3) for 10^{20} eV protons ($Z = 1$) (the dashed line refers to 10^{20} eV iron nuclei ($Z = 26$). Clearly, for

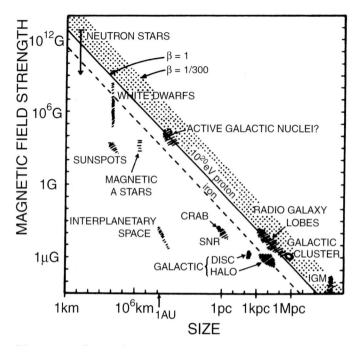

Fig. 16.16. Size and magnetic field strength of possible sites of particle acceleration. Objects below the **diagonal line** cannot accelerate protons to 10^{20} eV. From Hillas (1984 [227])

these high cosmic ray energies only very few sites remain as possibilities: either highly condensed objects with huge magnetic field strengths such as neutron stars, or enormously extended objects such as clusters of galaxies or radio galaxy lobes. Supernova remnant shock waves can only accelerate particles up to rigidities of $\simeq 10^{15}$ V.

17. Galactic Cosmic Rays

We have shown in Chap. 16 that the acceleration of galactic cosmic rays below a few 10^{15} eV/nucl. with power law momentum spectra can be well explained by diffusive shock wave acceleration near galactic supernova remnants, which also, by the overall power requirements (see Chap. 7), are the most probable sources of galactic cosmic rays. Here we address the subsequent propagation of cosmic rays in the Galaxy. We investigate in particular the relation of the cosmic ray momentum spectra near the solar system to the cosmic ray source spectra, and the influence of distributed interstellar re-acceleration on the momentum spectra.

Another topic of this section concerns the origin of ultrahigh energy cosmic rays above 10^{15} eV/nucl. in galaxies (for a review see Axford 1992 [20]). In the papers of Bryant et al. (1992 [78]), Ball et al. (1992 [27]) and Schneider (1993 [478]) the interest in second-order Fermi acceleration of this component by electromagnetic turbulence in the interstellar medium has been revived. Observationally it is well established that the interstellar medium is turbulent up to the largest scales, due to the motion of giant gas clouds, supernova explosions, stellar winds and the formation of superbubbles, loops, chimneys and galactic fountains. Measurements of the density fluctuation indicate the presence of a Kolmogorov-type fluctuation spectrum up to scales of 100 pc (Rickett 1990 [432]). Together with an interstellar magnetic field strength of $\simeq 5$ µG this would allow, according to (16.7.5), the acceleration up to particle energies of $5 \times Z\, 10^{17}$ eV, which in the case of iron nuclei would explain the acceleration of cosmic rays up to total energies of 10^{19} eV. Density fluctuations are associated with electromagnetic fluctuations, although the exact relation is quite complex (see Wu and Huba 1975 [567], Minter and Spangler 1997 [361]). We identify the electromagnetic fluctuations with parallel propagating Alfvénic and isotropic fast magnetosonic fluctuations for which the relevant particle acceleration rates have been calculated in Chap. 13.

17.1 Galactic Cosmic Ray Diffusion 2

Because of the geometry of our Galaxy as a highly flattened cylinder we may adopt the model discussed in Sect. 14.3.3, where the cosmic ray sources are distributed in a uniform ($Q_1(\boldsymbol{x}) = q_1 = $ const.), axisymmetric, cylindrical disk of radius R_0 and half-thickness z_0, and the confinement region is a concentric disk with radius $R_1 > R_0$ and half-thickness $z_1 > z_0$. At the edges of the confinement region, we apply the free-escape boundary conditions in the form (14.3.18). As before we neglect spatial convection against spatial diffusion, and we assume the spatial diffusion coefficient to be momentum dependent but independent of position, i.e.

$$\kappa_{zz}(\boldsymbol{x}, p) = K_0 \kappa(p) , \qquad (17.1.1)$$

with $\kappa(p)$ a dimensionless function. If the momentum operator (14.3.1b) is also independent of position (i.e. $h(\boldsymbol{x}) = 1$), the conditions for the scattering time method are fulfilled, and the steady-state cosmic ray distribution is given by (14.3.27),

$$M_a(R, z, p) = \sum_{l=0}^{\infty} \sum_{m=1}^{\infty} A_{lm}(R, z) N_{lm}(p) , \qquad (17.1.2)$$

where $A_{lm}(R, z)$ follows from (14.3.23) and (14.3.26). Each function $N_{lm}(p)$ obeys the leaky-box equation

$$\mathcal{O}_p N_{lm}(p) - \lambda_{lm}^2 \kappa(p) N_{lm}(p) = -Q_2(p) , \qquad (17.1.3)$$

where the spatial eigenvalues λ_{lm}^2 according to (14.3.24) enter, and where $\mathcal{O}_p$ denotes the momentum operator

$$\mathcal{O}_p N_{lm}(p) \equiv \frac{1}{p^2} \frac{\partial}{\partial p} \left(p^2 A_2 \frac{\partial N_{lm}(p)}{\partial p} \right.$$
$$\left. + \frac{p^3}{3} V_1 N_{lm}(p) - p^2 \dot{p}_{\text{loss}} N_{lm}(p) \right) - \frac{N_{lm}(p)}{T_c(p)} , \qquad (17.1.4)$$

where we use $\nabla \cdot \boldsymbol{V} = 3V_1 = \text{const}$.

17.2 Cosmic Ray Propagation in the Leaky-Box Approximation

We first investigate the case where no interstellar acceleration occurs $A_2 = 0$. For cosmic ray nuclei with energies above a few hundred MeV/nucleon we can also neglect any continuous momentum losses and adiabatic losses, so that (17.1.3) simplifies to

$$\frac{N_{lm}(p)}{T_c(p)} + \lambda_{lm}^2 \kappa(p) N_{lm}(p) = Q_2(p) , \qquad (17.2.1)$$

Under the leaky-box approximation one retains only the first term in the sum (17.1.2) so that (17.2.1) is equivalent to

$$\frac{M_a(p)}{T_c(p)} + \frac{M_a(p)}{T_{\text{esc}}(p)} = Q_a(p) , \qquad (17.2.2)$$

where

$$T_{\text{esc}}(p) = \left[\lambda_{01}^2 \kappa(p)\right]^{-1} \qquad (17.2.3)$$

is the escape time of cosmic rays from the Galaxy.

17.2.1 Secondary/Primary Ratio

We first apply (17.2.2) to a secondary cosmic ray element (e.g. B) that results from the spallation of primary elements (in the case of B predominantly C, N, and O). The secondary source therefore is given by $Q_B = \sigma_B v n_H M_{\text{CNO}}$, where n_H denotes the average interstellar gas density and σ_B the partial fragmentation cross-section for the production of B by CNO-nuclei, which is known from accelerator measurements. The transport equation (17.2.2) for boron nuclei reads

$$M_B \left[v n_H \sigma_{f,B} + T_{\text{esc}}^{-1}(p)\right] = \sigma_B v n_H M_{\text{CNO}} , \qquad (17.2.4)$$

where we express the catastrophic loss time due to spallation of B as $T_{c,B} = (v n_H \sigma_{f,B})^{-1}$. Equation (17.2.4) yields the secondary/primary ratio

$$\frac{M_B}{M_{\text{CNO}}} = \frac{\sigma_B}{\sigma_{f,B} + X^{-1}(p)} \qquad (17.2.5a)$$

in terms of the traversed grammage

$$X(p) \equiv v n_H T_{\text{esp}}(p) \qquad (17.2.5b)$$

which, for the homogenous cosmic ray confinement volume adopted in the leaky-box approximation, agrees with (3.4.1).

The power law decrease of the measured secondary/primary ratios $\propto p^{-b}$ with $b \simeq 0.5$ at relativistic particle energies shown in Fig. 3.25 indicates, according to (17.2.5), a corresponding decrease of the cosmic ray escape time $T_{\text{esc}}(p) \propto p^{-b}$ and, according to (17.2.3), a corresponding momentum variation of the spatial diffusion coefficient

$$\kappa(p) \propto p^b . \qquad (17.2.6)$$

According to (13.2.23), (13.3.77), and (13.4.10) such a momentum variation is possible for Kolmogorov turbulence when the turbulence spectral index $q = 2 - b \simeq 1.5$ is close to the Kraichnan value, regardless of whether interstellar turbulence consists of slab Alfvén waves, isotropic fast mode waves, or a mixture of both.

17.2.2 Secondary Cosmic Ray Clocks

Now we consider two isotopes of one secondary cosmic ray element (e.g. beryllium Be): one isotope (^{9}Be) is stable, and the other (^{10}Be) decays with lifetime (γT_d). However, because beryllium is a secondary cosmic ray element produced by spallation reactions of (C,N,O)-nuclei in the interstellar medium, their respective sources are

$$Q_9(p) = \sigma_9 v n_\mathrm{H} M_\mathrm{CNO}, \quad Q_{10}(p) = \sigma_{10} v n_\mathrm{H} M_\mathrm{CNO}, \qquad (17.2.7)$$

where the partial spallation cross-sections for the channels (CNO) $\to {}^9$Be and (CNO) $\to {}^{10}$Be are known from accelerator measurements.

Expressing the respective catastrophic loss times due to spallation as $T_{\mathrm{c},9} = (v n_\mathrm{H} \sigma_{\mathrm{f},9})^{-1}$ and $T_{\mathrm{c},10} = (v n_\mathrm{H} \sigma_{\mathrm{f},10})^{-1}$, the transport equation (17.2.2) for the stable ^{9}Be-isotope is

$$M_9 \left[v n_\mathrm{H} \sigma_{\mathrm{f},9} + T_\mathrm{esc}^{-1}(p) \right] = \sigma_9 v n_\mathrm{H} M_\mathrm{CNO}, \qquad (17.2.8)$$

while the transport equation for the unstable ^{10}Be-isotope reads

$$M_{10} \left[v n_\mathrm{H} \sigma_{\mathrm{f},10} + \frac{1}{\gamma T_\mathrm{d}} + T_\mathrm{esc}^{-1}(p) \right] = \sigma_{10} v n_\mathrm{H} M_\mathrm{CNO}. \qquad (17.2.9)$$

The ratio of the two isotopes becomes

$$\frac{M_{10}}{M_9} = \frac{\sigma_{10}}{\sigma_9} \frac{v n_\mathrm{H} \sigma_{\mathrm{f},9} + T_\mathrm{esc}^{-1}(p)}{v n_\mathrm{H} \sigma_{\mathrm{f},10} + (\gamma T_\mathrm{d})^{-1} + T_\mathrm{esc}^{-1}(p)}. \qquad (17.2.10)$$

All cross-sections and the decay lifetime appearing in this equation are known, so by measuring the ratio (17.2.10) at different energies one can infer the value of the gas density in the confinement region n_H, and the value and the momentum dependence of the escape time $T_\mathrm{esc}(p)$. By combining with inferences drawn on the traversed grammage (17.2.5) from secondary/primary measurements one can determine the values of n_H and $T_\mathrm{esp}(p)$ separately.

Table 17.1 summarizes the density and escape time measures from several cosmic ray isotopes. All measurements are consistent with a cosmic ray escape time of $\simeq 10^7$ yr at non-relativistic particle energies. Moreover, the average gas density seen by the cosmic rays is substantially lower than the Galactic disk average of 1 atom cm^{-3}, which indicates that cosmic rays have spent substantial parts of their lifetime in low-density regions of the interstellar medium like, for example, the halo of our Galaxy.

The estimates in Chap. 6 have shown that the time scale for continuous momentum losses of relativistic cosmic ray nucleons in the Galaxy are much longer than the cosmic ray escape time of 10^7 yr. Thus, a posteriori, neglect of these processes in the transport equation (17.2.1) is justified.

Table 17.1. Density and escape time measurements from cosmic ray clocks. After Connell et al. (1997 [105])

Isotope	ρ [atoms × cm^3]	CR lifetime [Myr]	Reference
^{10}Be	0.18(+0.18, −0.11)	17(+24, −8)	Garcia-Munoz, Mason, Simpson '77
	0.30(+0.12, −0.10)	8.4(+0.4, −2, 4)	Wiedenbeck, Greiner '80
	0.23(+0.13, −0.11)	14(+13, −5)	Garcia-Munoz, Simpson, Wefel '81
	0.24 ± 0.07	15(+7, −4)	Simpson, Garcia-Munoz '88
	0.28(+0.14, −0.11)	27(+19, −9)	Lukasiak et al. '94
	0.23 ± 0.04	18 ± 3	Connell '97
^{26}Al	0.28(+0.72, −0.19)	9(+20, −6.5)	Wiedenbeck '83
	0.52(+0.26, −0.20)	13.5(+8.5, −4.5)	Lukasiak, McDonald, Webber '94
	0.28(+0.72, −0.19)	15.6(+2.5, −2.6)	Connell, Simpson '97
^{36}Cl	limits	limits	Wiedenbeck '85
	0.39 ± 0.15	11 ± 4	This work
^{54}Mn	0.37(+0.16, −0.11)	14(+6, −4)	Duvernois '97

N.B.: Adapted from DuVernois 1997 [149]. These are the quoted, published, confinement times–using a different pathlength distribution (PLD) would alter these values somewhat. Mn confinement is for an assumed ^{54}Mn β^- partial half-life of 1 Myr

17.3 Cosmic Ray Propagation Including Interstellar Acceleration

We now investigate the influence of interstellar acceleration ($A_2 \neq 0$), but we again neglect any continuous momentum losses and adiabatic losses. In this case (17.1.3) becomes

$$\frac{1}{p^2}\frac{\partial}{\partial p}\left(p^2 A_2 \frac{\partial N_{lm}(p)}{\partial p}\right) - N_{lm}(p)\left[\frac{1}{T_c(p)} + \lambda_{lm}^2 \kappa(p)\right] = -Q_2(p) \ . \quad (17.3.1)$$

According to Chap. 13 (particularly (13.2.42) and (13.3.40)) the momentum diffusion coefficient $A_2(p)$ and the spatial diffusion coefficient $K_0\kappa(p)$ are closely related. Moreover, at relativistic momenta both diffusion coefficients are functions of the cosmic ray momentum per nucleon $\tilde{p} = p/A = R/\alpha$, where R is the particle rigidity and $\alpha \equiv A/|Z|$ the mass to charge number. Therefore it is convenient to work with the particle momentum per nucleon

$$\tilde{p} = p/A = R/\alpha \quad (17.3.2)$$

as a variable, and to introduce the particle phase space density

$$F_{lm}(\tilde{p}) = A^3 N_{lm}(p) \,, \tag{17.3.3a}$$

and

$$Q(\tilde{p}) = A^3 Q_2(p) \,, \tag{17.3.3b}$$

which refer to the momentum per nucleon instead of the total momentum. Equation (17.3.1) transforms to

$$\frac{1}{\tilde{p}^2}\frac{\mathrm{d}}{\mathrm{d}\tilde{p}}\left(\tilde{p}^2 D(\tilde{p})\frac{\mathrm{d}F_{lm}(p)}{\mathrm{d}\tilde{p}}\right) - F_{lm}\left[\frac{1}{T_c(\tilde{p})} + \lambda_{lm}^2 \kappa(\tilde{p})\right] = -Q(\tilde{p}) \,, \tag{17.3.4}$$

where $D(\tilde{p}) = A_2(p)/A^2$.

Representing the interstellar plasma turbulence by a mixture of slab Alfvén waves and fast magnetosonic waves with Kolmogorov-type power spectrum with spectral index $q < 2$ we obtain from Chap. 13 that

$$K_0 \kappa(\tilde{p}) = \frac{2}{3}\frac{\eta(q)c}{(2-q)(4-q)}(\alpha\epsilon)^{2-q}\beta\tilde{p}^{2-q} \,, \tag{17.3.5}$$

and

$$D(\tilde{p}) = \frac{3}{2}\frac{V_A^2}{q(q+2)c\eta(q)}\alpha^{q-2}\beta\tilde{p}^q \ln(v/V_A) \,, \tag{17.3.6}$$

where $\beta = v/c$ is the cosmic ray particle velocity, $\epsilon = v/V_A$. For abbreviation we introduce the parameter

$$\eta(q) \equiv \frac{3}{2\pi(q-1)}\left(\frac{B_0}{\delta B}\right)^2\left(\frac{c}{eB_0}\right)^{2-q} k_{\min}^{1-q} \; \mathrm{cm} \; (\mathrm{eV} \; \mathrm{c}^{-1})^{q-2} \,, \tag{17.3.7}$$

which is independent of cosmic ray properties.

For relativistic particles $\epsilon \simeq V_A/c = \mathrm{const.}$, $\beta \simeq 1$, so that

$$K_0 \kappa(\tilde{p}) = K_1 (\alpha\tilde{p})^{2-q} \tag{17.3.8}$$

and

$$D(\tilde{p}) = D_1 \alpha^{q-2}\tilde{p}^q \,, \tag{17.3.9}$$

with the constants

$$K_1 = \frac{2}{3}\frac{\eta(q)c}{(2-q)(4-q)}\left(\frac{V_A}{c}\right)^{2-q}, \tag{17.3.10}$$

$$D_1 = \frac{3}{2}\frac{V_A^2}{q(q+2)c\eta(q)}\ln(c/V_A) \,. \tag{17.3.11}$$

Finally, we introduce the dimensionless momentum variable

$$x = \tilde{p}/(m_p c) = p/mc \,, \tag{17.3.12}$$

17.3 Propagation Including Interstellar Acceleration

the corresponding phase space density

$$f(x) = (m_p c)^3 F(\tilde{p}), \quad q(x) = (m_p c)^3 Q(\tilde{p}), \quad (17.3.13)$$

and the two time scales

$$T_{\rm f} = (m_p c)^{2-q}/D_1, \quad \text{and} \quad T_{lm} = (m_p c)^{q-2} K_0 / \left[K_1 \lambda_{lm}^2\right], \quad (17.3.14)$$

which are characteristic for interstellar acceleration and escape from the Galaxy, respectively. Hence (17.3.4) becomes for relativistic particles

$$\frac{\alpha^{q-2}}{T_{\rm f} x^2} \frac{\rm d}{{\rm d}x} \left(x^{q+2} \frac{{\rm d} f_{lm}}{{\rm d}x}\right) - f_{lm} \left[T_{\rm c}^{-1} + \alpha^{2-q} T_{lm}^{-1} x^{2-q}\right] = -q(x). \quad (17.3.15)$$

17.3.1 Mathematical Solution

Multiplying by $T_{\rm f} \alpha^{2-q} x^2$ we can cast the transport equation (17.3.15) into the self-adjoint form

$$\frac{\rm d}{{\rm d}x} \left[x^{4-\chi} \frac{{\rm d} f_{lm}}{{\rm d}x}\right] - \left[\psi x^2 + \lambda x^{2+\chi}\right] f_{lm} = -T_{\rm f} x^2 \alpha^{\chi} q(x), \quad (17.3.16)$$

where for convenience we have introduced

$$\chi \equiv 2 - q, \quad (17.3.17)$$

$$\psi \equiv \frac{T_{\rm f}}{T_{\rm c}} \alpha^{2-q}, \quad (17.3.18)$$

and

$$\lambda \equiv \frac{T_{\rm f}}{T_{lm}} \alpha^{2(2-q)}. \quad (17.3.19)$$

According to Sect. 14.3.4 the self-adjoint equation for general source distribution $q(x)$ is solved in terms of the Green's function $G(x, x_0)$ as

$$f_{lm} = T_{\rm f} \alpha^{\chi} \int_0^{\infty} {\rm d}x_0 \, x_0^2 q(x_0) G(x, x_0), \quad (17.3.20)$$

where the Green's function satisfies the equation

$$\frac{\rm d}{{\rm d}x} \left[x^{4-\chi} \frac{{\rm d}G}{{\rm d}x}\right] - \left[\psi x^2 + \lambda x^{2+\chi}\right] G = -\delta(x - x_0). \quad (17.3.21)$$

17.3.1.1 Construction of the Green's function.
With the substitution $v = x^{\chi}/A_0$, where A_0 is a constant to be fixed later, the homogenous differential equation corresponding to (17.3.21) becomes

$$v^2 \frac{{\rm d}^2 g}{{\rm d}v^2} + \frac{3}{\chi} v \frac{{\rm d}g}{{\rm d}v} - \left[\frac{\psi A_0 v}{\chi^2} + \frac{\lambda A_0^2 v^2}{\chi^2}\right] g = 0. \quad (17.3.22)$$

Making the ansatz
$$g(v) = W(v)\exp(-kv) \qquad (17.3.23)$$
with the constant k, (17.3.22) yields
$$v\frac{d^2W}{dv^2} + \left[\frac{3}{\chi} - 2kv\right]\frac{dW}{dv}$$
$$+ \left[-\left(\frac{3k}{\chi} + \frac{\psi A_0}{\chi^2}\right) + v\left(k^2 - \frac{\lambda A_0^2}{\chi^2}\right)\right]W = 0. \quad (17.3.24)$$

Obviously, if we choose the constants to be
$$k = 1/2, \quad A_0 = \chi/\left(2\lambda^{1/2}\right), \qquad (17.3.25)$$

(17.3.24) reduces to the differential equation of the confluent hypergeometric functions $M(a,b,v)$ and $U(a,b,v)$ (Abramowitz and Stegun 1972 [3], Chap. 13)

$$v\frac{d^2W}{dv^2} + \left[\frac{3}{\chi} - v\right]\frac{dW}{dv} - \left[\frac{3}{2\chi} + \frac{\psi}{2\chi\lambda^{1/2}}\right]W = 0, \qquad (17.3.26)$$

and the solution of the homogenous differential equation corresponding to (17.3.21) becomes
$$H(x) = c_1 H_1(x) + c_2 H_2(x), \qquad (17.3.27)$$

where
$$H_1 = \exp\left[-\frac{\lambda^{1/2}x^\chi}{\chi}\right] M\left(\frac{3}{2\chi} + \frac{\psi}{2\chi\lambda^{1/2}}, \frac{3}{\chi}, \frac{2\lambda^{1/2}x^\chi}{\chi}\right), \qquad (17.3.28)$$

and
$$H_2 = \exp\left[-\frac{\lambda^{1/2}x^\chi}{\chi}\right] U\left(\frac{3}{2\chi} + \frac{\psi}{2\chi\lambda^{1/2}}, \frac{3}{\chi}, \frac{2\lambda^{1/2}x^\chi}{\chi}\right). \qquad (17.3.29)$$

The function $H_1(x)$ fulfills the momentum boundary condition at $x = 0$, and the function $H_2(x)$ fulfills the momentum boundary condition at $x = \infty$ (see also the discussion in Sect. 14.3.4.1). Hence we can construct the Green's function as (e.g. Arfken 1970 [17])

$$G(x, x_0) = -\frac{1}{\mathcal{D}}\begin{cases} H_1(x)H_2(x_0) & \text{for } 0 \le x \le x_0 \\ H_1(x_0)H_2(x) & \text{for } x_0 \le x \le \infty \end{cases}, \qquad (17.3.30)$$

where
$$\mathcal{D} = P(x_0)\left[H_1\frac{dH_2}{dx} - H_2\frac{dH_1}{dx}\right]_{x=x_0} \qquad (17.3.31)$$

is calculated from the Wronskian and the function $P(x_0) = x_0^{4-\chi}$ from the self-adjoint (17.3.16). Using again Abramowitz and Stegun (1972 [3]) for the Wronskian we obtain in terms of gamma functions

$$\mathcal{D} = -\frac{2\lambda^{1/2}}{(2\lambda^{1/2}/\chi)^{3/\chi}} \frac{\Gamma[3/\chi]}{\Gamma\left[(3/2\chi) + (\psi/2\chi\lambda^{1/2})\right]} \,. \qquad (17.3.32)$$

Collecting terms, we obtain for the Green's function (17.3.30)

$$G(x, x_0) = \frac{(2\lambda^{1/2})^{(3/\chi)-1}}{\chi^{3/\chi}} \frac{\Gamma\left[(3/2\chi) + (\psi/2\chi\lambda^{1/2})\right]}{\Gamma[3/\chi]} \exp\left[-\frac{\lambda^{1/2}(x^\chi + x_0^\chi)}{\chi}\right]$$

$$\begin{cases} M\left(\frac{3}{2\chi} + \frac{\psi}{2\chi\lambda^{1/2}}, \frac{3}{\chi}, \frac{2\lambda^{1/2}x^\chi}{\chi}\right) U\left(\frac{3}{2\chi} + \frac{\psi}{2\chi\lambda^{1/2}}, \frac{3}{\chi}, \frac{2\lambda^{1/2}x_0^\chi}{\chi}\right) & 0 \leq x \leq x_0 \\ U\left(\frac{3}{2\chi} + \frac{\psi}{2\chi\lambda^{1/2}}, \frac{3}{\chi}, \frac{2\lambda^{1/2}x^\chi}{\chi}\right) M\left(\frac{3}{2\chi} + \frac{\psi}{2\chi\lambda^{1/2}}, \frac{3}{\chi}, \frac{2\lambda^{1/2}x_0^\chi}{\chi}\right) & x_0 \leq x \leq \infty \end{cases}$$

(17.3.33)

17.3.2 Sources of Cosmic Ray Primaries

For cosmic ray primaries we assume, on the basis of the discussion in Chap. 16, that they are injected into the interstellar medium from individual supernova remnants with power law source spectra

$$q(x) = q_0 x^{-s-2} H\left[x - x_{\min}\right] H\left[x_{\max} - x\right], \qquad (17.3.34)$$

at rigidities above some non-relativistic minimum rigidity $R_{\min}$ and below $R_{\max} = 10^{15}$ V, corresponding to $x_{\min} = R_{\min}/(\alpha m_p c)$ and $x_{\max} = 10^6/\alpha$, respectively.

17.3.2.1 Supernova frequency.
In spiral galaxies like ours it is estimated that supernova explosions occur with a frequency of 1 event every 50 years per galaxy (e.g. Ostriker and McKee 1977 [379]). With a volume of our Milky Way (see Chap. 2) of $V_G = \pi(13 \text{ kpc})^2 \times 0.3 \text{ kpc} = 2 \times 10^{11} \text{ pc}^3$ this yields the supernova rate

$$\nu_{SN} = 10^{-13} \nu_{50} \text{ pc}^{-3} \text{ yr}^{-1}. \qquad (17.3.35)$$

According to Ostriker and McKee (1977 [379]) the maximum radius attained by an individual supernova remnant of initial explosion energy $10^{51} E_{51}$, until it adjusts to pressure equilibrium with the ambient interstellar medium of density n_H and pressure $P_4 = (n_0 T/10^4)$ K cm^{-3}, is

$$R_{\max} = 55 E_{51}^{0.32} n_H^{-0.16} P_4^{-0.20} \text{ pc}. \qquad (17.3.36)$$

During the typical residence time of cosmic rays in the Galaxy of $T_d \simeq 10^7$ yr (compare Sect. 17.2.2) therefore $N_s = \nu_{SN} T_d V_G = 2 \times 10^5 \nu_{50}$ supernova explosions occur. If we assume that these explosions occur randomly in the Galaxy, and each of them occupies a volume $V_{SN} = 4\pi R_{\max}^3/3$, with the maximum radius given by (17.3.36), then the probability that each volume element of our Galaxy is overrun by a supernova explosion during the typical cosmic ray residence time is

$$Q = \frac{N_{\rm s} V_{\rm SN}}{V_{\rm G}} = 0.7 \frac{\nu_{50} E_{51}^{0.96}}{n_{\rm H}^{0.48} P_4^{0.6}} \ . \tag{17.3.37}$$

Most of the interstellar medium is occupied by the coronal phase (compare Chap. 2) with densities $n_{\rm H} \simeq 10^{-2} n_{c,2}$ cm^{-3}, which is in pressure equilibrium with the other gas phases. (Such a small density of the effective cosmic ray residence volume is also suggested by the density estimate in Sect. 17.2.2.) For the probability (17.3.37) we then obtain

$$Q = 6.4 \frac{\nu_{50} E_{51}^{0.96}}{n_{c,2}^{0.48} P_4^{0.6}} \ . \tag{17.3.38}$$

Every point of the interstellar medium is overrun at least six times by a supernova explosion during the typical residence time of cosmic rays. Therefore on these time scales we may neglect the "granular" nature of the cosmic ray source distribution in the transport equation (17.3.15), and regard the cosmic ray source term $q(x)$ in (17.3.34) as a quasi-continuous galactic-averaged injection rate.

17.3.2.2 Dispersion of source spectral indices. We noticed in Sect. 6.1 that the observed radio spectral index values α_r from individual supernova remnants, in the Galaxy and in the Magellanic Clouds, exhibit a considerable dispersion $\sigma_\alpha \simeq 0.15$ around the mean value $\langle \alpha_r \rangle = 0.48$. Since the radio emission is non-thermal synchrotron radiation from relativistic electrons $s = 1 + 2\alpha_r$ (see (4.121)), this implies a corresponding dispersion in the source power law spectral index value s of cosmic rays in the injection term (17.3.34) that has to be taken into account (Brecher and Burbidge 1972 [75]).

If we represent the distribution of radio spectral indices by the Gaussian (Williams and Bridle 1967 [565], Milne 1970 [358])

$$n_\nu(\alpha_r) = \frac{n_\nu}{\sqrt{2\pi}\sigma_\alpha} \exp\left[-\frac{(\alpha_r - \langle \alpha_r \rangle)^2}{2\sigma_\alpha^2}\right] \ , \tag{17.3.39}$$

this yields with $s = 1 + 2\alpha_r$ and $\sigma_s = 2\sigma_\alpha$ for the corresponding distribution of the cosmic ray momentum distribution

$$n(s) = \frac{\sqrt{2} n_\nu}{\sqrt{\pi}\sigma_s} \exp\left[-\frac{(s - \langle s \rangle)^2}{2\sigma_s^2}\right] \ . \tag{17.3.40}$$

For the averaged injection spectrum we obtain

$$\begin{aligned}
\langle x^{-s} \rangle &= \int_0^\infty ds \ x^{-s} \frac{n(s)}{n_\nu} \\
&= \frac{\sqrt{2}}{\sqrt{\pi}\sigma_s} \int_0^\infty ds \ \exp\left[-s \ln x - \frac{(s - \langle s \rangle)^2}{2\sigma_s^2}\right] \\
&= \exp\left[-\langle s \rangle \ln x + \frac{1}{2}\sigma_s^2 \ln^2 x\right] \\
&= x^{-\langle s \rangle} \exp\left[\frac{1}{2}\sigma_s^2 \ln^2 x\right] \ , \tag{17.3.41}
\end{aligned}$$

17.3 Propagation Including Interstellar Acceleration

which, with increasing momentum, exhibits a flattening of the source spectrum. For the averaged spectral index we obtain

$$\bar{s} \equiv -\frac{\ln \langle x^{-s}\rangle}{\ln x} = \langle s \rangle - \frac{1}{2}\sigma_s^2 \ln x \ . \tag{17.3.42}$$

This flattening is a consequence of the fact that supernova remnants with the flattest power law spectra dominate in the superposition at large momenta. Hence if we want to account for this dispersion effect we should use, instead of (17.3.34), the modified source distribution

$$q(x) = q_1 x^{-\bar{s}-2}\, H\,[x - x_{\min}]\, H\,[x_{\max} - x]$$
$$= q_1 x^{-\langle s \rangle - 2 + (1/2)\sigma_s^2 \ln(x/x_*)}\, H\,[x - x_{\min}]\, H\,[x_{\max} - x]\ , \tag{17.3.43}$$

where x_* denotes some reference momentum.

17.3.3 Steady-State Primary Particle Spectrum Below $R_{\max}$

We first study the response of (17.3.20) to the single power law injection function (17.3.34) at particle momenta in the range $x_{\min} < x < x_{\max}$,

$$f_{lm}\,(x_{\min} < x < x_{\max}) = T_{\mathrm{f}} q_0 \alpha^\chi \int_{x_{\min}}^{x_{\max}} dx_0 x_0^{-s} G(x, x_0)$$
$$= \frac{q_0 T_{\mathrm{f}} \alpha^\chi g_0}{\chi} x^{1-s} \exp(-B/2) J\,(x, x_{\min}, x_{\max}, B)\ , \tag{17.3.44}$$

with

$$J\,(x, x_{\min}, x_{\max}, B) \equiv J_1 + J_2$$
$$= U(a, b, B) \int_{(x_{\min}/x)^\chi}^{1} dy\, y^{(1-s-\chi)/\chi}$$
$$\times\, e^{-By/2}\, M(a, b, By)$$
$$+ M(a, b, B) \int_{1}^{(x_{\max}/x)^\chi} dy\, y^{(1-s-\chi)/\chi}$$
$$\times\, e^{-By/2}\, U(a, b, By)\ , \tag{17.3.45}$$

$$a \equiv \frac{3}{2\chi} + \frac{\psi}{2\chi\lambda^{1/2}} = \frac{b}{2} + \frac{\psi}{2\chi\lambda^{1/2}}\ ,$$
$$b \equiv \frac{3}{\chi}\ ,\ g_0 \equiv \frac{\left(2\lambda^{1/2}\right)^{(3/\chi)-1}}{\chi^{3/\chi}}\frac{\Gamma[a]}{\Gamma[b]}\ , \tag{17.3.46}$$

and

$$B \equiv \frac{2\lambda^{1/2} x^\chi}{\chi} = \left(\frac{x}{x_c}\right)^\chi\ ,\quad \text{where}\quad x_c = \left(\frac{\chi}{2\lambda^{1/2}}\right)^{1/\chi}\ . \tag{17.3.47}$$

Using the asymptotic behavior of the confluent hypergeometric functions for small and large argument (Abramowitz and Stegun 1972 [3])

$$M(a,b,z) \simeq \begin{cases} 1 & \text{for } |z| \ll 1 \\ (\Gamma(b)/\Gamma(a))z^{a-b}\, e^z & \text{for } |z| \gg 1 \end{cases}, \quad (17.3.48a)$$

$$U(a,b,z) \simeq \begin{cases} (\Gamma(b-1)/\Gamma(a))\, z^{1-b} & \text{for } |z| \ll 1 \\ z^{-a} & \text{for } |z| \gg 1 \end{cases}, \quad (17.3.48b)$$

we can discuss the behavior of the approximate solution (17.3.44)–(17.3.45). Obviously, we have to consider the four cases $x < x_\text{max} \ll x_\text{c}$, $x \ll x_\text{c} < x_\text{max}$, $x_\text{min} \ll x_\text{c} < x < x_\text{max}$ and $x_\text{c} \ll x_\text{min} < x < x_\text{max}$. We discuss each in turn.

17.3.3.1 $x < x_\text{max} \ll x_\text{c}$. In this case the parameter $B \ll 1$ for all momentum values x, and the integral J_1 with (17.3.48) becomes

$$J_1 \simeq \frac{\Gamma(b-1)}{\Gamma(a)} B^{1-b} \int_{(x_\text{min}/x)^\chi}^{1} dy\, y^{(1-s-\chi)/\chi}$$

$$= \frac{\chi}{s-1} \frac{\Gamma(b-1)}{\Gamma(a)} \left(\frac{x}{x_\text{c}}\right)^{\chi-3} \left[\left(\frac{x}{x_\text{min}}\right)^{s-1} - 1\right]. \quad (17.3.49)$$

Likewise

$$J_2 \simeq \frac{\Gamma(b-1)}{\Gamma(a)} B^{1-b} \int_{1}^{(x_\text{max}/x)^\chi} dy\, y^{(1-s-b\chi)/\chi}$$

$$= \frac{\chi}{s+2-\chi} \frac{\Gamma(b-1)}{\Gamma(a)} \left(\frac{x}{x_\text{c}}\right)^{\chi-3} \left[1 - \left(\frac{x}{x_\text{max}}\right)^{s+2-\chi}\right]. \quad (17.3.50)$$

Collecting terms in (17.3.44) we obtain in this case

$$f_{lm}(x_\text{min} < x < x_\text{max} \ll x_\text{c}) \simeq \frac{q_0 T_\text{f}}{s-1} \frac{(2\lambda^{1/2})^{(3/\chi)-1}}{\chi^{3/\chi}\left(\frac{3}{\chi}-1\right)} \alpha^\chi x_\text{min}^{1-s} \left(\frac{x}{x_\text{c}}\right)^{\chi-3}$$

$$\left[1 - \frac{3-\chi}{s+2-\chi}\left(\frac{x_\text{min}}{x}\right)^{s-1} - \frac{s-1}{s+2-\chi} \frac{x^{3-\chi}x_\text{min}^{s-1}}{x_\text{max}^{s+2-\chi}}\right]$$

$$\simeq \frac{q_0 T_\text{f}}{s-1} \frac{(2\lambda^{1/2})^{(3/\chi)-1}}{\chi^{3/\chi}\left(\frac{3}{\chi}-1\right)} \alpha^\chi x_\text{min}^{1-s} \left(\frac{x}{x_\text{c}}\right)^{\chi-3}.$$

$$(17.3.51)$$

a very flat distribution due to the dominance of stochastic acceleration.

17.3.3.2 $x \ll x_\text{c} < x_\text{max}$. In this case again $B \ll 1$ and the integral J_1 remains unchanged from (17.3.49).

However, the integral J_2 becomes

17.3 Propagation Including Interstellar Acceleration 447

$$J_2 \simeq \frac{\Gamma(b-1)}{\Gamma(a)} B^{1-b} \int_1^{(x_c/x)^\chi} dy\, y^{(1-s-\chi-b\chi)/\chi}$$

$$+ B^{-a} \int_{(x_c/x)^\chi}^{(x_{\max}/x)^\chi} dy\, y^{(1-s-\chi-a\chi)/\chi} e^{-By/2}$$

$$= \frac{\chi}{s+2-\chi} \frac{\Gamma(b-1)}{\Gamma(a)} \left(\frac{x}{x_c}\right)^{\chi-3} \left\{ 1 - \left(\frac{x}{x_c}\right)^{s+2-\chi} \right.$$

$$+ \frac{\Gamma(a)}{2^\delta \Gamma(b-1)} \frac{s+2-\chi}{\chi} \left(\frac{x}{x_c}\right)^{s+2-\chi}$$

$$\left. \times \left[\Gamma\left(\delta, \frac{1}{2}\left(\frac{x_{\max}}{x_c}\right)^\chi\right) - \Gamma\left(\delta, \frac{1}{2}\right) \right] \right\}, \qquad (17.3.52)$$

in terms of the incomplete gamma function $\Gamma(\delta, z)$ and

$$\delta = \left(s + \frac{1}{2} + \frac{\psi}{2\lambda^{1/2}} \right) / \chi. \qquad (17.3.53)$$

To lowest order in the small parameters $(x/x_c) \ll 1$ and $(x_c/x_{\max}) \ll 1$ the distribution function (17.3.44) also approaches the limit (17.3.51), i.e.

$$f_{lm}(x_{\min} \ll x_c \ll x_{\max}) \simeq \frac{q_0 T_f}{s-1} \frac{(2\lambda^{1/2})^{(3/\chi)-1}}{\chi^{3/\chi} \left(\frac{3}{\chi} - 1\right)} \alpha^\chi x_{\min}^{1-s} \left(\frac{x}{x_c}\right)^{\chi-3}. \qquad (17.3.54)$$

17.3.3.3 $x_{\min} \ll x_c \ll x \ll x_{\max}$. Here the parameter $B \gg 1$ for all momentum values x, and the integral J_2 with (17.3.48) becomes

$$J_2 \simeq \frac{\Gamma(b)}{\Gamma(a)} B^{-b} \exp(B/2)\, j_3, \qquad (17.3.55)$$

where

$$j_3 = e^{B/2} \int_1^{(x_{\max}/x)^\chi} dy\, y^{(1-s-\chi-a\chi)/\chi} e^{-By/2}$$

$$= \int_0^{(x_{\max}/x)^\chi - 1} dt\, t^{(1-s-\chi-a\chi)/\chi} e^{-Bt/2}. \qquad (17.3.56)$$

Since $B \gg 1$ the main contribution to the integral (17.3.56) comes from small values of $t \leq 2/B \ll 1$, so that $j_3 \simeq 2/B$ yielding to lowest order in $B^{-1} \ll 1$ that

$$J_2 \simeq \frac{2\Gamma(b)}{\Gamma(a)} B^{-b-1} \exp(B/2). \qquad (17.3.57)$$

For the integral J_1 we obtain approximately

$$J_1 \simeq B^{-a} \left\{ \int_{(x_{\min}/x)^\chi}^{(x_c/x)^\chi} dy\, y^{(1-s-\chi)/\chi} \right.$$
$$\left. + \frac{\Gamma(b)}{\Gamma(a)} B^{a-b} \int_{(x_c/x)^\chi}^1 dy\, y^{[1-s+(a-b-1)\chi]/\chi}\, e^{By/2} \right\}$$
$$\simeq B^{-a} \left\{ \frac{\chi}{s-1} x^{s-1} x_{\min}^{1-s} \left[1 - (x_{\min}/x_c)^{s-1} \right] \right.$$
$$\left. + \frac{\Gamma(b)}{\Gamma(a)} B^{a-b} e^{B/2} \int_0^{1-(1/B)} dt\, (1-t)^{[1-s-\chi+(a-b)\chi]/\chi}\, e^{-Bt/2} \right\}$$
$$\simeq B^{-a} \left\{ \frac{\chi}{s-1} x^{s-1} x_{\min}^{1-s} \left[1 - (x_{\min}/x_c)^{s-1} \right] \right.$$
$$\left. + \frac{2\Gamma(b)}{\Gamma(a)} B^{a-b-1} e^{B/2} \right\}. \tag{17.3.58}$$

Because of the large exponential term, $\exp(B/2) \gg 1$, the second term in the bracket of (17.3.58) dominates, so that

$$J_1 \simeq J_2 = \frac{2\Gamma(b)}{\Gamma(a)} B^{-b-1} \exp(B/2). \tag{17.3.59}$$

For the phase space density (17.3.44) we obtain in this case

$$f_{lm}(x_{\min} \ll x_c \ll x \ll x_{\max}) \simeq \frac{4q_0 T_f \left(2\lambda^{1/2}\right)^{(3/\chi)-1}}{\chi^{3/\chi+1}} \alpha^\chi x_c^{\chi+3} x^{-(s+2+\chi)}, \tag{17.3.60}$$

a power law that is steepened by the factor χ as compared to the injection distribution (17.3.34).

17.3.3.4 $x_c \ll x_{\min} \ll x \ll x_{\max}$. In this case besides $B \gg 1$ we find that $By \gg 1$ for all values of $y \geq (x_{\min}/x)^\chi$. The integral J_2 again is approximated by (17.3.57), whereas the integral J_1 simply becomes

$$J_1 \simeq \frac{\Gamma(b)}{\Gamma(a)} B^{-b} \int_{(x_{\min}/x)^\chi}^1 dy\, y^{[1-s-\chi+(a-b)\chi]/\chi}\, e^{By/2}$$
$$= \frac{\Gamma(b)}{\Gamma(a)} B^{-b} e^{B/2} \int_0^{1-(x_{\min}/x)^\chi} dt\, (1-t)^{[1-s-\chi+(a-b)\chi]/\chi}\, e^{-Bt/2}$$
$$\simeq \frac{2\Gamma(b)}{\Gamma(a)} B^{-b-1} e^{B/2} = J_2. \tag{17.3.61}$$

Also in this case the phase space density approaches (17.3.60), i.e.

$$f_{lm}(x_c \ll x_{\min} \ll x \ll x_{\max}) \simeq \frac{4q_0 T_f \left(2\lambda^{1/2}\right)^{(3/\chi)-1}}{\chi^{3/\chi+1}} \alpha^\chi x_c^{\chi+3} x^{-(s+2+\chi)}. \tag{17.3.62}$$

17.3.3.5 Interlude.

Our analysis in the last subsection has shown that within the momentum range $x_{\min} \leq x \leq x_{\max}$ the steady-state particle distribution for single power law injection approaches the two power law limits

$$f_{lm}(x \ll x_c) \simeq \frac{q_0 T_f}{s-1} \frac{(2\lambda^{1/2})^{(3/\chi)-1}}{\chi^{3/\chi}((3/\chi)-1)} \alpha^\chi x_{\min}^{1-s} \left(\frac{x}{x_c}\right)^{\chi-3}, \quad (17.3.63a)$$

and

$$f_{lm}(x \gg x_c) \simeq \frac{4q_0 T_f (2\lambda^{1/2})^{(3/\chi)-1}}{\chi^{3/\chi+1}} \alpha^\chi x_c^{\chi+3} x^{-(s+2+\chi)}, \quad (17.3.63b)$$

depending on $x \ll x_c$ and $x \gg x_c$, respectively. The rigidity x_c characterizes the strength of stochastic acceleration by momentum diffusion. At rigidities below x_c stochastic acceleration is the dominant effect and generates a very flat particle power law spectrum with index $\chi - 3$, independent from the injection spectral index. At momenta above x_c stochastic acceleration is negligibly small resulting in a steady-state power law spectrum that is slightly steeper than the injection spectrum. This limit corresponds to (17.2.5) that has been derived neglecting interstellar acceleration completely, and that also exhibits the steepening of the secondary energy spectrum as compared to the primary source spectrum if the traversed grammage $X \ll \sigma_{f,B}^{-1}$.

According to (17.3.19) and (17.3.47) the characteristic rigidity x_c

$$x_c = \frac{(\chi/2)^{1/\chi}}{\alpha} \left(\frac{T_{lm}}{T_f}\right)^{1/2\chi} \quad (17.3.64)$$

is determined by the ratio of the escape time T_{lm} and the acceleration time T_f. Using (17.3.14)–(17.3.15) we obtain for this ratio

$$\frac{T_{lm}}{T_f} = \frac{1}{(m_p c)^{2\chi}} D_1 \left[K_1 \left(\frac{y_m^2}{R_1^2} + \frac{(2l+1)^2 \pi^2}{4z_1^2}\right) \right]^{-1}, \quad (17.3.65)$$

which, for a highly flattened disk $R_1 \gg z_1$, approaches

$$\frac{T_{lm}}{T_f} \simeq \frac{T_l}{T_f} = \frac{1}{(m_p c)^{2\chi}} D_1 \frac{4z_1^2}{(2l+1)^2 \pi^2 K_1} = \frac{1}{(m_p c)^{2\chi}} \frac{\tau_D}{\tau_F}. \quad (17.3.66)$$

Apart from the normalization factor $(m_p c)^{2\chi}$, (17.3.66) is the ratio of the cosmic ray escape time $\tau_D = 4z_1^2/(\pi^2(2l+1)^2 K_1)$ at nucleon rigidity $m_p c$ to the cosmic ray acceleration time $\tau_F = 1/D_1$ at nucleon energy $m_p c$ (compare with the discussion in Sect. 13.2.2.5). According to (17.3.7), (17.3.10) and (17.3.11) we find

$$\frac{\tau_D}{\tau_F} = \frac{\chi(2+\chi)}{(2-\chi)(4-\chi)} \left(\frac{V_A}{c}\right)^{2-\chi} \ln\left(\frac{c}{V_A}\right) \left[\frac{3z_1}{\pi(2l+1)\eta(q=2-\chi)}\right]^2, \quad (17.3.67)$$

which involves the ratio of the escape boundary z_1 and the characteristic length η (see (17.3.7)). Combining (17.3.64)–(17.3.67) we obtain

$$x_c = \frac{(\chi/2)^{1/\chi}}{\alpha m_p c} \left(\frac{\tau_D}{\tau_F}\right)^{1/2\chi}$$

$$= \frac{1}{\alpha m_p c} \left[\frac{\chi(2+\chi)}{(2-\chi)(4-\chi)} \ln\left(\frac{c}{V_A}\right)\right]^{1/2\chi} \left(\frac{V_A}{c}\right)^{(2-\chi)/2\chi}$$

$$\times \left[\frac{3\chi z_1}{2\pi(2l+1)\eta(q=2-\chi)}\right]^{1/\chi}. \quad (17.3.68)$$

Because the observed cosmic ray spectrum above 1 GV is certainly steeper than the very flat distribution (17.3.63a) we have to demand that the value of $x_c \leq 10$. With realistic parameters of the Galaxy in (17.3.68) and (17.3.7) this condition is fulfilled.

17.3.4 Influence of Source Spectral Index Dispersion on the Steady-State Primary Particle Spectrum Below $R_{\max}$

We investigate the influence of the dispersion in the source spectral indices on steady-state distribution function (17.3.20) by using the modified injection function (17.3.43), instead of (17.3.34),

$$N_{lm}(x_{\min} < x < x_{\max}) = 4\pi x^2 f_{lm}(x)$$

$$= 4\pi T_f q_1 \alpha^\chi x^2 \int_{x_{\min}}^{x_{\max}} dx_0 \, x_0^{-\bar{s}} G(x, x_0). \quad (17.3.69)$$

Equation (17.3.69) is solved by numerical integration for different parameter values, and then inserted in (17.1.2) to obtain the full rigidity spectrum at the position of the solar system. Figure 17.1 shows the normalized rigidity spectra for three different values of the mean source spectral index $\langle s \rangle = \beta$, turbulence spectral index $q = 1.8$, and $R_{\max} = 10^{15}$ V. We see that in all three cases slightly curved power laws result below $R_{\max}$, and that the rigidity spectrum cuts off exponentially at larger rigidities. The cutoff is strongly influenced by the value of q as Fig. 17.2 indicates. The closer the value of q lies to 2.0, the more moderate is the cutoff.

Significant deviations from a straight power law can be obtained by varying the characteristic rigidity x_c, as Fig. 17.3 indicates. For increasing x_c (corresponding to increasing τ_D/τ_F, see (17.3.68)) the rigidity spectrum become flatter in accord with (17.3.63).

17.3 Propagation Including Interstellar Acceleration

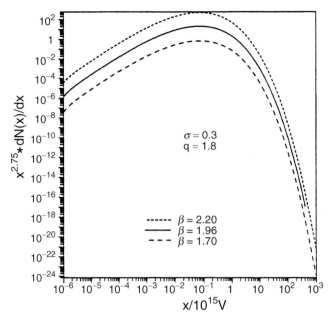

Fig. 17.1. Normalized cosmic ray rigidity spectrum for $q = 1.8$, $R_{\mathrm{max}} = 10^{15}$ V and different mean source spectral indices $\langle s \rangle = \beta$

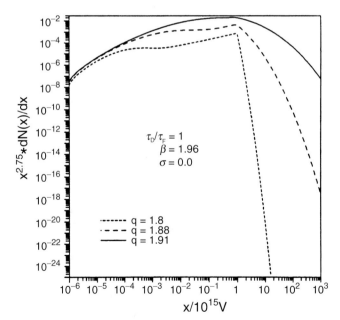

Fig. 17.2. Normalized cosmic ray rigidity spectrum for $R_{\mathrm{max}} = 10^{15}$ V, no source spectral index dispersion and different values of the turbulence spectral index q

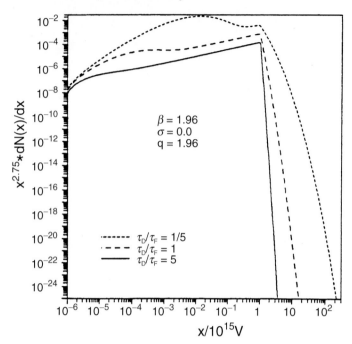

Fig. 17.3. Normalized cosmic ray rigidity spectrum for $q = 1.96$, $R_{\max} = 10^{15}$ V, no source spectral index dispersion, but different values of the characteristic rigidity x_c that reflect the ratio of escape (τ_D) to acceleration (τ_F) time scales at 1 GV

17.3.5 Steady-State Primary Particle Spectrum Above $R_{\max}$

Now we investigate the response of (17.3.20) to the single power law injection function (17.3.34) at particle momenta above $x \geq x_{\max}$. With the Green's function (17.3.33), and the same notation as in the last section, the steady-state situation between injection, fragmentation losses, escape by diffusion and second-order Fermi acceleration yields a cosmic ray spectrum above the highest injection momentum

$$f_{lm}(x \geq x_{\max}) = T_f q_0 \alpha^\chi \int_{x_{\min}}^{x_{\max}} dx_0 \, x_0^{-s} G(x \geq x_0)$$
$$= f_0 \alpha^\chi \, e^{-B/2} \, U(a, b, B) \,, \qquad (17.3.70)$$

with the constant

$$f_0 \equiv q_0 g_0 T_f \int_{x_{\min}}^{x_{\max}} dx_0 \, x_0^{-s} \, e^{-(\lambda^{1/2} x_0^\chi / \chi)} \, M\left(a, b, \frac{2\lambda^{1/2} x_0^\chi}{\chi}\right). \quad (17.3.71)$$

In the special case of negligible fragmentation losses ($\psi = 0$, $a = b/2 = (3-\chi)/2\chi$) we can express the confluent hypergeometric functions in terms of modified Bessel functions to obtain

17.3 Propagation Including Interstellar Acceleration

$$f_{lm}(x \geq x_{\max}; \psi = 0) = c_0 \alpha^\chi \left(\frac{x}{x_c}\right)^{(\chi-3)/2} K_{\left|\frac{3-\chi}{2\chi}\right|}\left[\frac{1}{2}\left(\frac{x}{x_c}\right)^\chi\right], \quad (17.3.72)$$

with the constant

$$c_0 \equiv q_0 g_0 T_f \, 2^{(\chi-3)/2\chi} x_c^{(3-\chi)/2}$$
$$\times \int_{x_{\min}}^{x_{\max}} dx_0 \, x_0^{-s+(\chi-3)/2} I_{\left|\frac{3-\chi}{2\chi}\right|}\left[\frac{1}{2}\left(\frac{x_0}{x_c}\right)^\chi\right]. \quad (17.3.73)$$

Now, it is essential to realize that the distributions (17.3.70) and (17.3.72) are in normalized particle rigidity x, which is simply energy per nucleon for relativistic particles. However, all measurements in this energy range refer to the total energy E of the cosmic ray particles. Adopting the measured cosmic-ray composition at lower energies (Swordy et al. 1990 [535]) we calculated the total energy spectrum of all cosmic ray species from their identical rigidity spectrum (17.3.72) for $R_{\max} = 8 \times 10^{15}$ V and for different values of the parameter $\chi = 2 - q$, which characterizes the rigidity dependence of both the escape (by spatial diffusion) time and the momentum diffusion time, by relating for all individual elements their total energy with their rigidity as $E = ZR$. The resulting all-particle total energy spectra are shown in Fig. 17.4 in comparison with the observed all particle cosmic ray energy spectrum.

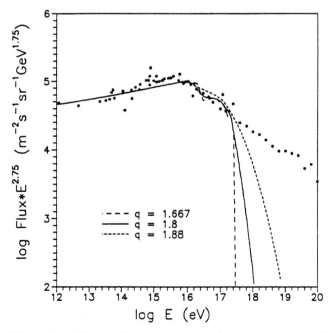

Fig. 17.4. All-particle cosmic ray total energy spectrum for different turbulence power law spectral indices and $R_{\max} = 8 \times 10^{15}$ V

One can clearly see that for values of q between 5/3 and 1.88, which are compatible with the observed Kolmogorov density turbulence spectrum, we achieve remarkable agreement with the measured energy spectrum of ultrahigh energy cosmic rays up to several $\sim 10^{18}$ eV.

In Fig. 17.5 we show the contributions from a few individual (protons, helium, carbon, oxygen and iron) elements in the case $q = 1.8$ to the all-particle energy spectrum. One notices the drastic change in chemical composition from total energies below 10^{16} eV (proton-dominated composition) to higher energies above 10^{16} eV (iron dominated-composition) implied by this model, which can be tested observationally by abundance measurements in this crucial energy range. Figures 17.4 and 17.5 also nicely illustrate how the observed "knee" and "dip" features in the all-particle spectra are reproduced by this in-situ acceleration model.

A very crucial parameter for this in-situ acceleration model is the maximum value $R_{\max}$ of cosmic ray injection in supernova remnant sources, as Fig. 17.6 demonstrates. Only for rather high values of $R_{\max} \geq 8 \times 10^{15}$ V can we account for the acceleration of $> 10^{16}$ eV cosmic rays.

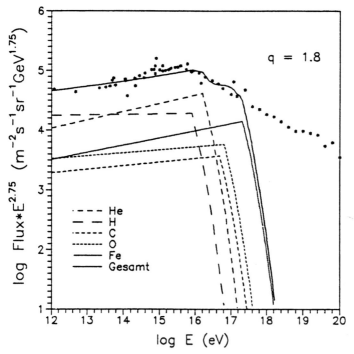

Fig. 17.5. Contribution of individual elements to the total (= Gesamt) all-particle cosmic ray total energy spectrum for $q = 1.8$ and $R_{\max} = 8 \times 10^{15}$ V

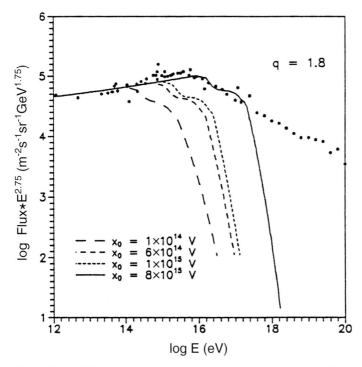

Fig. 17.6. All-particle cosmic ray total energy spectrum for different values of $R_{\max}$

17.4 Time-Dependent Calculation

We now investigate the following hypothesis: if, since the first generation of supernova explosions in our freshly formed galaxy, diffusive shock acceleration at supernova shocks has steadily accelerated cosmic rays up to rigidities of $R_{\max} = 10^{15}$ V, is it conceivable that, since then, over the $\leq 10^{10}$ yr of the lifetime of our Galaxy, the turbulence in the interstellar medium has accelerated some of these cosmic rays slowly but efficiently to ultrahigh energies by the second-order Fermi mechanism, although the mean escape time from the Galaxy is much shorter?

In Sect. 17.3.5 we demonstrated that in-situ second-order Fermi acceleration in a steady-state situation with injection and escape is capable of explaining the observed all-particle cosmic ray spectrum up to total energies of several 10^{18} eV. The crucial question is: can the steady-state spectrum (17.3.70) be reached within $\leq 10^{10}$ yr galactic lifetime?

17.4.1 General Values $\chi \neq 0$

To obtain an answer to this question we have to solve the corresponding (to 17.3.16) time-dependent diffusion-convection equation with steady injection since time $y = 0$ for general values of $\chi \neq 0$:

$$\frac{\partial f_{lm}}{\partial y} = \frac{1}{x^2}\frac{\partial}{\partial x}\left[x^{4-\chi}\frac{\partial f_{lm}}{\partial x}\right] - [\psi + \lambda x^\chi]f_m$$
$$+ Q_0 \alpha^\chi \delta(x - x_{\max}) H[y], \qquad (17.4.1)$$

where H denotes the Heaviside's step function, and

$$y = t/(T_{\text{acc}}\alpha^\chi) \qquad (17.4.2)$$

is the dimensionless time in units of the acceleration time at $R_{\max}$

$$T_{\text{acc}}(R_{\max}) = T_{\text{f}}\left(\frac{R_{\max}}{m_p c}\right)^\chi, \qquad (17.4.3)$$

and

$$T_{\text{f}} = \frac{(2-\chi)(4-\chi)}{\pi(1-\chi)}\left(\frac{B_0}{\delta B}\right)^2\left[\frac{\omega_{p,i}}{ck_{\min}}\right]^{1-\chi}\frac{(c/V_A)^{2+\chi}}{\ln(c/V_A)}\frac{1}{\omega_{p,i}}. \qquad (17.4.4)$$

At rigidities above the maximum injection rigidity $x > x_{\max}$ it suffices to solve (17.4.1) for δ-function injection at $x_{\max}$ since, as (17.3.70) indicates, the variation with rigidity x is independent of the specific injection distribution.

The solution of (17.4.1) for the case $\chi = 0$ is well-known (e.g. Kardashev 1962 [256]), see (17.5.2), and Ball et al. (1992 [27]) based their analysis on this limiting case. However, in the light of the interstellar power spectrum with $1.4 \leq q \leq 1.9$ corresponding to $0.1 \leq \chi \leq 0.6$, this case is of limited interest. The general solution of (17.4.1) is

$$f_{lm}(x,y) = Q_0\alpha^\chi \lambda^{1/2}(x_{\max}x)^{(\chi-3)/2}$$
$$\times \int_0^y d\tau I_{2\nu}\left[\frac{2\lambda^{1/2}(x_{\max}x)^{\chi/2}}{\chi\sinh(\chi\lambda^{1/2}\tau)}\right]$$
$$\times \frac{\exp\left[-\psi\tau - \lambda^{1/2}(x_{\max}^\chi + x^\chi)\coth(\chi\lambda^{1/2}\tau)/\chi\right]}{\sinh(\chi\lambda^{1/2}\tau)}, \qquad (17.4.5)$$

where $I_\nu(z)$ denotes the modified Bessel function of the first kind, and

$$\nu = \left|\frac{3-\chi}{2\chi}\right|. \qquad (17.4.6)$$

Figure 17.7 shows the particle spectra above $R_{\max} = 10^{14}$ V at different galactic ages as compared to the steady-state solution (17.3.70) for $\chi = 0.2$, in which case 10^{10} yr galactic lifetime corresponds to $y = 10^3$. For times $y \geq 5$ the time-dependent solution is indistinguishable from the steady-state solution for particle rigidities $x \leq 10^5$, so that it is justified to use the theoretical steady-state spectra for comparison with the observations.

For all values of χ within the indicated range the time-dependent solutions approach the steady-state solutions within the galactic lifetime, so that our analysis in Sect. 17.3.5 on the basis of steady-state solutions is justified.

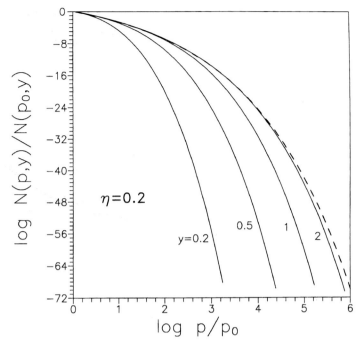

Fig. 17.7. Normalized momentum per nucleon spectrum for constant injection since time $y = 0$ at different later times y calculated from (17.4.5). At late times the **curves** approach the steady-state solution (17.3.70) shown as a **dashed curve**. The injection momentum is $p_0 = 10^{14}$ eV/c nucl. From Schlickeiser (1993 [461])

17.4.2 Special Case $\chi = 0$

For completeness we list the steady-state Green's function (17.3.33) and the time-dependent solution (17.4.5) in the special case of rigidity-independent escape and acceleration times, i.e. $\chi = 0$. The steady-state Green's function in this case is

$$G(x, x_0) = \frac{(xx_0)^{-3/2}}{2\sqrt{\psi + \lambda + (9/4)}} \begin{cases} (x/x_0)^{\sqrt{\psi+\lambda+(9/4)}} & 0 \le x \le x_0 \\ (x/x_0)^{-\sqrt{\psi+\lambda+(9/4)}} & x_0 \le x \le \infty \end{cases}, (17.4.7)$$

while the time-dependent solution is

$$f_{lm}(x, y) = \frac{Q_0 a^\chi \ln(x/x_{\max})}{4\pi^{1/2}\sqrt{\psi + \lambda + (9/4)}} (xx_{\max})^{-3/2} \int_0^y d\tau \, \tau^{-3/2}$$
$$\times \exp\left[-(\psi + \lambda + (9/4))\tau - \frac{\ln^2(x/x_{\max})}{4\tau}\right]. \quad (17.4.8)$$

With the integral representation (C26) of the modified Bessel function it is straightforward to demonstrate that in the limit $y \to \infty$ (17.4.8) approaches the steady-state solution (17.3.72) with $x_0 = x_{\max}$.

17.5 Secondary/Primary Ratio Including Interstellar Acceleration

The steady-state equilibrium distribution of secondary cosmic rays in this case is again given by the solution of the transport equation (14.2.7) where the secondaries source term

$$Q_s(\boldsymbol{x}, p) = v\sigma_{p \to s}(p) n_g(\boldsymbol{x}) M_p(R, z, p) \qquad (17.5.1)$$

is due to the fragmentation of primary cosmic rays with distribution M_p in collisions with interstellar gas. $\sigma_{p \to s}$ denotes the momentum dependence of the fragmentation cross-section, n_g the interstellar gas density, and v the primary cosmic ray velocity. For relativistic cosmic rays the fragmentation cross-section is a constant σ_0 (Letaw et al. 1983 [302]) so that the momentum dependence of the source term (17.5.1) is given by the momentum dependence of the primary cosmic ray equilibrium spectrum.

The solution of the secondary transport equation is given by an expression similar to (17.1.2),

$$M_s(R, z, p) = \sum_{l=0}^{\infty} \sum_{m=1}^{\infty} C_{lm}(R, z) S_{lm}(p) , \qquad (17.5.2)$$

which is quite involved since the expansion coefficients $C_{lm}(R, z)$, according to (14.3.7) and (14.3.8), depend both on the spatial distribution of the interstellar gas density and the cosmic ray primaries (see (17.5.1) above), respectively. Moreover, each function $S_{lm}(p)$ obeys the leaky-box equation

$$\mathcal{O}_p S_{lm}(p) - \Lambda_{lm}^2 \kappa(p) S_{lm}(p) = -Q_2(p) , \qquad (17.5.3)$$

where the secondaries spatial eigenvalues Λ_{lm}^2, in general, do not equal the primaries spatial eigenvalues.

Analytical (Cowsik 1980 [106], Lerche and Schlickeiser 1985 [297]) and numerical (Heinbach and Simon 1995 [224]) solutions of the system of equations (17.5.2)–(17.5.3) based on different simplifying assumptions have been obtained. The resulting secondary/primary ratio $R(p) = M_s(\boldsymbol{x}_s, p)/M_p(\boldsymbol{x}_s, p)$ at the position of the solar system ($\boldsymbol{x}_s$) decreases according to $R(p) \propto p^{-\chi}$ in the relativistic momentum range accessible to observations, which accounts for the measured ratios.

This result can be made plausible by the following argument: under the assumptions made for relativistic secondaries (17.5.3) reduces to an equation similar to (17.3.16)

$$\frac{d}{dx}\left[x^{4-\chi} \frac{dS_{lm}}{dx}\right] - \left[\psi_s x^2 + \lambda_s x^{2+\chi}\right] S_{lm} = -T_f x^2 \alpha^\chi c n_0 \sigma_0 P(x) , \qquad (17.5.4)$$

where $P(x)$ denotes the primaries momentum distribution (17.3.44).

17.5 Influence of Interstellar Acceleration

According to (17.3.63) the primaries spectrum is approximately a power law $P(x \gg x_c) \propto x^{-(s+2+\chi)}$ for momentum values $x \gg x_c$. Inserting this power law in (17.5.4) will yield, after repeating the calculations of Sect. 17.3.3 mutatis mutandis, an equilibrium secondary spectrum above x_c as $S(x \gg x_c) \propto x^{-(s+2+2\chi)}$. As a consequence, the ratio of the two becomes

$$R(x \gg x_c) = \frac{S(x)}{P(x)} \propto x^{-\chi}, \qquad (17.5.5)$$

as required by observations.

18. Formation of Cosmic Ray Momentum Spectra

The calculations in the preceding Chaps. 16 and 17 have indicated that near supernova shock waves, and in the Galaxy, the equilibrium cosmic ray momentum distributions result from the competition of particle acceleration and particle escape by diffusion and/or convection. The competition of these two processes results from the general cosmic ray transport equation (14.2.7) if all other momentum loss and gain processes operate on longer time scales. In this chapter we study the equilibrium spectra in cases where other elementary interaction processes (like radiation losses) are rapid compared to the acceleration and escape processes.

18.1 Fundamental Physical Picture

Figure 18.1 illustrates the basic physical picture we have in mind of the elementary processes operating in different cosmic systems containing cosmic rays. Either gravitational, nuclear or magnetic catastrophic activity provides free energy to the system that gives rise to shock waves and MHD turbulence. We have seen in Chaps. 14 and 16 that shock waves accelerate energetic charged particles both by adiabatic compression (first-order Fermi process) and stochastic acceleration (second-order Fermi process) due to transit-time dampingand cyclotron resonance damping the waves' electric fields.

On the other hand, the generated relativistic gas loses momentum in ionization (Sects. 4.5 and 5.3.8), bremsstrahlung (Sect. 4.4), radiation (Sects. 4.1 and 4.2), and spallation (Sect. 5.3.9) interactions with ambient matter and radiation fields. According to Sect. 14.1.1 these momentum loss processes are accounted for in the cosmic ray transport equation either by loss rates ($\dot{p}$) in the case of continuous losses or loss times (T_c) in the case of catastrophic losses. Mathematically the competition between the various elementary interactions, illustrated in Fig. 18.1, is described by the basic transport equation (14.2.7). This equation, and its time-dependent variant (14.2.3), holds under fairly general conditions, as its quasilinear derivation has indicated. It comprises three main features of any cosmic ray interaction process: regular change, stochastic behavior, escape.

In Chap. 14 we also discussed the appropriate spatial and momentum boundary conditions. Moreover, we demonstrated that for fairly general

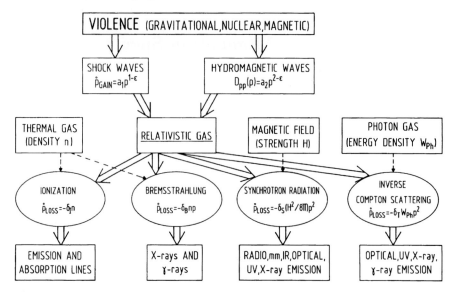

Fig. 18.1. Diagram showing schematically the basic physical model for the formation of cosmic ray momentum spectra. The **arrows** represent the direction of energy flow

assumptions, the formal mathematical solution of the transport equation (14.2.7) with the scattering time method leads to an effective decoupling of the spatial and the momentum transport. The steady-state phase space density $M_a(\boldsymbol{x}, p)$ can be expressed as an (infinite) sum of products of functions $R_i(p)$ and $A_i(\boldsymbol{x})$. The spatial functions $A_i(\boldsymbol{x})$ are determined by the cosmic ray age distribution at location $\boldsymbol{x}$ and depend on the geometry of the considered cosmic ray confinement region, the spatial boundary conditions and the spatial source distribution. Each momentum function $R_i(p)$ satisfies a leaky-box equation (see (14.3.15) and (17.1.3))

$$\mathcal{O}_p R_i(p) - \lambda_i^2 \kappa(p) R_i(p) = - Q(p) , \qquad (18.1.1)$$

where the corresponding spatial eigenvalues λ_i^2 enter in the form of an inverse catastrophic loss time. $\mathcal{O}_p$ denotes the momentum operator

$$\mathcal{O}_p R \equiv \frac{1}{p^2} \frac{\partial}{\partial p} \left(p^2 A_2 \frac{\partial R}{\partial p} + \frac{p^3}{3} V_1 R - p^2 \dot{p}_{\text{loss}} R \right) - \frac{R}{T_c(p)} . \qquad (18.1.2)$$

In the case of cosmic ray nucleons it is convenient to introduce, instead of the total momentum p, either

(i) the particle momentum per nucleon ($\tilde{p} = p/A$) and the corresponding phase space densities ($F_i(\tilde{p}) = A^3 R_i(A\tilde{p})$) and source distributions (see Sect. 17.3 for details), or

(ii) the dimensionless momentum $x = \tilde{p}/m_p c = p/mc$ with the corresponding phase space density and source distribution

18.1 Fundamental Physical Picture

$$f_i(x) = (m_p c)^3 F_i(\tilde{p}) = (mc)^3 R_i(p), \qquad (18.1.3)$$
$$q(x) = (m_p c)^3 Q(\tilde{p}) = (mc)^3 Q(p). \qquad (18.1.4)$$

Then the leaky-box equation (18.1.2) becomes

$$\frac{1}{x^2}\frac{\mathrm{d}}{\mathrm{d}x}\left(D(x)\frac{\mathrm{d}f_i}{\mathrm{d}x} + \frac{x^3}{3}V_1 f_i - x^2 \frac{\dot{p}_{\mathrm{loss}}}{mc} f_i\right)$$
$$- f_i \left[T_c^{-1} + \lambda_i^2 \kappa(x)\right] = -q(x), \qquad (18.1.5)$$

with (notation as in Sect. 17.3)

$$D(x) = A_2(p)/(mc)^2 = \frac{3 V_{\mathrm{A}}^2 (m_p c)^{q-2}}{2q(q+2) c\eta(q)} \alpha^{q-2} \beta x^q \, \ln(v/V_{\mathrm{A}}). \qquad (18.1.6)$$

Specializing to relativistic particles ($\beta \simeq 1$) we may express

$$D(x) = \frac{\alpha^{q-2} x^q}{T_{\mathrm{f}}}, \qquad \kappa(x) = \frac{(\alpha x)^{2-q}}{T_i \lambda_i^2} \qquad (18.1.7)$$

in terms of the basic acceleration and escape time scales (see (17.3.14))

$$T_{\mathrm{f}} = (m_p c)^{2-q}/D_1, \quad \text{and} \quad T_i = (m_p c)^{q-2} K_0 / \left[K_1 \lambda_i^2\right]. \qquad (18.1.8)$$

If we also express the continuous loss rates in (18.1.5) in the form

$$\dot{p}_{\mathrm{loss}} \equiv -\frac{p}{T_1(p)} = -\frac{mcx}{T_1}, \qquad (18.1.9)$$

and introduce the momentum independent adiabatic acceleration (if $V_1 < 0$) or adiabatic cooling (if $V_1 > 0$) time, $T_a = 3/V_1$, the basic leaky-box equation (18.1.5) becomes

$$\frac{1}{x^2}\frac{\mathrm{d}}{\mathrm{d}x}\left(\frac{\alpha^{q-2} x^{q+2}}{T_{\mathrm{f}}} \frac{\mathrm{d}f_i}{\mathrm{d}x} + x^3 \left[T_a^{-1} + T_1^{-1}(x)\right] f_i(x)\right)$$
$$- \left[T_c^{-1} + \frac{(\alpha x)^{q-2}}{T_i}\right] f_i = -q(x). \qquad (18.1.10)$$

Note that for electrons, positrons and protons, $\alpha = A/Z = 1$, and for heavier fully ionized cosmic ray nucleons $\alpha = 2$.

Equation (18.1.10) indicates that *the equilibrium cosmic ray spectrum $f_i(x)$ for given source distribution $q(x)$ is mainly determined by the value and the momentum variation of the five characteristic time scales $T_{\mathrm{f}}, T_a, T_1, T_c$ and T_i, which represent stochastic acceleration, adiabatic acceleration/cooling, continuous losses, catastrophic losses, and escape, respectively.* As we will illustrate in the following subsections rather different momentum spectra can form depending on which of the five time scales has the smallest value.

18.2 Some Exact Solutions for $q = 2$

For turbulence spectra with $q = 2$, the ratio of the escape to stochastic acceleration time scales is a constant, and (18.1.10) becomes

$$\frac{d}{dx}\left(x^4 \frac{df_i}{dx} + x^3 f_i(x)\left[a_1 + \left(T_f/T_1(x)\right)\right]\right) - \lambda x^2 f_i = -T_f x^2 q(x), \quad (18.2.1)$$

with the two constants

$$\lambda \equiv T_f \left[T_c^{-1} + T_i^{-1}\right], \quad (18.2.2)$$

and

$$a_1 \equiv T_f/T_a. \quad (18.2.3)$$

a_1 can be positive or negative depending on the sign of V_1.

In Chaps. 5–7 we have seen that each of the individual continuous loss rates varies proportional to $\dot{p}_{\rm loss} \propto -p^\delta$ with $\delta = 2$ for synchrotron and inverse Compton radiation losses, $\delta = 1$ for nonthermal bremsstrahlung losses, $\delta \simeq 0$ for ionization and Coulomb losses of relativistic particles, $\delta \simeq -2$ for ionization and Coulomb losses of non-relativistic particles above the Coulomb barrier. This implies, according to (18.1.9), for the corresponding loss times

$$T_1(p) \propto p^{1-\delta}, \quad (18.2.4)$$

so that over a wide momentum range the ratio of acceleration to loss times in (18.2.1) can be expressed as

$$\frac{T_f}{T_1(x)} = \theta x^{\delta - 1} \quad (18.2.5)$$

in terms of the constant θ. Without loss of generality it suffices to consider only the case of adiabatic acceleration ($a_1 = -a < 0$), since adiabatic cooling can be added to the nonthermal bremsstrahlung loss term due to its identical momentum dependence. Collecting terms (18.2.1) becomes

$$\frac{d}{dx}\left(x^4 \frac{df_i}{dx} - \left[ax^3 - \theta x^{2+\delta}\right] f_i(x)\right) - \lambda x^2 f_i = -T_f x^2 q(x). \quad (18.2.6)$$

Two linearly independent solutions $f_{1,2}$ of the corresponding homogeneous equation (i.e. $q(x) = 0$) of (18.2.6) are

$$f_{1,2}(x) = c_{1,2} x^k \exp\left(-\frac{\theta x^{\delta-1}}{\delta - 1}\right) \left\{ \begin{array}{c} M \\ U \end{array} \right. \left(\frac{\mu - (a+3)/2}{\delta - 1}, \frac{2\mu + \delta - 1}{\delta - 1}, \frac{\theta x^{\delta-1}}{\delta - 1}\right), \quad (18.2.7)$$

where

$$\mu = \sqrt{\lambda + \left(\frac{a+3}{2}\right)^2}, \qquad (18.2.8)$$

which allow us to construct the Green's function by the standard method described in Sect. 17.3.1. M and U in (18.2.7) denote confluent hypergeometric functions. Solution (18.2.7) is a generalization of the solutions of

(i) Schlickeiser (1984 [456]): shock wave acceleration of relativistic electrons undergoing synchrotron and inverse Compton radiation losses ($\delta = 2$), see also Sect. 18.2.1, and
(ii) Steinacker et al. (1988 [522]): shock wave acceleration of relativistic electrons suffering ionization and Coulomb losses ($\delta = 0$).

18.2.1 Shock Acceleration of Relativistic Electrons Undergoing Radiation Losses

In the case of $\delta = 2$ and a mono-momentum source $S(p) = q_0 \delta(p - p_0)$ the solution of (18.2.6) with the zero-momentum flux boundary condition (14.1.12) at $p = \infty$ and finite $f(p)$ as $p \to 0$ is (Schlickeiser 1984 [456]) ($c_3 =$ const.)

$$N(p) = 4\pi p^2 f(p) = c_3 p_0^{\mu+(1-a/2)} p^{\mu+(1+a/2)} \exp(-p/p_c)$$

$$\begin{cases} U\left(\mu - \frac{a+3}{2}, 1+2\mu, \frac{p_0}{p_c}\right) M\left(\mu - \frac{a+3}{2}, 1+2\mu, \frac{p}{p_c}\right) & \text{for } p \leq p_0 \\ M\left(\mu - \frac{a+3}{2}, 1+2\mu, \frac{p_0}{p_c}\right) U\left(\mu - \frac{a+3}{2}, 1+2\mu, \frac{p}{p_c}\right) & \text{for } p \geq p_0 \end{cases}. \quad (18.2.9)$$

where

$$p_c = 17.5 \, \frac{\text{MeV}}{c} \left(\frac{V_A}{100 \text{ km/s}}\right)^2 \left(\frac{K_\parallel}{10^{24} \text{ cm}^2 \text{ s}^{-1}}\right)^{-1} \left(\frac{w_H + w_{\text{ph}}}{10^{-5} \text{ erg cm}^{-3}}\right)^{-1}, \quad (18.2.10)$$

is that momentum value where the constant stochastic acceleration time scale T_f exactly equals the radiation loss time scale $T_l = c_4(w_H + w_{\text{ph}})^{-1} p^{-1}$, $c_4 =$ const., where $w_H = H^2/8\pi$ and w_{ph} denote the magnetic field and target photon energy densities, respectively. The interplay of these two time scales (and the constant escape time T_i through λ and μ) influences the behavior of solution (18.2.9):

(i) For $T_i \to 0$, yielding $\mu \propto \lambda^{1/2} \to \infty$, particles immediately leave the acceleration volume, and almost no particles are accelerated. The resulting equilibrium spectrum almost coincides with the source distribution (see Fig. 18.2a).
(ii) For $T_i \to \infty$, yielding $\lambda \to 0$ and $\mu \to (a+3)/2$, particles stay in the acceleration region for a long time, and attain an almost monoenergetic distribution ("pile-up distribution") at $p_M = (2+a)p_c$ with a low-energy

power law tail $N(p) \propto p^{-1}$. p_M is that momentum where the first-order Fermi acceleration time scale exactly balances the radiation loss time. The optically thin synchrotron spectrum from such an electron distribution is also illustrated in Fig. 18.2b: it is dominated by the "mono energetic" part of the electron distribution. It has been suggested that this behavior is observed in the nonthermal synchrotron radiation spectrum from our Galactic center (Lesch et al. 1988 [300]).

(iii) In the intermediate case, T_i comparable to T_f, an electron power law distribution with an exponential cutoff at p_M is formed (see Fig. 18.2c). In fact, it is precisely the ratio T_i/T_f which determines the spectral index of this power law. Depending on the actual value of the ratio all spectral indices between -1 (Fig. 18.2b) and $-\infty$ (Fig. 18.2a) can result.

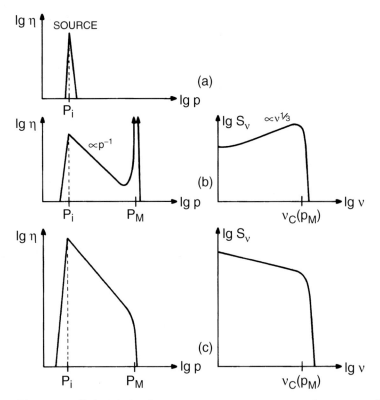

Fig. 18.2. Relativistic electron momentum spectra and corresponding optically thin synchrotron radiation frequency spectra resulting from the shock wave and stochastic acceleration of electrons undergoing synchrotron and inverse Compton radiation losses for three different ratios of the escape lifetime ($\tau_e = T_i$) to the stochastic acceleration time ($\tau_F = T_f$): **(a)** $\tau_e/\tau_f \to 0$, **(b)** $\tau_e/\tau_f \to \infty$, **(c)** $\tau_e/\tau_f \simeq 1$

18.2.2 Shock Acceleration of Relativistic Electrons Undergoing Ionization and Coulomb Losses

In the case $\delta = 0$ the Greent's function G constructed from (18.2.8) under the boundary conditions that

(i) G remains finite as $x \to 0$,
(ii) G approaches zero as $x \to \infty$, is (Steinacker et al. 1988 [522])

$$G(x, x_0) = \frac{\Gamma(\mu - (a + 1/2))}{\Gamma(1 + 2\mu)}(xx_0)^{(a-3/2)}$$

$$\begin{cases} \left(\frac{x}{x_0}\right)^{\mu} U\left(-\mu - \frac{a+1}{2}, 1 - 2\mu, \theta/x\right) M\left(\mu - \frac{a+1}{2}, 1 + 2\mu, \theta/x_0\right) & \text{for } x \leq x_0 \\ \left(\frac{x_0}{x}\right)^{\mu} U\left(-\mu - \frac{a+1}{2}, 1 - 2\mu, \theta/x_0\right) M\left(\mu - \frac{a+1}{2}, 1 + 2\mu, \theta/x\right) & \text{for } x \geq x_0 \end{cases}$$

(18.2.11)

Adopting the steady-state source distribution $Q(x) = q_0\delta(x - x_i)$ we obtain for the steady-state particle number density $N(x) = 4\pi x^2 f(x) = 4\pi x^2 q_0 G(x, x_i)$, at momenta $x > x_i$,

$$N(x > x_i) = C_1(x_i) x^{-\mu+(a+1/2)} M\left(\mu - \frac{a+1}{2}, 1 + 2\mu, \theta/x\right), \quad (18.2.12)$$

which in Fig. 3.15 is compared to the electron spectrum measured for the impulsive solar flare spectrum from 7 June 1980. Using the asymptotic expansion of the confluent hypergeometric function M at small and large arguments we find that (18.2.12) approaches the power law $N(x) \propto x^{-\mu+(a+1)/2}$ at large momenta $x \gg \theta$, and varies as $N(x) \propto x^{a+2} \exp(\theta/x)$ at small momenta $x \ll \theta$. $x = \theta$ corresponds to that momentum value where the stochastic acceleration time scale T_f exactly balances the ionization and Coulomb loss time scale $T_l(x) \propto x$. As Fig. 3.15 indicates, the measured electron spectrum of the impulsive flare from 7 June 1980 is well reproduced by the distribution (18.2.12) which results from the interplay of stochastic first- and second-order Fermi acceleration and Coulomb and ionization losses.

18.3 Bessel Function Solutions

Allowing for general and different power law dependences of the catastrophic escape/fragmentation time scale

$$T_{c,e} = T_0 x^{-b} \tag{18.3.1}$$

and the stochastic acceleration time scale

$$\frac{(\alpha x)^{q-2}}{T_{\mathrm{f}}} = \frac{1}{T_a x^\epsilon} , \qquad (18.3.2)$$

the leaky-box equation (18.1.10) is exactly solvable for vanishing adiabatic acceleration ($a = 0$) and continuous momentum losses ($\theta = 0$). For mono-momentum injection it reads

$$\frac{1}{x^2}\frac{\mathrm{d}}{\mathrm{d}x}\left(x^{4-\epsilon}\frac{\mathrm{d}f}{\mathrm{d}x}\right) - \Lambda x^b f = -q_0 \delta(x - x_0), \qquad (18.3.3)$$

where $\Lambda = T_a/T_0$ is the ratio of acceleration to catastrophic loss time. The solution of (18.3.3) is (Barbosa 1979 [28], Dröge and Schlickeiser 1986 [136])

$$f(x, x_0) = \frac{2q_0}{|b+\epsilon|} x_0^{(\epsilon+1/2)} x^{(\epsilon-3/2)}$$

$$\begin{cases} K_{\left|\frac{3-\epsilon}{b+\epsilon}\right|}\left(\frac{2\Lambda^{1/2}}{|b+\epsilon|}x_0^{(b+\epsilon/2)}\right) I_{\left|\frac{3-\epsilon}{b+\epsilon}\right|}\left(\frac{2\Lambda^{1/2}}{|b+\epsilon|}x^{(b+\epsilon/2)}\right) & \text{for } x \leq x_0 \\ I_{\left|\frac{3-\epsilon}{b+\epsilon}\right|}\left(\frac{2\Lambda^{1/2}}{|b+\epsilon|}x_0^{(b+\epsilon/2)}\right) K_{\left|\frac{3-\epsilon}{b+\epsilon}\right|}\left(\frac{2\Lambda^{1/2}}{|b+\epsilon|}x^{(b+\epsilon/2)}\right) & \text{for } x \geq x_0 \end{cases}, \qquad (18.3.4)$$

where $I_\nu(x)$, $K_\nu(x)$ are modified Bessel functions of the first and second kind, respectively.

Solution (18.3.4) is a generalization of the solutions found earlier by Pikelner and Tsytovich (1976 [400]) ($b = 0$) and Ramaty (1979 [416]) ($b = 0, \epsilon = 0, 1$). It also includes (17.3.72) as special case $b = \epsilon$. The special case $b = 0, \epsilon = 1$ provides an excellent fit to the measured particle momentum spectra of non-relativistic proton and α-particles from long-duration solar flares, as discussed in Sect. 3.3.1 (e.g. see Fig. 3.16).

18.4 Interlude

The special solutions discussed in Sects. 18.2 and 18.3 indicate that the simple model for the evolution of energetic particle spectra, described in Sect. 18.1, successfully accounts for the observed diversity of power law, exponential, inverse-exponential, Bessel-type and quasi-monoenergetic distribution functions. The model is based on the interplay of stochastic Fermi-type acceleration, continuous and catastrophic momentum losses due to ionization, radiation and fragmentation, and particle escape from the acceleration region. All these interaction processes are elementary, and should occur in all cosmic sources of nonthermal radiation, including solar flares, supernova remnants, galaxies, active galactic nuclei and clusters of galaxies. Since the global parameter values (such as the gas density, target photon density and ordered and turbulent magnetic field strengths, which determine the strength of the individual interaction processes) vary significantly from object to object, we can expect a wide diversity in the resulting nonthermal particle momentum spectra from this model.

19. Summary and Outlook

With Fig. 3.3 we indicated that most of the cosmic ray landscape is void. Of course, it will be very important for future cosmic ray particle observations to fill most of the existing gaps. Some important observational goals for the future are:

1. What are the highest energies and the fluxes of cosmic ray nucleons and cosmic ray electrons?
2. Is there cosmic ray antimatter with $Z \geq 2$?
3. What is the positron and antiproton fraction far above energies of 20 GeV? What is the isotope ratio of cosmic ray clocks like beryllium and aluminium at relativistic energies?
4. What is the elemental composition of cosmic ray nuclei at energies far above 10^{12} eV? And how does the cosmic ray anisotropy behave as a function of energy above 10^{14} eV ?

In Chap. 7 we have summarized the observational material on galactic cosmic rays as known today. We emphasized that a successful explanation of the origin of galactic cosmic rays has to explain the following key ingredients

1. an over 10^9 yr constant cosmic ray power of $\sim 10^{40}$ erg s^{-1};
2. a nearly uniform and isotropic distribution of cosmic ray nucleons with energies below 10^{15} eV and electrons with energies below 10^{12} eV over the Galaxy;
3. elemental and isotopic composition similar to solar flare particles;
4. electron/nucleon ratio in relativistic cosmic rays at the same Lorentz factor of about 0.01;
5. the formation of power law energy spectra for all species of cosmic rays over large energy ranges accounting for the systematic differences in the spectral index values of primary and secondary cosmic ray nucleons and cosmic ray electrons.

Now, it is time to compare the theory developed in Sects. B. and C. with these requirements. We discuss each point in turn.

19.1 Cosmic Ray Sources

We noted already in Chap. 7 that the global cosmic ray source energetics requirements of cosmic rays can be met if an efficient acceleration process for cosmic rays can be found that transforms the free energy involved in supernova explosions and stellar winds into energetic particles. In Chap. 16 we discussed the relevant acceleration processes near shock waves that form as a consequence of these powerful events. However, we treated the process in the test particle limit only. In this case, formally one hundred per cent efficiency is possible, although at large efficiencies one has to include self-consistently the backreaction of the accelerated particles on the explosive event. Existing work on this issue (Völk et al. 1984 [550], Kang and Jones 1991 [255], Jones 1993 [252]) is encouraging because it indicates efficiencies up to about 10%.

Instead of accelerating cosmic rays, the supernova energy may alternatively either be dissipated by

(1) heating the interstellar medium and/or
(2) by driving large galactic outflows (e.g. Habe and Ikeuchi 1980 [206]), superbubbles or large loops and spurs.

The heating of the interstellar medium most probably operates via collisionless heating processes that involve the efficient generation of the same plasma wave turbulence that also resonantly accelerates energetic charged particles. Whether this plasma turbulence primarily heats the background medium or accelerates relatively few particles out of the thermal pool to cosmic ray energies depends sensitively on the local plasma conditions, in particular the plasma beta.

If, indeed, supernova explosions are the main cosmic ray sources in the Galaxy we should detect the nonthermal radiation generated by the energetic particles in their immediate environment. We have noted in Sect. 6.1 that in radio surveys supernova remnants clearly stand out as powerful synchrotron radiation sources generated by relativistic electrons and positrons. Obviously in future we would like to have similar evidence for cosmic ray nucleons. Our discussion of the nucleonic radiation processes has shown that cosmic ray nucleons emit high-energy gamma rays and neutrinos from the decay of secondary pions. At most gamma-ray frequencies there is a non-negligible contribution of leptonic gamma-radiation produced by electron bremsstrahlung and inverse Compton interactions. The rather uncertain flux of cosmic ray electrons with energies below 1 GeV implies a large uncertainty in the bremsstrahlung contribution especially at gamma-ray energies below 300 MeV. And at photon energies above 10 GeV the leptonic inverse Compton contribution dominates over the pion-decay contribution (see Fig. 6.13). Therefore a sky map at gamma-ray energies between 0.5 and 5 GeV would reveal the celestial distribution of nucleonic gamma rays most clearly.

Even more significant would be a future sky map of the neutrino emission from the Galaxy because neutrinos are produced in the decay of the charged secondary pions. Unfortunately the atmospheric neutrino background is too large compared to the estimated diffuse galactic neutrino radiation but powerful galactic neutrino point sources may stand out.

Another helpful measurement concerns a sensitive sky survey at TeV photon energies. Here, inverse Compton interactions of high-energy ($E \simeq 15\,E_{15}$ TeV) cosmic ray electrons with the universal microwave background radiation is the dominating photon production process. Because these target photons are universal their intensity is the same everywhere in the Galaxy, so that the TeV gamma-ray intensities are directly proportional to the flux of the radiating electrons. The inverse Compton radiation loss time for these electrons is $T_R = 10^5 E_{15}^{-1}$ yr. Cosmic ray lifetime measurements indicate spatial diffusion coefficients of $\kappa \simeq 10^{28} \kappa_{28}$ cm^2 s^{-1}. Adopting a similar value for these TeV electrons we obtain for the average distance that the electrons diffuse from their sources

$$\langle x \rangle = \sqrt{2\kappa T_R} = 80 \, (\kappa_{28}/E_{15})^{1/2} \quad \text{pc} \,. \tag{19.1.1}$$

If the mean distance between cosmic ray sources is less than $\langle x \rangle$, the diffusion regions of these high-energy electrons will overlap and we will obtain a rather uniform distribution of TeV gamma rays over the Galaxy. Alternatively, if the mean source distance is much larger than $\langle x \rangle$, the TeV celestial distribution will be very patchy, and each patch will be associated with a cosmic ray source. Moreover, the electron energy spectrum at these energies will be very different within the diffusion regions and outside, which might explain some of the peculiarities detected lately in the energy spectrum of diffuse galactic gamma radiation above 1 GeV (Pohl and Esposito 1998 [402]).

19.2 Cosmic Ray Isotropy

The scattering of cosmic rays by low-frequency magnetohydrodynamic plasma waves guarantees the near isotropy of cosmic rays, as shown in Sect. 12.3, because the magnetic force of these waves is much larger than the electric force. Should these waves not be present in the Galaxy, the freely streaming cosmic ray particles along the ordered magnetic field efficiently generate these waves (Sect. 11.3) which leads to cosmic ray self-confinement.

From careful measurements of the interstellar turbulence Spangler (1991 [504]), as well as Minter and Spangler (1997 [361]), estimated the mean of the interstellar magnetic field fluctuations due to turbulence of $\delta B \simeq 0.9$ μG, which corresponds to an energy density of $\delta U \simeq 3.6 \times 10^{-14}$ erg cm^{-3}. According to Simonetti (1992 [490]) it is plausible that this turbulence is generated by supernovae and stellar winds. Moreover, the turbulent magnetic field strength $\delta B = 0.9$ μG is much less than the mean total magnetic field strength $B_t \simeq 4$ µG in the Galaxy so that the application of quasilinear theory (which requires smallness of the ratio $(\delta B/B_t)^2 \ll 1$) to cosmic ray transport and acceleration is well justified. Therefore, the observational evidence on interstellar plasma turbulence supports the picture of cosmic ray confinement by low-frequency magnetohydrodynamic plasma waves.

19.3 Cosmic Ray Isotopic and Elemental Composition

The similarity in the isotopic and elemental composition of solar flare particles and galactic cosmic rays is a key observation. The theory can account for these observations, implying that the basic particle acceleration processes are the same in both environments, i.e. first-order shock acceleration and second-order momentum diffusion by transit-time damping and cyclotron damping of the generated plasma turbulence. Our successful modeling of observed solar flare particle momentum spectra strongly supports this hypothesis.

It also implies that the fundamental and basic cosmic ray interaction processes can be, and should be, tested with future in-situ measurements of energetic particles and the ambient plasma condition in the heliosphere. However, there is one important difference in the study of solar and galactic cosmic ray particles. Whereas for the latter we have the cleanest observational results at relativistic energies (because of the solar modulation), the majority of solar cosmic ray particles have non-relativistic energies. As we have shown (Chap. 13), in gyroresonant transport and acceleration processes cosmic ray particles interact most efficiently with plasma waves of wavenumbers given by the inverse of the cosmic ray particles' gyroradius, $k \simeq R_L^{-1}$. Therefore to study the dynamics of the non-relativistic (i.e. small gyroradii) solar cosmic rays we have to know the plasma turbulence at very large wavenumbers corresponding to very large wave frequencies; this requires turbulence measurements with very small time resolution in the range of $\leq 10^{-3}$ s. Unfortunately, most of the existing plasma turbulence instruments flown in the interplanetary medium have time resolutions only of order seconds.

19.4 Electron/Nucleon Ratio

The observed electron/nucleon ratio in relativistic cosmic rays at the same Lorentz factor of about 0.01 is well accounted for by all electromagnetic acceleration processes. We have shown that for gyroresonant interactions the cosmic ray particle rigidity $R = p/Z$ is the basic variable in the fundamental diffusion-convection equation, which is simply the total particle momentum p for electrons and protons. Let us assume that electrons and protons are accelerated to the same power law spectrum for the differential number of particles, $N_e(p) = N_{0,e}\, p^{-q}$, $N_p(p) = N_{0,p}\, p^{-q}$, and that the total number of protons

19.4 Electron/Nucleon Ratio

and electrons above the non-relativistic kinetic energy $T_0 \simeq 10$ keV is the same, i.e.

$$n_0 = \int_{T_0}^{\infty} dT \, N_e(T) = \int_{T_0}^{\infty} dT \, N_p(T) \,. \tag{19.4.1}$$

Using the relation $T = \sqrt{m^2c^4 + p^2c^2} - mc^2$ we obtain

$$\frac{dp}{dT} = \left(\frac{T}{c^2} + m\right)\left[\frac{T^2}{c^2} + 2Tm\right]^{-1/2}, \tag{19.4.2}$$

so that

$$N(T) = N\left[p(T)\right] \frac{dp}{dT} = \frac{N_0}{c^2}\left(T + mc^2\right)\left[\frac{T^2}{c^2} + 2Tm\right]^{-(q+1)/2}. \tag{19.4.3}$$

Inserting (19.4.3) in (19.4.1) and integrating yields

$$N_0 = (q-1)n_0 \left[\frac{T_0^2}{c^2} + 2T_0 m\right]^{(q-1)/2}, \tag{19.4.4}$$

so that

$$\frac{N_{0,e}}{N_{0,p}} = \left[\frac{m_e + (T_0/2c^2)}{m_p + (T_0/2c^2)}\right]^{(q-1)/2} \simeq \left(\frac{m_e}{m_p}\right)^{(q-1)/2}, \tag{19.4.5}$$

since $T_0 \ll m_e c^2 \ll m_p c^2$.

With (19.4.5) we then obtain for the electron/proton ratio as a function of kinetic energy

$$\frac{N_e(T)}{N_p(T)} \simeq \left(\frac{m_e}{m_p}\right)^{(q-1)/2} \frac{(T + m_e c^2)}{(T + m_p c^2)} \left[\frac{T + 2m_e c^2}{T + 2m_p c^2}\right]^{-(q+1)/2}. \tag{19.4.6}$$

At non-relativistic kinetic energies $T \ll m_e c^2 = 511$ keV the electron/proton ratio (19.4.6) approaches unity, $N_e(T)/N_p(T) \simeq 1$, in the intermediate energy range $m_e c^2 \ll T \ll m_p c^2$ it becomes energy-dependent $\propto (T/m_p c^2)^{(1-q)/2}$, and at relativistic energies $T \gg m_p c^2 \simeq 1$ GeV it approaches the constant value

$$\frac{N_e \left(T \gg m_p c^2\right)}{N_p \left(T \gg m_p c^2\right)} \simeq \left(\frac{m_e}{m_p}\right)^{(q-1)/2}, \tag{19.4.7}$$

which is solely determined by the mass ratio of electrons and protons. Using a spectral index value of $q = 2.2$, as suggested by observations, the ratio (19.4.7) becomes 1.1×10^{-2}, which reproduces the observed electron/proton ratio extremely well.

19.5 Formation of Particle Momentum Spectra

In Chap. 18 we have demonstrated how the competition of elementary acceleration and loss processes can account for the diversity of the observed particle momentum spectra. If the continuous momentum loss processes are too slow compared to the acceleration and escape processes, momentum spectra close to power law spectra are formed near supernova shock waves and in the Galaxy.

19.6 Conclusion

Summing up our comparison of the theory with the observational requirements we note **that the quasilinear transport theory indeed can account for the basic observational requirements of galactic cosmic rays**. It seems that the theory presented for the origin of cosmic rays is general and flexible enough to explain nearly all observational facts known today. As demonstrated, the crucial cosmic ray acceleration and transport parameters depend on many input plasma parameters which, at least for all distant cosmic environments, very often are not well known. Besides applying this cosmic ray theory to astrophysical objects, where cosmic ray acceleration takes place, future refinements and improvements to the theory should therefore also be tested with cosmic ray populations in the local interplanetary medium where detailed in-situ measurements of many plasma parameters are possible.

Some important future improvements of the quasilinear theory are:

(a) the rigorous treatment of perpendicular cosmic ray diffusion;
(b) the calculation of quasilinear transport and acceleration parameters for all relevant magnetohydrodynamic plasma modes allowing for plasma wave dispersion, dissipation and oblique propagation;
(c) the combination of the test-wave and test-particle approaches to a fully self-consistent theory. This will require the solution of two coupled nonlinear kinetic equations, namely the diffusion-convection cosmic ray equation and the diffusion-type equation (11.3.6) for the plasma wave spectral density;
(d) the solutions of the kinetic particle equation (18.1.10) and their application to crucial cosmic ray observations, e.g. the acceleration of ultrahigh energy cosmic rays.

From this list it is clear that to a large extent the progress in cosmic ray transport is closely linked to our understanding of the nature and dynamics of cosmic plasma turbulence. Here again, detailed in-situ turbulence observations especially in the interplanetary medium will play a crucial role.

A. Radiation Transport

To interpret observations of electromagnetic radiation quantitatively, it is necessary to understand how radiation travels through a medium. As a first step we consider loss and gain mechanisms in the direction of the observer without regard to the details of atomic processes making up these mechanisms. We also assume all processes to be stationary, i.e. all parameters are time independent over the time scale of our observations, and we neglect scattering of radiation. We follow the treatment by Gordon (1988 [194]).

We consider the behavior of the total radiation intensity in a situation sketched in Fig. A.1.

The differential intensity dI contributed from an elemental volume of path-length dx is

$$dI = -I \sum_j \kappa_j dx + \sum_j \epsilon_j dx, \tag{A.1}$$

where κ_j denotes the absorption coefficient of the absorption process j operating in the elemental volume dx and ϵ_j denotes the corresponding emission process, each in the direction of the observer. Integrating from the far side of the cloud toward the telescope we find the observed intensity to be

$$I = I(0)\exp\left(-\sum_j \langle \kappa_j \rangle L\right) + \int_0^L \sum_j \frac{\epsilon_j}{\kappa_j} \exp\left(\kappa_j [x-L]\right) \kappa_j dx + I(x > L). \tag{A.2}$$

The first term on the RHS of (A.2) describes the attenuated background, the second term the contribution from the cloud, and the third term the foreground emission. L is the near edge of the cloud. The parameter $\langle \kappa_j \rangle$ is the mean of the absorption coefficient of process j over the physical length of the nebula. The radiation received by the telescope is simply the sum of

(1) the attenuated background radiation,
(2) the contribution from the cloud, and
(3) the contribution of foreground sources (if any). Equation (A.2) is important, because it relates the physical processes in a medium, as described by κ_j and ϵ_j, to the radiation intensity measured by the telescope. It remains to calculate the emission and absorption coefficients for the individual radiation processes.

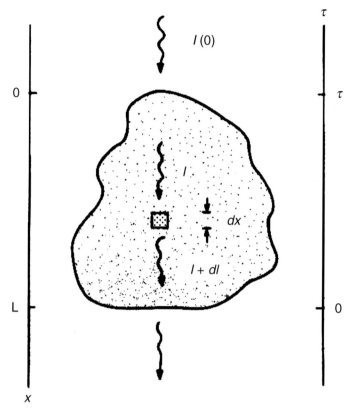

Fig. A.1. Background radiation I(0) travels through the gas cloud, where it is either strengthened or weakened, towards the telescope. From Gordon (1988 [194])

It is convenient to combine the absorption coefficient κ_j and L to a parameter τ_j called the *optical depth of process j*:

$$\tau_j \equiv \int_{x_1}^{x_2} \mathrm{d}x \kappa_j = \langle \kappa_j \rangle (x_2 - x_1) \,. \tag{A.3}$$

In a sense the optical depth may be thought of as the probability of a photon at point x_1 reaching point x_2. Its particular appeal is that it reduces the number of unknowns in (A.2), and it is more likely to be a direct observable. For most purposes we will assume the near limit of the path to be the telescope itself, $x_2 = 0$, and designate the optical path by the most distant point in the path, i.e. $\tau_j = \tau_j(x_1)$. In this notation the intensity transfer equation (A.2) becomes

A. Radiation Transport

$$I(x) = I(0) \exp\left(-\sum_j \tau_j(0)\right) + \sum_j \int_0^{\tau_j(0)} dt \, \frac{\epsilon_j(t)}{\kappa_j(t)} \exp(-t) + I(x > L) \,. \tag{A.4}$$

Depending on the value of the total optical depth

$$\tau = \sum_j \tau_j \tag{A.5}$$

we consider three different cases:

(a) in the *optically thin* case $\tau \ll 1$, (A.4) reduces to

$$I \simeq I(x > L) + I(0) + \sum_j \int_0^L dx \epsilon_j(x) \,, \tag{A.6}$$

which is just the sum of foreground and background emission plus the line-of-sight integrated spontaneous emission from the j radiation processes operating in the cloud;

(b) in the *optically thick* case $\tau \gg 1$ and $\tau > 0$, (A.4) reduces to

$$I \simeq I(x > L) + \sum_j \frac{\epsilon_j}{\kappa_j} \,. \tag{A.7}$$

The background emission $I(0)$ and most of the cloud emission, except emission from the short path-length from L to $L - (\sum_j \kappa_j)^{-1}$, are absorbed in the cloud;

(c) in the *maser* case $\tau \gg 1$ and $\tau < 0$, (A.4) reduces to

$$I \simeq I(x > L) + I(0) \exp(|\tau|) + \sum_j \int_0^{|\tau|} dt \, \frac{\epsilon_j(t)}{|\kappa_j|} \exp(t) \,. \tag{A.8}$$

Here the background and the intrinsic cloud emission are exponentially amplified since the stimulated emission exceeds the absorption in the cloud.

B. Conductivity Tensor in a Hot Magnetized Plasma

With the Fourier-Laplace representations (8.3.12) and (8.3.13) the linearized Vlasov equation (8.3.2) becomes

$$(-\imath\omega + \imath \boldsymbol{k}\cdot\boldsymbol{v})f_1(\boldsymbol{k},\omega,\boldsymbol{p}) + q_a \frac{\boldsymbol{v}\times\boldsymbol{B}_0}{c}\cdot\frac{\partial f_1(\boldsymbol{k},\omega,\boldsymbol{p})}{\partial \boldsymbol{p}}$$
$$= -q_a\left[\boldsymbol{E}_1(\boldsymbol{k},\omega) + \frac{\boldsymbol{v}\times(\boldsymbol{k}\times\boldsymbol{E}_1(\boldsymbol{k},\omega))}{\omega}\right]\cdot\frac{\partial f_a^{(0)}}{\partial \boldsymbol{p}}$$
$$+ f_a(\boldsymbol{k},0) - \frac{\imath q_a}{\omega c}[\boldsymbol{v}\times\boldsymbol{B}(\boldsymbol{k},0)]\cdot\frac{\partial f_a^{(0)}}{\partial \boldsymbol{p}}, \quad (\text{B.1})$$

where we also used (8.3.18). With our choice of coordinate system (see Fig. 8.3) it is convenient to introduce cylindrical momentum and velocity ($\boldsymbol{v} = \boldsymbol{p}/(m_a\gamma)$) coordinates,

$$p_x = p_\perp\cos\phi \quad p_y = p_\perp\sin\phi \quad p_z = p_\parallel, \quad (\text{B.2})$$

so that

$$q_a\frac{\boldsymbol{v}\times\boldsymbol{B}_0}{c}\cdot\frac{\partial f_1(\boldsymbol{k},\omega,\boldsymbol{p})}{\partial \boldsymbol{p}} = -\frac{q_a B_o}{\gamma m_a c}\frac{\partial f_1(\boldsymbol{k},\omega,\boldsymbol{p})}{\partial \phi}$$
$$= -\Omega_a\frac{\partial f_1(\boldsymbol{k},\omega,\boldsymbol{p})}{\partial \phi}, \quad (\text{B.3})$$

where we introduced the relativistic gyrofrequency (8.3.57). With (B.3) we obtain for (B.1)

$$\frac{\partial f_1(\boldsymbol{k},\omega,\boldsymbol{p})}{\partial \phi} - \frac{\imath\boldsymbol{k}\cdot\boldsymbol{v} - \imath\omega}{\Omega_a}f_1(\boldsymbol{k},\omega,\boldsymbol{p}) = \frac{\Phi(\phi)}{\Omega_a}, \quad (\text{B.4})$$

where we defined

$$\Phi(\phi) \equiv q_a Q - f_a(\boldsymbol{k},0) + \frac{\imath q_a}{\omega c}[\boldsymbol{v}\times\boldsymbol{B}(\boldsymbol{k},0)]\cdot\frac{\partial f_a^{(0)}}{\partial \boldsymbol{p}} \quad (\text{B.5a})$$

and

$$Q \equiv \left[\boldsymbol{E}_1(\boldsymbol{k},\omega) + \frac{\boldsymbol{v}\times(\boldsymbol{k}\times\boldsymbol{E}_1(\boldsymbol{k},\omega))}{\omega}\right]\cdot\frac{\partial f_a^{(0)}}{\partial \boldsymbol{p}}. \quad (\text{B.5b})$$

The factor $i\mathbf{k} \cdot \mathbf{v} - i\omega$ with (8.3.49) becomes $i(-\omega + k_\| v_\| + k_\perp v_\perp \cos\phi)$. Equation (B4) is a first-order linear equation for $f_1(\mathbf{k},\omega,\mathbf{p})$ and can be solved by the method of variation of the constant to yield

$$f_1(\mathbf{k},\omega,\mathbf{p}) = \frac{1}{\Omega_a} \int_{\infty\mathrm{sgn}(q_a\Gamma)}^{\phi} d\phi'$$
$$\times \Phi(\phi') \exp\left[\frac{i(k_\| v_\| - \omega)(\phi - \phi') + ik_\perp v_\perp (\sin\phi - \sin\phi')}{\Omega_a}\right]. \tag{B.6}$$

One notices that $f_1(\mathbf{k},\omega,\mathbf{p})$ is periodic in ϕ with a period 2π. The solution (B.6) converges for general values of the imaginary part of the complex frequency $\omega = \omega_R + i\Gamma$ in the limit $\phi' \to \infty\mathrm{sgn}(q_a\Gamma)$.

Introducing the electric field component $\mathbf{E}_1(\mathbf{k},\omega) = (E_x, E_y, E_z)$ expression (B5a) reduces to

$$Q(\phi) = E_x p_x \left[f_{0,\perp} + \frac{k_z}{\gamma m_a \omega} \left(\frac{\partial f_a^{(0)}}{\partial p_\|} - p_z f_{0,\perp}\right)\right]$$
$$+ E_y p_y \left[f_{0,\perp} + \frac{k_z}{\gamma m_a \omega} \left(\frac{\partial f_a^{(0)}}{\partial p_\|} - p_z f_{0,\perp}\right)\right]$$
$$+ E_z \frac{\partial f_a^{(0)}}{\partial p_\|} + E_z \left[\frac{p_x k_x + p_y k_y}{\gamma m_a \omega}\right] \left(p_z f_{0,\perp} - \frac{\partial f_a^{(0)}}{\partial p_\|}\right), \tag{B.7}$$

with the abbreviation

$$f_{0,\perp} \equiv \frac{\partial f_a^{(0)}}{\partial(p_\perp^2/2)}. \tag{B.8}$$

For the special case of *isotropic* distribution functions, i.e. $f_a^{(0)} = f_a^{(0)}(p^2) = f_a^{(0)}(p_\|^2 + p_\perp^2)$ we find

$$p_z f_{0,\perp} - \frac{\partial f_a^{(0)}}{\partial p_\|} = 0, \tag{B.9}$$

so that (B.7) reduces to

$$Q_\mathrm{isotropic}(\phi) = \left(E_x p_x + E_y p_y\right) f_{0,\perp} + E_z \frac{\partial f_a^{(0)}}{\partial p_\|}$$
$$= \left[\left(E_x \cos\phi + E_y \sin\phi\right) p_\perp + E_z p_\|\right] \frac{1}{p}\frac{\partial f_a^{(0)}}{\partial p}. \tag{B.10}$$

We introduce (B.2) and (B.7) into (B.6) and transform the integration variable from ϕ' to $\alpha = \phi - \phi'$ to obtain

$$f_1(\boldsymbol{k},\omega,\boldsymbol{p}) = -\frac{q_a}{\Omega_a}\left\{p_\perp\left[f_{0,\perp} + \frac{k_\|}{\gamma m_a \omega}\left(\frac{\partial f_a^{(0)}}{\partial p_\|} - p_z f_{0,\perp}\right)\right]\right.$$

$$\times \int_{-\infty\operatorname{sgn}(q_a\Gamma)}^0 \mathrm{d}\alpha\,[E_x\cos(\phi-\alpha) + E_y\sin(\phi-\alpha)]$$

$$\times \exp\left[\frac{\imath(k_\| v_\| - \omega)\alpha - \imath k_\perp v_\perp(\sin(\phi-\alpha) - \sin\phi)}{\Omega_a}\right]$$

$$+ E_z\frac{\partial f_a^{(0)}}{\partial p_\|}\int_{-\infty\operatorname{sgn}(q_a\Gamma)}^0 \mathrm{d}\alpha$$

$$\times \exp\left[\frac{\imath(k_\| v_\| - \omega)\alpha - \imath k_\perp v_\perp(\sin(\phi-\alpha) - \sin\phi)}{\Omega_a}\right]$$

$$+ \frac{E_z k_\perp p_\perp}{\gamma m_a \omega}\left(p_z f_{0,\perp} - \frac{\partial f_a^{(0)}}{\partial p_\|}\right)\int_{-\infty\operatorname{sgn}(q_a\Gamma)}^0 \mathrm{d}\alpha\cos(\phi-\alpha)$$

$$\left.\times \exp\left[\frac{\imath(k_\| v_\| - \omega)\alpha - \imath k_\perp v_\perp(\sin(\phi-\alpha) - \sin\phi)}{\Omega_a}\right]\right\} + I_1\,,$$

(B.11)

where the initial values from $f_a(\boldsymbol{k},0)$ and $\boldsymbol{B}(\boldsymbol{k},0)$ have been collected in the term I_1.

Substituting (B.11) into (8.3.16) for the Fourier-Laplace transform of the current density perturbation $\boldsymbol{J}_1(\boldsymbol{k},\omega)$ indeed provides Ohm's law (8.3.17), i.e. the relation between $\boldsymbol{J}_1(\boldsymbol{k},\omega)$ and $\boldsymbol{E}_1(\boldsymbol{k},\omega)$:

$$\sigma_{ij}(\boldsymbol{k},\omega)E_j = \sum_a q_a n_a \int_{-\infty}^\infty \mathrm{d}p_\| \int_0^\infty \mathrm{d}p_\perp p_\perp$$

$$\times \int_0^{2\pi} \mathrm{d}\phi(v_\perp\cos\phi, v_\perp\sin\phi, v_\|)f_1(\boldsymbol{k},\omega,\boldsymbol{p})_a\,. \quad (\text{B.12})$$

B.1 The General Case

Inspecting (B.12) and (B.11) we see that we have to do four integrations over the variables α, ϕ, $p_\|$ and $p_\perp$. In the case of general distribution functions $f_a^{(0)}(\boldsymbol{p})$ we do the ϕ-integral first by making use of the Bessel-function identities

$$\mathrm{e}^{\imath z\sin\phi} = \sum_{n=-\infty}^\infty J_n(z)\mathrm{e}^{\imath n\phi}\,, \tag{B.13a}$$

$$\mathrm{e}^{-\imath z\sin(\phi-\alpha)} = \sum_{m=-\infty}^\infty J_m(z)\mathrm{e}^{-\imath m(\phi-\alpha)}\,. \tag{B.13b}$$

From (B.13) we can derive

$$\int_0^{2\pi} d\phi\, e^{-iz[\sin(\phi-\alpha)-\sin\phi]} \begin{Bmatrix} \sin\phi\sin(\phi-\alpha) \\ \sin\phi\cos(\phi-\alpha) \\ \cos\phi\sin(\phi-\alpha) \\ \cos\phi\cos(\phi-\alpha) \\ 1 \\ \sin\phi \\ \cos\phi \\ \sin(\phi-\alpha) \\ \cos(\phi-\alpha) \end{Bmatrix} = 2\pi \sum_{n=-\infty}^{\infty} e^{in\alpha} \begin{Bmatrix} \left[J_n'\right]^2 \\ -inJ_nJ_n'/z \\ inJ_nJ_n'/z \\ n^2 J_n^2/z^2 \\ J_n^2 \\ -iJ_nJ_n' \\ nJ_n^2/z \\ iJ_nJ_n' \\ nJ_n^2/z \end{Bmatrix}. \tag{B.14}$$

The α-integrals can then also be carried out and contribute a factor

$$\int_{-\infty\,\mathrm{sgn}(q_a\Gamma)}^{0} d\alpha \exp\left[\frac{\alpha}{\Omega_a}(-i\omega + ik_\| v_\| + in\Omega_a)\right] = \frac{i\Omega_a}{\omega - k_\| v_\| - n\Omega_a}. \tag{B.15}$$

This reduces (B.12) to expression (8.3.53).

B.2 Isotropic Distribution Functions

For isotropic background distribution functions $f_a^{(0)}(\boldsymbol{p}) = f_a^{(0)}(p)$ use of (B.10) simplifies (B.11) to

$$f_1(\boldsymbol{k},\omega,\boldsymbol{p})_{\mathrm{iso}}$$
$$= -\frac{q_a}{\Omega_a}\frac{1}{p}\frac{\partial f_a^{(0)}}{\partial p}$$
$$\times \int_{-\infty\,\mathrm{sgn}(q_a\Gamma)}^{0} d\alpha \left[p_\perp \left[E_x\cos(\phi-\alpha) + E_y\sin(\phi-\alpha)\right] + p_\| E_z\right]$$
$$\times \exp\left[\frac{i(k_\| v_\| - \omega)\alpha - ik_\perp v_\perp(\sin(\phi-\alpha) - \sin\phi)}{\Omega_a}\right]. \tag{B.16}$$

Inserting (B.16) into (B.12) and introducing dimensionless spherical momentum coordinates

$$p_\| = m_a c x \cos\vartheta, \qquad p_\perp = m_a c x \sin\vartheta \tag{B.17}$$

as well as the plasma frequency (9.1.6) we obtain

$$\sigma_{ij}(\boldsymbol{k},\omega)$$
$$= -\frac{1}{4\pi}\sum_a \frac{\omega_{p,a}^2 (m_a c)^3}{\Omega_{0,a}} \int_0^\pi d\vartheta\, \sin\vartheta \int_0^{2\pi} d\phi$$

B.2 Isotropic Distribution Functions

$$\times \int_0^\infty \mathrm{d}x x^3 \frac{\partial f_a^{(0)}(x)}{\partial x} \int_{-\infty \mathrm{sgn}(q_a \Gamma)}^0 \mathrm{d}\alpha \exp\left[-\frac{\imath \omega \alpha \sqrt{1+x^2}}{\Omega_{0,a}}\right.$$

$$\left. + \frac{\imath c x}{\Omega_{0,a}} \left(k_\| \alpha \cos \vartheta + k_\perp (1-\cos\alpha) \sin\vartheta \sin\phi + k_\perp \sin\alpha \sin\vartheta \cos\phi\right)\right]$$

$$\begin{pmatrix} \sin^2\vartheta \cos\phi \cos(\phi-\alpha) & \sin^2\vartheta \cos\phi \sin(\phi-\alpha) & \sin\vartheta \cos\vartheta \cos\phi \\ \sin^2\vartheta \sin\phi \cos(\phi-\alpha) & \sin^2\vartheta \sin\phi \sin(\phi-\alpha) & \sin\vartheta \cos\vartheta \sin\phi \\ \sin\vartheta \cos\vartheta \cos(\phi-\alpha) & \sin\vartheta \cos\vartheta \sin(\phi-\alpha) & \cos^2\vartheta \end{pmatrix}.$$

(B.18)

The integrals over the momentum angles ϑ and ϕ are readily performed applying the general integral [Gradshteyn and Ryzhik 1965 [201], equation (4.624)]

$$J(A, B, D) \equiv \int_0^\pi \mathrm{d}\vartheta \sin\vartheta$$

$$\times \int_0^{2\pi} \mathrm{d}\phi F \left(A \cos\vartheta + B \sin\vartheta \cos\phi + D \sin\vartheta \sin\phi\right)$$

$$= 2\pi \int_{-1}^1 \mathrm{d}t F(Rt) = 4\pi \sinh(R)/R \qquad \text{(B.19)}$$

to $F(x) = \exp(x)$, where $R = \sqrt{A^2 + B^2 + D^2}$, and derivatives of (B.19) with respect to A, B, and D, respectively. Let M and L denote any of the variables A, B, and D; from (B.19) we then obtain

$$\frac{\partial^2 J}{\partial M \partial L} = \delta_{KL} \left[\frac{\cosh R}{R^2} - \frac{\sinh R}{R^3}\right]$$

$$+ \frac{ML}{R^3}\left[\left(1 + \frac{3}{R^2}\right)\sinh R - \frac{3}{R}\cosh R\right], \qquad \text{(B.20)}$$

where δ_{KL} denotes Kronecker's symbol. Equation (B.18) then reduces to

$$\sigma_{ij}(\boldsymbol{k},\omega) = -\sum_a \frac{\omega_{\mathrm{p},a}^2 (m_a c)^3}{\Omega_{0,a}} \int_0^\infty \mathrm{d}x x^3 \frac{\partial f_a^{(0)}(x)}{\partial x}$$

$$\times \int_{-\infty \mathrm{sgn}(q_a \Gamma)}^0 \mathrm{d}\alpha \exp\left[-\frac{\imath \omega \alpha \sqrt{1+x^2}}{\Omega_{0,a}}\right]$$

$$\times \left[\left(\frac{\cosh R}{R^2} - \frac{\sinh R}{R^3}\right)T_{ij}^{(1)} + R^{-1}V^{-2}\right.$$

$$\left. \times \left[\left(1 + \frac{3}{R^2}\right)\sinh R - \frac{3}{R}\cosh R\right]T_{ij}^{(2)}\right], \qquad \text{(B.21)}$$

with

$$R = \imath c x V/\Omega_{0,a}, \quad V = \sqrt{k_\|^2 \alpha^2 + 2 k_\perp^2 (1-\cos\alpha)} \qquad \text{(B.22)}$$

B. Conductivity Tensor

and where we defined the two tensors

$$T^{(1)}_{ij} \equiv \begin{pmatrix} \cos\alpha & -\sin\alpha & 0 \\ \sin\alpha & \cos\alpha & 0 \\ 0 & 0 & 1 \end{pmatrix}, \qquad (B.23)$$

and

$$T^{(2)}_{ij} \equiv \begin{pmatrix} k_\perp^2 \sin^2\alpha & -k_\perp^2 \sin\alpha\,(1-\cos\alpha) & k_\parallel k_\perp \alpha \sin\alpha \\ k_\perp^2 \sin\alpha\,(1-\cos\alpha) & -k_\perp^2 (1-\cos\alpha)^2 & k_\parallel k_\perp \alpha\,(1-\cos\alpha) \\ k_\parallel k_\perp \alpha \sin\alpha & -k_\parallel k_\perp \alpha(1-\cos\alpha) & k_\parallel^2 \alpha^2 \end{pmatrix}. \qquad (B.24)$$

C. Calculation of the Integral (10.3.10)

Our task is to evaluate the integral (10.3.10). We substitute $E = \sqrt{1+s^2}$ to derive

$$J(\mu, N) = y^2 \int_0^\infty ds \frac{\exp\left(-\mu\sqrt{1+s^2}\right)\left[1 + s^2 + (2\sqrt{1+s^2}/\mu) + \frac{2}{\mu^2}\right]}{\sqrt{1+s^2}\,(s^2 + y^2)}$$

$$= \frac{y}{2i} \int_{-\infty}^\infty ds \frac{\exp\left(-\mu\sqrt{1+s^2}\right)\left[1 + s^2 + (2\sqrt{1+s^2}/\mu) + \frac{2}{\mu^2}\right]}{\sqrt{1+s^2}\,(s - iy)}, \qquad (C.1)$$

where

$$y = \pm\sqrt{D}, \quad D = \frac{1}{1-N^2} = \frac{z^2}{z^2 - N_R^2}, \quad z = 1 + i\beta. \qquad (C.2)$$

We first note that the integral (C.1) is even in y,

$$J(\mu, -y) = J(\mu, y). \qquad (C.3)$$

Obviously, (C.1) can be calculated as

$$J = y^2 \left[\frac{2}{\mu^2} I(\mu, y) - \frac{2}{\mu}\frac{\partial I(\mu, y)}{\partial \mu} + \frac{\partial^2 I(\mu, y)}{\partial \mu^2}\right], \qquad (C.4)$$

where

$$I(\mu, y) = \int_0^\infty ds \frac{\exp\left(-\mu\sqrt{1+s^2}\right)}{\sqrt{1+s^2}\,(s^2 + y^2)} = \frac{1}{2iy} \int_{-\infty}^\infty ds \frac{\exp\left(-\mu\sqrt{1+s^2}\right)}{\sqrt{1+s^2}\,(s - iy)}. \qquad (C.5)$$

The further calculation of the integral (C.5) depends on the value of y defined in (C.2). We know already from our general discussion in Chap. 10 that β is negative throughout, so we write

$$\beta = -b, \quad z = 1 - ib, \qquad (C.6)$$

with $b \geq 0$. It is straightforward to show from (C.2) that in general

$$y = \frac{(1+b^2)^{1/2}}{\left[4b^2 + (N_R^2 + b^2 - 1)^2\right]^{1/4}} e^{i\psi}, \qquad (C.7)$$

where

$$\psi = \frac{\pi}{2}\theta(b_c - b) - \frac{1}{2}\arctan\frac{2bN_R^2}{(1+b^2)^2 - N_R^2(1-b^2)}$$

$$= \frac{\pi}{2}\theta(b_c - b) - \frac{1}{2}\arctan\frac{2b(1+b_c^2)^2}{(b_c^2 - b^2)[3 + b_c^2 + b^2(1-b_c^2)]}, \quad (C.8)$$

with

$$b_c = \left(\frac{N_R^2}{2}\left[\sqrt{1+\frac{8}{N_R^2}} - \left(1+\frac{2}{N_R^2}\right)\right]\right)^{1/2}. \quad (C.9)$$

For definiteness we have chosen the positive root in (C.2). It is straightforward to prove that $b_c \simeq \epsilon/3 \to 0$ for $N_R^2 = 1 + \epsilon$, $\epsilon \ll 1$ and $b_c \to 1$ from below for $N_R^2 \gg 1$.

In the case of superluminal waves $z = 1$ and $b = 0$ and we find from (C.7) and (10.2.28) that $y = \sigma$ is purely real with values between $1 \leq \sigma \leq \infty$.

In the case of subluminal ($N_R > 1$) waves we obtain from (C.7)

$$y = S + iQ, \quad (C.10a)$$

with

$$S = \frac{(1+b^2)^{1/2}}{\left[4b^2 + (N_R^2 + b^2 - 1)^2\right]^{1/4}}\cos\psi, \quad (C.10b)$$

and

$$Q = \frac{(1+b^2)^{1/2}}{\left[4b^2 + (N_R^2 + b^2 - 1)^2\right]^{1/4}}\sin\psi. \quad (C.10c)$$

According to (C.8), the angle ψ varies between 0 and $\pi/2$, so that the real part S of y is always non-zero provided $b \neq 0$.

In the case of subluminal ($N_R > 1$) waves in the weak-damping limit ($\beta = 0$) we find from (C.9) that $y = -i\alpha$ is purely imaginary, where

$$\alpha = -\left(N_R^2 - 1\right)^{-1/2} \quad (C.11)$$

attains values between $0 \leq |\alpha| < 1$.

C.1 The Integral J for Imaginary Values of y

The weak-damping limit of subluminal waves yields a purely imaginary value of $y = -i\alpha$ (see (C.11)), which needs special attention. Here (C.5) can be calculated as a limit from below

$$I(\mu,\alpha) = \frac{1}{2\alpha}\int_{-\infty}^{\infty}ds\,\frac{\exp\left(-\mu\sqrt{1+s^2}\right)}{\sqrt{1+s^2}(s-\alpha)}$$
$$= \frac{1}{2\alpha}\lim_{\epsilon\to 0}\int_{-\infty}^{\infty}ds\,\frac{\exp\left(-\mu\sqrt{1+s^2}\right)}{\sqrt{1+s^2}[s-(\alpha-\imath\epsilon)]}, \qquad (C.12)$$

Using the Plemelj formula we derive the integral in terms of its principal value as

$$I(\mu,\alpha) = \frac{1}{2\alpha}\left[P\int_{-\infty}^{\infty}ds\,\frac{\exp\left(-\mu\sqrt{1+s^2}\right)}{\sqrt{1+s^2}(s-\alpha)} - \pi\imath\frac{\exp\left(-\mu\sqrt{1+\alpha^2}\right)}{\sqrt{1+\alpha^2}}\right]. \qquad (C.13)$$

The principal value of the integral has been determined by Schlickeiser and Mause (1995 [468], Appendix B.3) as

$$-P\int_{-\infty}^{\infty}ds\,\frac{\exp\left(-\mu\sqrt{1+s^2}\right)}{\sqrt{1+s^2}(s-\alpha)}$$
$$= -2\alpha\int_{0}^{\infty}dt\,M\left(1,\frac{3}{2},-\alpha^2 t\right)\exp\left(-t-\frac{\mu^2}{4t}\right), \qquad (C.14)$$

so that we obtain

$$I(\mu,\alpha) = -\int_{0}^{\infty}dt\,M\left(1,\frac{3}{2},-\alpha^2 t\right)\exp\left(-t-\frac{\mu^2}{4t}\right)$$
$$-\frac{\imath\pi}{2\alpha}\frac{\exp\left(-\mu\sqrt{1+\alpha^2}\right)}{\sqrt{1+\alpha^2}}. \qquad (C.15)$$

It is straightforward to determine from (C.15) the first and second derivative with respect to μ and to derive according to (C.4) in this case that

$$J(\mu,\alpha) = W(\mu,\alpha) - \frac{\imath\pi}{2}\alpha\,\text{sgn}(\Im\alpha)$$
$$\times \sqrt{1+\alpha^2}e^{-\mu\sqrt{1+\alpha^2}}\left[1 + \frac{2}{\mu\sqrt{1+\alpha^2}} + \frac{2}{\mu^2(1+\alpha^2)}\right], \qquad (C.16)$$

where

$$W(\mu,\alpha) = \frac{2\alpha^2}{\mu^2}\int_{0}^{\infty}dt\left[1 + \frac{\mu^2}{4t} + \frac{\mu^4}{8t^2}\right]M\left(1,\frac{3}{2},-\alpha^2 t\right)\exp\left(-t-\frac{\mu^2}{4t}\right). \qquad (C.17)$$

The properties of the function W are discussed in detail by Schlickeiser and Mause (1995 [468]).

C.2 The Integral J for Non-imaginary Values of y

For superluminal waves and subluminal waves with finite damping the real part of $y \neq 0$. Using the identity

C. Calculation of the Integral J

$$\frac{1}{s-iy} = i\int_0^{\infty\Re(y)} dt\, e^{-it(s-iy)},\qquad \text{(C.18)}$$

we can cast (C.5) to the double integral

$$I(\mu,y) = \frac{1}{2y}\int_0^{\infty\Re(y)} dt\, e^{-yt}\int_{-\infty}^{\infty} ds\, \frac{\exp\left(-\mu\sqrt{1+s^2}-its\right)}{\sqrt{1+s^2}}.\qquad \text{(C.19)}$$

The s-integral is the Fourier-cosine integral

$$\int_{-\infty}^{\infty} ds\, \frac{\exp\left(-\mu\sqrt{1+s^2}-its\right)}{\sqrt{1+s^2}} = 2K_0\left(\sqrt{\mu^2+t^2}\right),\qquad \text{(C.20)}$$

so that (C.19) becomes

$$I(\mu,y) = \frac{1}{y}\int_0^{\infty\Re(y)} dt\, e^{-yt} K_o\left(\sqrt{\mu^2+t^2}\right).\qquad \text{(C.21)}$$

It is straightforward to determine from (C.21) the first and second derivatives with respect to μ and to derive, according to (C.4)

$$J = y\int_0^{\infty\Re(y)} dt\, e^{-yt}$$
$$\times\left[\frac{2K_0\left(\sqrt{\mu^2+t^2}\right)}{\mu^2} + \frac{K_1\left(\sqrt{\mu^2+t^2}\right)}{\sqrt{\mu^2+t^2}} + \frac{\mu^2 K_2\left(\sqrt{\mu^2+t^2}\right)}{\mu^2+t^2}\right].$$
$$\text{(C.22)}$$

Repeated partial integration of the integral (C.22) with respect to the variable t yields the alternating series

$$J = \sum_{i=0}^{\infty} \mu^{-i}\left[\left(1+\frac{2}{\mu^2}\right)K_i(\mu) + \frac{2i+3}{\mu}K_{i+1}(\mu)\right]\left(\frac{\partial}{\partial y}y^{-1}\right)^{(i)},\qquad \text{(C.23)}$$

where

$$\left(\frac{\partial}{\partial y}y^{-1}\right)^{(i)} \equiv \left(\frac{\partial}{\partial y}y^{-1}\right)\ldots \text{ i-times}\times \left(\frac{\partial}{\partial y}y^{-1}\right),\qquad \text{(C.24)}$$

implying

$$\left(\frac{\partial}{\partial y}y^{-1}\right)^{(0)} = 1,\quad \left(\frac{\partial}{\partial y}y^{-1}\right)^{(1)} = -y^{-2},\quad \left(\frac{\partial}{\partial y}y^{-1}\right)^{(2)} = 3y^{-4}.\qquad \text{(C.25)}$$

The series (C.22) converges rapidly for values of $\mu\Re(y^2)\gg 1$.

Using the integral representation of Bessel functions for $\Re(Z^2) > 0$ (Watson 1966 [556], p.183)

C.2 The Integral J for Non-imaginary Values of y

$$K_\nu(Z) = 2^{-(\nu+1)} Z^\nu \int_0^\infty dx \, x^{-\nu-1} \exp\left[-x - \frac{Z^2}{4x}\right]$$

$$= 2^{\nu-1} Z^{-\nu} \int_0^\infty dx \, x^{\nu-1} \exp\left[-x - \frac{Z^2}{4x}\right] \quad (C.26)$$

in (C.22) we derive

$$J = y \int_0^\infty dx \left[\frac{1}{\mu^2 x} + \frac{1}{4x^2} + \frac{\mu^2}{8x^3}\right] \exp\left[-x - \frac{\mu^2}{4x}\right]$$

$$\times \int_0^{\infty \Re(y)} dt \, \exp\left(-yt - \frac{t^2}{4x}\right). \quad (C.27)$$

In the case of subluminal waves we have, according to (C.10), $\Re(y^2) < 1$, and we find

$$J\left(\Re\left(y^2\right) < 1\right) = \frac{\pi^{1/2} y \operatorname{sgn}\left(\Re y\right)}{\mu^2} \int_0^\infty dx \left[\frac{1}{x^{1/2}} + \frac{\mu^2}{4x^{3/2}} + \frac{\mu^4}{8x^{5/2}}\right]$$

$$\times \exp\left[-x\left(1-y^2\right) - \frac{\mu^2}{4x}\right] - \frac{2y^2}{\mu^2} \int_0^\infty dx$$

$$\times \left[1 + \frac{\mu^2}{4x} + \frac{\mu^4}{8x^2}\right] M\left(1, \frac{3}{2}, xy^2\right) \exp\left[-x - \frac{\mu^2}{4x}\right]. \quad (C.28)$$

By using (C.26) and (C.17) we obtain for (C.28)

$$J\left(\Re\left(y^2\right) < 1\right) = W(\mu, \imath y) + \frac{\pi y \operatorname{sgn}\left(\Re y\right) e^{-\mu\sqrt{1-y^2}}}{\mu^2 \sqrt{1-y^2}}$$

$$\times \left[1 + \mu\sqrt{1-y^2} + \frac{\mu^2\left(1-y^2\right)}{2}\right]. \quad (C.29)$$

In the case of superluminal waves $y = -\sigma$ with $\sigma \geq 1$, and we find

$$\int_0^\infty dt \, \exp\left(-\sigma t - \frac{t^2}{4x}\right) = \pi^{1/2} x^{1/2} e^{\sigma^2 x} \operatorname{erfc}\left(\sigma x^{1/2}\right)$$

$$= (\pi x)^{1/2} w\left(\imath \sigma x^{1/2}\right), \quad (C.30)$$

in terms of the error function w of complex argument (Abramowitz and Stegun 1972 [3]). With the series expansion

$$w\left(\imath \sigma x^{1/2}\right) = \sum_{n=0}^\infty \frac{\left(-\sigma x^{1/2}\right)^n}{\Gamma\left(1 + \frac{n}{2}\right)} \quad (C.31)$$

we obtain in this case, from (C.27)

C. Calculation of the Integral J

$$J\left(\Re\left(y^2\right) \geq 1\right) = \frac{\pi^{1/2}\sigma}{\mu^2} \sum_{n=0}^{\infty} \frac{(-\sigma)^n}{\Gamma(1+(n/2))} \int_0^{\infty} dx$$

$$\times \left[x^{(n-1)/2} + \frac{\mu^2}{4} x^{(n-3)/2} + \frac{\mu^4}{8} x^{(n-5)/2} \right] \exp\left[-x - \frac{\mu^2}{4x}\right].$$
(C.32)

By using the integral representation (C.26) we derive for (C.32)

$$J\left(\Re\left(y^2\right) \geq 1\right) = \sqrt{\pi\mu\sigma^2/2} \sum_{n=0}^{\infty} \frac{\left(-\sqrt{\mu\sigma^2/2}\right)^n}{\Gamma(1+(n/2))}$$

$$\times \left[\left(1 + \frac{2}{\mu^2}\right) K_{(n+1)/2}(\mu) + \frac{2-n}{\mu} K_{(n-1)/2}(\mu) \right].$$
(C.33)

Collecting odd and even terms in the series separately we obtain

$$J\left(\Re\left(y^2\right) \geq 1\right) = \sqrt{\pi\mu\sigma^2/2} \left\{ \sum_{n=0}^{\infty} \frac{(\mu\sigma^2/2)^n}{\Gamma(1+n)} \right.$$

$$\times \left[\left(1 + \frac{2}{\mu^2}\right) K_{n+(1/2)}(\mu) - \frac{2(n-1)}{\mu} K_{n-(1/2)}(\mu) \right]$$

$$- \sqrt{\mu\sigma^2/2} \sum_{n=0}^{\infty} \frac{(\mu\sigma^2/2)^n}{\Gamma((3/2)+n)}$$

$$\left. \times \left[\left(1 + \frac{2}{\mu^2}\right) K_{n+1}(\mu) - \frac{2n-1}{\mu} K_n(\mu) \right] \right\}.$$
(C.34)

C.3 A Second Form of the Integral J

Expanding the exponential function in (C22) into an infinite series we derive

$$J = y \operatorname{sgn}(\Re y) \sum_{n=0}^{\infty} \frac{(-y \operatorname{sgn}(\Re y))^n}{\Gamma(n+1)} \int_0^{\infty} dt\, t^n$$

$$\times \left[\frac{2K_0\left(\sqrt{\mu^2+t^2}\right)}{\mu^2} + \frac{K_1\left(\sqrt{\mu^2+t^2}\right)}{\sqrt{\mu^2+t^2}} + \frac{\mu^2 K_2\left(\sqrt{\mu^2+t^2}\right)}{\mu^2+t^2} \right].$$
(C.35)

The integrals can be solved in closed form [Gradshteyn and Ryzhik 1965 [201], equation (6.596.3)] and we obtain, with the help of Bessel function identities,

$$J = y \operatorname{sgn}(\Re y) \sqrt{\pi\mu/2} \sum_{n=0}^{\infty} \frac{\left(-\operatorname{sgn}(\Re y) y \sqrt{\mu/2}\right)^n}{\Gamma(1+(n/2))}$$

$$\times \left[\left(1 + \frac{2}{\mu^2}\right) K_{(n+1)/2}(\mu) + \frac{2-n}{\mu} K_{(n-1)/2}(\mu) \right].$$
(C.36)

Collecting odd and even terms in the series separately we obtain

$$
\begin{aligned}
J = {} & \sqrt{\pi \mu y^2/2} \Biggl\{ \sum_{n=0}^{\infty} \frac{(\mu y^2/2)^n}{\Gamma(1+n)} \\
& \times \left[\left(1 + \frac{2}{\mu^2}\right) K_{n+(1/2)}(\mu) - \frac{2(n-1)}{\mu} K_{n-(1/2)}(\mu) \right] \\
& - \sqrt{\mu y^2/2} \sum_{n=0}^{\infty} \frac{(\mu y^2/2)^n}{\Gamma((3/2)+n)} \\
& \times \left[\left(1 + \frac{2}{\mu^2}\right) K_{n+1}(\mu) - \frac{2n-1}{\mu} K_n(\mu) \right] \Biggr\} ,
\end{aligned} \quad (\text{C.37})
$$

which for $y = \sigma$ agrees with (C.34). With (C.37) it is straightforward to approximate the integral J for non-relativistic ($\mu \gg 1$) and ultrarelativistic ($\mu \ll 1$) conditions by using the appropriate asymptotic expansions of the modified Bessel functions.

C.4 The Integral J for Non-relativistic Temperatures

In the non-relativistic limit $\mu \gg 1$ we obtain from (C.37) the ratio

$$
\begin{aligned}
\frac{J}{K_2(\mu)} & \simeq \sqrt{\pi \mu y^2/2} \left[\sum_{n=0}^{\infty} \frac{(\mu y^2/2)^n}{\Gamma(1+n)} - \sqrt{\mu y^2/2} \sum_{n=0}^{\infty} \frac{(\mu y^2/2)^n}{\Gamma((3/2)+n)} \right] \\
& = \sqrt{\pi \mu y^2/2} \left[\exp(\mu y^2/2) - \sqrt{\frac{2\mu y^2}{\pi}} M\left(1, \frac{3}{2}, \mu y^2/2\right) \right] \\
& \times \left[1 + O\left(\frac{1}{\mu}\right) \right] .
\end{aligned} \quad (\text{C.38})
$$

Expressing the confluent hypergeometric function in (C.38) in terms of the complementary error function we derive

$$
\frac{J(\mu \gg 1, N_R)}{K_2(\mu)} \simeq \sqrt{\pi \mu /2} \, y e^{\mu y^2/2} \operatorname{erfc}\left(\sqrt{\mu/2}\, y\right) \left[1 + O\left(\frac{1}{\mu}\right) \right] . \quad (\text{C.39})
$$

The well-known plasma dispersion function of Fried and Conte (1961 [177]) $Z(x)$ of complex argument x is related to the error function of complex argument $w(x)$ as

$$
Z(x) = \imath \sqrt{\pi} w(x) = \imath \sqrt{\pi} e^{-x^2} \operatorname{erfc}(-\imath x) , \quad (\text{C.40})
$$

provided $\Im(x) > 0$, and with $w(x)$ defined in (C.27). A representation of the plasma dispersion function valid for all arguments x is

C. Calculation of the Integral J

$$Z(x) = 2ie^{-x^2} \int_{-\infty}^{ix} dt \, e^{-t^2} \,. \tag{C.41}$$

From (C.41) one readily derives the relations

$$Z(-x) = 2i\pi^{1/2} e^{-x^2} - Z(x) \,, \tag{C.42}$$

$$Z^*(x) = -2i\pi^{1/2} e^{-(x^*)^2} + Z(x^*) \,, \tag{C.43}$$

$$Z(x^*) = -Z^*(-x) \,, \tag{C.44}$$

$$Z'(x) = -2\left[1 + xZ(x)\right] \,, \tag{C.45}$$

where the prime denotes differentiation of Z with respect to its argument.
The choice

$$x = -iy\sqrt{\mu/2} \,. \tag{C.46}$$

implies $\Im(x) < 0$ since, according to (C.10b), $\Re(y) > 0$. Equation (C.39) then becomes

$$\frac{J(\mu \gg 1, N_R)}{K_2(\mu)} \simeq i\pi^{1/2} x e^{-x^2} \operatorname{erfc}(ix) = i\pi^{1/2} x w(-x) \,. \tag{C.47}$$

After using the symmetry relation

$$w(-x) = 2e^{-x^2} - w(x)$$

we can make use of the identity (C.40) and, in terms of the plasma dispersion function, (C.47) then becomes

$$\frac{J(\mu \gg 1, N_R)}{K_2(\mu)} \simeq -xZ(x) + 2i\pi^{1/2} x e^{-x^2} \,. \tag{C.48}$$

The relativistic dispersion function is equivalent to the confluent hypergeometric function of the second kind (Lazzaro et al. 1982 [284])

$$F_q(v) = U(1, 2 - q, v) = v^{q-1} U(q, q, v) \,. \tag{C.49}$$

If one relates

$$x = i\sqrt{v} \,, \quad v = -x^2 \tag{C.50}$$

and uses the identity (Shkarofsky 1966 [482])

$$F_{1/2}(v) = \sqrt{\pi/v} \, e^v \operatorname{erfc}\left(\sqrt{v}\right) \,, \tag{C.51}$$

and the recurrence relation

$$(q - 1) F_q(v) = 1 - v F_{q-1}(v) \tag{C.52}$$

one can easily verify (e.g. Swanson 1989 [534]) the relations

C.4 The Integral J for Non-relativistic Temperatures

$$Z(x) = i\sqrt{v}\, F_{1/2}(v)\,, \tag{C.53}$$

$$xZ(x) = -v\, F_{1/2}(v) = \frac{1}{2} F_{3/2}(v) - 1\,, \tag{C.54}$$

and

$$-ix\left[1 + xZ(x)\right] = \frac{1}{2}\sqrt{v}\, F_{3/2}(v)\,. \tag{C.55}$$

where, according to the relations (C.46) and (C.50), one has

$$v = -\frac{\mu}{2}\, y^2\,. \tag{C.56}$$

References

Following the policy of the publisher some of the references are abbreviated with respect to the author list.

– **A** –

1. D.C. Abbott: In: *Proc. 6th Internat. Solar Wind Conference* ed. by V.J. Pizzo, T.E. Holzer, D.G. Sime (NCAR, Boulder, CO, 1988) Vol. **1**, p. 149
2. P.B. Abraham, K.A. Brunstein, T.L. Cline: Phys. Rev. **150**, 1088 (1966)
3. M. Abramowitz, I.A. Stegun: *Handbook of Mathematical Functions* (National Bureau of Standards, Washington, DC, 1972)
4. U. Achatz, R. Schlickeiser: Astron. Astrophys. **274**, 165 (1993)
5. U. Achatz, H. Lesch, R. Schlickeiser: Astron. Astrophys. **233**, 391 (1990)
6. U. Achatz, J. Steinacker, R. Schlickeiser: Astron. Astrophys. **250**, 266 (1991)
7. U. Achatz, W. Dröge, R. Schlickeiser, G. Wibberenz: J. Geophys. Res. **98**, 13261 (1993)
8. A. Achterberg: Astron. Astrophys. **97**, 259 (1981)
9. A. Achterberg, R.D. Blandford: Monthly Not. Royal Astron. Soc. **218**, 551 (1986)
10. F.A. Aharonian, L.O.C Drury, H.J. Völk: Astron. Astrophys. **285**, 645 (1994)
11. J.K. Alexander, L.W. Brown, T.A. Clark, R.G. Stone: Astron. Astrophys. **6**, 476 (1970)
12. C.W. Allen, C.S. Gum: Austr. J. Sci. Res. **3A**, 224 (1950)
13. H. Andernach, D. Schallwich, C.G.T. Haslam, R. Wielebinski: Astron. Astrophys. Suppl. **43**, 155 (1981)
14. J.E. Anderson: *Magnetohydrodynamic Shock Waves* (MIT Press, Cambridge, MA, 1963)
15. R. Antonucci: Ann. Rev. Astron. Astrophys. **31**, 473 (1993)
16. E.V. Appleton: J. Instr. Electr. Eng. **71**, 642 (1932)
17. G. Arfken: *Mathematical Methods for Physicists* (Academic Press, New York, 1970)
18. K. Asakimori, et al.: Astrophys. J. **502**, 278 (1998)
19. E.O. Aström: Arkiv Fysik **2**, 443 (1950)
20. W.I. Axford: In: *Particle Acceleration in Cosmic Plasmas*, ed. by G.P. Zank, T.K. Gaisser, AIP Conference Proceedings **264** (AIP, New York, 1992) p. 45
21. W.I. Axford, L.J. Gleeson: Astrophys. J. **149**, L115 (1967)
22. W.I. Axford, E. Leer, G. Skadron: *Proc. 15th Internat. Cosmic Ray Conf.* (Plovdiv, Bulgaria, 1977) **11**, 132

– B –

23. G. Backus: J. Math. Phys. **1**, 178 (1960)
24. G.D. Badhwar, S.A. Stephens: *Proc. 15th Internat. Cosmic Ray Conf.* (Plovdiv, Bulgaria, 1977) **1**, 198
25. M. Bailes, D.A. Kniffen: Astrophys. J. **391**, 659 (1992)
26. D.E. Baldwin, I.P. Bernstein, M.P.H. Weenink: In: *Advances in Plasma Physics*, ed. by A. Simon, W.B. Thompson (Wiley, New York, 1969) p. 1
27. L. Ball, D.B. Melrose, C.A. Norman: Astrophys. J. **398**, L65 (1992)
28. D.D. Barbosa: Astrophys. J. **233**, 383 (1979)
29. M.G. Baring: Monthly Not. Royal Astron. Soc. **253**, 388 (1991)
30. W.H. Barkas, M.J. Berger: *Tables of Energy Loss and Ranges of Heavy Charged Particles* (National Academy of Sciences NRC, 1964) Publ. No. 1133
31. A. Barnes: Phys. of Fluids **9**, 1483 (1967)
32. P.D. Barthel, J.E. Conway, S.T. Myers, T.J. Pearson, A.C.S. Readhead: Astrophys. J. **444**, L21 (1995)
33. R. Beck: IEEE Transactions on Plasma Science **PS-14**, 740 (1986)
34. W. Bednarek: Astron. Astrophys. **278**, 307 (1993a)
35. W. Bednarek: Astrophys. J. **402**, L29 (1993b)
36. J. Beeck, G.M. Mason, D.C. Hamilton, G. Wibberenz, H. Kunow, D. Hovestadt, B. Klecker: Astrophys. J. **322**, 1052 (1987)
37. M.C. Begelman, R.D. Blandford, M.J. Rees: Rev. Mod. Phys. **56**, 255 (1984)
38. M.C. Begelman, B. Rudak, M. Sikora: Astrophys. J. **362**, 38 (1990)
39. G. Bekefi: *Radiation Processes in Plasmas* (Wiley, New York, 1966)
40. J.W. Belcher, Leverett Davis, Jr., E.J. Smith: J. Geophys. Res. **74**, 2302 (1969)
41. A.R. Bell: Monthly Not. Royal Astron. Soc. **182**, 147 (1978)
42. V.S. Berezinsky, S.I. Grigoreva: Astron. Astrophys. **199**, 1 (1988)
43. V.S. Berezinsky, G.T. Zatsepin: Soviet Phys. Uspekhi **20**, 361 (1977)
44. E.M. Berkhuijsen: Astron. Astrophys. Suppl. **5**, 263 (1972)
45. E.M. Berkhuijsen: Astron. Astrophys. **40**, 311 (1975)
46. E.M. Berkhuijsen: Astron. Astrophys. **57**, 9 (1977)
47. I.B. Bernstein: Phys. Rev. **109**, 10 (1958)
48. H.A. Bethe, W. Heitler: Proc. R. Soc. London **A146**, 83 (1934)
49. K. Beuermann, G. Kanbach, E.M. Berkhuijsen: Astron. Astrophys. **153**, 17 (1985)
50. J. W. Bieber, W. H. Matthaeus: In: *Particle Acceleration in Cosmic Plasmas*, ed. by G.P Zank, T.K. Gaisser, AIP Conference Proceedings **264** (AIP, New York, 1992) p. 86
51. W.R. Binns: In: *Genesis and Propagation of Cosmic Rays*, ed. by M.M. Shapiro, J.P. Wefel (Reidel, Dordrecht, The Netherlands, 1988) p. 71
52. G.F. Bignami, P.A. Caraveo: Ann. Rev. Astron. Astrophys. **34**, 331 (1996)
53. G.F. Bignami, W. Hermsen: Ann. Rev. Astron. Astrophys. **21**, 67 (1983)
54. D.J. Bird, S.C. Corbato, H.Y. Dai, et al.: Astrophys. J. **441**, 144 (1995)
55. R.D. Blandford: In: *Active Galactic Nuclei*, ed. by T. Courvoisier, M. Mayor (Springer, New York, 1990) p. 161
56. R.D. Blandford: *Proc. Compton Symp.*, ed. by N. Gehrels, M. Friedlander (AIP, New York, 1993) p. 533
57. R.D. Blandford, A. Levinson: Astrophys. J. **441**, 79 (1995)
58. R.D. Blandford, J.P. Ostriker: Astrophys. J. **227**, L49 (1978)
59. R.D. Blandford, M.J. Rees: In: *Pittsburg Conference on BL Lac Objects*, ed. by A.M. Wolfe (Pittsburg Univ. Press, Pittburg PA, 1978) p. 328
60. H. Bloemen, R. Wijnands, K. Bennett, et al.: Astron. Astrophys. **281**, L5 (1994)

61. H. Bloemen, A.M. Bykov, S.V. Bozhokin, et al.: Astrophys. J. **475**, L25 (1997)
62. H. Bloemen, D. Morris, J. Knödlseder, et al.: Astrophys. J. **521**, L137 (1999)
63. J.B.G.M. Bloemen: Astron. Astrophys. **145**, 391 (1985)
64. J.B.G.M. Bloemen: Astrophys. J. **317**, L15 (1987)
65. J.B.G.M. Bloemen, P. Reich, W. Reich, R. Schlickeiser: Astron. Astrophys. **204**, 88 (1988)
66. S.D. Bloom, A.P. Marscher: *Proc. Compton Symp.*, ed. by N. Gehrels, M. Friedlander (AIP, New York, 1993) p. 578
67. G.R. Blumenthal: Phys. Rev. **D1**, 1596 (1970)
68. G.R. Blumenthal, R.J. Gould: Rev. Mod. Phys. **42**, 237 (1970)
69. M. Boezio, P. Carlson, T. Francke, et al.: Astrophys. J. **487**, 415 (1997)
70. M. Böttcher, R. Schlickeiser: Astron. Astrophys. **306**, 86 (1995)
71. T.J. Bogdan, M.A. Lee, P. Schneider: J. Geophys. Res. **96**, 161 (1991)
72. T.J. Bogdan, M.A. Lee, I. Lerche, G.M. Webb: *Proc. 19th Internat. Cosmic Ray Conf.* (San Diego, CA, 1985) **9**, 543
73. H.G. Booker: Proc. Royal Soc. **155A**, 235 (1936)
74. J.J. Brainerd, V. Petrosian: Astrophys. J. **320**, 703 (1987)
75. K. Brecher, G.R. Burbidge: Astrophys. J. **174**, 253 (1972)
76. W.N. Brouw, T.A.T. Spoelstra: Astron. Astrophys. Suppl. **26**, 129 (1976)
77. M.D. Brown, C.D. Moak: Phys. Rev. **B6**, 90 (1972)
78. D.A. Bryant, G.I. Powell, C.H. Perry: Nature **356**, 582 (1992)
79. S.V. Bulanov, V.A. Dogiel: Astrophys. Space Sci. **29**, 305 (1974)
80. S.V. Bulanov, V.A. Dogiel: *Proc. 14th Internat. Cosmic Ray Conf.* (Munich, Germany, 1975) **2**, 706
81. O. Bunemann: Phys. Rev. **115**, 503 (1959)
82. L.F. Burlaga: Space Sci. Rev. **12**, 600 (1971)
83. T.H. Burnett, S. Dake, M. Fuki, et al.: Phys. Rev. Lett. **51**, 1010 (1983)
84. W.B. Burton: In: *Galactic and Extragalactic Radio Astronomy*, ed. by G.L. Verschuur, K.I. Kellermann (Springer, Berlin, Heidelberg, 1988) p. 295
85. S.T. Butler, M.J. Buckingham: Phys. Rev. **126**, 1 (1962)

– C –

86. H. Cabannes: *Theoretical Magnetofluiddynamics* (Academic Press, New York, 1970)
87. M. Camenzind: Rev. Mod. Astron. **3**, 234 (1990)
88. A.G.W. Cameron: In: *Essays in Nuclear Astrophysics*, ed. by C.A. Barnes, D.D. Clayton, D.N. Schramm (Cambridge University Press, Cambridge, MA, 1981)
89. A. Campeanu, R. Schlickeiser: Astron. Astrophys. **263**, 413 (1992)
90. H.V. Cane: Austr. J. Phys. **31**, 561 (1978)
91. H.V. Cane, R.E. McGuire, T. von Rosenvinge: Astrophys. J. **301**, 448 (1986)
92. A. Carraminana: Astron. Astrophys. **264**, 127 (1992)
93. G. Case, D. Bhattacharya: Astron. Astrophys. Suppl. **120**, C437 (1996)
94. M. Casse, J.A. Paul: Astrophys. J. **258**, 860 (1982)
95. G. Cavallo, R.J. Gould: Nuovo Cim. **2B**, 77 (1971)
96. C.J. Cesarsky, J.A. Paul: Nuclear Sci. Applic. **1**, 191 (1981)
97. S. Chandrasekhar: *Hydrodynamic and Hydromagnetic Stability* (Oxford University Press, Oxford, 1961)
98. A. Chen, J. Dwyer, P. Kaaret: Astrophys. J. **463**, 169 (1996)
99. C.-C. Cheung: Space Sci. Rev. **13**, 3 (1972)
100. M.J. Chodorowski, A.A. Zdziarski, M. Sikora: Astrophys. J. **400**, 181 (1992)
101. D.F. Cioffi, T.W. Jones: Astron. J. **85**, 368 (1980)
102. D.H. Clark, J.L. Caswell: Monthly Not. Royal Astron. Soc. **174**, 267 (1976)

103. A.H. Compton, I.A. Getting: Phys. Rev. **47**, 817 (1935)
104. G.M. Comstock, K.C. Hsieh, J.A. Simpson: Astrophys. J. **173**, 691 (1972)
105. J. Connell, M.A. DuVernois, J.A. Simpson: *Proc. 25th Internat. Cosmic Ray Conf.* (Durban, South Africa, 1997) **3**, 397
106. R. Cowsik: Astrophys. J. **241**, 1195 (1980)
107. R. Cowsik, M.A. Lee: Astrophys. J. **228**, 297 (1979)
108. D.P. Cox, B.W. Smith: Astrophys. J. **189**, L105 (1974)
109. P. Cox, P.G. Mezger: Astron. Astrophys. Rev. **1**, 49 (1989)
110. R.M. Crutcher, T.H. Troland, I. Kazes: Astron. Astrophys. **181**, 119 (1987)
111. I. Cruz-Gonzalez, J.P. Huchra: Astron. J. **89**, 441 (1984)
112. A.C. Cummings, E.C. Stone: In: *Solar Wind VI Proc.*, ed. by V.J. Pizzo, T. Holzer, D.G. Sime (NCAR Technical Note, Boulder, CO, 1988)
113. A.C. Cummings, E.C. Stone: In: *The Heliosphere in the Local Interstellar Medium*, ed. by R. von Steiger, R. Lallement, M.A. Lee (Kluwer, Dordrecht, The Netherlands, 1996) p. 117
114. A.C. Cummings, E.C. Stone, R.E. Vogt: *Proc. 13th Internat. Cosmic Ray Conf.* (Denver, CO, 1973) **1**, 335

– D –

115. A. Dalgarno: In: *Atomic and Molecular Processes*, ed. by D.R. Bates, (Academic Press, New York, 1962)
116. A. Dalgarno, G.W. Griffing: Proc. R. Soc. London **A248**, 415 (1958)
117. J.M. Davila, J.S. Scott: Astrophys. J. **285**, 400 (1984)
118. H. Debrunner, E. Flückiger, H. Grädel, J.A. Lockwood, R.E. McGuire: J. Geophys. Res. **93**, 7206 (1988)
119. O.C. DeJager, A.K. Harding, P.F. Michelson, et al.: Astrophys. J. **457**, 253 (1996)
120. K.U. Denskat, H.J. Beinroth, F.M. Neubauer: J. Geophys. **54**, 60 (1983)
121. C.D. Dermer: Astron. Astrophys. **157**, 223 (1986)
122. C.D. Dermer: *Proc. Compton Symp.*, ed. by N. Gehrels, M. Friedlander (AIP, New York, 1993) p. 541
123. C.D. Dermer, N. Gehrels: Astrophys. J. **447**, 103 (1995)
124. C.D. Dermer, R. Schlickeiser: Astron. Astrophys. **252**, 414 (1991)
125. C.D. Dermer, R. Schlickeiser: Sience **257**, 1642 (1992)
126. C.D. Dermer, R. Schlickeiser: Astrophys. J. **416**, 484 (1993)
127. C.D. Dermer, R. Schlickeiser: Astrophys. J. Suppl. **90**, 945 (1994)
128. C.D. Dermer, R. Schlickeiser, A. Mastichiadis: Astron. Astrophys. **256**, L27 (1992)
129. J.M. Dickey: Astrophys. J. **273**, L71 (1983)
130. S.W. Digel, S.D. Hunter, R. Mukherjee: Astrophys. J. **441**, 270 (1995)
131. S.W. Digel, I.A. Grenier, A. Heithausen, S.D. Hunter, P. Thaddeus: Astrophys. J. **463**, 609 (1996)
132. V.A. Dogiel, M.J. Freyberg, G.E. Morfill, V. Schönfelder: *Proc. 4th Proc. Compton Symp.*, ed. by C.D. Dermer, M.S. Strickman, J.D. Kurfess (AIP, New York, 1997) p. 1069
133. D. Downes, R. Güsten: Mitt. Astron. Ges. **57**, 207 (1982)
134. W. Dröge: Astrophys. J. Suppl. **90**, 567 (1993)
135. W. Dröge, W. Priester: Z. Astrophys. **40**, 236 (1956)
136. W. Dröge, R. Schlickeiser: Astrophys. J. **305**, 909 (1986)
137. W. Dröge, I. Lerche, R. Schlickeiser: Astron. Astrophys. **178**, 252 (1987)
138. W. Dröge, P. Meyer, P. Evenson, D. Moses: Solar Phys. **121**, 95 (1989)
139. W. Dröge, U. Achatz, W. Wanner, R. Schlickeiser, G. Wibberenz: Astrophys. J. **407**, L95 (1993)

140. W. Dröge, K. Gibbs, J.M. Grunsfeld, P. Meyer, B.J. Newport, P. Evenson, D. Moses: Astrophys. J. Suppl. **73**, 279 (1990)
141. L.O.C. Drury: Rept. Progr. Phys. **46**, 973 (1983a)
142. L.O.C. Drury: Space Sci. Rev. **36**, 57 (1983b)
143. L.O.C. Drury, F.A. Aharonian, H.J. Völk: Astron. Astrophys. **287**, 959 (1994)
144. M. Dryer: In: *Solar Wind and the Earth*, ed. by S.-I. Akasofu, Y. Kamide (Reidel, Dordrecht, The Netherlands, 1987) p. 21
145. P. Duffy, J.G. Kirk, Y.A. Gallant, R.O. Dendy: Astron. Astrophys. **302**, L21 (1995)
146. G.A. Dulk: Ann. Rev. Astron. Astrophys. **23**, 169 (1985)
147. R. Dung, R. Schlickeiser: Astron. Astrophys. **237**, 504 (1990a)
148. R. Dung, R. Schlickeiser: Astron. Astrophys. **240**, 537 (1990b)
149. M.A. DuVernois: Astrophys. J. **481**, 241 (1997)

– E –

150. J.D. Elliot, S.L. Shapiro: Astrophys. J. **192**, L3 (1974)
151. R.S. Ellis, D.J. Axon: Astrophys. Space Sci. **54**, 425 (1978)
152. D.J. Ennis, G. Neugebauer, M. Werner: Astrophys. J. **262**, 460 (1982)
153. T. Erber: Rev. Mod. Phys. **38**, 626 (1966)
154. J. Esposito, A. Joseph D. Hunter Stanley, G. Kanbach, P. Sreekumar: Astrophys. J. **461**, 820 (1996)
155. R.D. Evans: *The Atomic Nucleus* (McGraw-Hill, New York, 1955)

– F –

156. C.Y. Fan, G. Glöckler, D. Hovestadt: Space Sci. Rev. **38**, 143 (1984)
157. G.G. Fazio, F.W. Stecker, J.P. Wright: Astrophys. J. **144**, 611 (1966)
158. E. Feenberg, H. Primakoff: Phys. Rev. **73**, 449 (1948)
159. D.E. Fegan: Nucl. Phys. B Suppl. **60B**, 37 (1998)
160. J.E. Felten, P. Morrison: Phys. Rev. Lett. **10**, 453 (1963)
161. J.E. Felten, P. Morrison: Astrophys. J. **146**, 686 (1966)
162. E. Fermi: Phys. Rev. **75**, 1169 (1949)
163. E. Fermi: Progr. Theor. Phys. **5**, 570 (1950)
164. E. Fermi: Astrophys. J. **119**, 1 (1954)
165. C.E. Fichtel, R.C. Hartman, D.A. Kniffen, et al.: Astrophys. J. **198**, 163 (1975)
166. C.E. Fichtel, D.L. Bertsch, J. Chiang, et al.: Astrophys. J. Suppl. **94**, 551 (1995)
167. H. Fichtner: Habilitationsschrift, Rheinische Friedrich-Wilhelms-Universität, Bonn (2000)
168. D. Fiebig, R. Güsten: Astron. Astrophys. **214**, 333 (1989)
169. G.B. Field: Mitt. Astron. Ges. **47**, 7 (1980)
170. L. Fisk: In: *High Energy Particles and Quanta in Astrophysics*, ed. by F.B. McDonald, C.E. Fichtel (MIT Press, Cambridge, MA, 1974) p. 170
171. L.A. Fisk: J. Geophys. Res. **81**, 4633 (1976)
172. L.A. Fisk, B. Koslovsky, R. Ramaty: Astrophys. J. **190**, L35 (1974)
173. J.W. Follin: Phys. Rev. **72**, 743 (1947)
174. S.E. Forbush: J. Geophys. Res. **59**, 525 (1954)
175. M.A. Forman, G.M. Webb: In: *Collisionless Shocks in the Heliosphere: A Tutorial Review*, ed. by P.G. Stone, B.T. Tsurutani, Geophys. Monogr. Ser. **34**, 91 (1985)
176. P.H. Fowler, R.A. Adams, V.G. Cowan, J.M. Kidd: Proc. Royal Soc. **A 301**, 39 (1967)

177. B.D. Fried, S.D. Conte: *The Plasma Dispersion Function* (Academic Press, New York, 1961)
178. B.D. Fried, R.W. Gould: Phys. of Fluids **4**, 139 (1961)
179. G.J. Fulks: J. Geophys. Res. **80**, 1701 (1975)

– G –

180. M. Garcia-Munoz, G.M. Mason, J.A. Simpson: Astrophys. J. **217**, 859 (1977)
181. M. Garcia-Munoz, P. Meyer, K.R. Pyle, J.A. Simpson, P. Evenson: J. Geophys. Res. **91**, 2858 (1986)
182. C.S. Gardner: Phys. of Fluids **6**, 839 (1963)
183. J. Geiss, P. Bochsler: In: *Proc. Intl. Conf. on Isotopic Ratios in the Solar system*, CNES (Paris, France, 1984)
184. H. Genzel, P. Joos, W. Pfeil: In: Landolt-Börnstein, Vol. **8**, ed. by H. Schopper (Springer, Berlin, Heidelberg, 1973) p. 16
185. V.L. Ginzburg, S.I. Syrovatskii: *The Origin of Cosmic Rays* (Pergamon Press, New York, 1964)
186. V.L. Ginzburg: *Elementary Processes for Cosmic Ray Astrophysics* (Gordon and Breach, New York, 1969)
187. I.M. Gioia, L. Gregorini, U. Klein: Astron. Astrophys. **116**, 164 (1982)
188. G. Glöckler: Adv. Space Res. **4**, 127 (1984)
189. W. Göbel, W. Hirth, E. Fürst: Astron. Astrophys. **93**, 43 (1981)
190. P.F. Goldsmith, W.D. Langer: Astrophys. J. **222**, 881 (1978)
191. M.L. Goldstein, W.H. Matthaeus: *Proc. 17th Internat. Cosmic Ray Conf.* (Paris, France, 1991) **3**, 204
192. M.L. Goldstein, R. Ramaty, L.A. Fisk: Phys. Rev. Lett. **24**, 1193 (1970)
193. J. Goodman: *Proc. Ultrahigh Energy Cosmic Ray Workshop* (Salt Lake City, UT, 1987) p. 253
194. M.A. Gordon: In: *Galactic and Extragalactic Radio Astronomy*, ed. by G.L. Verschuur, K.I. Kellermann (Springer, Berlin, Heidelberg, 1988) p. 37
195. M.A. Gordon, W.B. Burton: Astrophys. J. **208**, 346 (1976)
196. R.J. Gould: Phys. Rev. **185**, 72 (1969)
197. R.J. Gould: Physica **62**, 555 (1972)
198. R.J. Gould: Astrophys. J. **196**, 689 (1975)
199. R.J. Gould: Astrophys. J. **230**, 967 (1979)
200. R.J. Gould, G.R. Burbidge: Annales d' Astrophysique **28**, 171 (1965)
201. I.S. Gradshteyn, I.M. Ryzhik: *Tables of Integrals, Series and Products*, (Academic Press, New York, 1965)
202. K. Greisen: Phys. Rev. Lett. **16**, 748 (1966)
203. J.M. Grunsfeld, J. L'Heureux, P. Meyer, D. Müller, S.P. Swordy: Astrophys. J. **327**, L31 (1988)
204. J.M. Grunsfeld, J. L'Heureux, P. Meyer, D. Müller, S.P. Swordy: *Proc. 21st Internat. Cosmic Ray Conf.* (Adelaide, Australia, 1990) 3, 69
205. R. Güsten, P.G. Mezger: Vistas in Astronomy **26**, 159 (1982)

– H –

206. A. Habe, S. Ikeuchi: Progr. Theor. Phys. **64**, 1995 (1980)
207. R. Hagedorn: *Relativistic Kinematics* (Benjamin, Reading, UK, 1973)
208. D.E. Hall, P.A. Sturrock: Phys. of Fluids **10**, 2620 (1968)
209. A.K. Harding: Astrophys. J. **247**, 639 (1981)
210. J.I. Harnett, R. Beck, U.R. Buczilowski: Astron. Astrophys. **208**, 32 (1989)
211. G. Haro: In: *Stars and Stellar Systems* (University of Chicago Press, Chicago, IL, 1968) Vol. **7**, p. 141

212. R.C. Hartman, D.L. Bertsch, S.D. Bloom, et al.: Astrophys. J. Suppl. **123** 79 (1999)
213. T.W. Hartquist, M.A.J. Snijders: Nature **299**, 783 (1982)
214. D.R. Hartree: Proc. Camb. Phil. Soc. **27**, 143 (1931)
215. M. Harwit: *Cosmic Discovery* (Basics Books Inc., New York, 1981)
216. A.H. Hasegawa: *Plasma Instabilities and Nonlinear Effects* (Springer, Berlin, Heidelberg, 1975)
217. C.G.T. Haslam, C.J. Salter, H. Stoffel, W.E. Wilson: Astron. Astrophys. Suppl. **47**, 1 (1982)
218. K. Hasselmann, G. Wibberenz: Z. Geophys. **34**, 353 (1968)
219. E. Haug: Z. Naturforsch. **30a**, 1099 (1975)
220. E. Haug: Z. Naturforsch. **36a**, 413 (1981)
221. S. Hayakawa: *Cosmic Ray Physics* (Wiley, New York, 1969)
222. J.N. Hayes: Phys. of Fluids **4**, 1387 (1961)
223. J.N. Hayes: Nuovo Cim. **30**, 1048 (1963)
224. U. Heinbach, M. Simon: Astrophys. J. **441**, 209 (1995)
225. G. Henri, G. Pelletier, J. Roland: Astrophys. J. **404**, L41 (1993)
226. W. Heitler: *The Quantum Theory of Radiation* (Oxford University Press, Oxford, 1954)
227. A.M. Hillas: Ann. Rev. Astron. Astrophys. **22**, 425 (1984)
228. M. Hof, W. Menn, C. Pfeifer, et al.: Astrophys. J. **467**, L33 (1996)
229. F. de Hoffmann, E. Teller: Phys. Rev. **80**, 692 (1950)
230. G.D. Holman, J.A. Ionson, J.S. Scott: Astrophys. J. **228**, 576 (1979)
231. V.D. Hopper: *Cosmic Radiation and High Energy Interactions* (Logos Press, London, 1964)
232. E. Hummel, H. Lesch, R. Schlickeiser, R. Wielebinski: Astron. Astrophys. **197**, L29 (1988)
233. S. Hunter, D.L. Bertsch, J.R. Catelli, et al.: Astrophys. J. **481**, 205 (1997)

– I –

234. M. Israel: In: *Composition and Origin of Cosmic Rays*, ed. by M.M. Shapiro (Reidel, Dordrecht, The Netherlands, 1983) p. 47

– J –

235. J.D. Jackson: J. Nucl. Energy **C1**, 171 (1960)
236. U. Jaekel, R. Schlickeiser: Ann. Geophysicae **10**, 541 (1992)
237. U. Jaekel, W. Wanner, R. Schlickeiser, G. Wibberenz: Astron. Astrophys. **291**, L35 (1994)
238. C.E. Jäkel, K.O. Thielheim, H. Wiese: Astrophys. Space Sci. **51**, 329 (1977)
239. J.M. Jauch, F. Rohrlich: *The Theory of Photons and Electrons* (Addison-Wesley, Cambridge, MA, 1955)
240. S. Jarp, K.J. Mork: Phys. Rev. **D3**, 159 (1973)
241. A. Jeffrey, T Taniuti: *Non Linear Wave Propagation with Applications to Physics and Magnetohydrodynamics* (Academic Press, New York, 1964)
242. E.B. Jenkins: Comm. Astrophys. **7**, 121 (1978)
243. I.-G. Jiang, K.-W. Chan, C.-M. Ko: Astron. Astrophys. **307**, 903 (1996)
244. J.R. Jokipii: Astrophys. J. **146**, 480 (1966)
245. J.R. Jokipii: Space Sci. Rev. **36**, 27 (1983)
246. J.R. Jokipii, P.J. Coleman: J. Geophys. Res. **73**, 5495 (1968)
247. J.R. Jokipii, P. Meyer: Phys. Rev. Lett. **20**, 752 (1968)
248. J.R. Jokipii, E.N. Parker: Astrophys. J. **155**, 799 (1969)
249. F.C. Jones: Phys. Rev. **137**, B1306 (1965)

250. F.C. Jones: Phys. Rev. **167**, 1159 (1968)
251. T.W. Jones, S.L. O'Dell, W.A. Stein: Astrophys. J. **188**, 353 (1974)
252. T.W. Jones: Astrophys. J. **413**, 619 (1993)
253. N. Junkes, E. Fürst, W. Reich: Astron. Astrophys. Suppl. **69**, 451 (1987)

– K –

254. P. Kaaret, J. Cottam: Astrophys. J. **462**, L1 (1996)
255. H. Kang, T.W. Jones: Monthly Not. Royal Astron. Soc. **249**, 439 (1991)
256. N.S. Kardashev: Sov. Astron. J. **6**, 317 (1962)
257. D. Kazanas, D.C. Ellison: Astrophys. J. **304**, 178 (1986)
258. C.F. Kennel, F. Engelmann: Phys. of Fluids **9**, 2377 (1966)
259. F.J. Kerr, D. Lynden-Bell: Monthly Not. Royal Astron. Soc. **221**, 1023 (1986)
260. R. Kippenhahn, C. Möllenhoff: *Elementare Plasmaphysik* (BI-Verlag, Mannheim, 1975)
261. J.G. Kirk, A. Mastichiadis: Astron. Astrophys. **213**, 75 (1989)
262. J G. Kirk, R. Schlickeiser, P. Schneider: Astrophys. J. **328**, 269 (1988)
263. T. Kifune, K. Nishijima, T. Hara, et al.: Astrophys. J. **301**, 230 (1986)
264. U. Klein, D.T. Emerson: Astron. Astrophys. **94**, 29 (1981)
265. C.-M. Ko: Astron. Astrophys. **259**, 377 (1992)
266. G.E. Kocharov: Nucl. Phys. B Suppl. **22B**, 153 (1991)
267. G.E. Kocharov: In: *Particle Acceleration in Cosmic Plasmas*, ed. by G.P. Zank, T.K. Gaisser, AIP Conference Proceedings **264** (AIP, New York, 1992) p. 205
268. A.N. Konstantinov, G.E. Kocharov, V.A. Lerchenko: PAZh **16**, 799 (1990) (in Russian)
269. J. Kota, J.R. Jokipii: Astrophys. J. **265**, 573 (1983)
270. K. Koyama, R. Petre, E.V. Gotthelf, et al.: Nature **378**, 255 (1995)
271. B. Kozlovsky, R. Ramaty, R. Lingenfelter: Astrophys. J. **484**, 286 (1997)
272. N.A. Krall, A.W. Trivelpiece: *Principles of Plasma Physics* (McGraw-Hill, New York, 1973)
273. J.D. Kraus: *Radioastronomy* (McGraw-Hill, New York, 1967)
274. W.L. Kraushaar, G.W. Clark, G.P. Garmire, et al.: Astrophys. J. **177**, 341 (1972)
275. W. Krülls: Astron. Astrophys. **260**, 49 (1992)
276. G.F. Krymsky: Dok. Akad. Nauk. SSSR **234**, 1306 (1977)
277. R.M. Kulsrud, C.J. Cesarsky: Astrophys. Letters **8**, 189 (1971)
278. R.M. Kulsrud, W.P. Pearce: Astrophys. J. **156**, 445 (1969)
279. J.D. Kurfess: In: *Active Galactic Nuclei across the Electromagnetic Spectrum*, ed. by T. Courvoisier, A. Blecha (Kluwer, Dordrecht, The Netherlands, 1994) p. 39

– L –

280. L. Landau: J. Phys. USSR **10**, 25 (1946)
281. L. Landau, E. Lifshitz: *Electrodynamics of Continuous Media* (Pergamon Press, Oxford, 1960)
282. T.L. Landecker, R. Wielebinski: Austr. J. Phys. Suppl. **16**, 1 (1970)
283. K.R. Lang: *Astrophysical Formulae* (Springer, Berlin, Heidelberg, 1974)
284. E. Lazzaro, G. Ramponi, G. Giruzzi: Phys. of Fluids **25**, 1220 (1982)
285. M.A. Lee: J. Geophys. Res. **87**, 5063 (1982)
286. M.A. Lee, H.J. Völk: Astrophys. Space Sci. **24**, 31 (1973)
287. M.A. Lee, H.J. Völk: Astrophys. J. **198**, 485 (1975)
288. F. Le Guet, M. Stanton: Astron. Astrophys. **35**, 165 (1974)
289. C.E. Leith: Phys. of Fluids **10**, 1409 (1967)

290. I. Lerche: Phys. of Fluids **9**, 1073 (1966)
291. I. Lerche: Astrophys. J. **147**, 689 (1967)
292. I. Lerche: Astrophys. J. **190**, 165 (1974)
293. I. Lerche: Astron. Astrophys. **85**, 141 (1980)
294. I. Lerche, R. Schlickeiser: Astrophys. J. Suppl. **47**, 33 (1981)
295. I. Lerche, R. Schlickeiser: Astron. Astrophys. **107**, 148 (1982a)
296. I. Lerche, R. Schlickeiser: Monthly Not. Royal Astron. Soc. **201**, 1041 (1982b)
297. I. Lerche, R. Schlickeiser: Astron. Astrophys. **151**, 408 (1985)
298. I. Lerche, R. Schlickeiser: Astrophys. Space Sci. **145**, 319 (1988)
299. J.A. LeRoux, M.S. Potgieter: Astrophys. J. **361**, 275 (1990)
300. H. Lesch, R. Schlickeiser, A. Crusius: Astron. Astrophys. **200**, L9 (1988)
301. H. Lesch, A. Crusius, R. Schlickeiser: Astron. Astrophys. **209**, 427 (1989)
302. J.R. Letaw, R. Silberberg, C.H. Tsao: Astrophys. J. Suppl. **51**, 271 (1983)
303. G.G. Lichti, T. Balonek, T.J.-L. Courvoisier, et al.: Astron. Astrophys. **298**, 711 (1995)
304. R.P. Lin, R.A. Mewaldt, M.A.I. Van Hollebeke: Astrophys. J. **253**, 949 (1982)
305. J. Linsley: *Proc. 18th Internat. Cosmic Ray Conf.* (Bangalore, India, 1983) **12**, 135
306. J. Lloyd-Evans, R.N. Coy, A. Lambert, et al.: Nature **305**, 784 (1983)
307. B. Lovell: Phil. Trans. Royal Soc. London **A 277**, 489 (1974)
308. N. Lund: In: *Cosmic Rays in Contemporary Astrophysics*, ed. by M.M. Shapiro (Reidel, Dordrecht, The Netherlands, 1986) p. 1
309. A.G. Lyne, F.G. Smith: Monthly Not. Royal Astron. Soc. **237**, 533 (1989)
310. W.J. Lyuten: Astrophys. J. **109**, 532 (1949)

– **M** –

311. W.M. MacDonald, M.N. Rosenbluth, W. Chuck: Phys. Rev. **107**, 350 (1957)
312. J.M. MacLeod, J.P. Vallee, N.W. Broten: Astron. Astrophys. Suppl. **74**, 63 (1988)
313. W. Magnus, F. Oberhettinger, R.P. Soni: *Formulas and Theorems for the Special Functions Of Mathematical Physics* (Springer, Berlin, Heidelberg, 1966)
314. P. Mandrou, A. Bui-Van, G. Vedrenne, M. Niel: Astrophys. J. **237**, 424 (1980)
315. K. Mannheim: Space Sci. Rev. **75**, 331 (1996)
316. K. Mannheim, P.L. Biermann: Astron. Astrophys. **221**, 211 (1989)
317. K. Mannheim, R. Schlickeiser: Astron. Astrophys. **286**, 983 (1994)
318. S.P. Maran, A.G.W. Cameron: Phys. Today **21**, No. 8, 41 (1968)
319. L. Maraschi, G. Ghisellini, A. Celotti: Astrophys. J. **397**, L5 (1993)
320. S.H. Margolis, D.N. Schramm, R. Silberberg: Astrophys. J. **221**, 990 (1978)
321. E. Marsch: Rev. Mod. Astron. **4**, 145 (1991)
322. A.P. Marscher, R.L. Brown: Astrophys. J. **221**, 588 (1978)
323. A.P. Marscher, W.T. Vestrand, J.S. Scott: Astrophys. J. **241**, 1166 (1980)
324. A. Mastichiadis: Monthly Not. Royal Astron. Soc. **253**, 235 (1991)
325. A. Mastichiadis: Space Sci. Rev. **75**, 317 (1996)
326. A. Mastichiadis, A.P. Marscher, K. Brecher: Astrophys. J. **300**, 178 (1986)
327. A. Mastichiadis, K. Brecher, A.P. Marscher: Astrophys. J. **314**, 88 (1987)
328. D.S. Mathewson, D.K. Milne: Austr. J. Phys. **18**, 635 (1965)
329. D.S. Mathewson, V.L. Ford, M.A. Dopita, I.R. Tuohy, K.S. Long, D.J. Helfand: Astrophys. J. Suppl. **51**, 345 (1983)
330. J.S. Mathis, W. Rumpl, K.H. Nordsieck: Astrophys. J. **217**, 425 (1977)
331. J.S. Mathis, P.G. Mezger, N. Panagia: Astron. Astrophys. **128**, 212 (1984)
332. T. Matsumoto, S. Hayakawa, H. Matsuo, et al.: Astrophys. J. **329**, 567 (1988)
333. L.C. Maximon: J. Res. NBS **72B**, 79 (1966)

334. D. McCammon, D.N. Burrows, W.T. Sanders, W.L. Kraushaar: Astrophys. J. **269**, 107 (1983)
335. R.E. McGuire, T.T. von Rosenvinge: Adv. Space Res. **4**, 117 (1984)
336. J.K. McKenzie, K.O. Westphal: Planet. Space Sci. **17**, 1029 (1969)
337. K. Meisenheimer, H.-J. Röser, P.R. Hiltner, et al.: Astron. Astrophys. **219**, 63 (1989)
338. D.B. Melrose: Astrophys. J. **154**, 803 (1968)
339. D.B. Melrose, D.G. Wentzel: Astrophys. J. **161**, 457 (1970)
340. M. Meneguzzi, H. Reeves: Astron. Astrophys. **40**, 91 (1975)
341. W. Menn: private communication (1999)
342. R.A. Mewaldt: Rev. Geophys. Space Phys. **21**, 295 (1983)
343. R.A. Mewaldt: In: *Cosmic Abundances of Matter*, ed. by C.J. Waddington (AIP, New York, 1989) p. 124.
344. R.A. Mewaldt, E.C. Stone: Astrophys. J. **337**, 959 (1989)
345. P. Meyer: Nature **272**, 675 (1978)
346. P. Meyer: In: *Origin of Cosmic Rays, IAU Symp. 94*, ed. by G. Setti, G. Spada, A.W. Wolfendale (Reidel, Dordrecht, The Netherlands, 1980) p. 7
347. J.-P. Meyer: *Proc. 17th Internat. Cosmic Ray Conf.* (Paris, France, 1981) **3**, 145
348. J.-P. Meyer: Astrophys. J. Suppl. **57**, 173 (1985)
349. P. Meyer, E.N. Parker, J.A. Simpson: Phys. Rev. **104**, 768 (1956)
350. P.G. Mezger, A.P. Henderson: Astrophys. J. **147**, 471 (1967)
351. G. Michalek, M. Ostrowski: *Nonlinear Processes in Geophysics* **3**, 66 (1996)
352. A.B. Mikhailovski: Plasma Phys. **22**, 133 (1980)
353. J.A. Miller: Astrophys. J. **376**, 342 (1991)
354. J.A. Miller, D.A. Roberts: Astrophys. J. **452**, 912 (1995)
355. J.A. Miller, T.N. LaRosa, R.L. Moore: Astrophys. J. **461**, 445 (1996)
356. B.Y. Mills: In: *IAU Symp. No. 9 on Radio Astronomy*, ed. by R.N. Bracewell (Stanford University Press, Stanford, CA, 1959) p. 431
357. B.Y. Mills, A.J. Turtle, A.C. Little, J.M. Durdin: Austr. J. Phys. **37**, 321 (1984)
358. D.K. Milne: Austr. J. Phys. **23**, 425 (1970)
359. D.K. Milne: Austr. J. Phys. **32**, 83 (1979)
360. J. Milogradev-Turin, F.G. Smith: Monthly Not. Royal Astron. Soc. **161**, 269 (1973)
361. A.H. Minter, S.R. Spangler: Astrophys. J. **485**, 182 (1997)
362. E. Möbius, M. Scholer, D. Hovestadt, B. Klecker, G. Glöckler: Astrophys. J. **259**, 397 (1982)
363. D.C. Montgomery, D.A. Tidman: *Plasma Kinetic Theory* (McGraw-Hill, New York, 1964)
364. C. von Montigny, D.L. Bertsch, J. Chiang, et al.: Astrophys. J. **440**, 525 (1995)
365. M. Mori: Astrophys. J. **478**, 225 (1997)
366. K.J. Mork: Phys. Rev. **160**, 1065 (1967)
367. P. Morrison: Nuovo Cim. **7**, 858 (1958)
368. D. Moses, W. Dröge, P. Meyer, P. Evenson: Astrophys. J. **346**, 523 (1989)
369. D. Müller: Adv. Space Res. **9** (12) 31 (1989)
370. D. Müller, K.-K. Tang: Astrophys. J. **312**, 183 (1987)
371. H. Muraishi, T. Tanimori, S. Yanagita, et al.: Astron. Astrophys. **354**, L57M (2000)
372. R.J. Murphy, C.D. Dermer, R. Ramaty: Astrophys. J. Suppl. **63**, 721 (1987)

– N –

373. B.S. Newberger: J. Math. Phys. **23**, 1278 (1982)
374. C.E. Newman: J. Math. Phys. **14**, 502 (1973)
375. J. Nishimura, M. Fujii, T. Kobayashi, et al.: *Proc. 21st Internat. Cosmic Ray Conf.* (Adelaide, Australia, 1990) **3**, 213
376. L.C. Northcliffe: Ann. Rev. Nucl. Sci. **13**, 67 (1963)

– O –

377. J.F. Ormes, A.A. Moiseev, T. Saeki, et al.: Astrophys. J. **482**, L187 (1997)
378. L. Oster: Rev. Mod. Phys. **33**, 525 (1961)
379. J.P. Ostriker, C. McKee: Astrophys. J. **218**, 148 (1977)
380. M. Ostrowski, R. Schlickeiser: Astron. Astrophys. **268**, 812 (1993)

– P –

381. A.G. Pacholczyk: *Radio Astrophysics* (Freeman, San Francisco, CA, 1970)
382. A.G. Pacholczyk: *Radio Galaxies* (Pergamon Press, Oxford, 1977)
383. T.A.D. Paglione, A.P. Marscher, J.M. Jackson, D.L. Bertsch: Astrophys. J. **460**, 295 (1996)
384. R. Pallavicini, S. Serio, G. Vaiana: Astrophys. J. **216**, 108 (1977)
385. I.D. Palmer: Rev. Geophys. Space Phys. **20**, 355 (1982)
386. B.T. Park, V. Petrosian: Astrophys. J. **446**, 699 (1995)
387. E.N. Parker: Astrophys. J. **128**, 664 (1958)
388. E.N. Parker: *Interplanetary Dynamical Processes* (Interscience Publ., New York, 1963)
389. E.N. Parker: Planet. Space Sci. **13**, 9 (1965)
390. E.N. Parker, D.A. Tidman: Phys. Rev. **111**, 1206 (1958a)
391. E.N. Parker, D.A. Tidman: Phys. Rev. **112**, 1048 (1958b)
392. J. Paul, K. Bennett, G.F. Bignami, et al.: Astron. Astrophys. **63**, L31 (1978)
393. G. Pelletier, H. Sol: Monthly Not. Royal Astron. Soc. **254**, 635 (1992)
394. M. Perl: *High Energy Hadron Physics* (Wiley, New York, 1974)
395. G.C. Perola, L. Scarsi, G. Sironi: Nuovo Cim. **52B**, 455 (1967)
396. M.E. Pesses, J.R. Jokipii, D. Eichler: Astrophys. J. **246**, L85 (1981)
397. J. Pfleiderer, W. Priester, W. Köhnlein: *Lectures on Space Physics* **2**, ed. by A. Bruzek, H. Pilkuhn (Bertelsmann, 1973) p. 127
398. S. Philipps, S. Kearsey, J.L. Osborne, C.G.T. Haslam, H. Stoffel: Astron. Astrophys. **98**, 286 (1981a)
399. S. Philipps, S. Kearsey, J.L. Osborne, C.G.T. Haslam, H. Stoffel: Astron. Astrophys. **103**, 405 (1981b)
400. S.B. Pikelner, V.N. Tsytovich: Sov. Astron. J. **19**, 450 (1976)
401. M. Pohl: Astron. Astrophys. **287**, 453 (1994)
402. M. Pohl, J.A. Esposito: Astrophys. J. **507**, 327 (1998)
403. M. Pohl, R. Schlickeiser: Astron. Astrophys. **234**, 147 (1990)
404. M. Pohl, R. Schlickeiser: Astron. Astrophys. **252**, 565 (1991)
405. M. Pohl, G. Kanbach, S.D. Hunter, B.B. Jones: Astrophys. J. **491**, 159 (1997)
406. M. Pohl, W. Reich, T.P. Krichbaum, et al.: Astron. Astrophys. **303**, 383 (1995)
407. R.M. Price: Astron. Astrophys. **33**, 33 (1974)
408. T.A. Prince: Astrophys. J. **227**, 676 (1979)
409. R. Protheroe, A. Wolfendale: Astron. Astrophys. **92**, 175 (1980)
410. M. Punch, C.W. Akerlof, M.F. Cawley, et al.: Nature **358**, 477 (1992)
411. J.L. Puget, F.W. Stecker, J.H. Bredekamp: Astrophys. J. **205**, 638 (1976)

412. W.R. Purcell, L. Bouchet, W.N. Johnson, et al.: Astron. Astrophys. Suppl. **120**, C389 (1996)
413. K.R. Pyle, J.A. Simpson, A. Barnes, J.D. Mihalov: Astrophys. J. **282**, L107 (1984)

– R –

414. B.R. Ragot, J.G. Kirk: Astron. Astrophys. **327**, 432 (1997)
415. R. Ramaty: In: *High Energy Particles and Quanta in Astrophysics*, ed. by F.B. McDonald, C.E. Fichtel (MIT Press, Cambridge, MA, 1974) p. 122
416. R. Ramaty: In: *Particle Acceleration Mechanisms in Astrophysics*, ed. by J. Arons, C. Max, C. McKee (AIP, New York, 1979) p. 135
417. R. Ramaty, R.J. Murphy: Space Sci. Rev. **45**, 213 (1987)
418. R. Ramaty, N.J.Westergaard: Astrophys. Space Sci. **45**, 143 (1976)
419. R. Ramaty, B. Kozlovsky, R.E. Lingenfelter: Astrophys. J. Suppl. **40**, 487 (1979)
420. R. Ramaty, B. Kozlovsky, R. Lingenfelter: Astrophys. J. **456**, 525 (1996)
421. R. Ramaty, B. Kozlovsky, V. Tatischeff: *Proc. 4th Proc. Compton Symp.*, ed. by C.D. Dermer, M.S. Strickman, J.D. Kurfess (AIP, New York, 1997) p. 1049
422. D. Reames, T.T. von Rosenvinge, R.P. Lin: Astrophys. J. **292**, 716 (1985)
423. R.C. Reedy, J.R. Arnold, D. Lal: Sience **219**, 127 (1983)
424. W. Reich: Astron. Astrophys. Suppl. **48**, 219 (1982)
425. W. Reich: In: *The Center of the Galaxy*, IAU Symp. **136**, ed. by M. Morris (Reidel, Dordrecht, The Netherlands, 1989) p. 265
426. P. Reich, W. Reich: Astron. Astrophys. Suppl. **63**, 205 (1986)
427. P. Reich, W. Reich: Astron. Astrophys. **196**, 211 (1988)
428. W. Reich, Y. Sofue, R. Wielebinski, J.H. Seiradakis: Astron. Astrophys. **191**, 303 (1988)
429. W. Reich, H. Steppe, R. Schlickeiser, et al.: Astron. Astrophys. **273**, 65 (1993)
430. Y. Rephaeli: Comm. Astrophys. **12**, 265 (1988)
431. S. Reynolds: Astrophys. J. **459**, L13 (1996)
432. B.J. Rickett: Ann. Rev. Astron. Astrophys. **28**, 561 (1990)
433. E. Rieger, G.H. Share, D.J. Forrest, G. Kanbach, C. Reppin, E.L. Chupp: Nature **312**, 623 (1984)
434. G.H. Rieke, M.J. Lebofsky, W.Z. Wisniewski: Astrophys. J. **263**, 73 (1982)
435. E.I. Robson, W.K. Gear, P.E. Clegg, et al.: Nature **305**, 194 (1983)
436. E.C. Roelof: Can. J. Phys. **46**, 990 (1968)
437. H.-J. Röser: In: *Hot Spots in Extragalactic Radio Sources*, ed. by K. Meisenheimer, H.-J.Röser, Lecture Notes in Physics **327** (Springer, Berlin, Heidelberg, 1989) p. 91
438. H.-J. Röser, K. Meisenheimer: Astron. Astrophys. **154**, 15 (1986)
439. B.W. Roos: *Analytic Functions and Distributions in Physics and Engineering* (Wiley, New York, 1969)
440. M.N. Rosenbluth, W.M. MacDonald, D.L. Judd: Phys. Rev. **107**, 1 (1957)
441. G.B. Rybicki, A.P. Lightman: *Radioative Processes in Astrophysics* (Wiley, New York, 1979)

– S –

442. A.W. Saenz: J. Math. Phys. **6**, 859 (1963)
443. C.J. Salter, R.L. Brown: In: *Galactic and Extragalactic Radio Astronomy*, ed. by G.L.Verschuur, K.I. Kellermann (Springer, Berlin, Heidelberg, 1988) p. 1
444. M. Samorski, W. Stamm: Astrophys. J. **268**, L17 (1983)
445. M. Samorski, W. Stamm: Astrophys. J. **277**, 897 (1984)

446. B.D. Savage, K.S. DeBoer: Astrophys. J. **243**, 460 (1981)
447. M.P. Savedoff: Nuovo Cim. **13**, 12 (1959)
448. J.H. Scanlon, S.N. Milford: Astrophys. J. **141**, 718 (1965)
449. O.A. Schaeffer: *Proc. 14th Internat. Cosmic Ray Conf.* (München, Germany, 1975) **11**, 3508
450. P.A.G. Scheuer, A.C.S. Readhead: Nature **277**, 182 (1979)
451. R. Schlickeiser: Astrophys. J. **233**, 294 (1979)
452. R. Schlickeiser: Fortschr. der Physik **29**, 95 (1981)
453. R. Schlickeiser: Astron. Astrophys. **107**, 378 (1982a)
454. R. Schlickeiser: In: *Progress in Cosmology*, ed. by A.W. Wolfendale (Reidel, Dordrecht, The Netherlands, 1982b) p. 215
455. R. Schlickeiser: Astron. Astrophys. **106**, L5 (1982c)
456. R. Schlickeiser: Astron. Astrophys. **136**, 227 (1984)
457. R. Schlickeiser: Astron. Astrophys. **213**, L23 (1989a)
458. R. Schlickeiser: Astrophys. J. **336**, 243 (1989b)
459. R. Schlickeiser: Astrophys. J. **336**, 264 (1989c)
460. R. Schlickeiser: In: *Particle Acceleration in Cosmic Plasmas*, ed. by G.P Zank, T.K. Gaisser, AIP Conf. Proc. **264** (AIP, New York, 1992) p. 92
461. R. Schlickeiser: In: *Statistical Description of Transport in Plasma, Astro-, and Nuclear Physics*, ed. by J. Misguich, G. Pelletier, P. Schuck (Nova Science Publ., Commack, 1993) p. 175
462. R. Schlickeiser, U. Achatz: J. Plasma Phys. **49**, 63 (1993a)
463. R. Schlickeiser, U. Achatz: J. Plasma Phys. **50**, 85 (1993b)
464. R. Schlickeiser, A. Crusius: Astrophys. J. **328**, 578 (1988)
465. R. Schlickeiser, M. Kneller: J. Plasma Phys. **57**, 709 (1997)
466. R. Schlickeiser, I. Lerche: Astron. Astrophys. **151**, 151 (1985)
467. R. Schlickeiser, H. Lesch: Astrophys. J. **329**, 149 (1988)
468. R. Schlickeiser, H. Mause: Phys. of Plasmas **2**, 4025 (1995)
469. R. Schlickeiser, J.A. Miller: Astrophys. J. **492**, 352 (1998)
470. R. Schlickeiser, J. Steinacker: Solar Phys. **122**, 29 (1989)
471. R. Schlickeiser, K.O. Thielheim: Astrophys. Space Sci. **47**, 415 (1977)
472. R. Schlickeiser, A. Campeanu, I. Lerche: Astron. Astrophys. **276**, 614 (1993)
473. R. Schlickeiser, R. Dung, U. Jaekel: Astron. Astrophys. **242**, L5 (1991)
474. R. Schlickeiser, H. Fichtner, M. Kneller: J. Geophys. Res. **102**, 4725 (1997b)
475. R. Schlickeiser, A. Sievers, H. Thiemann: Astron. Astrophys. **182**, 21 (1987)
476. R. Schlickeiser, M. Pohl, R. Ramaty, J.G. Skibo: *Proc. 4th Proc. Compton Symp.*, ed. by C.D. Dermer, M.S. Strickman, J.D. Kurfess (AIP, New York, 1997a) p. 449
477. J.M. Schmidt, W. Dröge: *Proc. 25th Internat. Cosmic Ray Conf.* (Durban, South Africa, 1997) **1**, 221
478. P. Schneider: Astron. Astrophys. **269**, L13 (1993)
479. M. Scholer, J.W. Belcher: Solar Phys. **16**, 472 (1971)
480. E. Sedlmayr: Mitt. Astron. Ges. **63**, 75 (1985) (in German)
481. M.M. Shapiro, R. Silberberg: Ann. Rev. Nucl. Phys. **20**, 323 (1970)
482. I.P. Shkarofsky: Phys. of Fluids **9**, 561 (1966)
483. M. Sikora, M.C. Begelman, M. Rees: Astrophys. J. **421**, 153 (1994)
484. M. Sikora, J.G. Kirk, M.C. Begelman, P. Schneider: Astrophys. J. **320**, L81 (1987)
485. R. Silberberg, M.M. Shapiro: *Proc. 15th Internat. Cosmic Ray Conf.* (Plovdiv, Bulgaria, 1977) **6**, 237
486. R. Silberberg, C.H. Tsao: Astrophys. J. Suppl. **25**, 315 (1973)
487. R. Silberberg, C.H. Tsao: Phys. Rep. **191**, 351 (1990)

488. M. Simard-Normandin, P.P. Kronberg, S. Button: Astrophys. J. Suppl. **45**, 97 (1981)
489. M. Simon, A. Molnar, S. Roesler: Astrophys. J. **499**, 250 (1998)
490. J.H. Simonetti: Astrophys. J. **386**, 170 (1992)
491. J.A. Simpson: In: *Composition and Origin of Cosmic Rays*, ed. by M.M. Shapiro (Reidel, Dordrecht, The Netherlands, 1983) p. 1
492. J.A. Simpson: Adv. Space Res. **9**, 5 (1989)
493. J.A. Simpson, M. Garcia-Munoz: Space Sci. Rev. **46**, 205 (1988)
494. D.V. Sivukhin: In: *Rev. Plasma Physics*, ed. by M. Leontovich (Consultants Bureau, New York, 1965) Vol. **4**, p. 93
495. J.G. Skibo, R. Ramaty: Astron. Astrophys. Suppl. **97**, 145 (1993)
496. J.G. Skibo, R. Ramaty, W.R. Purcell: Astron. Astrophys. Suppl. **120**, C403 (1996)
497. J.G. Skibo, C.D. Derner, R. Schlickeiser: Astrophys. J. **483**, 56 (1997)
498. J. Skilling: Monthly Not. Royal Astron. Soc. **172**, 557 (1975)
499. P. Slane, S.M. Wagh: Astrophys. J. **364**, 198 (1990)
500. C.W. Smith, J.W. Bieber, W.H. Matthaeus: Astrophys. J. **363**, 283 (1990)
501. T.J. Sodroski, N. Odegard, E. Dwek, et al.: Astrophys. J. **452**, 262 (1995)
502. Y. Sofue, M. Fujimoto, R. Wielebinski: Ann. Rev. Astron. Astrophys. **24**, 459 (1986)
503. P. Sokolsky: *Introduction to Ultrahigh Energy Cosmic Ray Physics* (Addison-Wesley, Redwood City, 1979)
504. S.R. Spangler: Astrophys. J. **376**, 540 (1991)
505. L. Spitzer Jr.: Astrophys. J. **107**, 6 (1948)
506. L. Spitzer Jr., E.H. Scott: Astrophys. J. **158**, 161 (1969)
507. T.A.T. Spoelstra: Astron. Astrophys. Suppl. **5**, 205 (1972)
508. T.A.T. Spoelstra: Astron. Astrophys. **135**, 238 (1984)
509. T.A.T. Spoelstra, W. Hermsen: Astron. Astrophys. **135**, 135 (1985)
510. P. Sreekumar, et al.: Astrophys. J. **400**, L67 (1992)
511. P. Sreekumar, et al.: Phys. Rev. Lett. **70**, 127 (1993)
512. P. Sreekumar, et al.: Astrophys. J. **426**, 105 (1994)
513. I. Stakgold: *Boundary Value Problems of Mathematical Physics*, Vol. **I** (Wiley, New York, 1967)
514. F.W. Stecker: Phys. Rev. Lett. **21**, 1016 (1968)
515. F.W. Stecker: Phys. Rev. **180**, 1264 (1969)
516. F.W. Stecker: Astrophys. Space Sci. **6**, 377 (1970)
517. F.W. Stecker: Astrophys. J. **185**, 499 (1973)
518. F.W. Stecker: Phys. Rev. Lett. **35**, 188 (1975)
519. F.W. Stecker: Astrophys. J. **212**, 60 (1977)
520. F.W. Stecker: Astrophys. J. **228**, 919 (1979)
521. J. Steinacker, R. Schlickeiser: Astron. Astrophys. **224**, 259 (1989)
522. J. Steinacker, W. Dröge, R. Schlickeiser: Solar Phys. **115**, 313 (1988)
523. S.A. Stephens, G.D. Badhwar: Astrophys. Space Sci. **76**, 213 (1981)
524. R.M. Sternheimer: Phys. Rev. **88**, 851 (1952)
525. T.H. Stix: *The Theory of Plasma Waves* (McGraw-Hill, New York, 1962)
526. M. Stix: *The Sun* (Springer, Berlin, Heidelberg, 1989)
527. T.H. Stix: *Waves in Plasmas* (AIP, New York, 1992)
528. A.W. Strong, J.R. Mattox: Astron. Astrophys. **308**, L21 (1996)
529. A.W. Strong, K. Bennett, H. Bloemen, et al.: Astron. Astrophys. Suppl. **120**, C381 (1996)
530. S.J. Sturner, C.D. Dermer: Astron. Astrophys. **293**, L17 (1995)
531. S.J. Sturner, C.D. Dermer: Astrophys. J. **461**, 872 (1996)
532. B.N. Swanenburg, K. Bennett, G.F. Bignami, et al.: Nature **275**, 298 (1978)

533. B.N. Swanenburg, K. Bennett, G.F. Bignami, et al.: Astrophys. J. **243**, L69 (1981)
534. D.G. Swanson: *Plasma Waves* (Academic Press, New York, 1989)
535. S.P. Swordy, D. Müller, P. Meyer, J.L'Heureux, J.M. Grunsfeld: Astrophys. J. **349**, 625 (1990)
536. S.I. Syrovatskii: Sov. Astron. J. **3**, 22 (1959)

– T –

537. E. Tademaru: Astrophys. J. **158**, 959 (1969)
538. K. Takayanagi: Publ. Astr. Soc. Japan **25**, 327 (1973)
539. M. Takeda, N. Hayashida, K. Honda, et al.: Phys. Rev. Lett. **81**, 1163 (1998)
540. K.-K. Tang: Astrophys. J. **278**, 881 (1984)
541. T. Tanimori, Y. Hayami, S. Kamei, et al.: Astrophys. J. **497**, L 25T (1998)
542. T. Terasawa, M. Scholer: Sience **244**, 1050 (1989)
543. W.B. Thompson: *An Introduction to Plasma Physics* (Pergamon Press, Oxford, 1962)
544. I.N. Toptygin: Space Sci. Rev. **26**, 157 (1980)

– U –

545. I.H. Urch: Astrophys. Space Sci. **46**, 389 (1977)
546. C.M. Urry, P. Padovani: Publ. Astr. Soc. Pacific **107**, 803 (1995)

– V –

547. R. Vainio, R. Schlickeiser: Astron. Astrophys. **331**, 793 (1998)
548. J.P. Vallee: Astron. J. **95**, 750 (1988)
549. G.L. Verschuur: In: *Interstellar Magnetic Fields: Observations and Theory*, ed. by R. Beck, R. Gräve (Springer, Berlin, Heidelberg, 1987) p. 154
550. H. Völk, L. Drury, J. McKenzie: Astron. Astrophys. **130**, 19 (1984)
551. V. Votruba: Phys. Rev. **73**, 1468 (1948)

– W –

552. G. Wallis: In: *IAU Symp. No. 9 on Radio Astronomy*, ed. by R.N. Bracewell (Stanford University Press, Stanford, CA, 1959) p. 595
553. Y.-M. Wang, R. Schlickeiser: Astrophys. J. **313**, 200 (1987)
554. A.A. Watson: Adv. Space Res. **4**, 35 (1984)
555. A.A. Watson: Nucl. Phys. B Suppl. **22B**, 116 (1991)
556. G.N. Watson: *Theory of Bessel Functions* (Cambridge University Press, Cambridge, MA, 1966)
557. G.M. Webb: Astrophys. J. **270**, 319 (1983)
558. W.R. Webber: Austr. J. Phys. **21**, 845 (1968)
559. W.R. Webber: In: *Cosmic Abundances of Matter*, ed. by C.J. Waddington (AIP, New York, 1989) p. 100
560. W.R. Webber, S.M. Yushak: Astrophys. J. **275**, 391 (1983)
561. A.S. Webster: Astrophys. Letters **5**, 189 (1970)
562. T.C. Weekes: *High Energy Astrophysics* (Chapman and Hall, London, 1972)
563. G. Westerhout, C.L. Seeger, W.N. Brouw, J. Tinbergen: Bull. Astron. Inst. Neth. **16**, 187 (1962)
564. W. Whaling: In: *Handbuch der Physik*, ed. by S. Flügge (Springer, Berlin, Heidelberg, 1958) Vol. **34**, p. 193
565. P.J.S. Williams, A.H. Bridle: Observatory **87**, 280 (1967)

566. A.W. Wolfendale: Quart. J. Royal Astron. Soc. **24**, 122 (1983)
567. C.S. Wu, J.D. Huba: Astrophys. J. **196**, 849 (1975)

– **Y** –

568. I.-A. Yadigaroglu, R.W. Romani: Astrophys. J. **449**, 211 (1995)
569. N.Y. Yamasaki, T. Ohashi, F. Takahara, et al.: Astron. Astrophys. Suppl. **120**, C393 (1996)
570. N.Y. Yamasaki, T. Ohashi, F. Takahara, et al.: Astrophys. J. **481**, 821 (1997)
571. D.G. York: Ann. Rev. Astron. Astrophys. **20**, 221 (1982)

– **Z** –

572. G.T. Zatsepin, V.A. Kuzmin: Soviet Phys. JETP **14**, 1294 (1962)
573. G.T. Zatsepin, V.A. Kuzmin: Soviet Phys. JETP Letters **4**, 78 (1966)
574. Y. Zhou, W.H. Matthaeus: J. Geophys. Res. **95**, 14881 (1990)

Illustration Credits

All figures and tables listed below are reprinted with explicit permission from the authors and publishers.

Reprinted with permission from The American Astronomical Society, 1630 Connecticut Ave NW, Suite 200, Washington, DC 20009, USA
Fig. 2.11 – J.P. Vallee: Astron. J. **95**, 750 © (1988)

Reprinted with permission from Annual Review of Astronomy and Astrophysics, Editorial Board, 4139 El Camino Way, PO Box 10139, Palo Alto CA, 94303-0139, USA
Fig. 16.16 – A.M. Hillas: Ann. Rev. Astron. Astrophys. **22**, 425 © (1984)

Reprinted with permission from Elsevier Science, PO Box 800, Oxford OX5 1DX, UK
Fig. 3.29 – D. Müller: Adv. Space Res. **9**, (12) 31 © (1989)
Fig. 3.30 – D. Müller: Adv. Space Res. **9**, (12) 31 © (1989)
Fig. 3.39 – D. Müller: Adv. Space Res. **9**, (12) 31 © (1989)
Fig. 3.35 – A.A. Watson: Adv. Space Res. **4**, 35 © (1984)
Fig. 3.36 – A.A. Watson: Adv. Space Res. **4**, 35 © (1984)

Reprinted with permission from Gordon & Breach Publishers, World Trade Center, CH-1000 Lausanne, Switzerland
Fig. 2.7 – E.B. Jenkins: Comm. Astrophys. **7**, 121 © (1978)

Reprinted with permission from Kluwer Academic Publishers, PO Box 17, NL-3300 AA Dordrecht, The Netherlands
Fig. 3.3 – N. Lund: In: *Cosmic Rays in Contemporary Astrophysics*, ed. by M.M. Shapiro, (Reidel, Dordrecht, The Netherlands, © 1986) p. 1
Fig. 3.5 – J.A. Simpson: In: *Composition and Origin of Cosmic Rays*, ed. by M.M. Shapiro, (Reidel, Dordrecht, The Netherlands, © 1983) p. 1
Fig. 3.14 – W. Dröge, P. Meyer, P. Evenson, D. Moses: Solar Phys. **121**, 95 © (1989)
Fig. 3.15 – W. Dröge, P. Meyer, P. Evenson, D. Moses: Solar Phys. **121**, 95 © (1989)

Fig. 3.23 – P. Meyer: In: *Origin of Cosmic Rays, IAU Symp. 94*, ed. by G. Setti, G. Spada, A.W. Wolfendale, (Reidel, Dordrecht, The Netherlands, ©) 1980) p. 7

Fig. 3.27 – M. Israel: In: *Composition and Origin of Cosmic Rays*, ed. by M.M. Shapiro, (Reidel, Dordrecht, The Netherlands, © 1983) p. 47

Table 4.2 – R. Schlickeiser: In: *Progress in Cosmology*, ed. by A.W. Wolfendale, (Reidel, Dordrecht, The Netherlands, © 1982b) p. 215

Fig. 3.32 – J.A. Simpson, M. Garcia-Munoz: Space Sci. Rev. **46**, 205 © (1988)

Fig. 5.5 – C.-C. Cheung: Space Sci. Rev. **13**, 3 © (1972)

Fig. 5.9 – R. Ramaty, N.J.Westergaard: Astrophys. Space Sci. **45**, 143 © (1976)

Fig. 5.13 – R. Schlickeiser, J. Steinacker: Solar Phys. **122**, 29 © (1989)

Fig. 16.1 – L.F. Burlaga: Space Sci. Rev. **12**, 600 © (1971)

Fig. 16.2 – L.F. Burlaga: Space Sci. Rev. **12**, 600 © (1971)

Fig. 16.3 – L.F. Burlaga: Space Sci. Rev. **12**, 600 © (1971)

Fig. 16.4 – L.F. Burlaga: Space Sci. Rev. **12**, 600 © (1971)

Reprinted with permission from Macmillan Magazines Ltd, Nature Editorial Board, Porters South, 4-6 Crinan Street, London N1 9XW, UK

Fig. 2.8 – T.W. Hartquist, M.A.J. Snijders: Nature **299**, 783 © (1982)

Reprinted with permission from MIT Press, 77 Massachusetts Avenue, Cambridge, MA 02139-4307, USA

Fig. 5.6 – R. Ramaty: In: *High Energy Particles and Quanta in Astrophysics*, ed. by F.B. McDonald, C.E. Fichtel (MIT Press, Cambridge, MA, © 1974) p. 122

Fig. 5.8 – R. Ramaty: In: *High Energy Particles and Quanta in Astrophysics*, ed. by F.B. McDonald, C.E. Fichtel (MIT Press, Cambridge, MA, © 1974) p. 122

Index

γ-ray
– lines, 114, 128, 163
π^0-decay, 125, 155, 157
^{3}He-rich events, 44
"pile-up" distribution, 466

absorption, 102, 171
– coefficients, 76–78, 144
– free-free, 143, 144
– photoelectric, 15, 164, 175
abundances
– elemental, 15, 16, 43, 46, 48
– isotopic, 44, 57
acceleration, 311
– Fermi, 366
– – first-order, 366, 379, 420
– – second-order, 366, 420, 435
– interstellar, 181, 439
– particle, 188, 402, 433
– process, 3, 5
– shock wave, 391, 461
– stochastic, 357, 419, 446, 461
accretion disk, 166, 167, 171, 172
actinides, 48, 56
active galactic nuclei, 165, 166
– jets of, 172
adiabatic
– cooling, 389, 463
– deceleration, 310, 366, 373, 463
airshower
– experiment, 25
Alfvén shock, 402
Alfvén wave, 221, 266, 280, 313, 386
– transmission, 403
amplification
– total wave, 410
analytic continuation, 200
anisotropy, 3, 64, 70, 307
antimatter, 28, 469
antinuclei, 28
antiproton, 29, 469
approach

– test particle, 4, 190
– test wave, 4, 190
approximation
– cold plasma, 193, 212
– leaky-box, 436, 462
– monochromatic, 76, 83
– weak damping, 262, 275
astronomy
– γ-ray, 97, 154, 158
– cosmic ray, 1
– radio, 143
atmosphere
– stellar, 2
atmospheric electricity, 1
atmospheric neutrino background, 471

background radiation
– microwave, 17
Bessel function, 40, 74, 206, 242, 274, 299, 374, 452, 468, 481, 488
black hole accretion, 169
blackbody distribution, 17, 18
blazars, 165, 168, 171, 174
Breit-Wigner
– type resonance function, 303, 338
bremsstrahlung, 95
– Bethe-Heitler cross-section, 93
– electron source function, 96
– energy loss, 97, 367, 461
– nonthermal, 2
– nonthermal electron, 96

case
– maser, 477
– optically thick, 477
– optically thin, 102, 477
Cauchy
– principle value integral, 200, 276
center, 7
– galactic, 146, 153, 466
characteristic
– method of, 296

coefficient
- absorption, 76, 78
- emission, 76, 79
- Fokker-Planck, 297, 313
-- fast mode wave, 344
- reflection, 403
- transmission, 403
collision
- frequency, 183
- inelastic, 106
- particle-photon, 114
compactness, 166, 168, 171
component
- anomalous cosmic ray, 38, 46
- high-energy particle, 2
composition, 42, 70, 179, 469
- chemical, 2
- isotopic, 3, 44, 57, 63, 180
Compton
- inverse energy loss, 88, 149, 367
- inverse scattering, 2, 80, 155, 162, 174, 464
Compton-Getting effect, 309
conductivity tensor, 197, 204, 211, 241, 479
contact discontinuity, 395
convective motion, 365
correlation function, 297
cosmic ray
- clocks, 58, 61, 180, 438, 469
- energetics, 182
- injection of the, 54
- isotropy, 65, 182, 471
- power, 180, 469
- scattering, 182, 471
- self-confinement, 471
cosmic rays
- primary, 25, 48, 52, 63, 70, 71
- secondary, 25, 50, 52, 58, 62, 437, 458
Coulomb, 130
- barrier, 131
- collisional length, 4
- energy loss, 98, 134, 367
- interaction, 73, 183
- logarithm, 130
- loss, 464
cross helicity, 323, 345
cross-section
- Bethe-Heitler, 93
- fragmentation, 50, 135, 437
- Klein-Nishina, 82
- stopping, 133
- Thomson, 80, 105

- total, 118
curvature drift, 389
curvature radiation, 2
cutoff
- Greisen-Zatsepin-Kuzmin, 71
- Klein-Nishina, 85, 88, 162
CYG X-3, 66

damping
- cyclotron, 5, 290, 337, 387, 461
- ion-neutral, 291
- ion-neutral collision, 287
- Landau, 203, 277
-- nonlinear, 287
- transit-time, 347, 353, 461
- weak approximation, 262, 275
de Hoffmann and Teller frame, 392
Debye length, 184
decay, 115
- π^0, 114, 115
- radioactive, 58, 62, 438
density
- charge, 189
- current, 4, 189
- fluctuation, 435
depth
- optically, 476
diffusion
- galactic cosmic ray, 373, 436
- momentum, 333, 336, 349, 366, 439
- perpendicular, 366, 474
diffusion approximation, 306, 365
diffusion-convection equation, 4, 311, 365, 415
diffusion-convection operator, 378
diffusion-convection transport, 367
discontinuity
- contact, 395
- magnetohydrodynamic
-- shock, 394
- rotational, 397
- tangential, 395
dispersion function
- plasma, 254, 491
dispersion measure, 20
dispersion relation, 201, 203, 242
dissipation, 203, 336, 338
distribution
- "pile-up", 466
- age, 372
- function, 3, 187
-- ensemble of, 295
- of radio spectral index, 444
Doppler

– boosting, 171, 174
– factor, 171
drift
– curvature, 389
– gradient, 388

Eddington luminosity, 169
electron/proton ratio, 473
Elliot-Shapiro relation, 168, 169, 171
emission
– coefficients, 76, 78, 79, 83, 143
– electromagnetic, 1
– free-free, 143
– measure, 144
– neutrino, 471
energy loss rate, 80, 88–90, 92, 97, 99
energy spectrum, 3, 179
equation
– Boltzman, 4
– Fokker-Planck, 297
– leaky-box, 436, 462
– linearization of kinetic plasma, 195
– Maxwell, 4, 188
– moment, 191
– Rankine-Hugoniot, 392
– transport, 5, 293, 367
– Vlasov, 188, 194
external galaxy, 13, 21, 22

Faraday
– rotation, 77, 239
Faraday effect, 19, 20
Faraday rotation, 20, 23
fast shock, 402
fast shock waves, 5
Fermi
– first-order acceleration, 366, 418, 420
– second-order acceleration, 366, 420, 435
– theory of pion production, 118
Fermi acceleration, 366
first ionization potential, 43, 53, 55
flare stars, 181
flares, 22
– impulsive, 42, 467
– long-duration, 39, 468
– solar, 5, 38, 48, 387
fluctuation density, 435
Fokker-Planck
– coefficient, 313
– – fast mode wave, 344
Fourier-Laplace
– transform, 198

fragmentation cross-section, 51, 437
frame
– de Hoffmann and Teller, 392
frequency
– collision, 183
– electron plasma, 74, 184
– gyro-, 74, 186, 218
– supernova, 443
– turnover, 144

galactic center, 7
galactic nuclei
– active, 165, 166
– – jets of, 172
galaxy, 7
– active, 149
– clusters of, 23, 150
– edge-on, 23
– halo of our, 438
– Seyfert, 144, 150
Geminga CG, 177
grammage, 437
graybody distribution, 17
Green's function, 376, 426, 441, 465
group velocity, 216
guiding center, 293
gyrofrequency, 74, 186
gyrophase, 306

harmonics, 309
– first, 64
– second, 64
helicity
– cross, 323
– magnetic, 316
heliosphere, 8, 9
Hubble flow, 23

inertial frame, 106
instability, 203
– cosmic ray induced, 273, 283
– precursor streaming, 413
– streaming, 282
– two-stream, 272
interaction
– Coulomb, 73, 183
– nuclear, 116
intercloud medium, 10
intergalactic matter, 10
intergalactic medium, IX, 23, 24
intergalactic plasma, 23
interplanetary medium, 8
interstellar
– acceleration, 181, 439

– gas, 10
interstellar dust, 17
interstellar gas, 11
interstellar hydrogen, 11
interstellar matter, 10
interstellar medium, 7
invariance, 106
ionization, 99, 130, 139, 367
– secondary, 140
isotopes, 44, 57, 180
– radioactive, 62, 438
isotropy, 182

kinematics
– relativistic, 105
Klein-Nishina
– formula, 82
– limit, 85, 89
Klein-Nishina cutoff, 85, 88, 103, 162
knee, 454
Kolmogorov, 410, 435
– type turbulence spectra, 317

Landau
– contour, 200, 245
– damping, 203, 248, 262, 277
– – nonlinear, 287
– principal mode, 258
Larmor radius, 30, 182
leaky-box approximation, 373, 436, 462
Lorentz
– force, 187
– system, 106
– transformation, 105

Mach number, 399
Magellanic Clouds, 147, 151, 155, 444
magnetic field, IX, 19, 20, 24, 29
– irregularities, 29, 31, 182
magnetohydrodynamic, 190
– shock discontinuity, 392
– turbulence, 36
magnetohydrodynamics
– ideal, 187
magnetosonic waves, 229, 282, 342, 388
maser case, 477
Maxwell
– operator, 198
Maxwell's
– equation, 188
Maxwellian
– velocity distribution, 187
mean free path, 31, 327, 349, 383
mean residence time, 52, 58, 443

– of $\sim$ 1 GeV, 180
measure
– dispersion, 217
– rotation, 240
medium
– intergalactic, 2
– interplanetary, 383
– interstellar, 2, 10
metagalactic origin, 154, 162
meteorites, 69
MHD turbulence, 461
Milky Way, 7, 149
mode
– longitudinal, 243, 270
– transverse, 243, 270
momentum diffusion, 310, 333, 336, 349, 366, 415, 439
multiplicity, 116, 118
muon, 68, 70, 114, 115

neutrino, 67, 114
neutrino emission, 125, 471
neutron, 67, 112, 124
Newcomb–Gardner theorem, 249
nuclei, 29
– cosmogenic, 69, 180
– excited, 128
– secondary, 52, 58, 437
nucleosynthesis, 54, 57, 63

OB association, 156
observation
– X-ray, 12
Ohm's law, 197
optical depth, 143, 476
optically thick case, 144, 173, 477
optically thin case, 102, 144, 477

pair annihilation radiation, 174
pair attenuation, 169, 175
pair production, 109
– triplet, 89
pairs
– electron-positron, 89
parallel shock wave, 398
particle
– exotic, 67
– rigidity, 28, 30, 433
particle spectrum
– primary, 71
pathlength, 58, 61, 439
Peclet number shock wave, 428
perpendicular diffusion, 366, 474

perpendicular shock wave, 400
photo
- disintegration, 112
- hadron production, 111
- pair production, 111
- pion production, 109
pion
- production, 28, 108, 367
-- Fermi's theory, 118
- production spectra, 115
plasma, 1
- cold, 193, 212
- cutoff, 214
- dispersion function, 254, 491
- hot, 205, 241, 479
- kinetic description of, 187
- kinetic plasma equation
-- linearization of, 195
- magnetohydrodynamics of, 190
- parameter, 185
- resonance, 214, 222, 225
- waves, 182, 203, 212, 226
Plemlj formulae, 200
point source, 147, 156
polarization, 19, 203
- circular, 220
- degree of linear, 77
- elliptical, 76, 236
- position angle, 77
positron, 29, 46, 114, 469
positron/electron ratio, 57, 124
positrons
- secondary, 115, 124
power law, 39, 40, 52, 56, 179
precursor streaming instability, 413
pressure tensor, 190
process
- catastrophic loss, 136, 139, 367
- continuous loss, 367
- continuum radiation, 102
- interaction, 73
- r-process, 54, 56
- radiation, 73
- s-process, 54
production
- secondary, 28, 114, 115, 123, 437
propagation
- super-Alfvénic, 290
pulsar, 20, 144, 156, 181
- Crab, 177
- Vela, 177
pulsars, 22

quasi-longitudinal, 231

quasi-transverse, 231
quasilinear theory, 293, 295, 415

radiation transfer, 2, 76, 78, 144
radiation transport, 76, 475
radio
- continuum survey, 3, 145
- polarization, 147
radio polarization, 21–23
radio spectral index, 149
- distribution of, 151, 444
radionuclides, 69
Rankine-Hugoniot equation, 392
rate
- ionization, 2
ratio
- compression, 399
- electron/proton, 39, 473
- primary, 52
- scattering center compression, 418
- secondary, 52
Razin-Lorentz factor, 80
redshift expansion, 111, 113
reflection coefficient, 403
refraction
- index of, 205, 229
relativistic beaming, 176
- hypothesis, 166
residence time, 3, 52, 180, 443
resonance, 214
- electron cyclotron, 222
- function
-- Breit-Wigner type, 303, 338
- hybrid, 225
-- lower, 226
-- upper, 226
- proton cyclotron, 223
rest system, 107, 108
rigidity, 31, 39
- particle, 28, 30, 383, 433
rotation measure, 20, 23
rotational discontinuity, 396

scattering time method, 375, 436
Schwarzschild radius, 169
secondary
- cosmic rays, 25, 50, 52, 58
- electrons, 114
- ionization, 139
- neutrinos, 114, 125
- nuclei, 58, 437
- positrons, 114, 469
- production, 28, 115, 437

self-confinement, 286
Seyfert galaxy, 144, 150
shielding, 95, 184
shock, 46, 166
– Alfvén, 402
– classification of, 402
– fast, 402
– slow, 402
– switch-off, 402
– switch-on, 402
shock discontinuity
– magnetohydrodynamic, 392
shock wave, 5, 36
– acceleration, 391, 461
– parallel, 402
– Peclet number, 428
– perpendicular, 402
– supernova, 391
slab turbulence, 387
slow shock, 402
solar
– corona, 29
– cosmic rays, 29
– flare events, 38, 383, 387, 468
– modulation, 29, 32, 179, 388
– system, 30, 388
– wind, 29, 384
solar corona, 12
solar physics, 8
solar system, 7
solar vicinity, 12
solar wind, 8
source
– function, 84, 94, 115, 119, 121
spallation, 48, 50, 54, 57, 58, 367, 437
spectral index, 40, 52, 149, 151, 179, 429
states
– charge, 45
– ionization, 45
stellar wind, 164, 391
stochastic acceleration, 5, 357, 446, 461
Stokes parameters, 76
Sturm-Liouville, 372
– eigenfunction theory, 372, 376
superluminal source, 165, 166
supernova
– explosion, 7, 181
– frequency, 443
– rate, 13, 181, 443
– remnant, 144, 468
–– shell-type, 150
– shock waves, 13, 391

supernova rate, 13
supernova remnant, 13
survey
– linear polarization, 147
– radio continuum, 145
synchrotron
– energy loss, 80, 464
– monochromatic approximation, 76
– power, 74
– radiation, 74, 143, 147, 367
– Razin-Lorentz factor, 80
– self-absorption process, 78
– self-Compton process, 173
synchrotron radiation, 2, 19
system
– center-of-momentum, 106
– Lorentz, 106

tangential discontinuity, 395
tensor
– conductivity, 197, 204, 241, 479
– dielectric, 198
–– non-Hermitian, 209
– pressure, 190
test particle approach, 4, 190
test wave approach, 4, 190
Thomson
– cross-section, 80, 105, 168
– limit, 83, 88
threshold energy, 106, 108, 110
time variability, 67, 166, 169
total wave amplification, 410
transmission
– Alfvén wave, 403
transmission coefficient, 405
transport
– equation, 4, 5, 293, 367
turbulence
– dynamical magnetic, 302
– isotropic, 316
– magnetohydrodynamic, 36
– measurement, 472
– MHD, 461
– plasma wave, 4, 302
– slab, 316, 387
– spectra
–– Kolmogorov-type, 317, 435
– tensor, 315

ultraheavy elements, 55, 56
unified scenario, 166

vacuum approximation, 75
vector

- energy-momentum, 106
- four-vector, 105
- polarization, 203, 215, 229

Vela pulsar, 177

velocity
- Alfvén, 219
- group, 216

wave
- Alfvén, 221, 313, 386
-- compressional, 226
- cascading, 287
- electron cyclotron, 222
- extraordinary, 224
- high frequency, 227
- hybrid
-- lower, 227
-- upper, 227
- ion cyclotron, 221
- longitudinal, 215
- low frequency, 229
- magnetohydrodynamic, 4, 320
- magnetosonic, 229, 342, 388
- ordinary, 224
- shock, 391
- total wave amplification, 410
- transverse, 215
- whistler, 222

Zeeman effect, 20
Zeeman splitting, 19

Printing (Computer to Film): Saladruck Berlin
Binding: Stürtz AG, Würzburg